FSC
www.fsc.org
MIX
Papier | Fördert gute Waldnutzung
FSC® C083411

Kolibri 360 GmbH
Ferdinand-Jühlke-Straße 7
99095 Erfurt, Deutschland
produktsicherheit@kolibri360.de

Organoselenium Compounds in Biology and Medicine
Synthesis, Biological and Therapeutic Treatments

Organoselenium Compounds in Biology and Medicine
Synthesis, Biological and Therapeutic Treatments

Edited by

Vimal Kumar Jain
Bhabha Atomic Research Centre, India
Email: jainvk@barc.gov.in

and

K. Indira Priyadarsini
Bhabha Atomic Research Centre, India
Email: kindira@barc.gov.in

ROYAL SOCIETY OF CHEMISTRY

Print ISBN: 978-1-78801-029-0
PDF eISBN: 978-1-78801-190-7
EPUB eISBN: 978-1-78801-223-2

A catalogue record for this book is available from the British Library

The Royal Society of Chemistry is a charity, registered in England and Wales, Number 207890, and a company incorporated in England by Royal Charter (Registered No. RC000524), registered office: Burlington House, Piccadilly, London W1J 0BA, UK, Telephone: +44 (0) 207 4378 6556.

Visit our website at www.rsc.org/books

Printed in the United Kingdom by CPI Group (UK) Ltd, Croydon, CR0 4YY, UK

Preface

Selenium, like the moon from which its name has been derived, has different facets. It is a micronutrient for humans and animals. Its deficiency leads to several diseases, while at marginally higher concentration it becomes toxic. It is an important element for the development of semiconductors. The year 2017 is the bicentenary of its discovery by the Swedish scientist J. J. Berzelius. To commemorate this occasion, the publication of this multi-authored volume on "Organoselenium Compounds in Biology and Medicine", of interest to postgraduate students and researchers in those multidisciplinary areas that have selenium at centre stage, is a fitting tribute.

To date, the majority of books available in the literature on selenium biology are focussed on inorganic selenium compounds. During the past two decades or so there have been significant advances in the design and synthesis of organoselenium compounds exhibiting potential biological activities. Having realized the high toxicity of inorganic forms of selenium (*e.g.* sodium selenite), which limits the therapeutic window, efforts have been directed to develop a variety of organoselenium compounds that are considerably less toxic. The redox properties, toxicity and specificity of organoselenium compounds can be further tuned by the organic substituents as well as by the presence of non-bonding interactions between selenium and a hetero atom. For example, ebselen, the first synthetic organoselenium compound with glutathione peroxidase-mimicking activity is in clinical trials as an antioxidant drug for heart disease and bipolar disorders, while selenocystine, the 21st amino acid, and its derivatives are very promising compounds in cancer therapy. Several other compounds are in advanced stages of research for therapeutic usage.

The research on selenium has emerged as a multidisciplinary science through the interactions of researchers working in different areas such as

Organoselenium Compounds in Biology and Medicine: Synthesis, Biological and Therapeutic Treatments
Edited by Vimal Kumar Jain and K. Indira Priyadarsini

Published by the Royal Society of Chemistry, www.rsc.org

organic/inorganic/bio-chemistry, materials science, biology, pharmacology, medicine, agriculture and environmental science. This growing and sustained interest can be witnessed in the publication of several reviews and monographs over the past two or three decades, and the series of international symposia organized globally on a regular basis. These include the International Symposium on Industrial Uses of Selenium and Tellurium, the International Symposium on Selenium in Biology and Medicine, the International Conference on the Chemistry of Selenium and Tellurium (ICCST) and the International Conference on Selenium in the Environment and Human Health. Such events are organized in different countries manifesting the wide geographical distribution of selenium research.

With this background, this volume aims to cover the recent advancements of organoselenium compounds, with emphasis on their biological relevance as antioxidants, antibacterial agents, radioprotectors and anticancer agents, *etc.* The book, comprising 16 chapters, is therefore representative of the current status of research and developments in the synthesis of organoselenium compounds and their applications in biology and medicine. The first chapter provides an overview of organoselenium chemistry with an obvious bias towards biologically relevant compounds. The remaining 15 chapters have been divided into three sections, each including five chapters. The first section of the book dicusses synthetic advances on bioactive selenium compounds, providing details of the synthetic aspects of biologically relevant novel organoselenium compounds. The second section covers biophysics and chemistry and effects on oxidative stress of organoselenium compounds, where redox properties of selenium compounds, reactions with reactive oxygen species, non-bonding interactions in glutathione peroxidase mimics and important selenoproteins in maintaining cellular redox balance are discussed. The third section deals with therapeutic applications of organoselenium compounds. In this section, recent advances in the research related to the applications of selenium and organoselenium, in particular in radiotherapy, cancer therapy and cancer prevention, are discussed.

In the preparatory stage of our editorial work, we decided that all chapters should be written in such a way that they covered most of the recent advances in the field and should be easy for the beginners to follow. The selection of chapters should be such that the repetition and overlap of their content could be minimal. As editors we have been fortunate to have worked with authors experienced in their specialized fields from different parts of the world for the individual chapters. Their wholehearted support and co-operation in preparing the chapters is gratefully acknowledged.

We are indebted to all the reviewers for their critical comments and constructive suggestions, which enabled us to improve the content of the chapters. We acknowledge the help of our junior colleague Mr Amey Wadawale with computing during editing. We are thankful to the Royal Society of Chemistry (RSC) for giving us this opportunity. Special thanks to Dr Michelle Carey, senior commissioning editor at the RSC, for many useful suggestions while preparing the book proposal. We thank the authorities of our institute, Bhabha Atomic Research Centre, for encouraging us to take up this project with the RSC. And last, but not least, we appreciate all the efforts of Catriona Clarke and the team at the RSC for the editorial assistance in bringing out the book in time.

Vimal K. Jain and K. Indira Priyadarsini

Contents

Organoselenium Compounds in Biology and Medicine: Synthesis, Biological and Therapeutic Treatments
Edited by Vimal Kumar Jain and K. Indira Priyadarsini

Published by the Royal Society of Chemistry, www.rsc.org

CHAPTER 1

An Overview of Organoselenium Chemistry: From Fundamentals to Synthesis

VIMAL K. JAIN

Chemistry Division, Bhabha Atomic Research Centre,
Mumbai-400 085, India
Email: jainvk@barc.gov.in

1.1 Introduction

Selenium is a member of the group 16 elements (O, S, Se, Te and radioactive Po), collectively known as chalcogens. It was discovered in 1817 by J.J. Berzelius in the reddish deposits that formed in the lead chambers at his sulfuric acid plant at Gripsholm in Sweden. He named the element selenium in the honour of Greek goddess 'Selene' meaning moon.[1]

The chemistry of selenium compounds was neglected for more than a century; the entire literature[2] comprised only ~200 papers until 1920 and it remained an arcane field of investigation until 1970. This slow development can be attributed to the malodorous reputation of its compounds, toxicity, the instability of certain derivatives as well as the general belief that the chemistry of selenium, due to its proximity to sulfur, would be more or less similar to that of sulfur compounds. However, such beliefs and

Organoselenium Compounds in Biology and Medicine: Synthesis, Biological and Therapeutic Treatments
Edited by Vimal Kumar Jain and K. Indira Priyadarsini

Published by the Royal Society of Chemistry, www.rsc.org

perceptions for organoselenium compounds were defied by an exponential growth of organoselenium chemistry during past three decades or so. The following three major, interdependent factors have contributed to this rapid development of the field.

(i) Role in organic chemistry: since its discovery in the early 1930s as an oxidizing agent for organic compounds,[3] selenium dioxide (SeO_2) was used predominantly in organic synthesis until the early 1970s. However, around this time several useful reactions and processes were discovered[4–6] and the interest in organoselenium compounds was further catalysed with the publication of a monograph by Klayman and Günther.[7] Since then, the number of reactions as well as the variety of selenium compounds have grown dramatically.[8] Selenium can be introduced to a myriad organic substrates as an electrophile, nucleophile or even as a radical in a chemo-, regio- and stereo-selective manner.

(ii) Organometallic chemistry and materials science: although metal complexes of seleno ligands (*e.g.* $[PtCl_2(R_2Se)_2]$; R = Me, Et, Pr^n, Ph) were first synthesized more than a century ago,[9] reports on organoselenium complexes appeared only sporadically until the early 1990s,[10] possibly due to poorly developed synthetic processes for the desired organoselenium compounds. Selenium ligands quite often show unusual reactivity that differs from their sulfur counterparts.[11] Platinum group metal complexes with seleno ligands were developed as catalysts for various reactions since the 1990s,[12–14] and in some cases exhibit even better catalytic activity than the corresponding thio derivatives.[12] Further impetus to selenium chemistry comes from recent interest in semiconductor metal selenide nano-materials.[15–17] Metal selenolates have emerged as versatile single-source molecular precursors for the synthesis of nano-particles and deposition of thin films of metal selenides.

(iii) Selenium in biology:[18] selenium was long considered a poison until 1957 when Schwarz and Foltz identified it as an essential micronutrient.[19] Fifteen years later selenocysteine, the 21st amino acid, was discovered at the active site of glutathione peroxidase (GPx), establishing the role of selenium in mammals.[20] Since then, approximately 40 seleno enzymes, exhibiting a range of functions have been identified.[21] GPx has been extensively investigated due to its diversity of biological roles. To mimic the functions of GPx and related enzymes, several organoselenium compounds have been designed and developed. Ebselen is a promising candidate as an oxidant.[22] The mechanistic aspects of the antioxidant activity of GPx have been worked out and the formation of several selenium species has been proposed during the catalytic cycle.[23]

In the light of the above, an overview of organoselenium chemistry is presented in this chapter.

1.2 General Considerations

Selenium is a trace element occurring at an average level of $9\times10^{-5}\%$ (0.09 ppm) in the earth's crust. There is substantial geographical variations in agricultural soils, giving selenium-deficient, -adequate and -excess (toxic) regions.[24] In general, selenium-deficient and -adequate regions are much more widespread than the selenium-excess regions. Selenium exists in various chemical forms in soil, which influences the availability of the element to plants. Selenium in food grains, legumes and vegetables occurs primarily in organic form (such as MetSe, cysSeMe, cysSeSecys, *etc.*) and is often referred as dietary selenium.[25] Keshan disease, Kashin–Beck disease and several dangerous viral infections (H1N1 influenza, SARS, HIV/AIDS, Ebola, *etc.*) are associated with selenium deficiency and outbreaks of them have originated either in bio-geo-chemically selenium-poor regions of China or selenium nutrient-depleted sub-Saharan Africa.[26] Acute toxicities in grazing animals were first reported in South Dakota in the 1930s, which was locally known as 'alkali disease'.[27] More recently, high selenium content (750–5000 $\mu g\,day^{-1}$ per person) has been reported in food grown in the Punjab province of India.[28] Selenium intake varies worldwide, ranging from 7 $\mu g\,day^{-1}$ to 4990 $\mu g\,day^{-1}$.[25] In fact, there is a very narrow window between dietary deficiency (<40 $\mu g\,day^{-1}$) and toxic levels (>400 $\mu g\,day^{-1}$) in humans, and the recommended dietary intake of selenium is 50–70 $\mu g\,day^{-1}$.[29] For this reason, selenium is referred as an 'essential poison'.[30]

Selenium occurs in sulfide ores of heavy non-ferrous metals—copper, copper–nickel and multi-metallic sulfides. There is a wide variation in the selenium content of these ores. As there is no primary mineral/ore that contains an economically significant amount of selenium, it is recovered as a by-product, mostly from the anode slimes generated from the electrolytic refining of copper. The selenium content of the slimes vary from 2% to 55%, the average being ~10%. Although selenium was discovered in 1817, its commercial production in appreciable quantities began only about a century later. Global production of selenium is hard to estimate, because it is a by-product of copper refineries. Nevertheless, it is estimated that in 2013 the global production (excluding USA and China) was >2170 metric tonnes (Figure 1.1a).

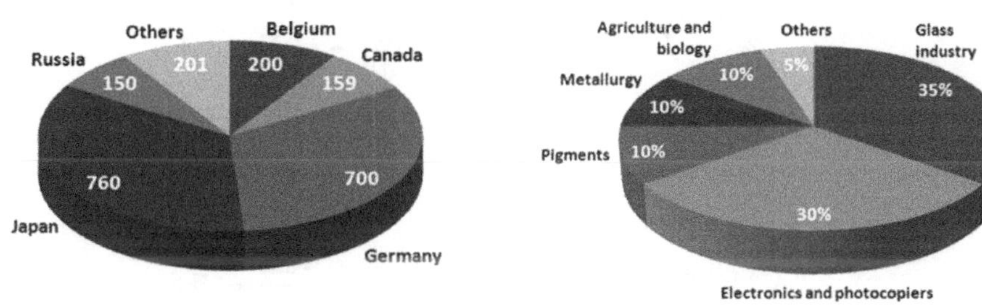

(a) (b)

Figure 1.1 (a) Global production (excluding USA and China) of selenium in 2013; (b) consumption of selenium in different industries.

The first practical application of selenium was realized in 1873 when photoconducting properties were reported, which were exploited for the development of photo-cells. Since then a wide range of applications have been developed which include (i) electronics and photocopier machines (~30%); (ii) the glass industry (~35%); (iii) pigments (~10%); (iv) metallurgy (~10%); (v) agriculture and biology (~10%); and (vi) others (~5%) (Figure 1.1b).

Selenium has several allotropic forms in both the amorphous and crystalline states at room temperature and adopts either helical polymeric chain or Se_8 ring structures with Se–Se distances varying between 2.32 and 2.37 Å.[1,31,32] The density of selenium varies between 4.20 and 4.81 $g\,cm^{-3}$. The following modifications of selenium are now well recognized: (i) amorphous or α-selenium (red and black forms); (ii) vitreous or glassy selenium (ordinary commercial form); (iii) crystalline monoclinic or β-selenium (red α-form and deep-red β-form); and (iv) trigonal grey or γ-selenium. The latter form (grey selenium), comprising of helical polymeric chains, is a p-type semiconductor and shows appreciable photoconductivity, whereas other modifications are insulators.

Selenium can adopt a range of integral and fractional formal oxidation states. The main integral oxidation states are −2 (*e.g.* sodium selenide, Na_2Se), −1 (*e.g.* disodium diselenide, Na_2Se_2), 0 (*e.g.* Se_8), +1 (*e.g.* diselenium dichloride, Se_2Cl_2), +2 (*e.g.* selenium dichloride, $SeCl_2$), +4 (*e.g.* sodium selenite, Na_2SeO_3) and +6 (*e.g.* sodium selenate, Na_2SeO_4); the latter oxidation state being less stable than the corresponding known sulfur compounds. The fractional oxidation states are reported in polyselenium-cations (Se_n^{2+}; $n=4$ (oxidation state $+\frac{1}{2}$) and $n=8$ (oxidation state $+\frac{1}{4}$))[33–35] and -anions (Se_n^{2-}; $n=3$–11, 16).[34–36] The hypervalent nature of selenium based on 3c–4e interactions is also encountered in its compounds.[37] In the divalent state, selenium can either weakly donate its electron pair (a Lewis base) to a metal centre (M) or partially accept a lone pair (a Lewis acid) from another atom (Y, such as H, N, O, S, Cl, I, *etc.*) (Scheme 1.1). The latter interaction results in a linear C–Se···Y bond, which is thought to be responsible for the bioactivity of these compounds.[37] Redox cycling of selenium between various oxidation states occurs readily (Scheme 1.2).[38] Accordingly, selenium

Scheme 1.1 Non-bonding interactions involving divalent selenium.

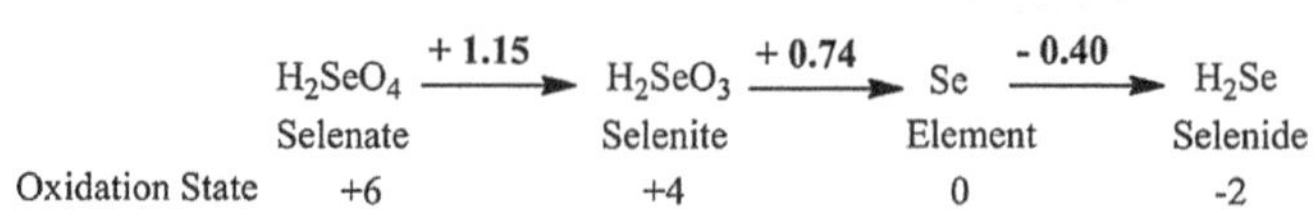

Scheme 1.2 Standard redox potentials (in volts at 25 °C) in acid solution (pH = 0).

can act as an oxidant as well as reductant in many reactions. Various single-bond energies involving selenium, such as Se–H (66 kcal mol^{-1}), Se–C (56 kcal mol^{-1}), Se–Se (46 kcal mol^{-1}), are intermediate between corresponding sulfur and tellurium compounds.[39]

1.3 NMR Spectroscopy

Out of six naturally occurring isotopes of selenium, only the ^{77}Se isotope has spin quantum number $\frac{1}{2}$ with natural abundance of 7.58%. It has favorable nuclear magnetic resonance (NMR) properties that include a positive magnetogyric ratio (5.101) and 5.26×10^{-4} relative receptivity with respect to proton.[40] The nuclear Overhauser effects are absent, while longitudinal relaxation times (T_1) are usually a few seconds (1–30 s), are influenced by spin-rotation (small molecules) and chemical shift anisotropy (larger molecules) mechanisms.[41] Accordingly, ^{77}Se NMR spectroscopy has emerged as a powerful diagnostic tool in organoselenium chemistry and its popularity is growing,[42,43] although initial progress was sluggish—only ~300 articles were published before 1985.[42] Different materials have been used as a reference; dimethylselenide (Me_2Se) is now universally accepted, but being malodorous and volatile, a secondary reference, diphenyldiselenide (Ph_2Se_2; δ ^{77}Se = 463 ppm) in C_6D_6 is commonly used.

Like any other heavy nuclei, the ^{77}Se NMR chemical shifts cover a large spectral window of ~3300 ppm, bridging selenides being most shielded (*e.g.* [{CpW(CO)$_2$}$_2$(μ-Se)] δ ^{77}Se = −900 ppm), while selenoaldehydes are most deshielded (*e.g.* 2,4,6-Bu$^t_3$C$_6$H$_2$–C(H)=Se; δ ^{77}Se = 2398 ppm).[42] The large chemical shift range is advantageous in the dispersion of resonances of closely related species; even a small chemical shift difference of diastereomeric diselenides can be resolved. For instance, for a mixture of regio-isomeric diselenides (Scheme 1.3), closely spaced (~1 ppm) resonances for two distinct selenium centers for each isomeric diselenide have been reported.[44] The ^{77}Se NMR chemical shifts are highly sensitive to oxidation state, the stereochemistry of selenium and its local environment.[45–48] Larger δ values (deshielding) are usually associated with a decrease in electron density of selenium. A wide variation in ^{77}Se NMR chemical shifts with respect to the chemical state of selenium can be noted in biologically important compounds: $H_3N^+CH_2CH_2SeH$ (δ ^{77}Se = −81.6 ppm), $H_3N^+CH_2CH_2Se^-$ (δ ^{77}Se = −245.6 ppm), $(H_3N^+CH_2CH_2Se)_2$ (δ ^{77}Se = 251.3 ppm), $H_3N^+CH_2CH_2Se\text{-}SCH_2CH_2NH_3^+$ (δ ^{77}Se = 322.7 ppm)[48] and $H_2NCH_2CH_2SeO_2H$ (δ ^{77}Se = 1226 ppm).[47] The effect of the local

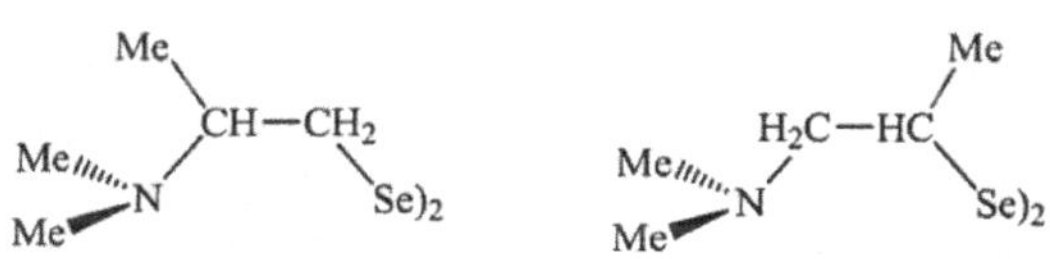

Scheme 1.3 Regio-isomeric diselenides.

environment on ^{77}Se NMR chemical shifts, as an example, is evident in *o*-carbonyl benzeneselenenyl compounds,[45] 2-RC_6H_4SeX. With a given R the shifts are spread over >800 ppm on changing X (X = Cl, Br, SCN, CN, Me) (R = Ac, X = Cl (δ, 1087 ppm); X = Me (δ, 282 ppm)), while this variation is ~100 ppm upon varying R and keeping the X group reserved. The intramolecular non-bonding Se···X interaction, resulting from a n_x-σ*Se orbital interaction leads to a downfield shift of the ^{77}Se NMR resonance.[49] Approximate linear correlation between the ^{77}Se NMR shifts and the strength of non-bonding Se···X interaction has been found using theoretical calculations.[50] ^{77}Se NMR spectroscopy is gaining momentum for understanding various process, like conformational mobility, molecular interactions of selenocysteine (sec) in biological macromolecules *etc.*, involving selenoproteins in biological samples.[51] Site-specific pK_a values of multiple sec-incorporated peptides have been determined using ^{77}Se NMR spectroscopy, which differ (3.3 and 4.3) depending on its position in the polypeptide, and are significantly lower than the values for free sec (5.2–5.6).[52] Peroxidase activity of selenosubtilisin and selenonicotinamide has been investigated using ^{77}Se NMR spectroscopy and the involvement of selenol (RSeH), selenenic acid (RSeOH), selenenyl sulfide (RSeSR′) and seleninic acid (RSe(O)OH), have been identified in the catalytic cycle.[53,54]

The presence of other nuclear spin 1/2 nuclei results in spin–spin couplings which appear as satellite peaks and provide invaluable information about the structure and stereochemistry of the molecule. Almost all coupling constants $^nJ(^{77}Se–X)$ are now reported. The $^1J(^{77}Se–^1H)$ coupling constants for selenol range between 44 and 65 Hz. The $^1J(^{77}Se–^{13}C)$ couplings in organoselenium compounds vary in the range 45–90 Hz,[55,56] whereas these values are much larger (127–250 Hz) in fluorinated selenium compounds and selenocynates.[57] The $^2J(^{77}Se–^{13}C)$ couplings in alkylseleno-ethers and dialkyldiselenides are 4–15 Hz.[55] Selenium–phosphorus couplings have been extensively investigated. The magnitude of $^1J(^{77}Se–^{31}P)$ couplings falls in the range 200–500 Hz and 500–~1000 Hz for formal selenium–phosphorus single and double bonds, respectively. The $^2J(^{77}Se–^{31}P)$ values vary from few hertz to a few tens of hertz and have great diagnostic importance in determining the stereochemistry of the molecule[58] (*e.g.* $^2J(^{77}Se–^{31}P)$ values for *cis* and *trans* isomers of $[Pt(SePh)_2(PPh_3)_2]$ are 45 Hz and 7 Hz, respectively[58a]). The $^1J(^{77}Se–^{77}Se)$ couplings in diselenides and inorganic seleno-cations fall in the range of 11–400 Hz; cations and cyclic diselenides showing larger coupling constants.[56,59] Single bond coupling with heavier nuclei can run in several hundreds of hertz. The $^1J(^{119}Sn–^{77}Se)$ in diorganotin selenolates (*e.g.* $[Me_2Sn\{SeC_4H(Me\text{-}4,6)_2N_2\}_2]$, $^1J(^{119}Sn–^{77}Se)$ = 725 Hz) varies in the range 595–1000 Hz.[17,60] Similarly, the $^1J(^{195}Pt–^{77}Se)$ coupling constants range from ~100 Hz to several hundred hertz (*e.g.* $^1J(^{195}Pt–^{77}Se)$ for *trans*-$[Pt(SeC_5H_4N\text{-}4)_2(PEt_3)_2]$ is 81 Hz and for $[Pt_2Cl_2(\mu\text{-}SeBz)_2(PPr_3)_2]$ is 134 229 Hz).[61]

1.4 Mass Spectrometry

Naturally occurring selenium is a mixture of six isotopes: ^{74}Se (0.87%), ^{76}Se (9.02%), ^{77}Se (7.58%), ^{78}Se (23.52%), ^{80}Se (49.82%) and ^{82}Se (9.19%). This distribution gives rise to characteristic isotopic patterns in mass spectra of selenium compounds. Mass spectrometry has therefore emerged as one of the principal techniques for characterization of selenium-containing molecules. Besides its routine use in synthetic organic chemistry of selenium, its use for the characterization of selenium-containing metabolites in Se-rich yeast, as well as in biological samples is growing.[62,63]

1.5 X-ray Crystallography

In addition to the ^{77}Se NMR and mass spectral investigations on selenium compounds, a rapid progress can also be attributed to X-ray crystallography. A search of the Cambridge Structural Database (2015) revealed that there were ~300 structural data on selenium compounds before 1980, which increased rapidly after 2000. The database now covers >11 000 structures (Figure 1.2). The structures of selenium compounds in all covalencies from one to six, as well as cyclic ring compounds are reported. Structures, in general, are consistent with the valence shell electron pair repulsion model where a stereochemically active lone pair of electrons is involved. The C–Se distances fall in the region 1.89–1.98 Å with C_{alkyl}–Se being longer than the C_{aryl}–Se distances. Compounds containing hydroxyl, carboxyl and amino groups are associated in the solid state through intra- and/or inter-molecular hydrogen bonding, resulting in dimeric[64] to infinite[65] chains. The stereochemistry of selenium compounds is quite diverse (Table 1.1).[66–74]

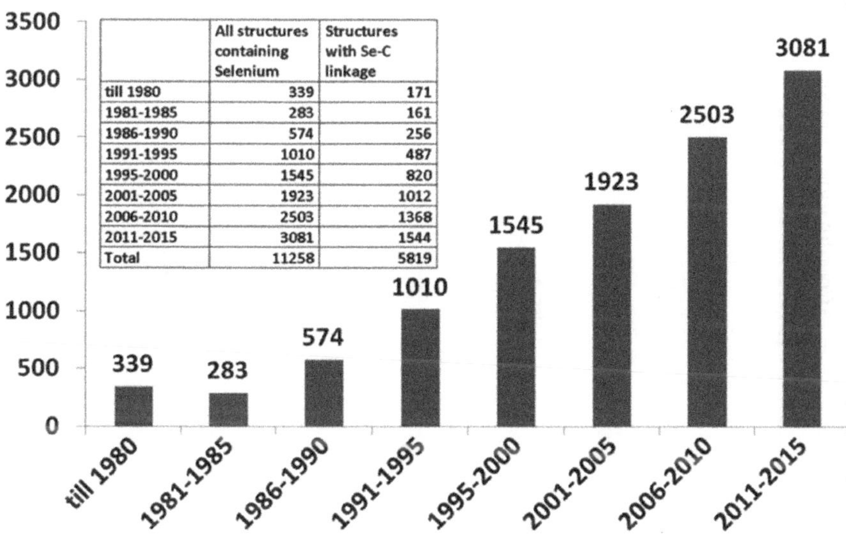

	All structures containing Selenium	Structures with Se-C linkage
till 1980	339	171
1981-1985	283	161
1986-1990	574	256
1991-1995	1010	487
1995-2000	1545	820
2001-2005	1923	1012
2006-2010	2503	1368
2011-2015	3081	1544
Total	11258	5819

Figure 1.2 Selenium compounds characterized using X-ray crystallography.

Table 1.1 Organoselenium compounds with selenium coordination numbers in different oxidation states.[66–74]

Oxidation state of selenium	Coordination number of selenium	Examples
One	Two	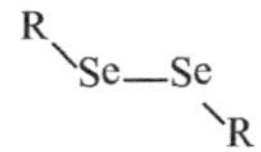 (diorganodiselenides)
	Three	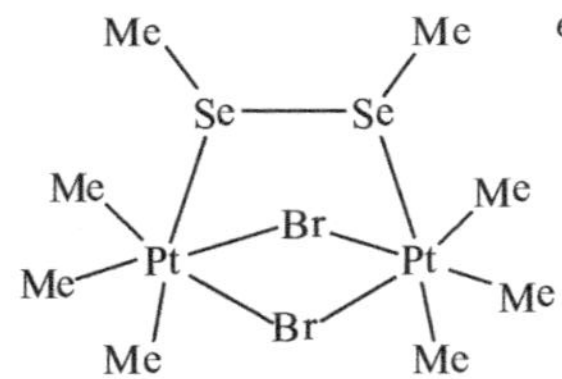 66
Two	Two	ReSeH (selenols); Re-Se-R′ (diorganoselenides); R-Se-X [X = Cl or Br] (seleneyl halides); OH (selenenic acid); OR (selenenic acid ester); CN (selenocyanate) 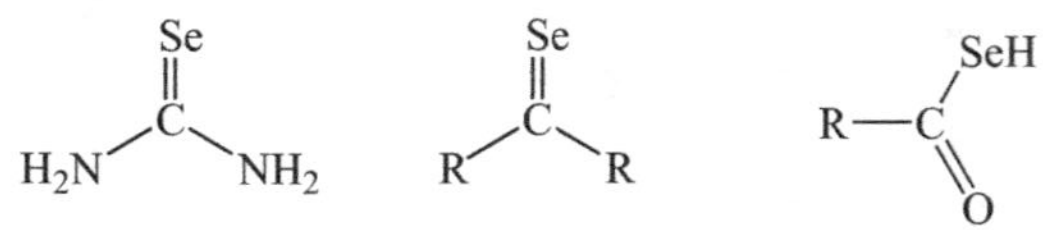 selenourea selenoketone selenocarboxylic acids Metal selenolates (*e.g.* $[Pt(SeC_4H_3N_2)_2(PPh_3)_2]$[67]
	Three	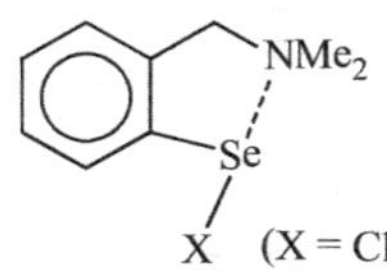 68 (X = Cl or I) Metal selenolates (*e.g.* $[Pd_2Cl_2(\mu\text{-}SeR)_2(PR'_3)_2]$)[69]
	Four	Metal selenolates (*e.g.* $[Pd_3Cl_2(\kappa^2\text{-}Se,N\text{-}SeCH_2CH_2NMe_2)(\eta^3\text{-}C_3H_5)_3]$[70]

Four	Two	O=Se=O (selenium dioxide); RN=Se=NR (selenodiimides)
	Three (pyramidal)	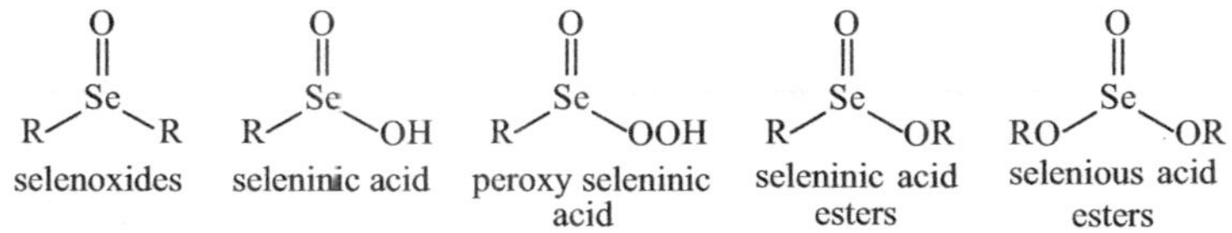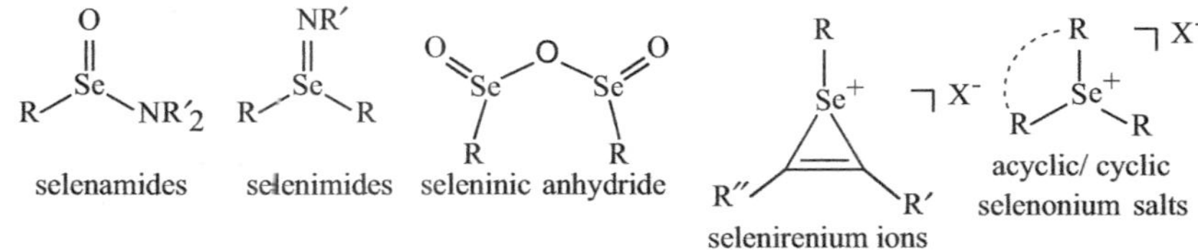
	Four (trigonal bipyramidal)	R_2SeX_2 (X = Cl or Br; R = Me, Ph, *etc.*),[71] $PhSeCl_3$, $[Se(OR)_4]$ 
	Five	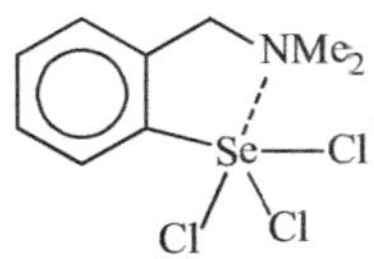[68]; $[PPh_4][Se(N_3)_5]$[72]

Table 1.1 (*Continued*)

Oxidation state of selenium	Coordination number of selenium	Examples
Six	Four	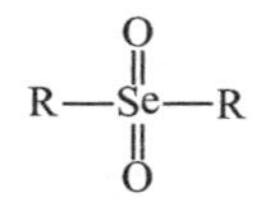 selenones 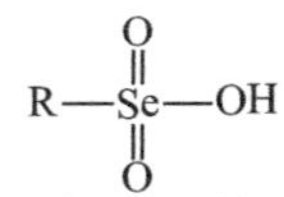 selenonic acid 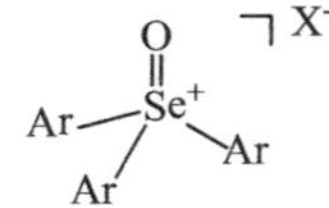 selenonium salt oxide
		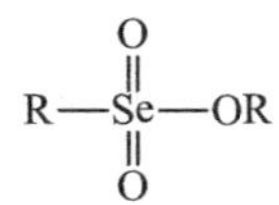 elenonic acid esters 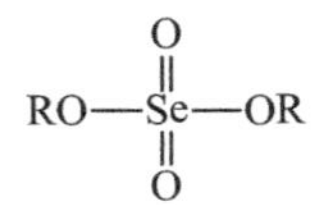 seleninic acid ester 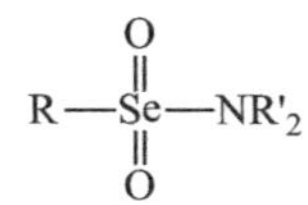 selenonic amides 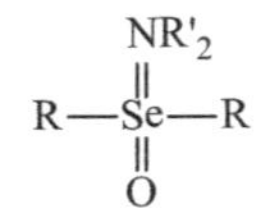 selenoximines
	Five	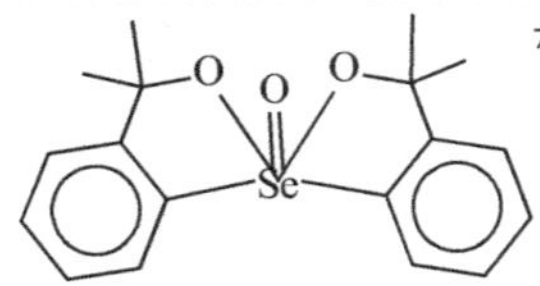 73
	Six	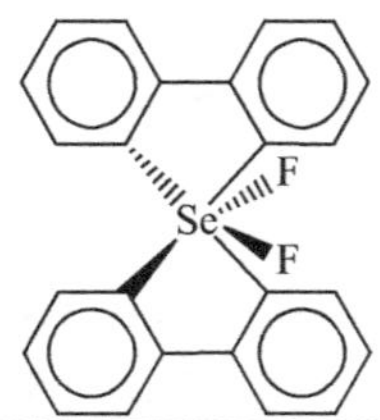 74

Diorganodiselenides, like other dichalcogenides, have skewed structures in solution and in the solid state, and both the enantiomeric forms display chiral *P* and *M* helical configurations (Scheme 1.4a). Racemization of these forms is quite facile due to the low barrier of rotation of the Se–Se bond (*e.g.* the energy barrier in the cisoid and transoid forms of Ph_2Se_2 is 8.2 kcal mol^{-1} and 5.2 kcal mol^{-1}, respectively[75]). The Se–Se distance in diselenides ranges from 2.24 to 2.39 Å and is dependent on the nature of the substituent,[16] the bond being shorter for more electronegative R groups (*e.g.* for Se_2Br_2 the Se–Se = 2.241 Å[76]). Shorter (2.265(1) Å) and quite long (2.974(1) Å) Se···Se distances between the two molecular halves of dimeric rectangular dication $[RSe]_4^{2+}$ (Scheme 1.4c), obtained by one-electron oxidation of diorganodiselenide by NO^+OTF^- or $XeF_2/BF_3 \cdot OEt_2$, have been reported.[77] The dihedral angle (Scheme 1.4b) in diorganodiselenides deviates significantly from the idealized value of 90°, depending on the nature of the organic groups and on the existence of other secondary interactions. In most cases the C–Se–Se–C angle[16] lies in the range 70–100°, while intra-molecular Se···N interactions (as in $[(3,5\text{-}Me_2C_5H_2N)_2Se_2]$ and $[\{2\text{-}NC_5H_3(3\text{-}CONHPh)\}_2Se_2]$[78]) or sterically bulky groups (as in $[\{(Me_3Si)_3C\}_2Se_2]$[79]) result in the opening of the angle to 180°. Polymorphism in diorganodiselenides is also encountered, and examples include $[2\text{-}py_2Se_2]$[80] and $[\{2\text{-}(3\text{-}H_2NCOC_5H_3N)\}_2Se_2]$.[81]

A large number of symmetrical and unsymmetrical diorganoselenides (Scheme 1.5) have been structurally characterized. These compounds adopt a V-shaped geometry with C–Se–C angles[16] varying in the range 95–104°. Diorganoselenoxides adopt a trigonal pyramidal shape, but considering the role of a stereochemically active lone pair of electrons, structures can be defined as distorted tetrahedral. The Se–O distances are in the range

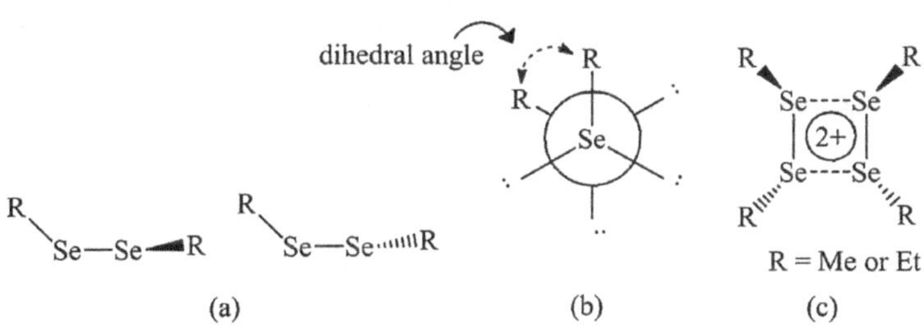

Scheme 1.4 (a) Chiral forms of diselenide; (b) view down the Se–Se bond showing the dihedral angle; and (c) schematic representation of the diselenide cation.

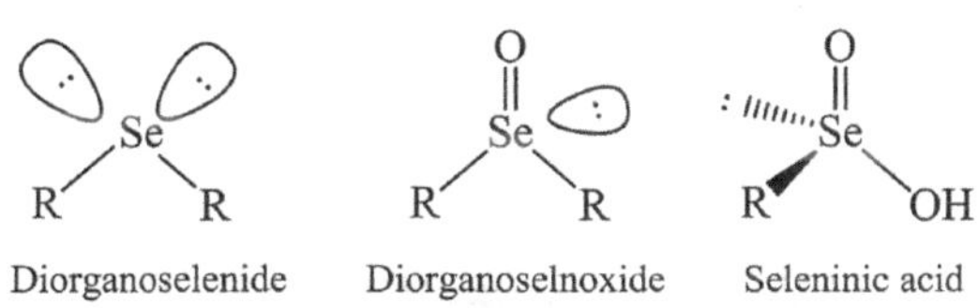

Scheme 1.5 Diorganoselenide, diorganoselenoxide and seleninic acid.

1.62–1.69 Å and the C–Se–C angles are similar to parent selenides, while the C–Se–O angles lie at ~102°. These compounds are often associated in the solid state through weak inter-molecular hydrogen C–H···O or Se···O interactions to give dimers,[82] star-shaped hexamers[83] or infinite strings.[84] The hydrated selenoxides ($R_2SeO \cdot H_2O$, misrepresented in older literature as $R_2Se(OH)_2$), show extensive hydrogen bonding with water molecules (Se–O···H–O).[84] The Se–O bond lengths are little influenced by such secondary bonding interactions. The Se–O distances in seleninic acids (RSe(O)OH) are similar to selenoxides; however the Se–OH distance (1.74 Å), being the single bond, is longer.[85]

1.6 Synthesis of Organoslenium Compounds

1.6.1 Diorganodiselenides

Diorganodiselenides are yellow to orange-red, shelf-stable compounds and are important precursors in synthetic chemistry; diphenyldiselenide is the most extensively used reagent. Several synthetic approaches are in practice for the preparation of diorganodiselenides. The most commonly used method is the reaction of Li_2Se_2 or Na_2Se_2, prepared in several different ways (Scheme 1.6), with an appropriate alkyl or aryl halide followed by aerial oxidation.[86–95] Diazonium salts, in place of an organic halide, can also be used for the synthesis of diaryldiselenides.[96] Most of these reactions are quite often accompanied by the formation of selenides (R_2Se) in variable amounts, and hence the purification of diselenides is essential.

Sodium borohydride reduces elemental selenium to NaSeH or Na_2Se_2, depending on the molar ratio of the reactants. Sodium hydrogen selenide (NaSeH), generated either by $NaBH_4$ reduction of selenium[90] or by the reaction of H_2Se with NaOEt in ethanol,[97] has also been used for the synthesis of diselenides. For instance, the reaction of NaSeH with 6-tosyl

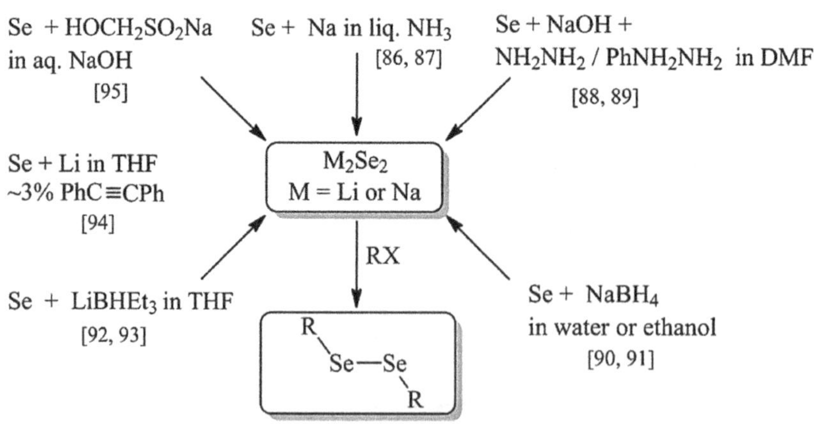

Scheme 1.6 Synthesis of diorganodiselenides from M_2Se_2.[86–95]

β-cyclodextrin in phosphate buffer followed by aerial oxidation yields cyclodextrin-based diselenides (Scheme 1.7, **I**), which shows 4.3-fold higher GPx activity than that of ebselen.[98] Other cylcodextrin-based diselenides have also been synthesized and characterized.[21]

Another promising approach for the synthesis of various aromatic diselenides has been heteroatom-directed aromatic lithiation.[99–101] *Ortho* deprotonation of the aromatic substrate by alkyllithium reagent, usually ^{n}BuLi, followed by treatment with selenium and aerial oxidation results in the corresponding diselenides (Scheme 1.8). A variety of substrates, such as *N*-substituted benzylamines, 2-phenyloxazolines, 1-dimethylamino naphthalene, 1-bromobenzaldehyde, aromatic Schiff bases, ferrocenes and 3-substituted thiophenes, have been used successfully in these reactions.

Acid (*e.g.* HCl) and base (*e.g.* KOH, NaOMe) hydrolysis of selenosulfates ($RSeSO_3K$)[48] and selenocyanates (RSeCN),[102,103] respectively, has also been employed for the preparation of diselenides (Scheme 1.9).

Other, less frequently employed methods for the synthesis of diorgano-diselenides include the reduction of selenium with carbon monoxide and aromatic aldehyde as substrates,[104] and use of CuO nano-powder as a catalyst in the reaction of selenium in KOH solution with an organic halide (Scheme 1.10).[105]

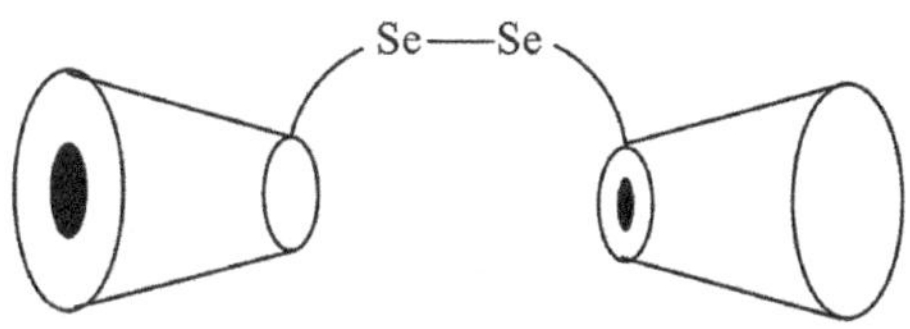

(**I**) 6,6′-bis(cyclodestrin)diselenide

Scheme 1.7

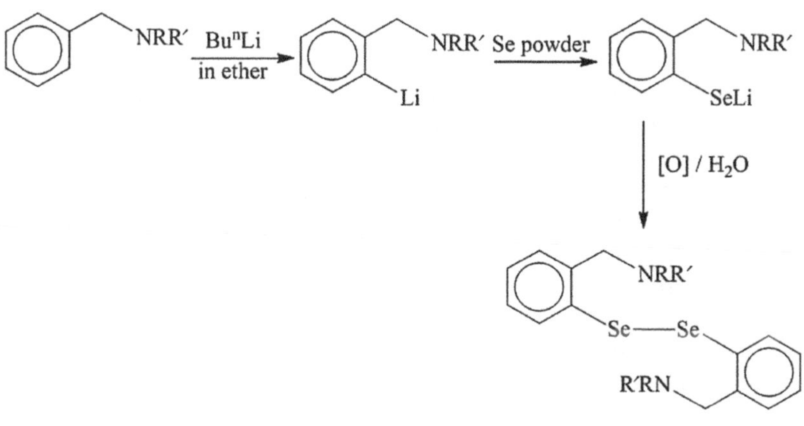

Scheme 1.8 Synthesis of diorganodiselenides from ArSeLi.

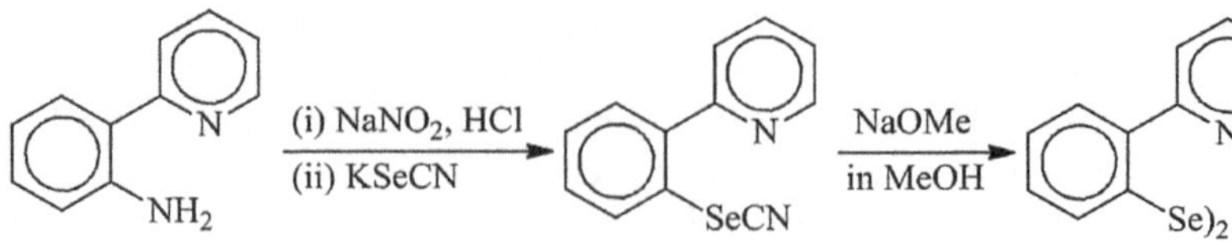

Scheme 1.9 Preparation of diselenides using selenocyanates.

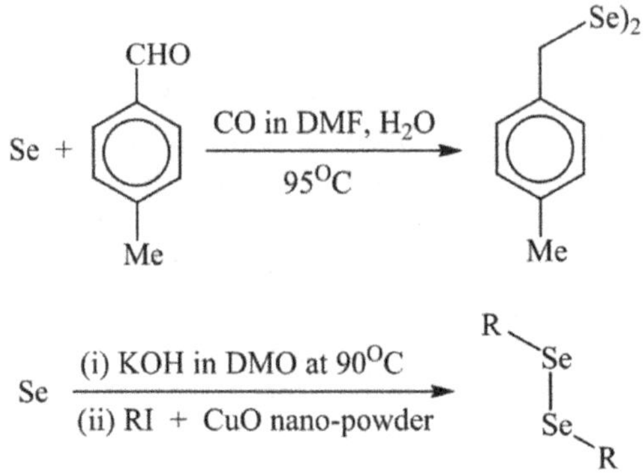

Scheme 1.10

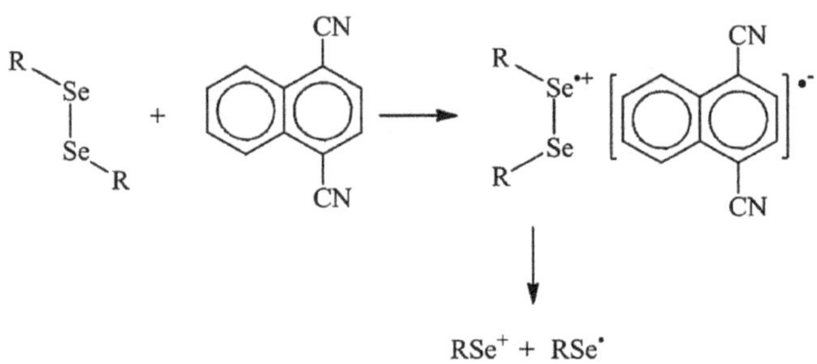

Scheme 1.11 Formation of radical cations.

Diorganodiselenides show remarkable redox behaviour. They can conveniently be oxidized to electrophilic states (*e.g.* RSeX or $RSeX_3$) and reduced to nucleophilic selenolate ions by the cleavage of the Se–Se bond. Cyclic voltammetric measurements of diselenides reveal that reversible electron transfer is typical for selenol/diselenide couples and is more negative than those for the corresponding thiol/disulfide couples.[106] Single-electron oxidation results in a radical cation intermediate species (Scheme 1.11).[107] Diselenides, like $(HOOCCH_2CH_2Se)_2$, react with one-electron oxidants (*e.g.* hydroxyl radicals produced during pulse radiolysis) to give diselenide radical cations with an absorption maximum at 560 nm.[108] However, the radical cations with simple alkyl derivatives, $[Me_2Se_2]^{\bullet+}$, dimerize as a rectangular dicationic species, $[Me_2Se_2]_2^{2+}$ by π^*–π^* interaction with long Se–Se bonds.[77]

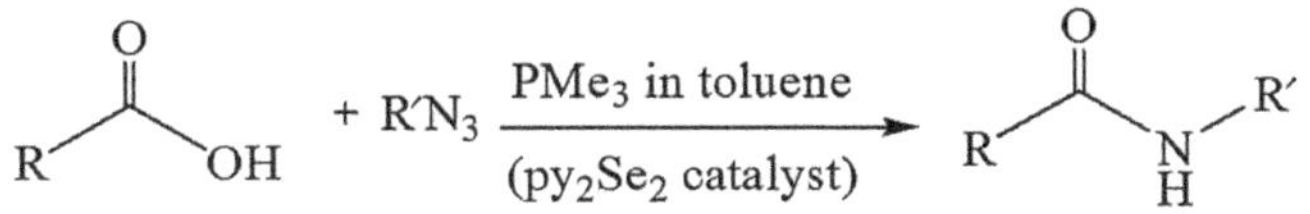

Scheme 1.12

Deselenization of diselenides can take place under photolytic or thermolytic conditions. Under photolytic conditions in the presence of tertiary phosphine, they yield the selenides (R_2Se) and R_3PSe.[109] Selenium is extruded from dibenzyldiselenide under photolytic as well as thermolytic conditions.[110]

Diorganodiselenides are used as a catalyst in a number of organic transformations.[111] For example, dipyridyldiselenide is used as a catalyst in the Staudinger–Vilarrasa reaction between carboxylic acids and azides in the presence of PMe_3 at room temperature (Scheme 1.12).[112] Diorganodiselenides are versatile synthons for the preparation colloidal semiconductor nanocrystals.[113]

1.6.2 Selenols

Selenols, selenium analogues of alcohols and thiols, are stronger acids than the corresponding thiols (pK_a 5.9 and 6.5 for PhSeH and PhSH, respectively). They are readily oxidized by atmospheric oxygen to diselenides and are therefore usually generated *in situ* for a chemical reaction. There are several methods to prepare selenols (Scheme 1.13);[114–117] the general routes for their synthesis are as follows.

(i) Reaction of Grignard reagent or organolithium with elemental selenium followed by hydrolysis with dilute acid yields RSeH. The reaction is often accompanied with the formation of selenides, diselenides and H_2Se as by-products. For instance, carboraneselenol, $B_{10}H_{11}C_2SeH$, is prepared by the reaction of 1,2-dicarba-*closo*-dodecaborane with ^{n}BuLi in DME followed by treatment with selenium powder and subsequent hydrolysis by HCl.[118] In this reaction diselenide is also formed as a by-product.

(ii) Reaction of organic halides[80b,119] or aromatic diazonium bromide[120] with $NaSeH/Na_2Se_2$ is another common route for the synthesis of selenols. In general, organic bromides and iodides are employed in this reaction.

(iii) Diselenides, being stable compounds, are reduced conveniently to afford the corresponding selenols. Several reducing agents are used for the reduction of diselenides; among them sodium borohydride[121] and hypophosphorus acid (H_3PO_2) in hydrochloric acid/organic solvents[120,122] are common reducing agents. Thiols (such as PhSH, dithiothreitol)[123,124] and Bu_3SnH[125] have also been used for reduction of diselenides, but in these reactions selenenyl sulfides (RSeSR′) and $Bu_3SnSeAr$, respectively, are also formed.

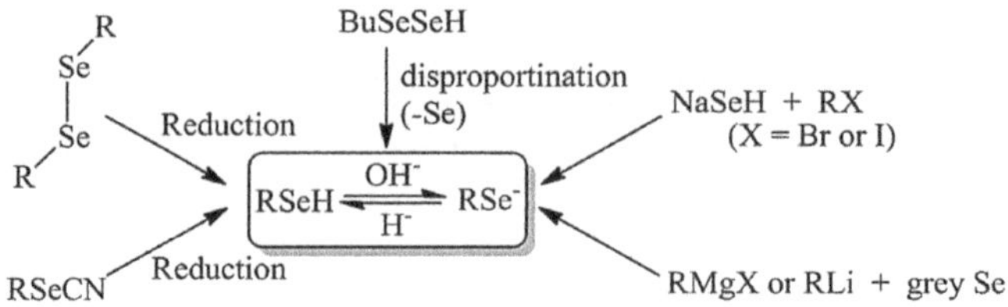

Scheme 1.13 Preparation of selenols.

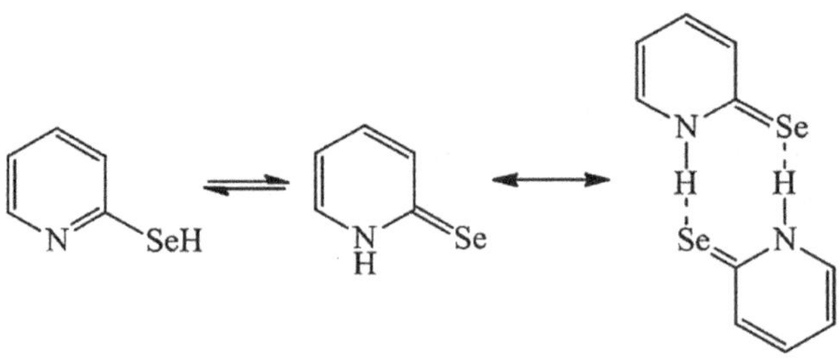

Scheme 1.14 Selenol–selone tautomerism.

(iv) Basic hydrolysis or reduction of selenocyanates has been employed for the synthesis of selenols.[116,126] Aliphatic and aromatic selenocyanates, RSeCN, which are easily obtained by the reaction of aliphatic halides or aromatic diazonium salts with KSeCN, are reduced by zinc in acidic medium.

(v) The reaction of alkylsulfate with potassium selenide and reaction between selenourea and *tert*-butylmalonodialdehyde[119] in refluxing ethanol are other protocols used for the synthesis of selenols.

Selenols are readily deprotonated in alkaline solution to give selenolate ions. Selenols containing unsaturated/N-heterocyclic rings tautomarize to the corresponding selone (C=Se linkage).[127,128] The selenol and selone can exist as discrete species, *e.g.* 2-pridyl selenol and 2-pryridylselone (Scheme 1.14).[127]

Selenols and their conjugate bases, selenolates (RSe^-) have high energy highest occupied molecular orbitals and are weak bases but powerful nucleophiles and react readily with electrophilic inorganic and organic compounds.[114,115] Selenols are remarkable reagents for the introduction of organoselenium groups to organic molecules either through addition to carbon–carbon multiple bonds or by reaction with organic halides, carboxylic acid chlorides, epoxides, *etc.* Selenols undergo addition of the Se–H bond to alkynes to give α,β-unsaturated compounds.[129] The non-catalytic (radical or base initiated) addition proceeds non-stereoselectively to give a mixture of anti-Markovnikov products, while palladium-catalyzed reactions result in Markovnikov products (Scheme 1.15).[130]

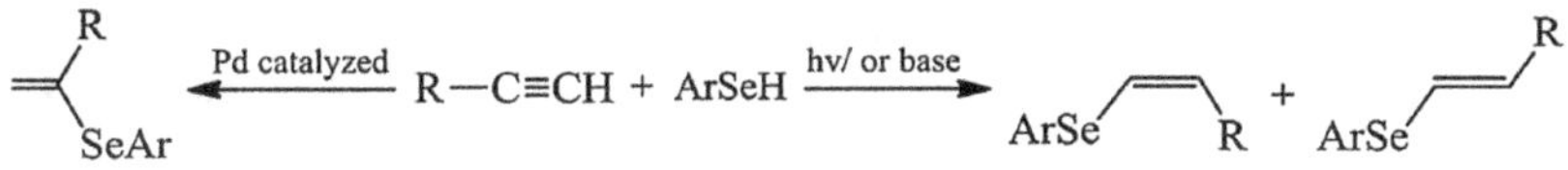

Scheme 1.15 Reactions of selonol with alkynes.

1.6.3 Diorganoselenides

Diorganoselenides are the oldest organoselenium compounds, prepared as early as 1836 by Löwig, who described the synthesis of diethylselenium.[131] These compounds are the most prevalent organoselenium derivatives and are also called as selenoethers. A myriad symmetrical and unsymmetrical selenides,[8a,116,132] which include acyclic alkyl, aryl, cyclic with varying ring sizes,[133] functionalized,[16,134] macrocyclic,[135] bi- and multi-selenium centred (*e.g.* $RSe(CH_2)_nSeR$), have been synthesized. These compounds are extensively used as ligands in coordination and organometallic chemistry[11c,128] and as a nucleophile in numerous organic reactions.[8a,116]

Although several protocols have been adopted for the synthesis of diorganoselenides, the most commonly employed methods are depicted in Scheme 1.16. The reaction of sodium selenide (Na_2Se), generated in several ways, with an organic halide in the appropriate solvent has been used to prepare a wide range of symmetrical as well as functionalized selenides.[8a,11c,65a,116,132,134–136] Both symmetrical and unsymmetrical selenides are conveniently synthesized by treatment of selenolate ion with non-activated organic halides (aromatic, hetero-aromatic, vinyl, *etc.*).[14b,65a,137] The selenolate ion is usually generated by reductive cleavage of the Se–Se bond of a diselenide. Sodium borohydride[14b] and hydrazine in the presence of NaOH in DMF are the common reducing agents. Diarylselenides can also be obtained by the reaction of selenolates with diazonium compounds.[7] Unsymmetrical aromatic selenides have been prepared by the reaction of ArSeX (X = H, Br, CN) with an organolithium compound or a Grignard reagent.

Several synthetic approaches have been adopted for the preparation of selenides using selenocyanates.[138,139] Functionalized selenides are efficiently prepared by ring opening of epoxides, aziridines and related molecules with selenols.[116,140] For example, β-hydroxyaryl selenides are conveniently obtained by regio-selective ring opening of epoxide with selenophenol (Scheme 1.17).[140] Electrophilic (RSeX; X = Cl or Br) (Scheme 1.17)[139,141] and nucleophilic (RSeH) selenium compounds add to alkynes, alkenes and carbonyl compounds to give functionalized selenides. Redox transmetalation of $Ag(C_5H_4N)$ with red selenium in EtCN at room temperature yields $(NH_4C_5)_2Se$, but at 50 °C diselenide is formed.[142]

Diorganoselenides undergo a variety of reactions. They are readily oxidized by one- and two-electron oxidants. Oxidation of Ph_2Se by one-electron oxidants (*e.g.* $OH^{\bullet}$, $Br_2^{\bullet-}$, $N_3^{\bullet}$) generated on pulse radiolysis, gives radical cation,[143] $Ph_2Se^{\bullet+}$. The redox potential of the $Ph_2Se^{\bullet+}/Ph_2Se$ couple is 1.37 V. One-electron oxidation of 1,8-bis(phenylselenyl)naphthalene, $napSe_2Ph_2$ by $NO[Al(OR)_4]$ ($R = C(CF_3)_3$) gives a blue-colored paramagnetic

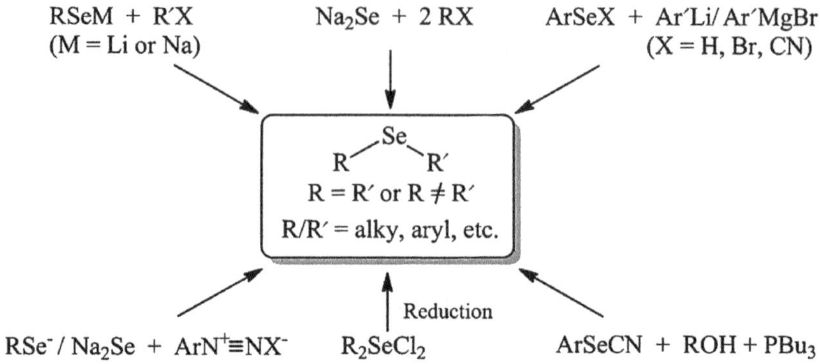

Scheme 1.16 Synthesis of monoselenides.

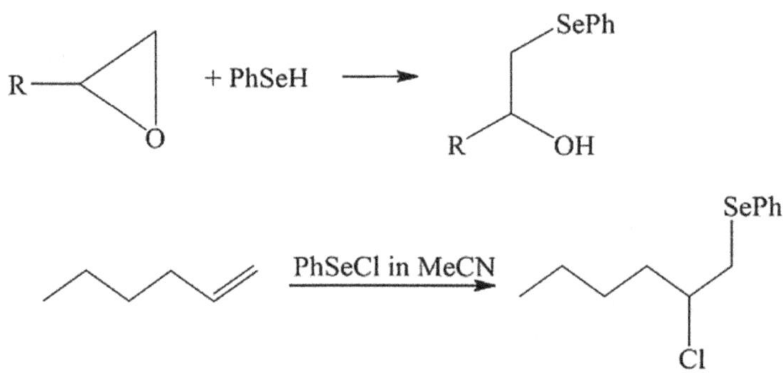

Scheme 1.17

radical cation, $[napSe_2Ph_2]^{\bullet+}$ containing a Se∴Se three-electron σ-bond which dimerizes to a brown diamagnetic species in the solid state.[144] Oxidation by two-electron oxidants results in Se(IV) compounds. Oxidation by H_2O_2 and alkyl halides yields selenoxides (R_2SeO) and selenonium salts ($[R_2SeR']^+X^-$), respectively. Selenides containing α-hydrogen on treatment with strong bases like lithium diisopropylamide (LDA) results in deprotonation to give the corresponding carbanions (RCH–SeAr).

1.6.4 Diorganoselenoxides

Diorganoselenoxides (R_2SeO; R = alkyl or aryl) are among the best-known organoselenium compounds[145] and find extensive applications in chemical synthesis as oxygen transfer agents in organic[146] and organometallic synthesis[147] and as oxygen donor ligands in coordination chemistry.[148]

Selenoxides are conveniently obtained by the oxidation of selenides. Several oxidizing agents, such as hydrogen peroxide,[149] peroxy acids (*e.g.* *m*-chloroperbenzoic acid[150]), ozone, sodium hypochlorite (NaOCl) in DMF[151] or *tert*-butylhypochlorite[150] are generally employed for the oxidation of

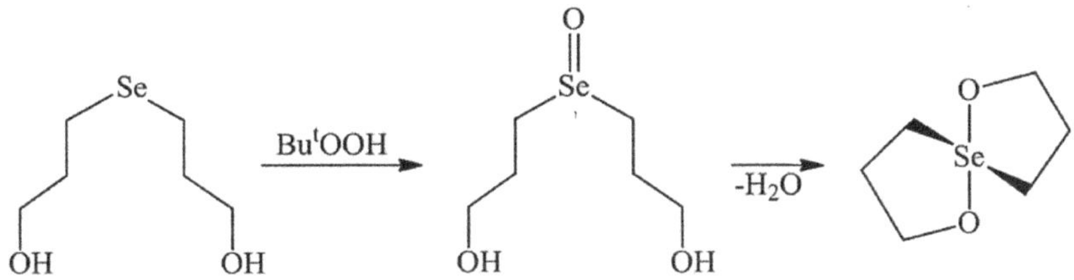

Scheme 1.18 Cyclization of selenoxide *via* dehydration.

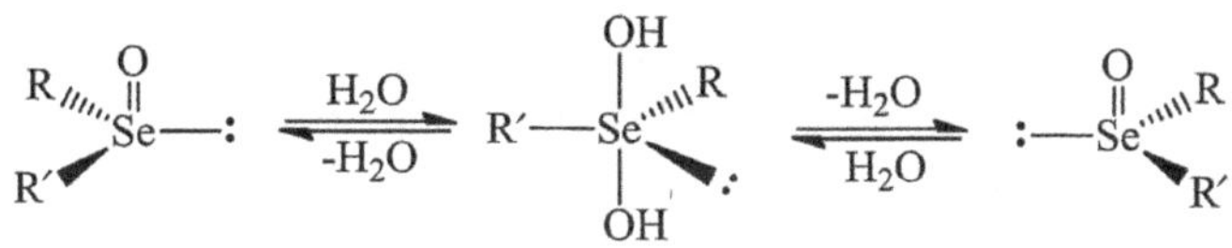

Scheme 1.19 Pyramidal inversion in unsymmetrical selenoxides.

selenides. Alternatively, selenoxides can be prepared by hydrolysis of di-organoselenium dichlorides using alkali metal hydroxides, sodium acetate and silver oxide in aqueous medium. Rearrangement in selenoxides containing protic organic groups (*e.g.* OH, COOH) may take place during oxidation.[152,153] For example oxidation of di(3-hydroxypropyl)selenide with *tert*-butylhydroperoxide results in to the formation of expected selenoxide, di(3-hydroxypropyl)selenoxide, which readily undergoes intra-molecular dehydration to yield a spiro-selenurane (Scheme 1.18).[152] Similarly 3,3′-selenodipropionic acid on oxidation readily yields a dehydrated cyclized product, whereas the homologous 4,4′-selenodibutyric acid and 5,5′-seleno divaleric acid yield corresponding selenoxides.[153] Further oxidation of selenoxides is a slow process and requires strong oxidizing agents like peroxy acids. Unsymmetrical selenoxides are chiral molecules and are stable towards pyramidal inversion at room temperature, but racemization is facilitated in the presence of an acid (Scheme 1.19).[154]

1.6.5 Selenenyl Sulfides

Diorganoselenenyl sulfides (RSeSR′) are considered to be important intermediates during catalytic cycle of GPx. The high reactivity of diselenides with thiols has been exploited for the synthesis of selenenyl sulfides.[155] This method has been employed successfully for the preparation of slenenyl sulfide glycopeptides[156] and selenenyl sulfide-bearing lipids[157] which have been characterized using mass spectroscopic techniques. The reaction of ebselen and its analogues with aromatic thiols yields selenenyl sulfides (Scheme 1.20), which exhibit strong Se···O intra-molecular interactions.[158] These interactions prevent the regeneration of catalytically active selenol species; as a consequence, low catalytic activity of ebselen analogues in the presence of thiols is observed and such species are considered as the dead-end products.[158] This type of hypervalent T-shaped selenenyl sulfides,

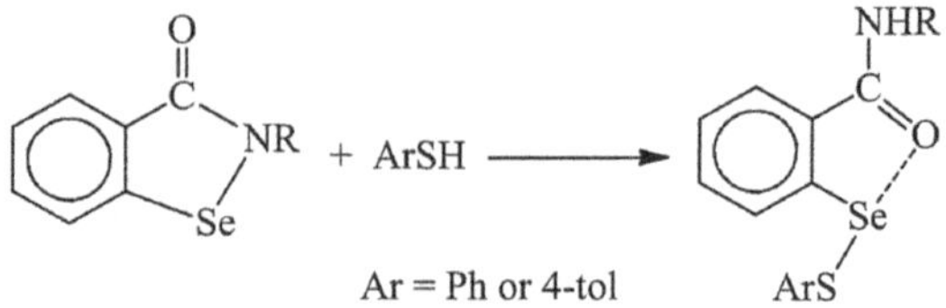

Scheme 1.20 Synthesis of selenenyl sulfides.

[{2–$R_2NCH_2C_6H_4$}SeSP(S)R'_2], showing weak intra-molecular Se···N (~2.67 Å) interactions,[159] have been isolated by redistribution reactions between diorganodiselenides and bis(diorganophosphinothioyl)disulfanes, [R'_2P(S)S]$_2$. Diorganoselenenyl sulfides have also been synthesized by the reaction of RSeX with the sodium salt of a thiol.

1.6.6 Organoselenium Halides

Organoselenium halides find extensive applications as electrophilic reagents to introduce selenium into organic molecules, and are found in +2 (RSeX) and +4 ($RSeX_3$; R_2SeX_2 and R_3SeX) oxidation states. The stability of a particular compound depends on the nature of X as well as the organic group attached to selenium. Compounds in the +4 oxidation state are less stable. For instance, oxidative addition of bromine to R_2Se results in to diorganoselenium dibromide, R_2SeBr_2, which tends to decompose on standing (lose bromine and/or bromoalkanes) and is therefore used *in situ* for organic synthesis.[160] Similarly, $PhSeCl_3$, obtained by the reaction of Ph_2Se_2 with SO_2Cl_2 in 1:3 ratio in chloroform, is used for selenium insertion in to the α-position of a ketonic substrate.[161] The trihalides ($RSeX_3$) upon heating in a vacuum above their melting points decompose to selenenyl halides and halogens. Selenenyl halides are relatively stable and are primarily used as electrophilic reagents for selenenylation.

Selenenyl halides (RSeX) are prepared by the reactions of diorganodiselenides with halogens or halogenating agents such as sulfuryl chloride (Scheme 1.21). The products are isolated in quantitative yields. Selenocyanates (RSeCN) can be cleaved by chlorine or bromine in chloroform solution to yield selenenyl halides.

Selenenyl halides have low-lying lowest unoccupied molecular orbitals and are therefore powerful electrophiles. Phenyl selenenyl halides (PhSeX; X = Cl or Br) are the common electrophilic reagents used for selenenylation of olefins and carbonyl compounds.[162,163] However, the nucleophilicity of the halide in RSeX often gives rise to side reactions in organic synthesis. Accordingly, several non-halide compounds have been developed which show similar behaviour to RSeX; the trifilate (RSeOTf) being the most preferred reagent. These compounds are isolated by the reaction of RSeX with appropriate silver salts (Scheme 1.22).[164–168]

The reaction of selenium electrophiles with alkenes takes place in a stereospecific manner and proceeds *via* a seleniranium ion intermediate. The

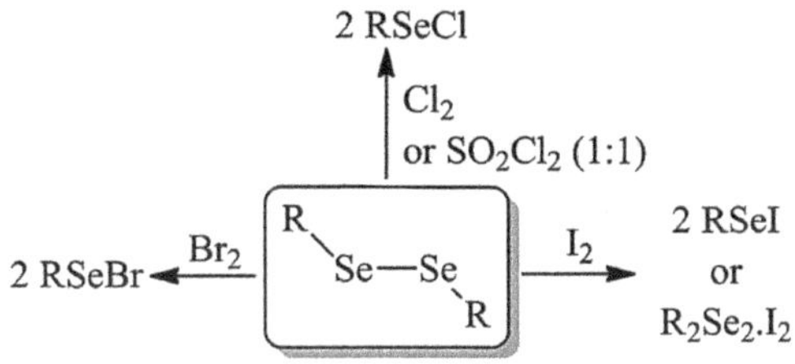

Scheme 1.21 Synthesis of selenenyl halides.

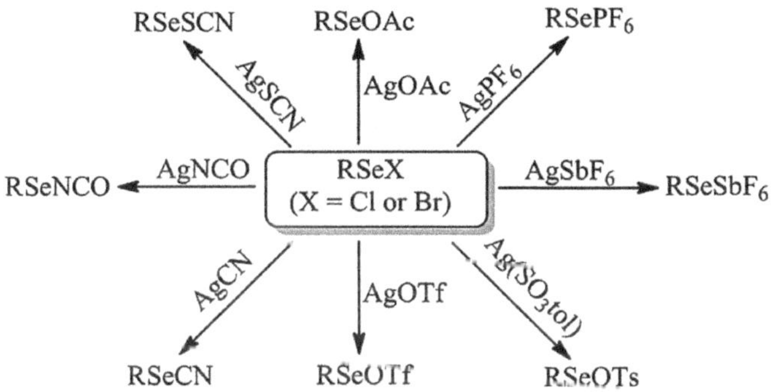

Scheme 1.22 Synthesis of selenenyl compounds.

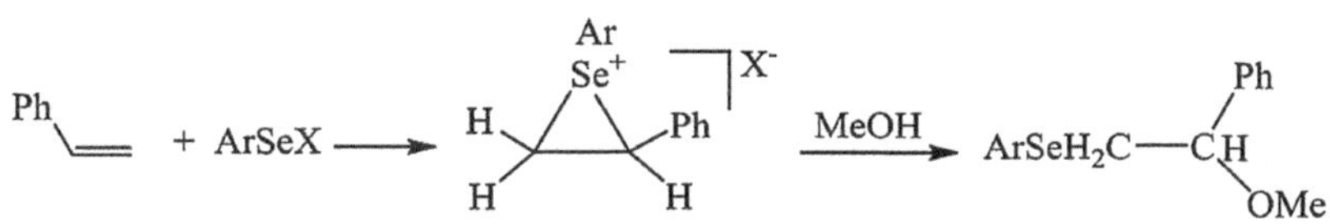

Scheme 1.23

latter is opened in the presence of a nucleophile to give addition products (Scheme 1.23).[163] α-Phenylselenocarbonyl compounds are widely used to give olefins *via* selenoxide elimination. The 2- and 4-pyridyl seleno group is a better leaving group than the phenyl seleno group in selenoxide elimination (Scheme 1.24).[169–171] The pyridylseleno compounds afford enones in excellent yields, even in cases where satisfactory results are not obtained with α-phenylselenocarbonyl compounds.

The selenenyl halides, RSeX (X = Cl or Br), are discrete molecules, but the corresponding iodo compounds exhibit considerable structural diversity. Compounds derived from bulky organic groups[172] or aryl groups containing hetero-atoms (O or N) on the pendant arms[173,174] usually exist as RSeI with a covalent Se–I bond. The majority of iodo compounds are charge-transfer (CT) species in which the Se–Se bond of the diselenide remains intact. The I_2 molecule interacts either with one selenium atom (*e.g.* $Me_2Se_2 \cdot I_2$, $(4\text{-}FC_6H_4)_2Se_2 \cdot I_2$)[175,176] to give a three-coordinate spoke structure or binds with both the selenium atoms to result in a centrosymmetric dimer

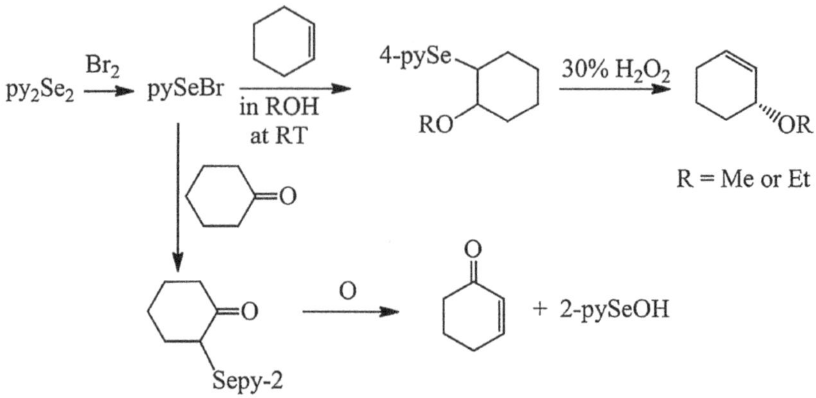

Scheme 1.24 Synthesis of olefins using selenenyl bromide.

(*e.g.* $(Ph_2Se_2 \cdot I_2)_2$).[177] The electron-donating R groups which increase the basicity of Se in general results in CT compounds.[178] A 1:1 CT compound containing a covalent Se–I bond, *p*-$ClC_6H_4SeI \cdot I_2$ is isolated by the reaction of (*p*-$ClC_6H_4Se)_2$ with I_2 in a 1:3 molar ratio.[176] The existence of Se···X (X = N, O, S) non-bonding interactions in a number of organoselenenyl halides has been reported by several authors and has been probed by X-ray crystallography, NMR and density functional theory calculations.[37]

1.6.7 Selenenic, Seleninic and Selenonic Acids

Selenenic acids (RSeOH) (Scheme 1.25; selenenic (**II**), seleninic (**III**) and selenonic (**IV**) acids) are highly reactive intermediates and are generated during the oxidation of selenols and diselenides. They are postulated in a number of reactions such as selenoxide *syn* elimination, *etc.* and are thought to be the active species in the catalytic cycle of GPx. These compounds are highly unstable and undergo disproportionation, possibly involving **V** and **VI** intermediates, to the corresponding diselenides and seleninic acids or anhydrides (Scheme 1.26).

The existence of several arylselenenic acids stabilized by the coordination of nitro, carbonyl or amine groups in solution has been reported,[179] but in no case has alkylselenenic acid been detected. With bulky aryl groups,[180] *e.g.* 2,4,6-$Pr^i_3C_6H_2$ and triptycyl (trip),[181] ArSeOH has been characterized using NMR spectroscopy, and even the redox behavior of tripSeOH has been investigated.[182] A stable selenenic acid derived from a bowl-shaped organic group, BmtSeOH (Scheme 1.27, **VII**), has been isolated by the direct oxidation of selenol with H_2O_2 and structurally characterized. The compound shows remarkable stability both in solution and in the solid state.[183]

Hydrolysis of selenenyl halides or the reduction of seleninic acids/anhydrides also yields selenenic acids. The reaction has been used to generate selenenic acids *in situ*. 2-Nitrobenzene seleninic acid on reduction with NaH_2PO_2 yields the corresponding selenenic acid.[179a] Hydrolysis of

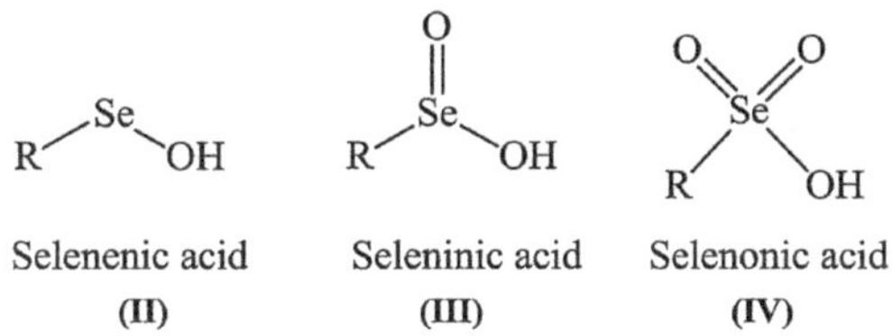

Scheme 1.25

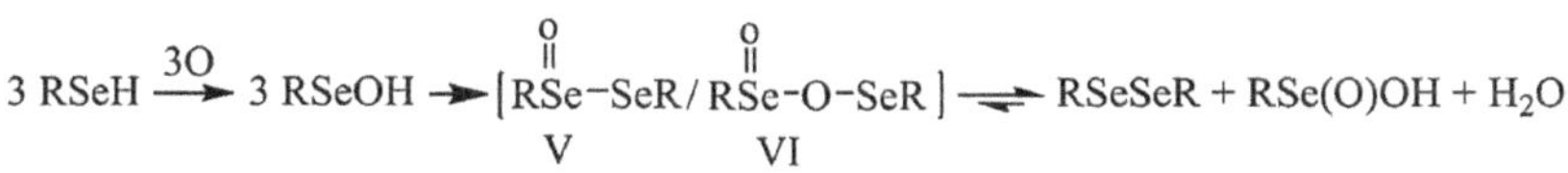

Scheme 1.26

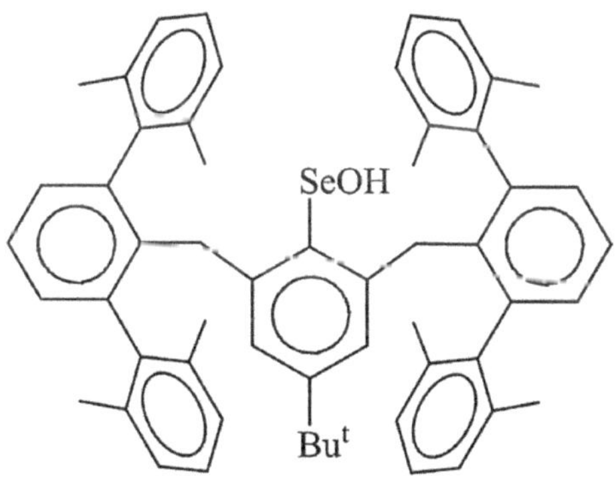

BmtSeOH (**VII**)

Scheme 1.27

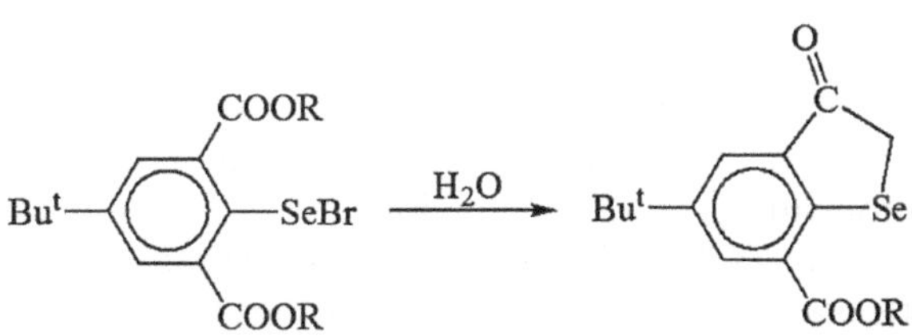

(**VIII**) (**IX**) (R = Me or Pr^i)

Scheme 1.28

selenenyl bromide (**VIII**) during crystallization leads to the formation of selenenate esters (**IX**) (Scheme 1.28).[184] Oxidation of secondary and tertiary amide based diselenides with hydrogen peroxide has been investigated.[185] In these reactions the involvement of selenenic acid has been suggested. In the case of the secondary amide group, rapid cyclization takes place to produce seleneyl amides which on oxidation with excess H_2O_2 yield seleninic

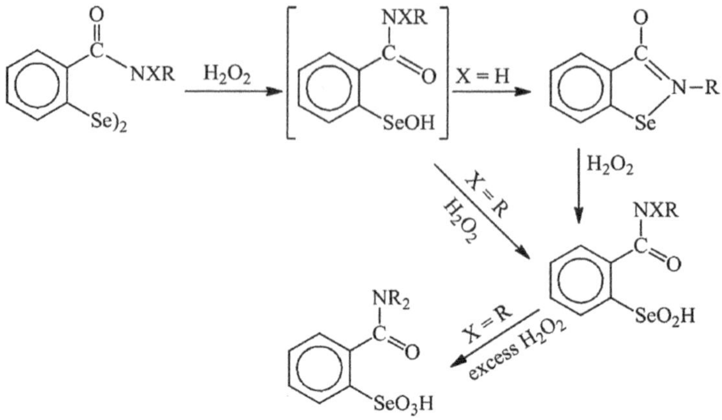

Scheme 1.29 Oxidation of seleninic acids to selenonic acids.

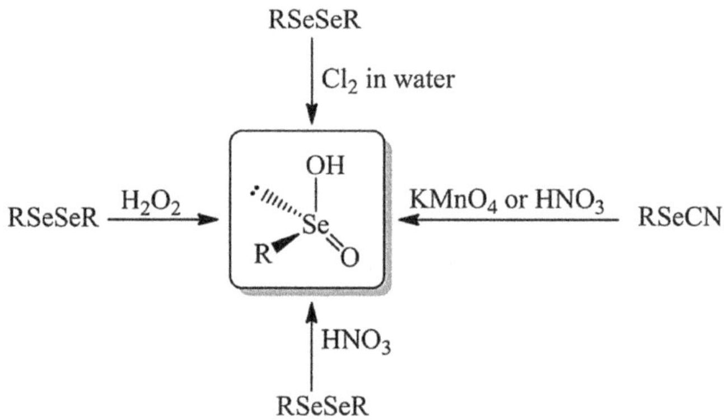

Scheme 1.30 Synthesis of seleninic acids.

acids. The seleninic acids of tertiary amides undergo further oxidation with an excess of H_2O_2 to afford selenonic acids (Scheme 1.29). The seleninic acids show intra-molecular Se···O interactions.[185]

Unlike selenenic acids, seleninic and selenonic acids are stable, isolable, colorless and odorless solids. They are used as oxidants in organic synthesis. Both aliphatic and aromatic seleninic acids are obtained by oxidation of diselenides or selenocyanates with concentrated HNO_3, 30% H_2O_2, potassium permanganate in acetic acid, chlorine in aqueous medium (Scheme 1.30). Nitration of aromatic groups can be observed on oxidation of aromatic selenium compounds with HNO_3.[186] Hydrolysis of organoseleniumtrihalides yields seleninic acids. Isolation of trihalides can be avoided if chlorination or bromination of diselenides is performed in aqueous medium.[187] Other oxidants, such as dimethyldioxirane (DMD)[181,188] and ozone,[189] have also been used for conversion of selenium compounds to seleninic acids. Oxidation of tripSeOH with DMD affords the corresponding seleninic acid, tripSe(O)OH quantitatively.[181] Similarly, polyfunctionalseleno esters, such as *Se*-glucopyranosylphenyl seleno acetate, are oxidized by DMD to seleninic acid.[188] Further oxidation of seleninic acid, *e.g.* PhSe(O)OH, with H_2O_2 results in peroxyseleninic acids, RSe(O)OOH, which find applications in Baeyer–Villiger oxidation of various carbonyl compounds.[190]

Seleninic acids are optically active molecules, although isolation of optically pure seleninic acid is a tedious process due to their facile racemization. Kamigata and co-workers obtained enantiomeric pure forms of seleninic acids by using either a chiral column in medium-pressure liquid chromatography or chiral crystallization (*e.g.* MeSe(O)OH).[75,191]

Oxidation of seleninic acids with strong oxidizing agents, like $KMnO_4$ in aqueous KOH, results into the formation of potassium salts of selenonic acid, $RSe(O)_2OK$. They are strong oxidizing agents and can be reduced to seleninic acids in concentrated HCl.

References

1. R. A. Zingaro and W. C. Cooper, *Selenium*, van Nostrand Reinhold Company, New York, 1974.
2. J. V. Comasseto, *J. Braz. Chem. Soc.*, 2010, **21**, 2027.
3. G. R. Waitkins and C. W. Clark, *Chem. Rev.*, 1945, **36**, 235.
4. D. Seeback and N. Peleties, *Angew. Chem., Int. Ed. Engl.*, 1969, **8**, 450.
5. K. B. Sharpless and R. F. Lauer, *J. Am. Chem. Soc.*, 1972, **94**, 7154.
6. T. G. Back, D. H. R. Barton, M. R. Britten-Kelly and F. S. Guziec, Jr., *J. Chem. Soc., Perkin Trans. 1*, 1976, 2079.
7. *Organic Selenium Compounds: Their Chemistry and Biology*, ed. D. L. Klayman and W. H. H. Günther, John Wiley & Sons Inc., Chichester, 1973.
8. (a) A. Krief, Selenium, in *Comprehensive Organometallic Chemistry-II*, ed. E. W. Abel, F. G. A. Stone and G. Wilkinson, Elsevier, Oxford, 1995, vol. 11, ch. 13, pp. 515–569; (b) T. Wirth, Selenium, in *Comprehensive Organometallic Chemistry-III*, volume ed. P. Knochel, ed. R. H. Crabtree and D. M. P. Mingos, Elsevier, Oxford, 2007, vol. 9, ch. 11, pp. 457–499; (c) *The Chemistry of Organic Selenium and Tellurium Compounds*, ed. Z. Rappoport, John Wiley & Sons, Ltd., 2014.
9. J. E. Fritzmann, *Z. Anorg. Allg. Chem.*, 1912, **73**, 239.
10. E. G. Hope and W. Levason, *Coord. Chem. Rev.*, 1993, **122**, 109; J. Arnold, *Prog. Inorg. Chem.*, 1995, **43**, 353.
11. (a) K. D. Karlin and J. Arnold, *Prog. Inorg. Chem.*, 2007, **43**, 353; (b) R. S. Chauhan, G. Kedarnath, A. Wadawale, A. M. Z. Slawin and V. K. Jain, *Dalton Trans.*, 2013, **42**, 259; (c) S. Kolay, A. Wadawale, D. Das, H. K. Kisan, R. B. Sunoj and V. K. Jain, *Dalton Trans.*, 2013, **42**, 10828.
12. Q. Yao, E. P. Kinney and C. Zheng, *Org. Lett.*, 2004, **6**, 2997.
13. A. Kumar, G. K. Rao, F. Saleem and A. K. Singh, *Dalton Trans.*, 2012, **41**, 11949.

14. (a) D. K. Paluru, S. Dey, K. R. Chaudhari, M. V. Khedkar, B. M. Bhange and V. K. Jain, *Tetrahedron Lett.*, 2014, **55**, 2953; (b) D. K. Paluru, S. Dey, A. Wadawale, D. K. Maity, N. Bhuvanesh and V. K. Jain, *Eur. J. Inorg. Chem.*, 2015, 397.
15. (a) M. Bochmann, *Chem. Vap. Deposition*, 1996, **2**, 85; (b) M. A. Malik, M. Afzaal and P. O'Brien, *Chem. Rev.*, 2010, **110**, 4417; (c) S. Schulz, *Coord. Chem. Rev.*, 2015, **297–298**, 49.
16. G. Kedarnath and V. K. Jain, *Coord. Chem. Rev.*, 2013, **257**, 1409.
17. (a) R. K. Sharma, A. Wadawale, G. Kedarnath, D. Manna, T. K. Ghanty, B. Vishwanadh and V. K. Jain, *Dalton Trans.*, 2014, **43**, 6525; (b) A. Tyagi, G. Kedarnath, A. Wadawale, A. Y. Shah, V. K. Jain and B. Vishwanadh, *RSC Adv.*, 2016, **6**, 8367.
18. H. J. Reich and R. J. Hondal, *ACS Chem Biol.*, 2016, **11**, 821.
19. K. Schwarz and C. M. Foltz, *J. Am. Chem. Soc.*, 1957, **79**, 3292.
20. (a) L. Flohe, W. A. Günzler and H. H. Schock, *FEBS Lett.*, 1973, **32**, 132; (b) J. T. Rotruck, A. L. Pope, H. E. Ganther, A. B. Swansen, D. G. Hafeman and W. G. Hoekstra, *Science*, 1973, **179**, 588.
21. X. Huang, X. Liu, Q. Luo, J. Liu and J. Shen, *Chem. Soc. Rev.*, 2011, **40**, 1171.
22. H. Sies and H. Masumoto, *Adv. Pharmacol.*, 1997, **38**, 229.
23. T. Wirth, *Angew. Chem., Int. Ed.*, 2015, **54**, 10074.
24. J. E. Oldfield, *Selenium World Atlas*, Selenium-Tellurium Development Association, Grimbergen, Belgium, 1999.
25. C. M. Weekley and H. H. Harris, *Chem. Soc. Rev.*, 2013, **42**, 8870.
26. M. Harthill, *Biol. Trace Elem. Res.*, 2011, **143**, 1325.
27. E. P. Painter, *Chem. Rev.*, 1941, **28**, 179.
28. C. Reilly, *Selenium in Food and Health*, 2nd edn, Springer, New York, 2006, p. 206.
29. L. H. E. Winkel, C. A. Johnson, M. Lenz, T. Grundl, O. X. Leupin, M. Amini and L. Charlet, *Environ. Sci. Technol.*, 2012, **46**, 571.
30. T. H. Jukes, *J. Appl. Biochem.*, 1983, **5**, 233.
31. D. M. Chizhikov and V. P. Shchastlivyi, *Selenium and Selenides*, (translated from Russian to English by E. M. Eltein), Collet's Publishers Ltd., London, 1968.
32. Z. Cernosek, A. Ruzicka, J. Holubova and O. E. Cernoskova, *Main Group Met. Chem.*, 2007, **30**, 231.
33. R. J. Gillespie and J. Passmore, *Acc. Chem. Res.*, 1971, **4**, 413.
34. M. A. Ansari and J. A. Ibers, *Coord. Chem. Rev.*, 1990, **100**, 223.
35. B. Krebs, S. Bonmann and I. Eidenschink, Selenium: Inorganic Chemistry, in *Encyclopedia of Inorganic Chemistry*, ed. R. B. King, John Wiley & Sons, 1994, vol. 7, pp. 3667–3689.
36. L. C. Roof and J. W. Kolis, *Chem. Rev.*, 1993, **93**, 1037.
37. A. J. Mukherjee, S. S. Zade, H. B. Singh and R. B. Sunoj, *Chem. Rev.*, 2010, **110**, 4357.
38. D. Banerjea, *Inorganic Chemistry - A Modern Treatise*, Asian Books Pvt. Ltd, New Delhi, 2012.

39. T. Chivers and R. S. Laitinen, *Chem. Soc. Rev.*, 2015, **44**, 1725.
40. R. K. Harris and B. E. Mann, *NMR and Periodic Table*, Academic Press, London, 1978.
41. J. D. Odom, W. H. Dawson and P. D. Ellis, *J. Am. Chem. Soc.*, 1979, **101**, 5815.
42. H. Duddeck, *Prog. Nucl. Magn. Reson. Spectrosc.*, 1995, **27**, 1.
43. (a) T. M. Klapötke and M. Broschag, *Compilation of Reported ^{77}Se NMR Chemical Shifts*, John Wiley & Sons Inc., New York, 1996; (b) H. Duddeck, *Annu. Rep. NMR Spectrosc.*, 2004, **52**, 105; (c) B. A. Demko and R. E. Wasylishen, *Prog. Nucl. Magn. Reson. Spectrosc.*, 2009, **54**, 208–238; (d) A. Panda and H. B. Singh, *Chemistry of Organic Selenium and Tellurium Compounds*, ed. Z. Rappoport, John Wiley & Sons Ltd, 2014, vol. 4, pp. 277–445.
44. S. Dey, L. B. Kumbhare, V. K. Jain, T. Schurr, W. Kaim, A. Klein and F. Belaj, *Eur. J. Inorg. Chem.*, 2004, 4510.
45. G. Llabrès, M. Baiwir, J. L. Piette and L. Christiaens, *Org. Magn. Reson.*, 1981, **15**, 152.
46. D. H. O'Brien, N. Dereu, C. K. Huang and K. J. Irgolic, *Organometallics*, 1983, **2**, 305.
47. D. Dowd and P. Gettins, *Magn. Reson. Chem.*, 1988, **26**, 1.
48. K. S. Tan, A. P. Arnold and D. L. Rabenstein, *Can. J. Chem.*, 1988, **66**, 54.
49. (a) M. Iwaoka and S. Tomoda, *J. Am. Chem. Soc.*, 1994, **116**, 2557; (b) H. Komatsu, M. Iwaoka and S. Tomoda, *Chem. Commun.*, 1999, 205.
50. (a) M. Iwaoka and S. Tomoda, *Phosphorus, Sulfur, Silicon Relat. Elem.*, 2005, **180**, 755; (b) S. Hayashi and W. Nakanishi, Hypervalent chalcogen compounds, in *Handbook of Chalcogen Chemistry: New perspectives in Sulfur, Selenium and Tellurium*, ed. F. A. Devillanova and W. W. Du Mont, 2nd edn, RSC Publishing, 2013, ch. 12.3.
51. J. Liu and S. Rozovsky, ^{77}Se NMR Spectroscopy of selenoproteins, in *Selenium in Molecular Biology & Human Health*, ed., D. L. Hatfield, *et al.*, Springer, 2016, pp. 187–200.
52. M. Mobli, D. Morgenstern, G. F. King, P. F. Alewood and M. Muttenthaler, *Angew. Chem., Int. Ed.*, 2011, **50**, 11952.
53. K. L. House, R. B. Dunlap, J. D. Odom, Z. P. Wu and D. Hilvert, *J. Am. Chem. Soc.*, 1992, **114**, 8573.
54. P. Prabhu, B. G. Singh, M. Noguchi, P. P. Phadnis, V. K. Jain, M. Iwaoka and K. I. Priyadarsini, *Org. Biomol. Chem.*, 2014, **12**, 2404.
55. W. McFarlane, D. S. Rycroft and C. J. Turner, *Bull. Soc. Chim. Belg.*, 1977, **86**, 457.
56. J. A. Anderson and J. D. Odom, *Organometallics*, 1988, **7**, 267.
57. (a) W. Gombler, *Magn. Reson. Chem.*, 1991, **29**, 73; (b) A. Sorensen, B. Rasmussen and M. Pittelkov, *J. Org. Chem.*, 2015, **80**, 3852.
58. (a) V. K. Jain, S. Kannan and E. R. T. Teikink, *J. Chem. Res.*, 1994, 501; (b) T. Niksch, H. Görls, M. Friedrich, R. Oilenkaniemi, R. Laitinen and W. Weigand, *Eur. J. Inorg. Chem.*, 2010, 74.
59. I. Johannsen, L. Henriksen and H. Eggert, *J. Org. Chem.*, 1986, **51**, 1657.

60. R. K. Sharma, G. Kedarnath, A. Wadawale, C. A. Betty, B. Vishwanadh and V. K. Jain, *Dalton Trans.*, 2012, **41**, 12129.
61. (a) S. Dey, V. K. Jain and B. Verghese, *J. Organomet. Chem.*, 2001, **623**, 48; (b) L. B. Kumbhare, A. P. Wadawale, V. K. Jain, S. Kolay and M. Nethaji, *J. Organomet. Chem.*, 2009, **694**, 3892; (c) K. V. Vivekananda, S. Dey, A. Wadawale, N. Bhuvanesh and V. K. Jain, *Dalton Trans.*, 2013, **42**, 14158.
62. M. Dernovics, J. Far and R. Lobinski, *Metallomics*, 2009, **1**, 317.
63. (a) M. Haratake, M. Hongoh, M. Miyauchi, R. Hirakawa, M. Ono and M. Nakayama, *Inorg. Chem.*, 2008, **47**, 6273; (b) Y. Yamashita and M. Yamashita, *J. Biol. Chem.*, 2010, **285**, 18134; (c) S. A. Schaefer, M. Dong, R. P. Rubenstein, W. A. Wilkie, B. J. Bahnson, C. Thorpe and S. Rozovsky, *J. Mol. Biol.*, 2013, **425**, 222.
64. K. I. Dondin, R. K. Berge, K. J. Borve, J. Songstad and K. W. Törnroos, *J. Mol. Struct.*, 2000, **554**, 149.
65. (a) A. S. Hodage, C. P. Prabhu, P. P. Phadnis, A. Wadawale, K. I. Priyadarsini and V. K. Jain, *J. Organomet. Chem.*, 2012, **720**, 19; (b) A. S. Hodage, P. P. Phadnis, A. Wadawale, K. I. Priyadarsini and V. K. Jain, *X-Ray Struct. Anal. Online*, 2009, **25**, 101.
66. E. W. Abel, A. R. Khan, K. Kite, K. G. Orrell, V. Šbik, T. S. Cameron and R. Cordes, *J. Chem. Soc., Chem. Commun.*, 1979, 713.
67. R. S. Chauhan, R. K. Sharma, G. Kedarnath, D. B. Cordes, A. M. Z. Slawin and V. K. Jain, *J. Organomet. Chem.*, 2012, **717**, 180.
68. M. Kulcsar, A. Silvestru, C. Silvestru, J. E. Drake, C. L. B. MacDonald, M. E. Hursthouse and M. E. Light, *J. Organomet. Chem.*, 2005, **690**, 3217.
69. L. B. Kumbhare, V. K. Jain, P. P. Phadnis and M. Nethaji, *J. Organomet. Chem.*, 2007, **692**, 1546.
70. N. Ghavale, S. Dey, A. Wadawale and V. K. Jain, *Organometallics*, 2008, **27**, 3297.
71. (a) S. M. Godfrey, C. A. McAuliffe, R. J. Pritchard and S. Sarwar, *J. Chem. Soc., Dalton Trans.*, 1997, 1031; (b) A. M. Toma, A. Nicoara, A. Silvestru, T. Rüffer, H. Lang and M. Mehring, *J. Organomet. Chem.*, 2016, **810**, 33.
72. T. M. Klapötke, B. Krumm, M. Scherr, R. Haiges and K. O. Christe, *Angew. Chem., Int. Ed.*, 2007, **46**, 8686.
73. A. G. Petrovic, P. I. Polavarapu, J. Drabowicz, Y. Zhang, O. J. McConnel and H. Duddeck, *Chem. - Eur. J.*, 2005, **11**, 4257.
74. N. Furukawa and S. Sato, *Heteroat. Chem.*, 2002, **13**, 406.
75. N. Kamigata, *Phosphorus, Sulfur Silicon Relat. Elem.*, 2005, **180**, 667.
76. M. ElHamdi, J. Poater, M. F. Bickelhaupt and M. Sola, *Inorg. Chem.*, 2013, **52**, 2458.
77. (a) B. Mueller, H. Poleschner and K. Seppelt, *Dalton Trans.*, 2008, 4424; (b) H. Poleschner and K. Seppelt, *Angew. Chem., Int. Ed.*, 2013, **52**, 12838.
78. (a) J. S. Dhau, A. Singh and R. Dhir, *J. Organomet. Chem.*, 2011, **696**, 2008; (b) C. P. Prabhu, P. P. Phadnis, A. Wadawale, K. I. Priyadarsini and V. K. Jain, *J. Organomet. Chem.*, 2012, **713**, 42.

79. I. Wagner, W. W. Du Mont, S. Pohl and W. Saak, *Chem. Ber.*, 1990, **123**, 2325.
80. (a) S. Dunne, E. von Nagy Felsobuki and M. Mackay, *Acta Crystallogr.*, 1998, **C 54**, 887; (b) C. O. Kienitz, C. Thöne and P. G. Jones, *Inorg. Chem.*, 1996, **35**, 3990.
81. P. P. Phadnis, A. Kunwar, R. K. Mishra, M. Kumar, A. Wadawale, K. I. Priyadarsini and V. K. Jain, unpublished results.
82. A. S. Filatov, E. Block and M. A. Petrukhina, *Acta Crystallogr.*, 2005, **C 61**, o596.
83. T. M. Klapötke, B. Kurmm, P. Mayer, H. Piotrowski and O. P. Ruscitti, *Z. Naturforsch.*, 2002, **57B**, 145.
84. J. Beckmann and A. Duthie, *Z. Anorg. Allg. Chem.*, 2005, **631**, 1849.
85. T. M. Klapötke, B. Kurmm and K. Polborn, *Eur. J. Inorg. Chem.*, 1999, 1359.
86. L. Brandsma and H. E. Wijers, *Recl. Trav. Chim. Pays-Bas*, 1963, **82**, 68.
87. S. Dey, V. K. Jain, S. Chaudhury, A. Knoedler, F. Lissner and W. Kaim, *J. Chem. Soc., Dalton Trans.*, 2001, 723.
88. (a) K. K. Bhasin and J. Singh, *J. Organomet. Chem.*, 2002, **658**, 71; (b) K. K. Bhasin, E. Arora, K. Kaur, S. Kang, M. Gobel, T. M. Klapoetke and S. K. Mehta, *Tetrahedron*, 2009, **65**, 247.
89. J. Q. Li, W. L. Bao, P. Lue and X. J. Zhau, *Synth. Commun.*, 1991, **21**, 799.
90. D. L. Klayman and T. S. Griffin, *J. Am. Chem. Soc.*, 1973, **95**, 197.
91. K. K. Bhasin, E. Arora and T. M. Klapoetke, *J. Organomet. Chem.*, 2011, **696**, 835.
92. J. A. Gladysz, J. L. Hornby and J. E. Garbe, *J. Org. Chem.*, 1978, **43**, 1204.
93. A. S. Hodage, P. P. Phadnis, A. Wadawale, K. I. Priyadarsini and V. K. Jain, *Org. Biomol. Chem.*, 2011, **9**, 2992.
94. L. Syper and J. Mlochowski, *Tetrahedron*, 1988, **44**, 6119.
95. L. B. Agenas, *Acta Chem. Scand.*, 1962, **16**, 1809.
96. M. Vafai and M. Renson, *Bull. Soc. Chim. Belg.*, 1966, **75**, 145.
97. D. S. Margolies and R. W. Pittman, *J. Chem. Soc.*, 1957, 799.
98. J. Q. Liu, S. J. Gao, G. M. Luo, G. L. Yan and J. C. Shen, *Biochem. Biophys. Res. Commun.*, 1998, **247**, 397.
99. Y. Nishibayashi, K. Segawa, J. D. Singh, S. Fukuzawa, K. Ohe and S. Uemura, *Organometallics*, 1996, **15**, 370.
100. R. Kaur, H. B. Singh and R. P. Patel, *J. Chem. Soc., Dalton Trans.*, 1996, 2719.
101. G. Mugesh and H. B. Gingh, *Acc. Chem. Res.*, 2002, **35**, 226.
102. W. H. H. Günther and H. G. Mautner, *J. Am. Chem. Soc.*, 1960, **82**, 2762.
103. H. Ehara, M. Noguchi, S. Sayama and T. Onami, *J. Chem. Soc., Perkin Trans. 1*, 2000, 1429.
104. F. Tan, Z. Yu and S. Lu, *J. Org. Chem.*, 2004, **69**, 4520.
105. D. Singh, A. M. Deobald, L. R. S. Camargo, G. Tabarelli, O. E. D. Rodrigues and A. L. Braga, *Org. Lett.*, 2010, **12**, 3288.
106. C. A. Collins, F. H. Fry, A. L. Holme, A. Yiakouvaki, A. Al-Qenaei, C. Pourzand and C. Jacob, *Org. Biomol. Chem.*, 2005, **3**, 1541.

107. G. Pandey, V. Jayathirtha Rao and U. T. Bhalerao, *J. Chem. Soc., Chem. Commun.*, 1989, 416.
108. (a) A. Kunwar, B. Mishra, A. Barik, L. B. Kumbhare, R. Pandey, V. K. Jain and K. I. Priyadarsini, *Chem. Res. Toxicol.*, 2007, **20**, 1482; (b) B. Mishra, L. B. Kumbhare, V. K. Jain and K. I. Priyadarsini, *J. Phys. Chcm. B*, 2008, **112**, 4441.
109. (a) T. G. Back and M. V. Krishna, *J. Org. Chem.*, 1988, **53**, 2533; (b) A. Ogawa, H. Yokoyama, K. Yokoyama, T. Masawaki, N. Kambe and N. Sonoda, *J. Org. Chem.*, 1991, **56**, 5721.
110. J. Y. C. Chu, D. G. Marsh and W. H. H. Günther, *J. Am. Chem. Soc.*, 1975, **97**, 4905.
111. F. V. Singh and T. Wirth, in *Organoselenium Chemistry: Synthesis and Reactions*, ed. T. Wirth, Wiley VCH Verlag GmbH, 2012, pp. 321–360.
112. J. Brues, M. Martin, F. Urpi and J. Vilarrasa, *J. Org. Chem.*, 2009, **74**, 2203.
113. R. L. Brutchley, *Acc. Chem. Res.*, 2015, **48**, 2918.
114. T. G. Back, Selenium: organoselenium chemistry, in *Encyclopedia of Inorganic Chemistry*, ed. R. B. King, John Wiley & Sons, Ltd., 1994, vol. 7, pp. 3690–3700.
115. M. Iwaoka, Nucleophilic selenium, in *Organoselenium Chemistry: Synthesis and Reactivity*, ed. T. Wirth, Wiley-VCH Verlag, GmbH, 2012, ch. 2, pp. 53–109.
116. S. S. Zade and H. B. Singh, Synthesis of organoselenium compounds in *The Chemistry of Organic Selenium and Tellurium Compunds*, ed. Z. Rappoport, John Wiley & Sons Ltd, 2014, vol. 4, ch. 1, pp. 1–180.
117. A. Krief, T. van Wemmed, M. Redon, W. Dumont and C. Delmotte, *Angew. Chem., Int. Ed.*, 1999, **38**, 2245.
118. S. Canales, O. Crespo, M. C. Gimeno, P. G. Jones, A. Laguna and P. Romero, *Dalton Trans.*, 2003, 4525.
119. C. J. Donahue, V. A. Martin, B. A. Schoenfelner and E. C. Kosinski, *Inorg. Chem.*, 1991, **30**, 1588.
120. (a) J. Ashaks, Yu. Bankovsky, D. Zoruma, I. Shestakova, I. Domracheva, A. Nesterova and E. Lukevics, *Chem. Heterocycl. Compd.*, 2004, **40**, 776; (b) E. Lukevics, D. Zoruma, J. Ashaks, I. Shestakova, I. Domracheva, V. Bridane and E. Yashchenko, *Chem. Heterocycl. Compd.*, 2009, **45**, 182.
121. (a) K. B. Sharpless and R. F. Lauer, *J. Am. Chem. Soc.*, 1973, **95**, 2697; (b) J. V. Comasseto, L. W. Ling, N. Petragnani and H. A. Stefani, *Synthesis*, 1997, 373; (c) D. Manna and G. Mugesh, *Angew. Chem., Int. Ed.*, 2010, **49**, 9246.
122. N. Metanis, E. Keinan and P. E. Dawson, *J. Am. Chem. Soc.*, 2006, **128**, 16684.
123. K. P. Bhabak and G. Mugesh, *Chem. – Eur. J.*, 2008, **14**, 8640.
124. S. Yoshida, F. Kumakura, I. Komatsu, K. Arai, Y. Onuma, H. Hojo, B. G. Singh, K. I. Priyadarsini and M. Iwaoka, *Angew. Chem., Int. Ed.*, 2011, **50**, 2125.

125. D. Crich, J. T. Hwang, S. Gastaldi, F. Recupero and D. J. Wink, *J. Org. Chem.*, 1999, **64**, 2877.
126. J. Gosselck, *Chem. Ber.*, 1958, **91**, 2345.
127. J. Laube, S. Jäger and C. Thöne, *Eur. J. Inorg. Chem.*, 2001, 1983.
128. (a) V. K. Landry, M. Minoura, K. Pang, D. Buccella, B. V. Kelly and G. Perkin, *J. Am. Chem. Soc.*, 2006, **128**, 12490; (b) L. J. Guziec and F. S. Guziec Jr., *J. Org. Chem.*, 1994, **59**, 4691.
129. G. Perkin, E. J. Lenardo, R. G. Jacob and R. B. Panatieri, *Chem. Rev.*, 2009, **109**, 1277.
130. (a) I. P. Beletskaya and V. P. Ananikov, *Pure Appl. Chem.*, 2007, **79**, 1041; (b) V. P. Ananikov, N. V. Orlov, S. S. Zalesskiy, I. P. Beletskaya, V. N. Khrustalev, K. Morokuma and D. G. Musaev, *J. Am. Chem. Soc.*, 2012, **134**, 6637.
131. C. J. Löwig, *Pogg. Ann.*, 1836, **37**, 552.
132. M. Renson, *The Chemistry of Organic Selenium and Tellurium Compounds*, ed. S. Patai and Z. Rappoport, Wiley, Chicheter, 1987, vol. 1, ch. 13.
133. (a) M. Elsherbini, W. S. Hamama and H. H. Zoorob, *Coord. Chem. Rev.*, 2016, **312**, 149; (b) M. Elsherbini, W. S. Hamama and H. H. Zoorob, *Coord. Chem. Rev.*, 2017, **330**, 110.
134. M. D. Milton, S. Khan, J. D. Sigh, V. Mishra and B. L. Khandelwal, *Tetrahedron Lett.*, 2005, **46**, 755.
135. A. Panda, *Coord. Chem. Rev.*, 2009, **253**, 1056.
136. S. G. Murray and F. R. Hartley, *Chem. Rev.*, 1981, **81**, 365; W. Levason, S. D. Orchad and G. Reid, *Coord. Chem. Rev.*, 2002, **225**, 159.
137. A. Bhalla, Y. Nagpal, R. Kumar, S. K. Mehta, K. K. Bhasin and S. S. Bari, *J. Organomet. Chem.*, 2009, **694**, 179.
138. D. L. J. Clive, *Aldrichimica Acta*, 1978, **11**, 43.
139. L. H. Andrade, A. V. Silva, P. Milani, D. Koszelewski and W. Kroutil, *Org. Biomol. Chem.*, 2010, **8**, 2043.
140. S. Berlin, C. Ericsson and L. Engman, *Org. Lett.*, 2002, **4**, 3.
141. D. Liotta, *Acc. Chem. Res.*, 1984, **17**, 28.
142. S. Aboulkacem, D. Naumann, W. Tyrra and I. Patenburg, *Organometallics*, 2012, **31**, 1559.
143. L. Engman, J. Lind and G. Merenyi, *J. Phys. Chem.*, 1994, **98**, 3174.
144. S. Zhang, X. Wang, Y. Su, Y. Qiu, Z. Zhang and X. Wang, *Nat. Commun.*, 2014, DOI: 10.1038/ncomms 5127.
145. F. Krafft and W. Vorster, *Chem. Ber.*, 1893, **26**, 2813.
146. A. Ariyoshi, Y. Aso, T. Otsubo and F. Ogura, *Chem. Lett.*, 1984, 891.
147. A. M. Kelly, G. P. Rosini and A. S. Goldman, *J. Am. Chem. Soc.*, 1997, **119**, 6115.
148. E. V. Dikarev, M. A. Petrukhina, X. Li and E. Bock, *Inorg. Chem.*, 2003, **42**, 1966.
149. A. Krief, A. Destree, V. Durisotti, N. Morean, C. Smal and C. Colaux-Castillo, *Chem. Commun.*, 2002, 1558.
150. J. M. Khurana, B. M. Kandpal and Y. K. Chauhan, *Phosphorus, Sulfur Silicon Relat. Elem.*, 2003, **178**, 1369.

151. T. Sama, T. Shimizu, K. Hirabayashi and N. Kamigata, *Heteroat. Chem.*, 2007, **18**, 301.
152. T. G. Back, Z. Moussa and M. Parvez, *Angew. Chem., Int. Ed.*, 2004, **43**, 1268.
153. L. B. Agenäs and G. Lindgren, *Acta Chem. Scand.*, 1970, **24**, 3301.
154. (a) T. Shimizu, K. Kikuchi, Y. Ishikawa, I. Ikemoto, M. Kobayashi and N. Kamigata, *J. Chem. Soc., Perkin Trans. 1*, 1989, 597; (b) F. A. Davis, J. M. Billmers and O. D. Stringer, *Tetrahedron Lett.*, 1993, **24**, 3191.
155. L. A. Wessjohann, A. Schneider, M. Abbas and W. Brandt, *Biol. Chem.*, 2007, **338**, 997.
156. (a) D. P. Gamblin, P. Garnier, S. van Kasteren, N. J. Oldham, A. J. Fairbanks and B. G. Davis, *Angew. Chem., Int. Ed.*, 2004, **43**, 828; (b) O. Boutureira, G. J. L. Bernardes, M. F. Gonzalez, D. C. Anthony and D. G. Davies, *Angew. Chem., Int. Ed.*, 2012, **51**, 1432.
157. M. Haratake, Y. Tachibana, Y. Emaya, S. Yoshida, T. Fuchigami and M. Nakayama, *ACS Omega*, 2016, **1**, 58.
158. K. P. Bhabak and G. Mugesh, *Chem. – Eur. J.*, 2007, **13**, 4594.
159. E. Duhamel, A. Pöllnitz, A. Stegarescu and A. Silvestru, *Z. Anorg. Allg. Chem.*, 2011, **637**, 1355.
160. S. Kumar, J. C. P. Helt, J. Autschbach and M. R. Detty, *Organometallics*, 2009, **28**, 3426.
161. L. Engman, *Tetrahedron Lett.*, 1985, **26**, 6385.
162. G. Holzle and W. Jenny, *Helv. Chim. Acta*, 1958, **41**, 594.
163. C. Santi and S. Santoro, Electrophilic selenium, in *Organoselenium Chemistry: Synthesis and Reactivity*, ed. T. Wirth, Wiley-VCH Verlag, GmbH, 2012, ch. 1, pp. 1–51.
164. W. P. Jackson, S. V. Ley and A. J. Whittle, *J. Chem. Soc., Chem. Commun.*, 1980, 1173.
165. F. A. Davis, R. Zhou and L. G. Sankar, *Tetrahedron Lett.*, 1990, **31**, 1653.
166. T. G. Back and K. R. Murlidharan, *J. Org. Chem.*, 1991, **56**, 2781.
167. S. Murata and T. Suzuki, *Tetrahedron Lett.*, 1987, **28**, 4415.
168. I. I. Lapkin, A. N. Dedugov and N. N. Pavlova, *Zh. Org. Khim.*, 1985, **21**, 345.
169. A. Toshimitsu, H. Owada, K. Terao, S. Uemura and M. Okano, *J. Org. Chem.*, 1984, **49**, 3796.
170. B. Boduszek and R. Gancarz, *J. Prakt. Chem.*, 1996, **338**, 186.
171. T. Andreou, J. Bures and J. Vilarrasa, *Tetrahedron Lett.*, 2010, **51**, 1863.
172. (a) W. W. du Mont, S. Kubiniok, K. Peters and H. G. von Schnering, *Angew. Chem., Int. Ed.*, 1987, **26**, 780; (b) W. W. du Mont, A. Marlens, S. Pohl and W. Saak, *Inorg. Chem.*, 1990, **29**, 4847; (c) K. Goto, D. Sonoda, K. Simada, S. Sare and T. Kawashima, *Angew. Chem., Int. Ed.*, 2010, **49**, 545.
173. K. Kandasamy, S. Kumar, H. B. Singh, R. J. Butcher and K. T. Holman, *Eur. J. Inorg. Chem.*, 2004, 1014.
174. G. Mugesh, H. B. Singh and R. J. Butcher, *Tetrahedron: Asymmetry*, 1999, **10**, 237.

175. B. Mueller, T. T. Takaluoma, R. S. Laitinen and K. Seppelt, *Eur. J. Inorg. Chem.*, 2011, 4970.
176. N. A. Barnes, S. M. Godfrey, J. Hughes, R. Z. Khan, I. Mushtaq, R. T. A. Ollerenshaw, R. G. Pritchard and S. Sarwar, *Dalton Trans.*, 2013, **42**, 2735.
177. S. Kubniok, W. W. du Mont, S. Pohl and W. Saak, *Angew. Chem., Int. Ed.*, 1988, **27**, 431.
178. N. Kaltsoyannis, *J. Chem. Soc., Dalton Trans.*, 1997, 4759.
179. (a) H. Rheinboldt and E. Giesbrecht, *Chem. Ber.*, 1955, **88**, 1974; (b) W. Jenny, *Helv. Chim. Acta*, 1958, **41**, 317; (c) H. Reich, C. A. Hoeger and W. W. Willis, Jr., *J. Am. Chem. Soc.*, 1982, **104**, 2936.
180. H. J. Reich and C. P. Jasperse, *J. Org. Chem.*, 1988, **53**, 2389.
181. A. Ishii, S. Matsubayashi, T. Takahashi and J. Nakayama, *J. Org. Chem.*, 1999, **64**, 1084.
182. Z. Zielinski, N. Presseau, R. Amorati, L. Valgimigli and D. A. Pratt, *J. Am. Chem. Soc.*, 2014, **136**, 1570.
183. K. Goto, M. Nagahara, T. Mizushima, K. Shimada, T. Kawashima and R. Okazaki, *Org. Lett.*, 2001, **3**, 3569.
184. K. Selvakumar, H. B. Singh and R. J. Butcher, *Chem. - Eur. J.*, 2010, **16**, 10576.
185. (a) K. P. Bhabak and G. Mugesh, *Chem. - Asian J.*, 2009, **4**, 974; (b) K. P. Bhabak and G. Mugesh, *Chem. - Eur. J.*, 2009, **15**, 9846.
186. V. V. Kozlov and S. E. Suvorova, *Zh. Obshch. Khim.*, 1961, **31**, 3034.
187. L. Pichat, M. Herbert and M. Thiers, *Tetrahedron*, 1961, **12**, 1.
188. M. Abdo and S. Knapp, *J. Am. Chem. Soc.*, 2008, **130**, 9234.
189. T. Shimizu, Y. Nakashima, I. Watanabe, K. Hirabayashi and N. Kamigata, *J. Chem. Soc., Perkin Trans. 1*, 2002, 2151.
190. (a) G. Brink, J. M. Vis, I. W. C. E. Arends and R. A. Sheldon, *J. Org. Chem.*, 2001, **66**, 2429; (b) D. Crich and Y. Zou, *J. Org. Chem.*, 2005, **70**, 3309.
191. Y. Nakashima, T. Shimizu, K. Hirabayashi, M. Yasui, M. Nakazato, F. Iwasaki and N. Kamigata, *Bull. Chem. Soc. Jpn.*, 2005, **78**, 710.

Synthetic Advances on Bioactive Selenium Compounds

CHAPTER 2

Looking Beyond the Traditional Idea of Glutathione Peroxidase Mimics as Antioxidants

C. SANTI,*[a] F. MARINI[a] AND E. J. LENARDÃO[b]

[a] Department of Pharmaceutical Sciences, University of Perugia, Via del Liceo 1, 06100, Perugia, Italy; [b] Centre of Chemical, Pharmaceuticals and Food Sciences, Federal University of Pelotas, P.O. Box 354, CEP 96010-900, Pelotas/RS, Brazil
*Email: claudio.santi@unipg.it

2.1 Introduction

For a long time, the main biological activity of selenium (and its derivatives) was supposed to be connected to its toxicity. In 1285, Marco Polo reported that, crossing the Shanxi province of Western China, the horses on his ships were affected by an unexpected loss of hair due to unknown reasons. After the discovery of selenium by the chemist Jöns Jacob Berzelius in 1817, someone supposed that this disease was a consequence of an exceptionally high concentration of the element in forage growing in that region. The accumulation of this element in the livestock probably caused an acute intoxication, nowadays known as "selenosis". In 1860, a US Army surgeon documented the first selenium poisoning in livestock, suggesting a fascinating theory that the horse illness that slowed General Custer's cavalry relief

Organoselenium Compounds in Biology and Medicine: Synthesis, Biological and Therapeutic Treatments
Edited by Vimal Kumar Jain and K. Indira Priyadarsini

Published by the Royal Society of Chemistry, www.rsc.org

may have been due to selenium intoxication. Beside myths and legends, several studies have confirmed that some disorders in animals[1–3] and humans[4] are consequence of selenium bioaccumulation, and a role in the genesis of some cancers has also been postulated.[5]

In contrast, following the discovery that selenium is an essential micronutrient for mammals, 30 years ago, it was demonstrated that the reduction of peroxides in living systems is a catalytic cycle mediated by the selenoenzyme glutathione peroxidase (GPx).[6] This enzyme catalyzes the reduction of harmful peroxides and has a crucial role in controlling the damage produced by the reactive species of oxygen (ROS), protecting the lipid membranes and other cellular components against oxidation. More than 25 selenium-containing proteins are expressed by humans and are distributed in a broad range of tissue and cells.[7] Beside GPx, other well characterized selenoproteins have been linked to defense against ROS damage, both in terms of detoxification (thioredoxin reductase (TrxR) and selenoprotein P), and repairing (some isoforms of methionine sulfoxide reductase). Selenoproteins K, M, N and H with several roles in maintaining the redox homeostasis in living systems, have been also identified.[8] The common catalytic core of all these families of selenoproteins is an atom of selenium incorporated into the 21st amino acid, named selenocysteine. It has been demonstrated that the presence of selenium is mandatory for good enzymatic activity, especially for GPx. In fact, catalytic efficiency drops by two or three orders of magnitude when selenium is replaced by a sulfur. Engineered selenium-containing glutaredoxin[9] and glutathione transferase[10] in eukaryotic cells showed a GPx activity higher than the native sulfur-containing enzyme, demonstrating that the replacement of a cysteine with a selenocysteine resulted in exceptional improved antioxidant activity. This is correlated to several chemical differences between the two chalcogen atoms. Selenium, in comparison with sulfur, establishes longer bonds with both carbons and hydrogens, and this affects the reactivity. As an example, it corresponds to a lower pK_a for selenol with respect to thiol (5.2 for selenocysteine *vs.* 8.4 for cysteine), allowing the former to be deprotonated at physiological pH (~7).[11] The increased nucleophilicity corresponds to a better reactivity with oxidant species.

In the catalytic cycle, two molecules of glutathione, (2*S*)-2-amino-4-{[(1*R*)-1-[(carboxymethyl)carbamoyl]-2-sulfanylethyl]carbamoyl}butanoic acid (GSH) are oxidized to the corresponding disulfide, acting as a co-factor during the reduction of peroxides into the safer alcohol or water. The first step of the catalytic cycle, summarized in Figure 2.1, is the oxidation of the selenol (or the selenolate) to the corresponding selenenic acid, that is rapidly reduced by the free thiol of GSH, forming a selenenyl sulfide intermediate subjected to a second thiol-mediated reduction. This latter reaction regenerates the catalytic site of the enzyme producing a molecule of oxidized glutathione (GSSG) which, through a reduction catalyzed by NADPH-dependent glutathione reductase (GR) is converted back to free GSH, enabling the redox homeostasis. Looking at the catalytic cycle, an additional evolutionary advantage of using

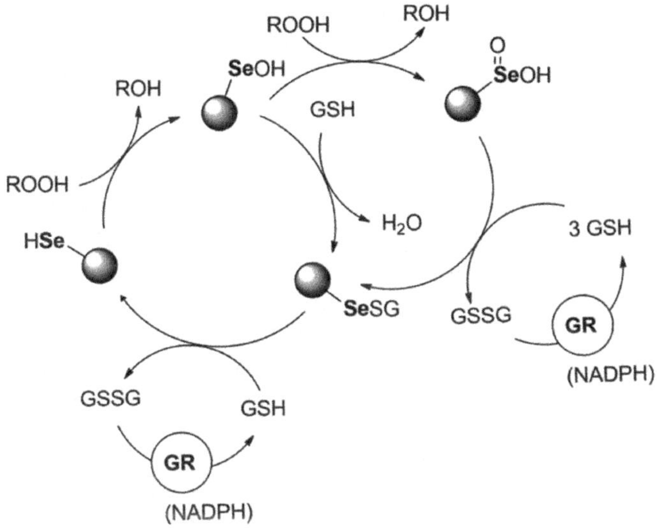

Figure 2.1 Catalytic cycle of glutathione peroxidase (GPx). GSH: reduced form of glutathione; GSSG: oxidized dimeric form of glutathione; GR: glutathione reductase.

selenocysteine in redox protein in the place of cysteine is evident.[12] The catalysis will be possible only if the chalcogen atom can be readily reduced back to the corresponding nucleophilic form and only if non-reversible overoxidations are prevented. When sulfenic acid is overoxidized to sulfinic or sulfonic derivatives, they cannot be easily reduced back to thiols. Sulfonic acid formation is irreversible[13] and sulfinic acid was demonstrated to be reduced only in a few cases by sulfiredoxin.[14] In contrast, when the selenium is embedded in a selenoprotein it is more resistant to overoxidation and selenenic and seleninic species are readily susceptible to the thiol-mediated reduction.[15] Experimental evidence confirmed the currently accepted catalytic GPx mechanism[16] and, in some cases, the selenenic acid was isolated and characterized. It has been observed, by crystallographic analysis, that in the bovine cellular enzyme, selenium is in the form of selenol, whereas in GPx isolated from human plasma, selenium is in the form of selenenic acid.[17]

Selenium concentration in mammalian serum ranges between 7 and 14 $\mu g\,dL^{-1}$,[18] and normally, it is acquired from food as inorganic (selenite and selenate) and/or organic species (selenomethionine and selenocysteine). An intake <20 $\mu g\,day^{-1}$ is considered a deficiency and has been correlated to some cancers, infections, neurological disorders (Alzheimer's and Parkinson's diseases), male infertility and thyroid dysfunction.[19] As a consequence of the evidence, many new organoselenium compounds, having the ability to mimic the catalytic activity of GPx, have been synthesized and evaluated as antioxidants in recent decades. Review articles summarized these new findings in the definition of the catalytic mechanisms as well as in the study of structure–activity correlations. Catalytic activities were studied, attempting to reproduce the reduction of a peroxide (hydrogen or *tert*-butyl hydroperoxide) in an organic solvent or water using different thiol cofactors [GSH, thiophenol (PhSH), $PhCH_2SH$ and dithiotreitol (DTT)].[20–24] The kinetics of the reduction is monitored by ultraviolet-spectrophotometry at 366 nm when the couple NADPH/GSH is used as probe, or at 305 nm when the cofactor is benzenethiol, which is oxidized to the corresponding disulfide. The reaction in the presence of DTT can be conveniently monitored in different solvents using nuclear magnetic resonance (NMR). MeOD-*d*3 was used to obtain slow and more easily measurable kinetics, while in deuterium oxide the kinetics are very fast and difficult to measure, but closer to biological conditions. The combination of all these variables affords a complex panorama that can be misleading and difficult to interpret or evaluate, because a lot of the reported data on the GPx-like activity cannot be directly compared due to different conditions in which the mimetic process was conducted. An additional point of difference is that the reaction mechanism proposed for the enzyme is different for the different class of mimetic derivatives. In the native enzyme, the hydrogen bonding between selenium and the amino acidic environment considerably affects the physicochemical properties of the catalytic center. X-ray analysis demonstrated that in human GPx, the Sec45 is in the proximity of a glutamine (Gln 80) and a tryptophan (Trp 158) residue, forming a triad that improves the catalytic activity based on (i) increased nucleophilicity of selenium; (ii) enhanced C–Se stability, preventing the poisoning effect of selenium; and (iii) reduced possibility of selenium overoxidation. Different classes of small-sized GPx mimetics were synthesized and functionalized in the attempt to reproduce the effects of the catalytic triads, leading to interesting results and considerations for different classes of compounds and the different reaction mechanisms.

In consideration of several reviews and book chapters covering the topic, our aim is not a complete and comprehensive overview of the state of the art.[25–32] We have tried to distil and collect the recently reviewed information for different topics, including selected current updates for the main classes of organoselenium derivatives claimed as GPx mimics: (i) ebselen and its derivatives or analogues; (ii) diselenides; and (iii) selenides and spiroselenuranes. For ebselen derivatives, due to the large and general interest, the chemistry of the preparation is briefly discussed and for all the classes, the GPx-like mechanism is detailed, bearing in mind the idea that it is strongly dependent on the experimental conditions used for its evaluation and that it is always necessary to take into account the pro-oxidant activity that can be associated to some of these compounds. The control of these aspects is crucial to prevent the toxicity and some unwanted side-biological reactions toward relevant physiological thiols. Furthermore, these reactions can be considered new and still unexplored weapons for specific targets and pathologies or to reinterpret the role of organoselenium compounds in a biological environment and for potential therapeutic and analytical scopes. These latter aspects are discussed in the conclusive part of the chapter.

2.2 Chemistry of Organoselenium GPx-mimetics

2.2.1 Ebselen and Its Derivatives or Analogues

Although ebselen (2-phenyl-1,2-benzisoselenazol-3(2*H*)-one) **1** (Figure 2.2) was synthesized by Lesser and Weiss almost a century ago,[33] its GPx-like activity was first described in the 1980s, by Sies and colleagues[34] and Wendel *et al.*[35] Since then, several analogues have been developed, with new straightforward assays to evaluate the GPx-like *in vitro* activity. Different combinations of thiol cofactors and peroxides as mimics of the GSH and peroxides, respectively, allowed the possibility of exploring in depth the GPx-like mechanism of ebselen and novel related structures.[24] In this section, we present some recent advances in the synthesis of new ebselen analogues with increased GPx-like activity, focusing the attention on the mechanistic information collected for the reaction of old and new compounds in the catalytic reduction of peroxides in the presence of free thiols.

2.2.1.1 *Synthesis of Ebselen Analogues: Boosting the GPx-like Catalysis*

Strategies to modify ebselen essentially involve the replacement of the phenyl group **B** at the nitrogen, combined or not with the insertion of different substituents in the aromatic ring **A** of the benzoselenazolone (Figure 2.2). These modifications have gained the important support of computer calculations, which have enabled the optimization of the preparation of molecules in a more rational way.[36–40]

There is a plethora of methods of preparing the benzoselenazolone core of ebselen and its analogues and we report some of those more recently used to prepare derivatives subsequently evaluated for the GPx-like activity. These approaches can be classified in three main groups, according to the key intermediate involved: benzamides **2a** (the Engman and Hallberg approach),[41] *ortho*-iodobenzamides **2b** or 2-(chloroseleno)benzoyl chloride **3a** (Figure 2.2), with a preferential use of this latter synthetic strategy. This is probably a consequence of chloride **3a** being easily obtained in three steps from the commercially available anthranilic acid.[42]

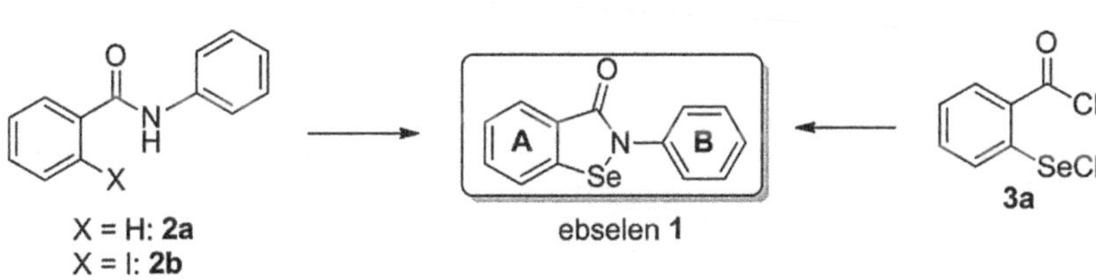

Figure 2.2 Ebselen **1**: synthetic approaches.

2.2.1.2 Replacement of the **B** Ring of Ebselen

Modifications of the aromatic **B** ring bonded at the nitrogen atom of the benzoselenazolone has been largely explored in the attempt to improve the catalytic activity of ebselen. Recently, Scianowski and co-workers prepared several *N*-alkyl and *N*-aryl ebselen analogues in quite good yields reacting alkyl and aryl *o*-iodobenzamides **2b** with Li_2Se_2 (Scheme 2.1).[43] Compounds **4–6** (5 mol%) oxidized DTT in the presence of 30% H_2O_2 more efficiently than ebselen, as observed using proton NMR using MeOD-*d3* as the solvent. Furthermore, *N*-allyl derivative **7** was prepared in 91% yield starting from **3a** by reaction with allylamine and it was evaluated using the same NMR-based test. Compound **5** showed the highest GPx-like activity, with total oxidation of DTT_{red} to DTT_{ox} after 60 min, followed by **6**, **4** and ebselen **1**. The GPx activity of **7** was essentially the same as ebselen, while the ring-opened diselenide **8**, easily prepared from **7** by a reduction–oxidation sequence was considerably less reactive, oxidizing only 12% of DTT after 15 min in comparison with the 64% oxidized by ebselen.[44]

Sterically hindered six-membered nitroxides were combined with the ebselen scaffold by reacting **3a** with alicyclic nitroxides **9** (Scheme 2.2). The authors observed that compounds with generic structures **10** and **11** could act as peroxide and nitroxide "sponges", due to the synergistic effect of ebselen and the redox-active pyrroline group in the GPx-like activity.[45] A total of 14 derivatives were prepared and evaluated for their GPx-like activities by reduction of ^{t}BuOOH using GSH as a cofactor. All of them showed initial rates higher than ebselen, compounds **12, 13** and **14** being the most efficient. In these cases, the corresponding ring-opened diselenides showed a GPx-like activity comparable to that of the corresponding cyclic precursors.

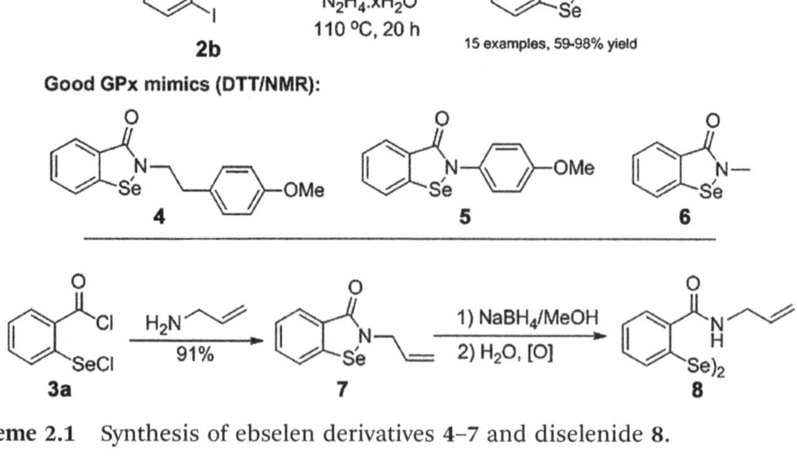

Scheme 2.1 Synthesis of ebselen derivatives **4–7** and diselenide **8**.

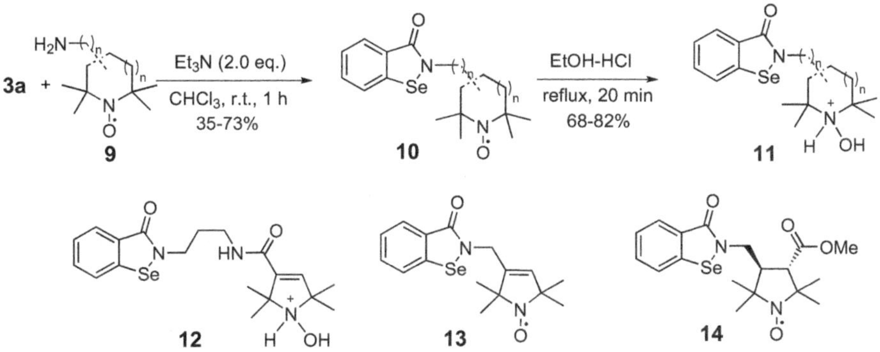

Scheme 2.2 Synthesis of ebselen functionalized with six- and five-membered nitroxides.

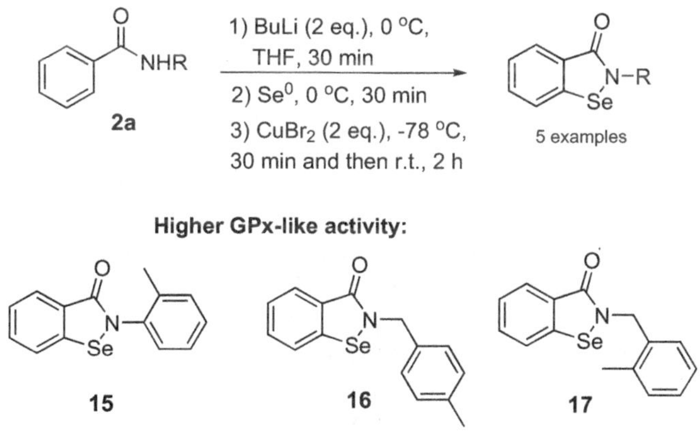

Scheme 2.3 Synthesis of the GPx mimetics **15–17**.

Hsu and co-workers prepared five new ebselen derivatives with methylated aryl and benzyl groups at the nitrogen atom of the benzoselenazolone.[46] The authors started from benzamide **2a**, which after *o*-lithiation and reaction with elemental selenium, afforded the desired ebselen analogues (Scheme 2.3). Ebselen **1** was prepared using the same procedure, although only at 42% yield. The GPx-like activity of these compounds was determined by measuring the consumption of NADPH (at 340 nm) in the presence of the couple glutathione/GR. Among the analyzed panel of compounds, 2-tolyl- **15** and benzyl-substituted **16** and **17** were 60% more active than ebselen.

Looking for new redox-modulators active against hepatocellular carcinoma, Wessjohann and co-workers designed and synthesized the ebselen-selenocyanate **20** and diselenide **21** analogues, reacting **3a** with 4-selenocyanoaniline **18** and *p*-aminophenyl diselenide **19**, respectively (Scheme 2.4). The authors prepared other symmetrical diselenides and tested them for their GPx-like activity and cytotoxicity against Hep G2 cancer

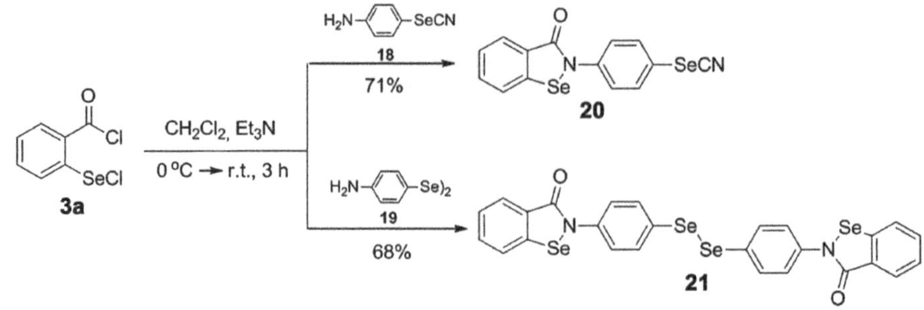

Scheme 2.4 Synthesis of ebselen-based selenocyanate **20** and diselenide **21**.

cells. Compounds **20** and **21** were not active in the GPx assay (NADPH/GSH/GR) and in the anticancer test, the IC_{50} values were 64 and 40 μM, respectively.[47]

Bhabak and Mugesh[48] prepared a series of ebselen analogues by using different primary amines in the reaction with **3a**, with the aim of studying the effect of different substituents in the nitrogen atom of the benzoselenazolone (Scheme 2.5). Density functional theory calculations showed that the presence of a phenyl group in the nitrogen is important for antioxidant activity. The authors tested C_6H_5SH, GSH and 4-Me-C_6H_4SH as thiol cofactors and three different peroxides (H_2O_2, ^{t}BuOOH and Cum-OOH). The outcomes indicate that the nature of the thiol cofactor drastically influences the catalytic activity exhibited by the selenium compounds, while the peroxide was not important, despite the initial reaction rate for reduction of H_2O_2 being higher than for other peroxides. As demonstrated before for ebselen **1**, analogues **22–26** also showed a low GPx-like activity when aromatic thiols were used, and this behavior was attributed to the presence of strong Se···*O* noncovalent interactions in the selenenyl sulfide intermediates formed after attack of the thiol to the selenium atom (Scheme 2.5). This work reported the first clear evidence that the *in vitro* GPx-like assays are strongly influenced by the choice of thiols (more than of peroxides), leading to appreciable differences in the comparison of the same group of compounds using different assays.

Haan and co-workers compared *m*-hydroxy-ebselen **24**, prepared according Scheme 2.5, with ebselen **1** in their efficacy in reduction of oxidative stress, atherogenic lesions and the ability to lessen endpoints of kidney diseases in a diabetic model *in vivo*. Compound **24** was more effective than ebselen **1** in attenuating the activation of hydrogen peroxide-induced pro-inflammatory mediators in human aortic endothelial cells, showing a direct connection between GPx mimics and beneficial effects in the treatment of diabetic complications.[49]

Satheeshkumar and Mugesh[50] evaluated the GPx-like activity of di- and tripeptide analogues of ebselen *in vitro* using H_2O_2, ^{t}BuOOH and Cum-OOH as substrates and GSH as a cosubstrate. Compounds **27–34** were prepared by reaction of 2-(chloroseleno)benzoil chloride **3a** with free NH_2-containing peptides (Scheme 2.6). The authors observed that, despite the presence of

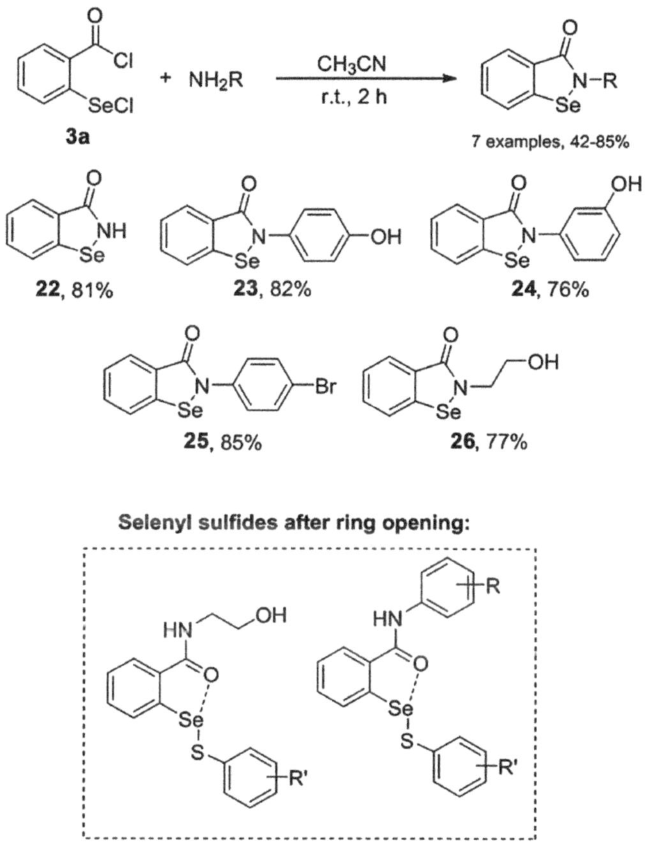

Scheme 2.5 Ebselen derivatives **22–26**; in the box, two examples of non-covalent Se–heteroatom interactions.

benzoselenazolone ring in all compounds, their GPx activity was highly dependent on the nature of the peptide.

Compound **33**, with aliphatic amino acids valine and alanine (Val-Ala) connected to the ebselen core, was the most active of the derivatives **27–34** tested toward the reduction of ^{t}BuOOH and Cum-OOH in the presence of glutathione. In contrast, compounds **28**, **30** and **34** exhibited lower activity. The authors suggested that the aliphatic dipeptide Val-Ala facilitates the formation of selenol **35** (Scheme 2.7), the active catalyst, while aromatic amino acids, like phenylalanine (Phe), inhibit its formation. Besides, Phe-containing ebselen prevents the cyclization of selenenic acids to produce selenenyl amides. Therefore, selenenic and seleninic acids can be over-oxidized to afford the inactive selenonic acids *in vitro*. However, these species are unlikely to be formed *in vivo*, where there is generally an excess of thiols compared to peroxides.

The mechanism depicted in Scheme 2.7 is similar to that recently demonstrated by Sarma and Mugesh for ebselen **1**[51] and starts with the fast

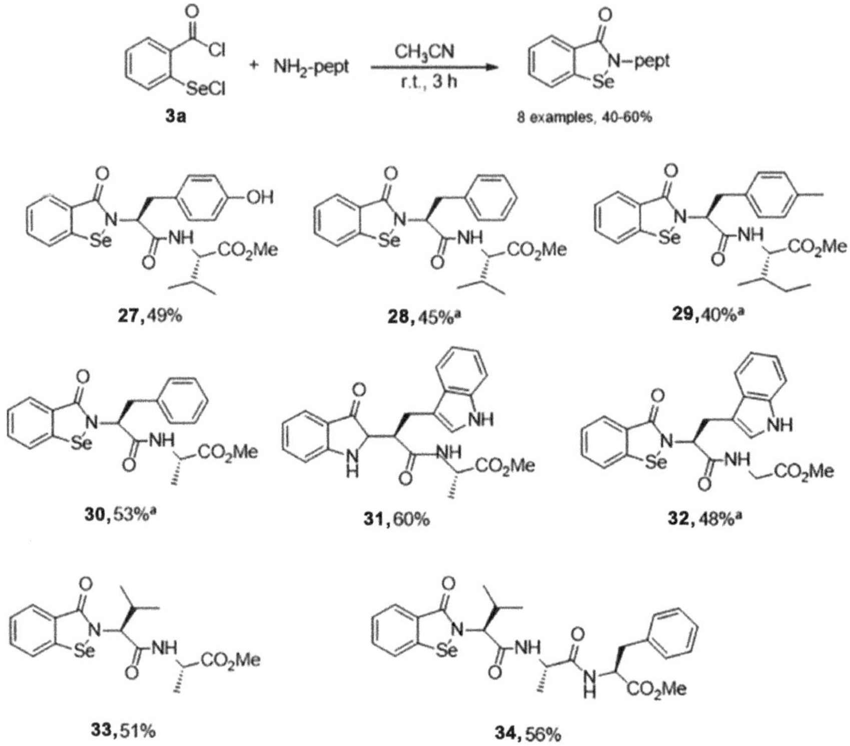

a The respective ring-opened diselenides were also obtained.

Scheme 2.6 Di- and tripeptide analogues of ebselen.

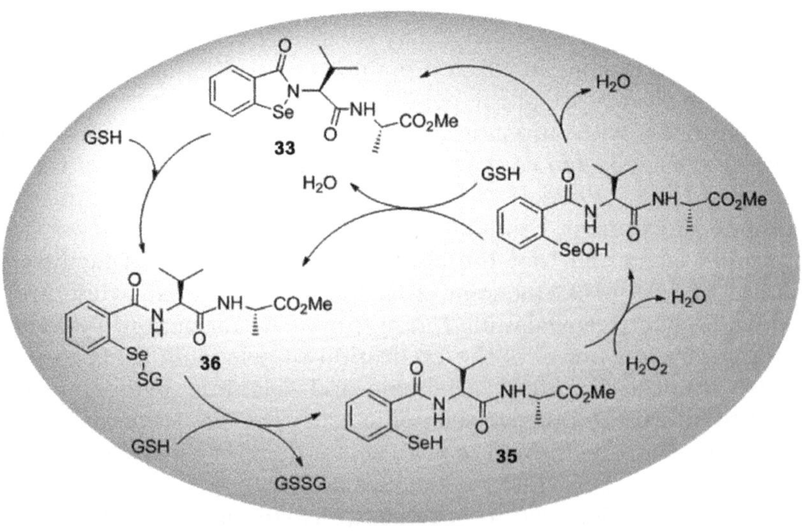

Scheme 2.7 GPx-like catalytic cycle of derivatives **27–34**.

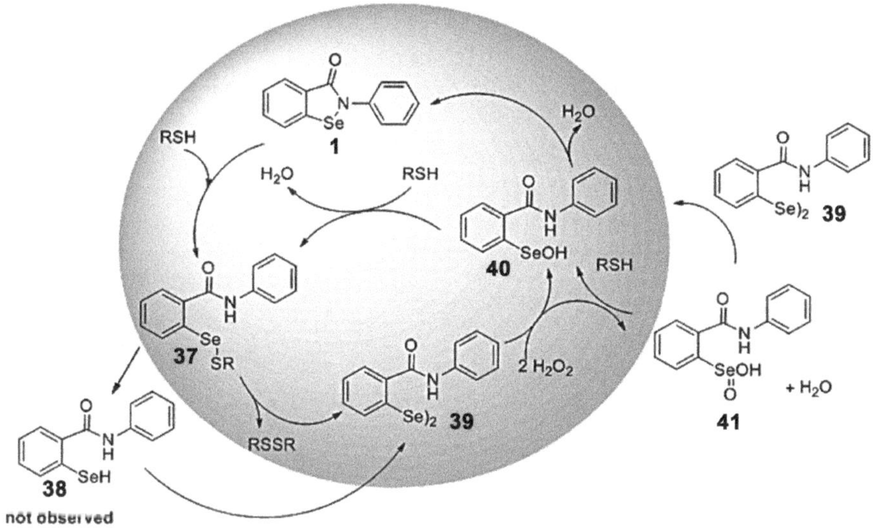

Scheme 2.8 GPx-like catalytic cycle of ebselen **1**.

formation of the selenenyl sulfide **36** by the reaction with GSH, which is subsequently transformed into the actual catalyst **35** by the consumption of a second equivalent of glutathione. Using ^{77}Se NMR and high-resolution mass spectrometry (HRMS) techniques, it was found that selenenyl sulfide **37** originated by **1** (Scheme 2.8) is the predominant species at high concentrations of thiol (RSH), while seleninic acid **41** is the main component in the presence of large amounts of peroxide. At the same time, starting from ebselen (**1**), the authors were not able to detect and demonstrate the formation of selenol **38** or the ebselen-selenoxide, in contrast with the previously proposed catalytic cycle. The rate-determining step is the formation of diselenide **39**, and it depends on the thiol used in the reaction. When dihydrolipoic acid or DTT were used as cofactors, selenol could be involved in the formation of **39**, while in the presence of aromatic thiols, diselenide **39** is mainly formed by the disproportionation reaction of selenenyl sulfide **37**, once the formation of selenol **38** is suppressed by thiol exchange reactions.

New water-soluble ebselen derivatives were prepared by Liu *et al.*,[52] reacting amino-functionalized β-cyclodextrins (β-CD) **42a–d** with 2-(chloroseleno)benzoyl chloride **3a**. Using this strategy, ebselen-hosted β-CDs **43–46** were prepared at 55–60% yields after 5 h. Similarly, ebselen carboxylic acid **47** was converted to the ester **48** and amide **49**, by reaction with β-CD and **42a**, respectively. These ebselen derivatives, containing modified β-CD in the place of the **B** ring, despite the interesting solubility in natural media (with the only exception of **48**), presented GPx-like activities (in GSH/NADPH assay) inferior respect to the unmodified structure **1** (Scheme 2.9).

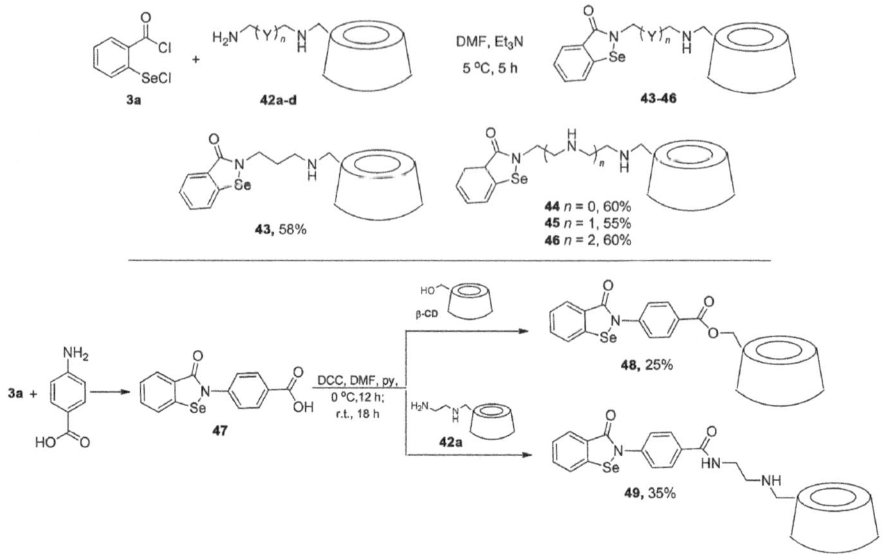

Scheme 2.9 Cyclodextrin-functionalized ebselen derivatives.

Yang *et al.*[53] also prepared some derivatives of ebselen linked to β-CD claiming their GPx-like activity, but no details were published about the assays used for this evaluation.

2.2.1.3 *Introduction of Substituents in the* **A** *Ring of Ebselen*

Since the observation that the introduction of a nitro group in the *ortho* position to the selenium atom in the **A** ring of ebselen increases its GPx-like activity 10-fold,[54] several differently substituted derivatives have been described and evaluated as catalytic antioxidants. The functionalization of the **A** ring is often combined with modification in the group attached to the nitrogen atom (the **B** ring) of the original core.

Singh and co-workers prepared several differently substituted benzisochalgenazolones (S, Se and Te) with a bulky group in the *para* position respect to the chalcogen. All these derivatives were tested for their GPx-like activities using the NADPH-GSH assay. Three of them are reported in Scheme 2.10, **50** ($R = CH_3$), **51** ($R = {}^{i}Pr$) and benzisotellurazolone **52** ($R = CH_3$) and showed higher activity in the reduction of ^{t}BuOOH in comparison to ebselen (**1**): 3-, 1.5- and 5-times, respectively. The key starting material used in the synthesis was *N,N′*-diorganyl-2-bromo-5-*tert*-butylisophtalimide **49**, which was reacted with the respective dilithium dichalcogenide similarly to the reaction shown in Scheme 2.1.[55] Alternatively, **51** could be prepared by the *o*-lithiation of **49**, followed by treatment with dimethoxymethyl diselenide and bromine.

The molecular hybridization of ebselen having a bulky group in the **A** ring and quinine in the place of the **B** ring was described by Kumar and

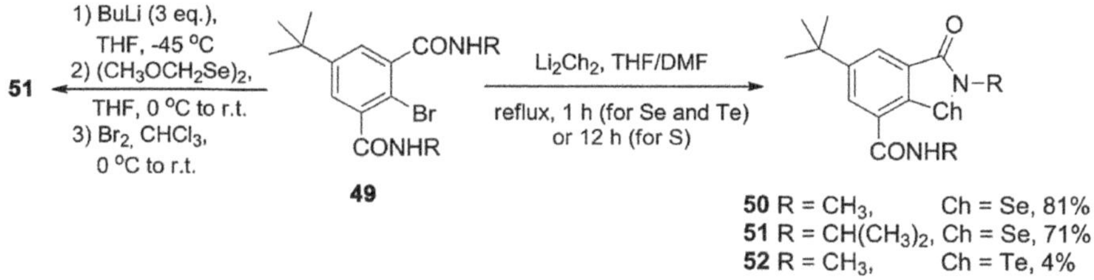

Scheme 2.10 Synthesis of benzisochalgenazolones **50–52**.

co-workers[56] to prepare potent GPx mimics, up to 10^3-fold more active than ebselen **1** in the oxidation of PhSH with H_2O_2. The key intermediate for the synthesis of the quinine–ebselen hybrids like **53** is a properly functionalized *o*-iodobenzamide bearing the quinine scaffold on the nitrogen atom. The construction of the corresponding benzoselenazolone ring can be achieved using KSeO^{t}Bu as the selenium source. The authors monitored the reaction of **53** with PhSH combining ^{77}Se NMR in CD_3OD and HRMS analysis. The expected selenenyl sulfide **54** was observed only as transient intermediate, and a ^{77}Se NMR peak near to zero (author described it in as −0.2 ppm and 4.8 ppm) evidenced the presence of a selenol (**55**), which is probably stabilized by the interaction with the neighbor heteroatoms and by the presence of a steric hindrance in the *ortho* position of the **A** ring respective to the selenium atom. The authors demonstrated that the simple selenium–heteroatom (*N* or *O*) interaction is not sufficient to properly stabilize the selenol, allowing the powerful efficiency observed for **53**, and that the catalytic activity follows a first-order dependence on the catalyst concentration. Increasing the peroxide concentration, the initial rate gains a constant value, indicating the formation of an oxidized specie (**56**) that is subsequently reduced through a rate-limiting step regenerating the cyclic selenoamide **53**. Formation of **56** was also confirmed by ^{77}Se NMR evidence (Scheme 2.11).

Engman and Kumar and colleagues[57] described the synthesis and GPx-like activity of a series of ebselenols **59a–b**, which were prepared by the reduction of the corresponding methoxylated precursors **58** with BBr_3, easily accessible by the reaction of *o*-chloro and *o*-bromobenzamides **57** with elemental selenium in the presence of CuI and phenantroline (Scheme 2.12). The authors compared the GPx-like activity of synthesized compounds **59a–b** with ebselen **1** in the GSH/H_2O_2 and GSH/^{t}BuOOH assays; all compounds were more active than ebselen and all are more active when the oxidant is H_2O_2. Using PhSH as thiol cofactor and hydrogen peroxide as ROS mimic, ebselenol **59a** was >10-fold faster than the others and ebselen and this can be ascribed to the proximal position of the hydroxyl group with respect to the selenium center. Interestingly, these compounds evidenced a low level of toxicity.

Singh and co-workers prepared a series of amino acid-containing ebselen derivatives and evaluated their GPx activity using as cofactors both GSH and PhSH.[58] The synthesis of the derivatives **62a–e** was performed using different

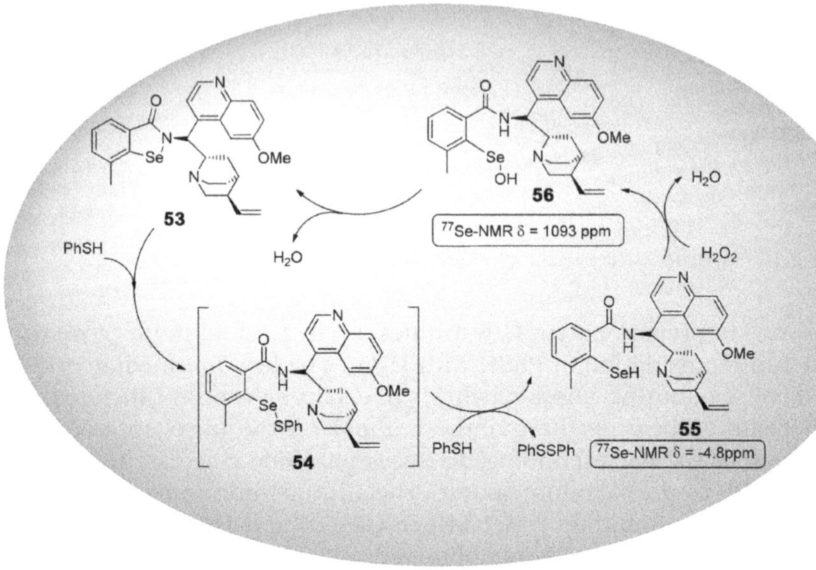

Scheme 2.11 GPx-like catalytic cycle of **53**.

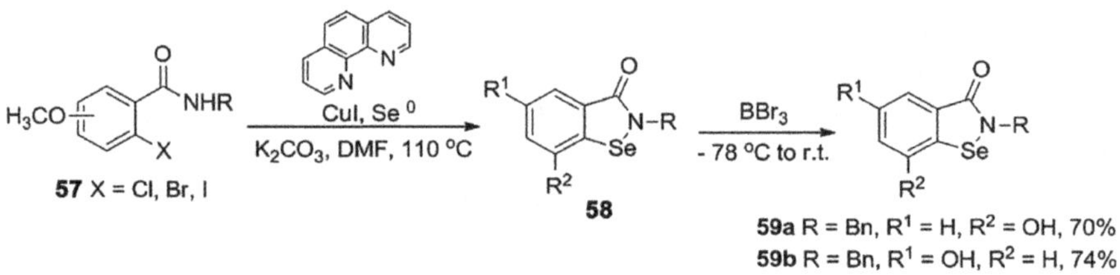

Scheme 2.12 Synthesis of ebselenols **59a–d**.

approaches and the best procedures are summarized in Scheme 2.13: (i) starting from the aryl bromide **60** through the intermediate formation of selenide **61** to afford **62a**; (ii) starting from the diselenide **63** to afford **62a–d**; and (iii) starting from the classical intermediate **3a** to obtain **62e**.

In some studies, focused on evaluating the structure–activity correlation, authors observed that intramolecularly stabilized ebselen analogues like **62a–b** showed maximal initial V_{max} slightly superior to ebselen **1**, but they were not active at low GSH concentrations, at which **1** and related diselenides are active. A detailed mechanistic study using ^{77}Se NMR spectroscopy was performed to understand this low activity using glycine derivatives **62f** (Scheme 2.14). In the reaction of **62a** with PhSH (in a 2 : 1 mixture of $CDCl_3/CD_3OH$) the respective selenenyl sulfide was not observed, probably due the low nucleophilicity of PhSH compared to GSH. In view of the low solubility of GSH in $CDCl_3/CD_3OH$ (2 : 1), the reaction of **62a** with GSH failed completely. To circumvent this problem, **62a** was hydrolyzed to the carboxylic acid **62f**, which was treated with $NaHCO_3$ affording *in situ* the

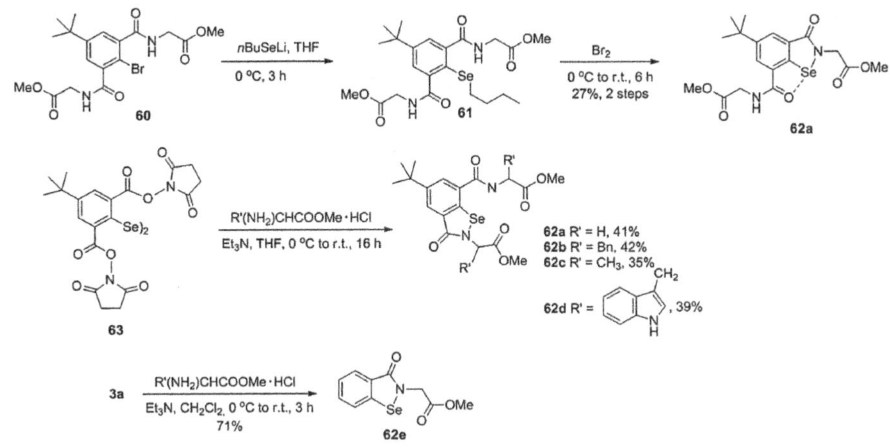

Scheme 2.13 Synthesis of amino acid-containing ebselen derivatives.

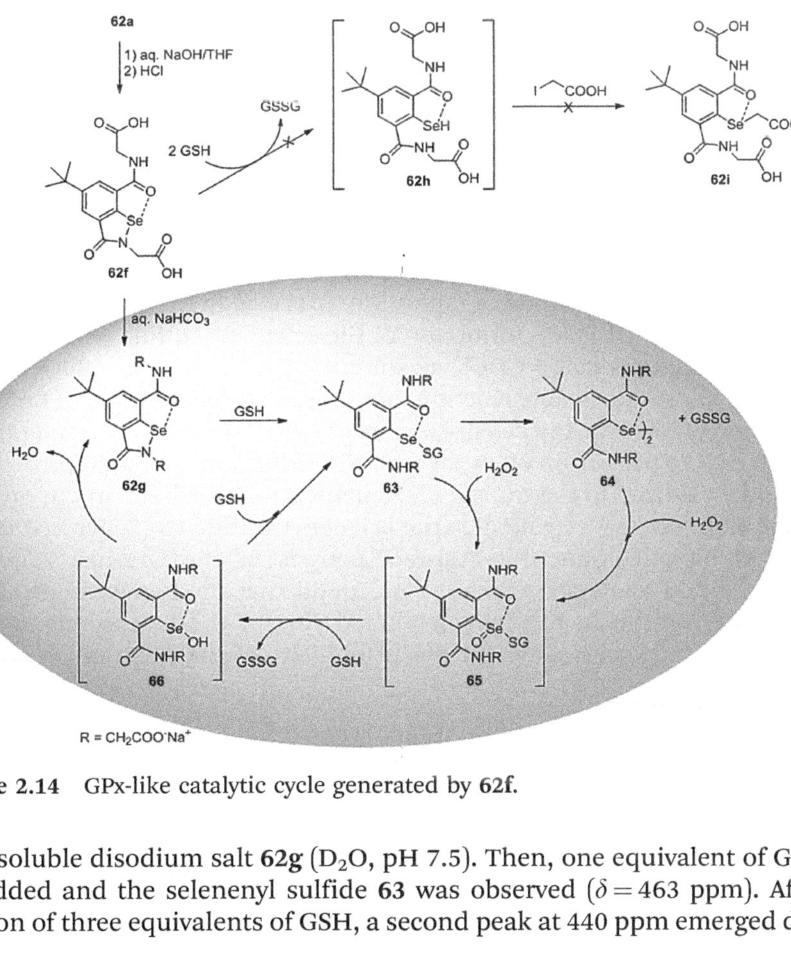

Scheme 2.14 GPx-like catalytic cycle generated by **62f**.

water-soluble disodium salt **62g** (D_2O, pH 7.5). Then, one equivalent of GSH was added and the selenenyl sulfide **63** was observed ($\delta = 463$ ppm). After addition of three equivalents of GSH, a second peak at 440 ppm emerged due to diselenide **64**, and the peak corresponding to selenol **62h** was not observed, which was consistent with the unsuccessful trapping as the carboxylic acid **62i** by reaction with iodoacetic acid. The addition of H_2O_2 to the mixture of **63**, **64** and GSH afforded the ebselen analogue **62**. The most likely intermediates for the cyclization of **63** to **62** are selenoxide **65** and selenenic acid **66**. These results were supported by computational studies and indicate that ebselen analogues with an amide group at the *ortho* position catalyze the reduction of peroxides by a mechanism similar to that of unmodified ebselen.

2.2.1.4 Ebselen-related Heterocycles

As can be observed through this chapter, the search for new small molecules with GPx activity is most often accompanied by mechanistic insights which help to explain how the new organoselenium compounds are acting and, importantly, drives the rational design of new GPx-mimetics. Since the seminal work by Reich and Jasperse in 1987,[59] which synthesized several *bis* selenamides with GPx-like activity, work has been described based on the replacement of both the aromatic **B** ring and the carbonyl group in the benzoselenazolone skeleton. Back and Dyck described the synthesis of the potent GPx-mimetic camphor-derived selenenamide twenty years ago, (Scheme 2.15, **67**).[60]

Mechanistic studies using BnSH/^{t}BuOOH system were performed to evaluated the catalytic activity of **67**. In this technique, the amount of BnSSBn generated in the oxidation of thiol is monitored using high-performance liquid chromatography. The first step in the reaction of **67** with BnSH and ^{t}BuOOH is the formation of the selenenyl sulfide **68** (observed in ^{77}Se NMR analysis using $CDCl_3$ as solvent). The addition of ^{t}BuOOH to **68** did not give any detectable reaction; however, after the addition of BnSH, the formation of BnSSBn and selenol **69** was observed. Even if it was not possible to observe the peak from **69** in the ^{77}Se NMR, the authors could prove that it is rapidly oxidized to the transient selenenic acid **70**, which, in the presence of additional BnSH, regenerates the selenenyl sulfide **68** (Scheme 2.16).

Based on a computation-assisted analysis of the GPx-like activity of ebselen **1** and analogues, Singh *et al.* found that the replacement of C=O with a CH_2 group in the five-membered ring of the heterocycle activates the Se–*N* bond, decreasing the electrophilicity of the selenium atom and,

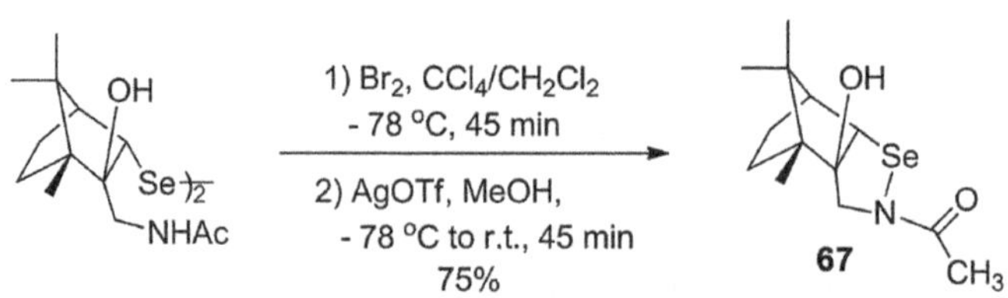

Scheme 2.15 Synthesis of the camphor-derived selenenamide **67**.

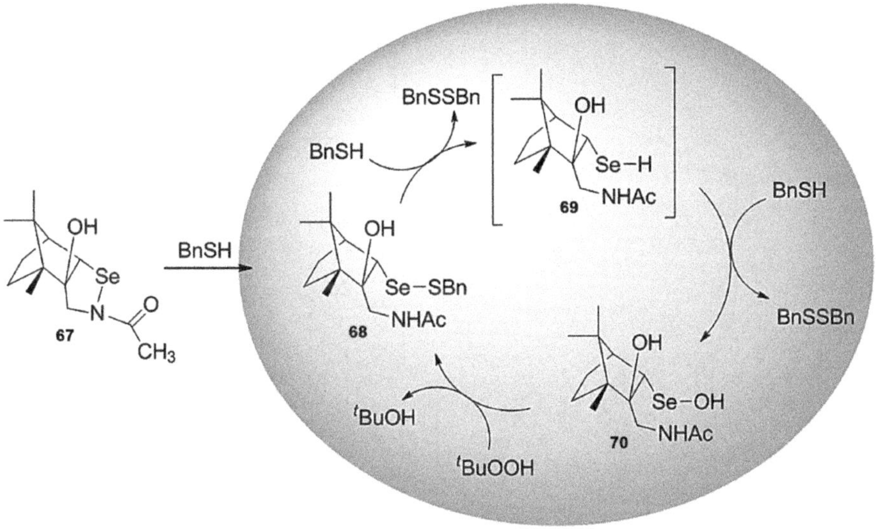

Scheme 2.16 GPx-like catalytic cycle of the camphor-derived selenenamide **67**.

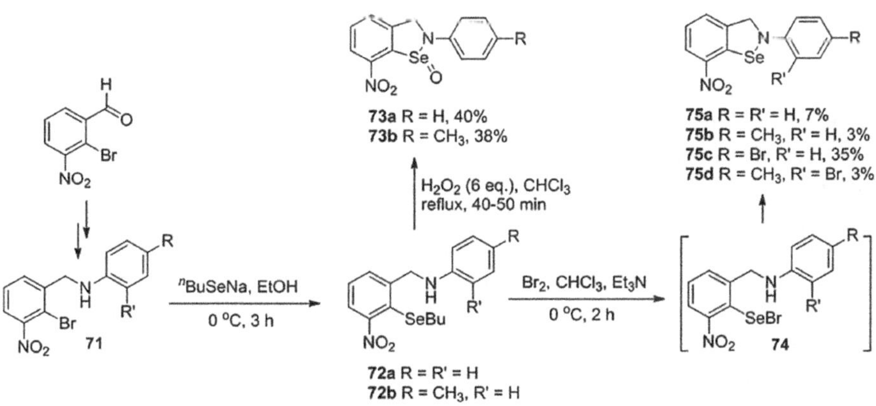

Scheme 2.17 Synthesis of isoselenazolines **75** and isoselenazoline *Se*-oxides **73**.

consequently, increases the rate of the reaction with GSH in the catalytic cyclic.[61] To prove these findings, a series of isoselenazolines **75** and isoselenazoline *Se*-oxides **73** were prepared and their GPx-like activities were evaluated and compared to the corresponding ebselen analogues. The key intermediate in the synthesis of isoselenazolines **75** and the respective oxides **73** is the benzyl amine **71**, which was prepared from 2-bromo-3-nitrobenzaldehyde in four steps (Scheme 2.17). Reaction of **71** with ^{n}BuSeNa, followed by bromination, afforded the desired *o*-nitro isoselenazolines **75** in modest yields. The cyclization step passes by the arylselenenyl bromide intermediate **74**. After treatment with an excess of H_2O_2, butylselenides **72a–b** were converted to the corresponding selenoxides **73a–b** in few minutes.

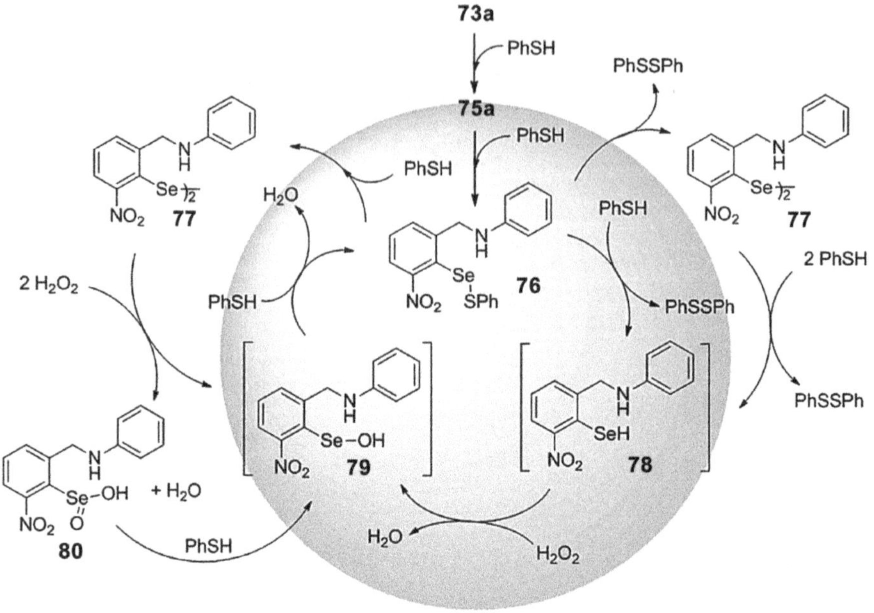

Scheme 2.18 GPx-like catalytic cycle of **73a** *via* **75a**.

Compounds **73b** and **75b** showed excellent GPx-like activity in the coupled reductase assay (GSH/NADPH and H_2O_2). The mechanism of the reaction between **75a** and PhSH in the presence of H_2O_2 in DMSO-*d*6 was investigated using ^{77}Se NMR and authors concluded that the mechanism is different to that reported by Back and Dyck[47,60] for the camphor-derivate selenamide **67**, in that no diselenide was observed in the first case. However, close examination of the mechanisms depicted in Schemes 2.16 and 2.18 reveals that they are similar to that of GPx, involving a selenol, selenenic acid and selenenyl sulfide as intermediates in the primary catalytic cycles (Scheme 2.18).[61]

The second equivalent of PhSH affords the selenenyl sulfide **76** and the diselenide **77**. A further reductive step promoted by thiol generates selenol **78**, which is now in a condition to reduce one equivalent of peroxides, forming the selenenic acid **79** that can be readily reactivated by reaction with PhSH, regenerating **76**. From the other side, diselenide **77** can enter in the cycle in two different manners; it can be reduced to selenol **78** (by reaction with the PhSH) or oxidized to selenenic (**79**) and seleninic acids (**80**) (by reaction with peroxide).

2.2.2 Diselenides

The GPx-like activity of diaryl diselenides was first reported in 1989 by Spector and co-workers,[62] five years after the discovery of the GPx-like activity of ebselen. These compounds were stable, easier to prepare and catalytically more active than ebselen. In the following years, a number of diselenides have been synthesized and the *in vitro* antioxidant activities evaluated mainly by measuring the ability to catalyze peroxide mediated thiol/disulfide oxidations (thiol assay).[38,63–72] H_2O_2, Cum-OOH and ^{t}BuOOH are the most common peroxides, and PhSH or GSH the most used thiol cofactors. Illustrative examples of the studied diaryl- and diheteroaryl diselenides are reported in Figure 2.3. Ferrocenyl diselenides **84** with redox-active groups were also investigated.

Great efforts have been devoted to the determination of kinetic parameters and in the comprehension of the GPx-like catalytic mechanism, not only as a guide for understanding the chemistry at the enzyme active site, but also for designing new GPx functional mimics as potential drugs. Computational calculations, ^{77}Se NMR experiments and X-ray structural studies have provided support to the mechanistic hypotheses and occurrence of key intermediates.[24,73,74] The analysis of the collected results evidenced the following general trends.

(1) Diselenides usually act as pro-catalysts that express their activity after cleavage of the Se–Se bond and generation of the actual catalyst.
(2) In most of the catalytically active diselenides, the catalytic triad of the GPx active site is imitated by incorporating a coordinating heteroatom (usually nitrogen or oxygen) near the selenium atom (see Figure 2.3).

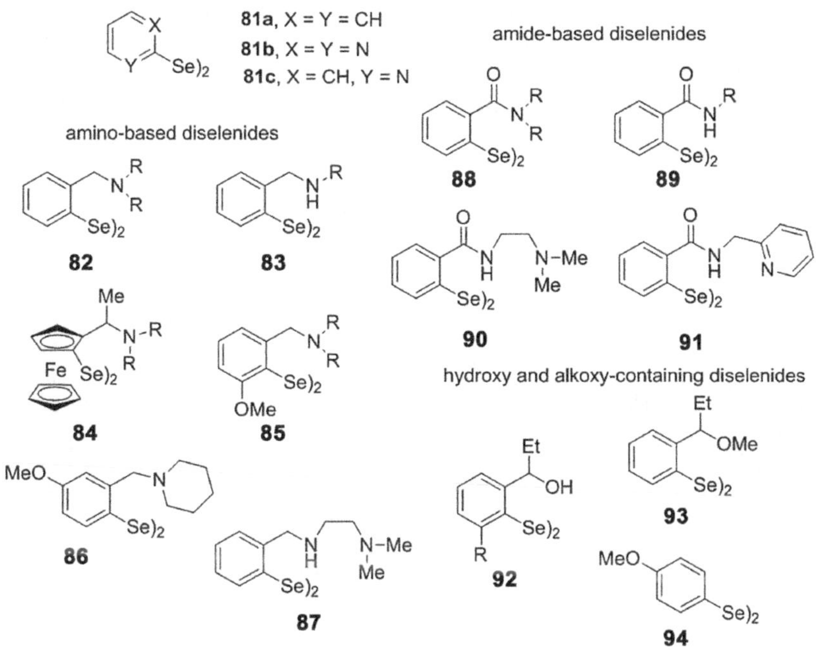

Figure 2.3 Selected classes of diaryl and diheteroaryl diselenides with glutathione peroxidase (GPx)-like activity.

The formation of intramolecular Se···heteroatom non-covalent interactions plays a key role in modulating the redox properties of the diselenide.

(3) Different diselenides can operate through distinct catalytic mechanisms. In fact, the nature of the selenium environments not only modulates the GPx-like activity by steric and electronic effects, but also affects the chemical steps in the reaction mechanism and the formation of intermediates.

(4) The same diselenide can operate through different catalytic cycles, depending on the nature and concentration of the peroxides and/or thiols used in the study.

(5) De-activation pathways, dependent on the nature of the substituents, may reduce the catalytic efficiency of diselenides.

Several diaryl diselenides containing an oxygen near the selenium atom were prepared in 1998 by Wirth.[64] Compound **92**, particularly if unsubstituted at the *ortho* position, showed good GPx-like activities, but they were poorer than those of structurally related amine-derivatives.

Scheme 2.19 shows the generally accepted catalytic cycle for the reaction of amine-based diaryl diselenides, first proposed by Iwaoka and Tomoda[63] and revisited by Mugesh and colleagues.[24,67]

In the initial step, the thiol acts as a nucleophile and produces a mixture of the selenol **95** and the selenenyl sulfide **96**. The catalytically active species **95** is more properly represented as a zwitterion, due to the selenol deprotonation by the amino moiety. Next, **95** is converted by the peroxide into the short-lived selenenic acid intermediate **97**, which further reacts with a thiol molecule to produce **96**. Finally, the active catalyst **95** is regenerated by

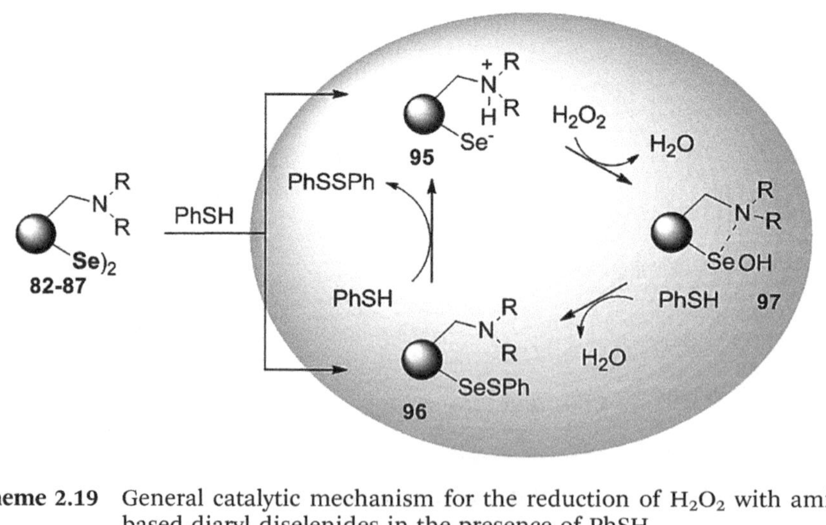

Scheme 2.19 General catalytic mechanism for the reduction of H_2O_2 with amine-based diaryl diselenides in the presence of PhSH.

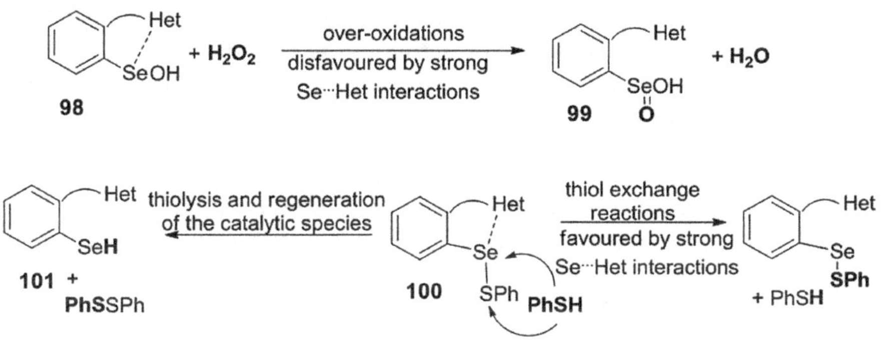

Scheme 2.20 Possible deactivation pathways in GPx mimics.

nucleophilic attack of another thiol to the sulfur atom of the selenenyl sulfide **96**. Over-oxidation of selenenic species to seleninic acids **99** (or even selenonic acid) observed in the presence of an excess of peroxide or thiol depletion, and thiol exchange reactions by nucleophilic attack of the incoming thiol at the selenium center of the selenenyl sulfide **100**, are possible deactivation steps, as illustrated in Scheme 2.20. These processes, slowing or preventing the regeneration of the catalytically active species **101** reduce the catalytic efficiency of the diselenides *in vitro*. As mentioned before, in biological systems the amount of thiol surpasses that of peroxides and over-oxidation of selenium species probably does not occur. The presence of an amino group in benzylamine-derived diselenides **82–87** beneficially affect the catalytic cycle beneficially, not only because of deprotonation of the nucleophile by the basic nitrogen, but also through of intramolecular non-covalent Se···*N* interactions. These coordination effects increase the stability of **98** and prevent over-oxidation. Conversely, too-strong Se···*N* interactions in the selenenyl sulfides **100** should be avoided, because they increase the electrophilic character of the selenium and speed up the deactivation pathway by thiol exchange.

These considerations clearly indicate that a change in the substitution in proximity of the selenium atom can modulate the GPx-mimicking activity by controlling the catalyst deactivation, and alter the catalytic cycle. For example, the absence of Se···*N* interactions in selenenyl sulfides from ferrocenyl diselenides **84** is crucial to explain their excellent catalytic activity.[65] Moreover, any substituent that can enhance the nucleophilic attack of thiol at sulfur rather than selenium in the selenenyl sulfide **100** increases the catalytic activity, preventing the thiol exchange reaction. Bhabak and Mugesh[67] reported interesting results on the possibility of improving the catalytic performances of diselenides through the combined effects of amine and methoxy groups **85**. The introduction of a methoxy group at the 6-position of the aromatic ring, which is able to interfere with the Se···*N* interaction, enhances the zwitterionic character in the selenol and hampers the over-oxidation to seleninic acid, mainly by steric protection. Moreover, it prevents both thiol exchange reaction and selenenyl sulfide re-oxidation (reverse-GPx cycle). Interestingly, when the assays were performed with larger thiols, such as GSH, the catalytic activities decreased by steric hindrance effect. In such cases, a methoxy group at the 4-position **86** significantly improved the catalytic activity.[69]

The replacement of *tert*-amino groups with *sec*-amino ones in diselenides **83** has been used to improve the catalytic activity.[67] *N*-propyl and *N*-isopropylamino-substituted diselenides are between eight and 18 times more active than the corresponding dipropyl and diisopropyl derivatives. This behavior is due to negligible thiol exchange reactions.

Mono- and disubstituted amide-based diselenides **88–89**[68] exhibited good GPx-like activities, but completely different catalytic mechanisms were involved. Mechanistic investigations showed that these diselenides are practically unreactive toward PhSH. The catalytic cycle starts with the oxidation of diselenides by the peroxide to generate the corresponding selenenic (**102**) or seleninic (**103**) acids. Then, both species react with thiols generating selenenyl sulfide **104**, which disproportionates into the corresponding diselenides and disulfides. The poorer catalytic efficiency in respect to amine-derived diselenides is due to strong Se···*O* interactions that cause undesired thiol exchange reactions in selenenyl sulfides **104**, particularly in the case of mono-substituted amides (Scheme 2.21).

A strategy that further enhances the GPx-like activity of amine- or amide-based diselenides consists of the insertion of an additional amino moiety as in **87** and **90** (Figure 2.3).[70] These compounds mimic the structural features of the GPx active site with glutamine and tryptophan residues located within

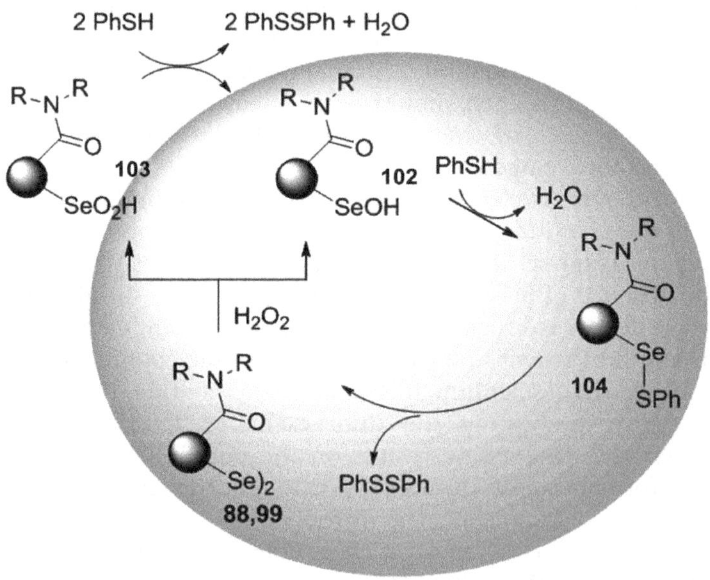

Scheme 2.21 Proposed catalytic mechanism for the reduction of H_2O_2 by amide-based diaryl diselenides in the presence of PhSH.

hydrogen bonding distances to the selenium atom of a selenocysteine. Interestingly, the catalytic mechanism is the same in both cases and involves the thiol-mediated formation of selenenyl sulfides and selenols, as described in the general mechanism for amine-based diselenides. Discrepancies in comparisons of the measured catalytic activity were observed when PhSH was replaced by GSH or in assays undertaken with different peroxides.

Press and Back demonstrated that conformationally restricted naphthalene *peri*-diselenides **105** have superior GPx-like activity to typical diaryl diselenides.[75] The C–Se–Se–C dihedral angle constraint and the long Se–Se bond destabilize the diselenide moiety and facilitate its oxidation by peroxides. Interestingly, as described in Scheme 2.22, the catalytic cycle starts with the oxidation of the diselenide into the corresponding selenolseleninate **106** and not with a ring-opening by the thiol.

In recent years, several diselenides were synthesized starting from biologically relevant molecules searching for new GPx mimics with enhanced catalytic activity and lower toxicity. Among them, nicotinoyl-[76,77] and pyridoxine-like diselenides **107–109**,[78,79] which have shown a notable peroxidase activity (Figure 2.4). In particular, GPx activities of pyridoxine-derived

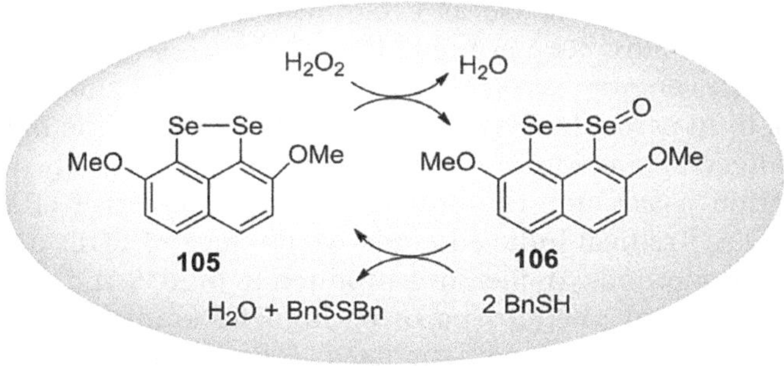

Scheme 2.22 Catalytic cycle for conformationally constrained naphthalene *peri*-diselenides **105**.

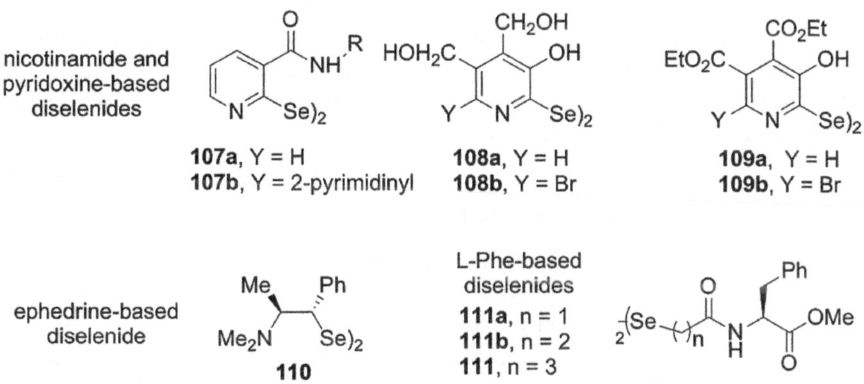

Figure 2.4 Diselenides derived from biologically relevant scaffolds.

diselenides were monitored in aqueous phosphate buffer (pH 7.5) by NADPH consumption in a NADPH–GSSG-coupled reductase assay. Compound **109b** gave the best results, demonstrating the beneficial effect of the insertion of the bromine atom into the parent compound. The authors suggested that the bromine converts the pyridylseleno group into a better leaving group. Thus, the nucleophilic attacks of thiol at selenium in the diselenide, and at sulfur in the selenenyl sulfide, are facilitated. However, conflicting results have been observed in the evaluation of the peroxidase activity by comparison with the PhSH assay.

Braga and co-workers investigated chiral diselenides from ephedrine,[80] or commercially available L-aminoacids.[81] In the latter case, the modular synthetic approach allowed the synthesis of compounds with controlled variations of structure and chain length. The evaluation of the GPx-like activity by thiol assays of L-phenyl alanine derivatives demonstrated that compounds **111b** and **111c**, with longer chains, showed higher catalytic activity than diphenyl diselenide.

The biological response to oxidative stress is not limited to the GPx activity, since other enzymatic pathways can limit the adverse effects of ROS. Thus, for a more efficient assessment of the antioxidant properties, and in view of a potential use as drugs in diseases related to oxidative stress, diselenides have been evaluated in multiple test models.[72,78,79,82–86] Free radical-scavenging assays (DPPH radical-scavenging and ROS-scavenging), lipid peroxidation models for the determination *in vitro* of the protective effect against lipid peroxidation and low-density lipoprotein (LDL) oxidation assays for the evaluation of a protective effect against copper or peroxyl radical-induced oxidative damage of LDLs are the most frequently used methods. In fact, overproduction of ROS and lipid and LDL oxidations are typical phenomena of pathologies traditionally related to oxidative stress such as cancer, diabetes, inflammation, atherosclerosis and stroke.

Finally, it was demonstrated that diphenyl diselenide **81a** and simple analogues serve as substrates for TrxR.[83–86] At present, the structural requirements for this activity are unknown, but interestingly, diselenides that are good GPx mimics are not necessarily good substrates for TrxR. TrxR is an enzyme with a key role in oxidative stress protection and diselenides would express antioxidant properties not only by mimicking the activity of GPx, but also *via* reduction in the TrxR pathway.

2.2.3 Selenides and Spiroselenuranes

In selenides, the selenium atom is stably bonded to two carbons and, differently from selenazoles and diselenides, in this case it is not possible to envision a mechanism involving the direct formation of a selenenic acid. In some cases, selenides were demonstrated to act as pre-catalysts. As an example, Back and Moussa demonstrated that functionalized allyl selenides containing in a suitable position a hydroxyl group (**112a–c**) are rapidly oxidized by ^{t}BuOOH into the corresponding selenoxides. The latter, following a [2,3]-sigmatropic rearrangement, afford cyclic seleninate esters **113a–c** that are the actual GPx-mimetic catalysts in the reduction of the ^{t}BuOOH in the presence of benzenethiol. The catalytic cycle starts with the thiolysis of **113** and the formation of the seleninate thioester **114**, which is reduced to selenenic acid **115** by the consumption of a second molecule of thiol. The selenenic acid **115** rapidly reduces a molecule of peroxide closing the catalytic cycle with the formation of **113** (Scheme 2.23). The screening of different substrates demonstrated that the best catalytic efficiency can be obtained when the hydroxyl group is spaced by three carbon atoms with respect to selenium, affording a five-membered cyclic seleninate ester **113a.**[87,88] A few years later, Singh and co-workers reported the synthesis of a seven-membered cyclic selenenilate ester derived from 2-phenoxyethanol, demonstrating an interesting GPx-like activity.[89]

These evidences have led different researchers to prepare cyclic seleninate esters fused to a functionalized benzenic ring (selected examples **116–118** are reported in Figure 2.5). Simple derivative **116** showed a considerably reduced catalytic activity with respect to **113**. The introduction of an electron-donating group (**117**) enhances the activity, and the mechanism depicted in Scheme 2.23 has been confirmed for all these compounds demonstrating that, in the presence of an excess of thiol, takes place an inactivation pathway that leads the formation of a selenenyl sulfide.[90–92] This latter is slowly reactivated by reaction with ^{t}BuOOH and, only in the

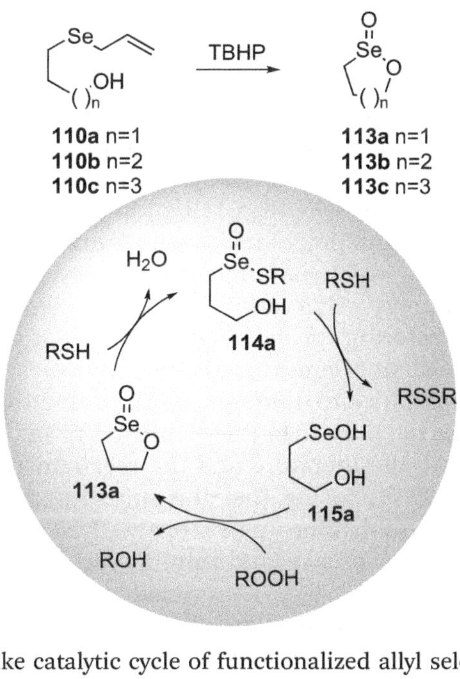

Scheme 2.23 GPx-like catalytic cycle of functionalized allyl selenides **112a–c**.

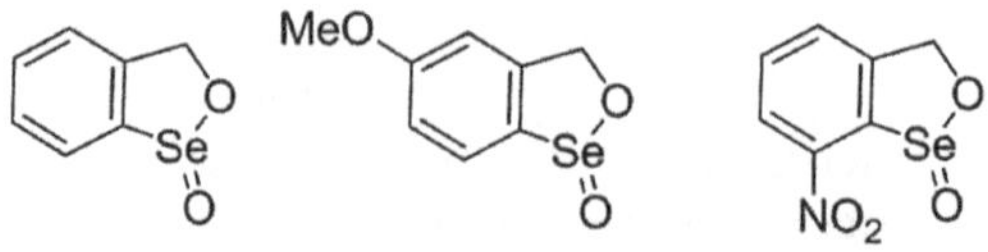

Figure 2.5 Examples of a cyclic seleninate ester fused with a functionalized benzenic ring.

case of **118**, it can be transformed into the corresponding diselenide, demonstrating the occurrence of different reaction mechanisms among analogue compounds. This difference is reasonably due to the presence of the nitro group in a suitable position to stabilize the selenenyl sulfide intermediate through a non-covalent Se···*O* interaction.[93,94] Saito *et al.* demonstrated the reductant activity of *N*-phenyl-(2-selenoallyl)benzoylamide against peroxides, claiming a GPx-like activity several times higher than selenomethionine. This allyl selenide, after oxidation to the corresponding selenoxide, according the mechanism proposed by Back and Moussa, could be considered to be the precursor of ebselen. Nevertheless, experimental evidences demonstrate that its antioxidant activity is only about one-tenth that of ebselen, indicating that, in this case, a different mechanism from that proposed in Scheme 2.23 should be involved, even if the first step is the formation of an allyl selenoxide intermediate.[95]

Selenides symmetrically substituted with 3-hydroxy propyl groups have been demonstrated to be highly efficient GPx mimics when compared to ebselen. This activity involves an interesting and novel mechanism, which passes through a spirodioxaselenurane intermediate, formed subsequently to the direct oxidation of selenide into selenoxide.[96] This demonstrates the importance of the Se–*O* bond, which can be conveniently used in the place of the well-known Se–*N* bond already discussed, for the preparation of GPx-like mimics. A similar mechanism was reported by Braverman *et al.* for a novel class of water-soluble vinyl selenides **119** endowed with high catalytic antioxidant activity.[97] The reduction of hydrogen peroxide leads to the formation of the selenoxide **120** which is subjected to a cyclization, affording the spirodioxaselenurane **121**. This latter reacts with benzenethiol and closes the catalytic cycle, restoring the selenide **119** through the intermediate formation of the hypervalent selenium derivative **122a** (Scheme 2.24). Given the propensity of selenium(IV) to undergo facile reductive eliminations, the formation of species such as **122a** (never proved to be formed) is superfluous, and according to the literature and the mechanism described by Back *et al.*,[96] is reasonable supposing that the spiro intermediate undergoes a single substitution by thiol, affording **122b**, and then a reductive elimination at selenium by the attack of a second thiol molecule.

Recently, Back and co-workers demonstrated that the water-soluble selenides **123–124** have high catalytic activity for the reduction of peroxides in the presence of glutathione as a reducing cofactor.[98] They demonstrated that

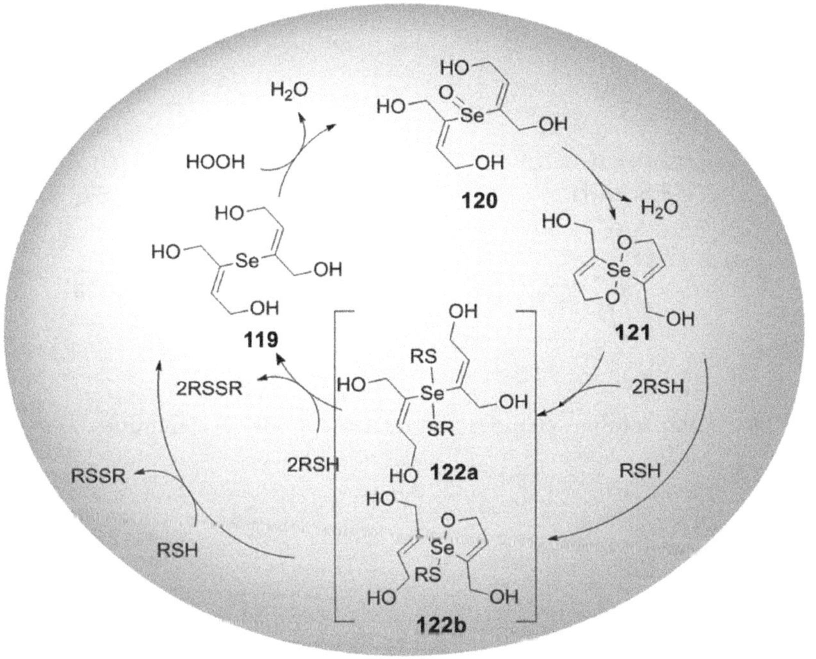

Scheme 2.24 GPx-like catalytic cycle of vinyl selenide **119** *via* the intermediate formation of the spirodioxaselenurane **121**.

this activity is strongly dependent on the nature as well as on the position of the substituents in the aromatic ring. All substituents able to increase the positive charge on the selenium atom as a consequence of resonance stabilization produce a considerable increment of the catalytic rate. It was demonstrated that in the presence of hydrogen peroxide, **124** and **125** are converted into the spirocyclic derivatives **126** and **127**, passing through the intermediate formation of the corresponding selenoxides, demonstrating that the oxidation is the rate-determining step. The derivatives **126** and **127**, after isolation in the presence of two equivalents of benzenethiol (and in absence of peroxides), afforded the original selenides **124** and **125** (Figure 2.6). In the same figure, structure **128** illustrates a generic skeleton of a spirodiazaselenurane, a class of compounds introduced by Back and co-workers[99] and extensively studied by Mugesh and co-workers.[100–102] Opportunely, decorated diaryl selenides, by oxidation into the corresponding selenoxides, afford the spirocyclic derivative *via* the condensation of the amidic nitrogen with the selenium atom, concomitant to the loss of a molecule of water. The authors observed that the catalytic thiol peroxidase activity is enhanced when a tellurium is in the place of selenium. Considering the substituent on the nitrogen atom for selenium derivatives, the better results were achieved when aromatic groups are present. The introduction of an electron-donating group on the aromatic ring enhances

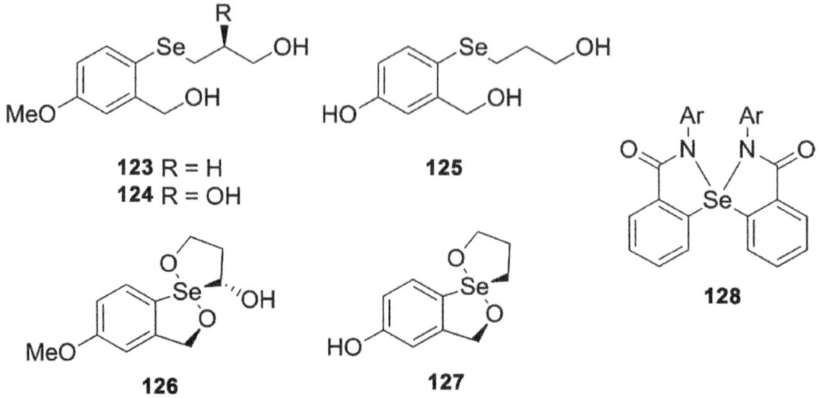

Figure 2.6 Water-soluble selenides and related spirodioxaselenuranes.

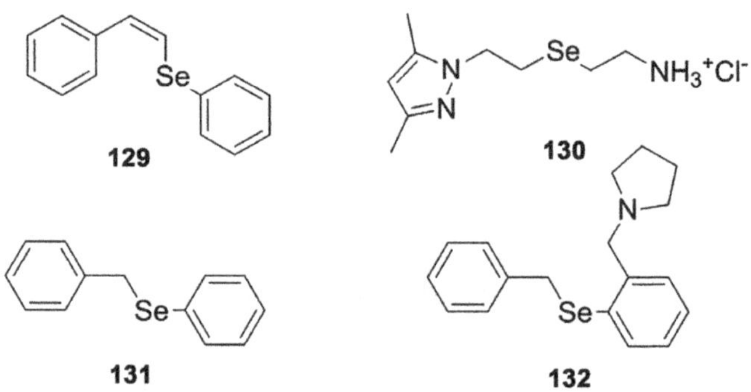

Figure 2.7 Selected selenides having glutathione peroxidase (GPx)-like activity *via* a selenoxide intermediate.

the GPx activity, whereas an electron-withdrawing group showed a detrimental effect.

Figure 2.7 depicts selected examples, among a series of other selenides, that act as Gpx-mimics passing simply through the formation of the corresponding selenoxide.[103,104] In the case of vinyl selenide **129** and dimethyl pyrazol derivative **130**, the formation of a selenoxide and its reduction to the starting selenide in the presence of DTT was confirmed by ^{77}Se NMR experiments. Braga and co-workers deeply studied the mechanism that allowed selenoxides of **131** and **132** to react with peroxide and benzenethiol, exploiting the redox catalytic activity.[105] They demonstrated that the GPx-like cycle does not follow a simple Se(II)–Se(IV) interchange. In the presence of hydrogen peroxide and water, the selenoxide **133** is converted into the more efficient oxidizing derivative **136**, reasonably involving an equilibrium with the hydrated form **134** and the selenonium ion **135**. The hydroxy perhydroxyselenane **136** oxidizes PhSH into the corresponding disulfide (PhSSPh) 10^4-fold faster than selenoxide **133**, regenerating the intermediate **135** with the transient formation of **137** (Scheme 2.25).

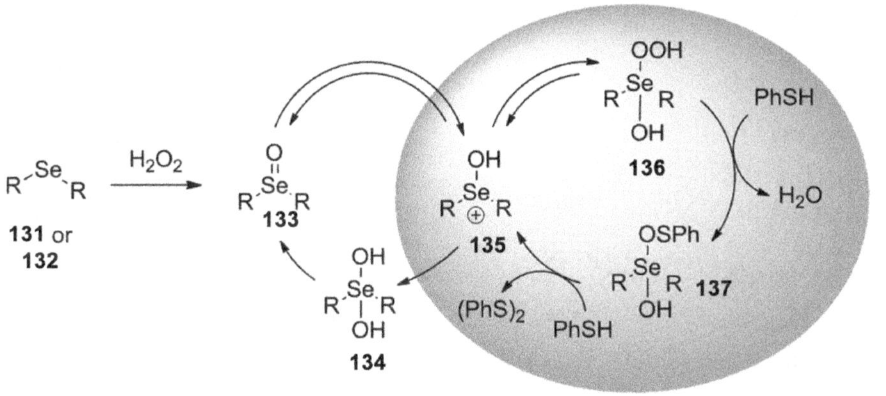

Scheme 2.25 GPx-like catalytic cycle of selenides **131** and **132**.

Recently, Iwaoka and co-workers studied the GPx-like activity of a series of water-soluble cyclic selenides, making some comparisons with the corresponding open-chain derivatives and making some considerations concerning the ring size and the presence of additional functionalization on the ring. Using ^{77}Se NMR spectroscopy, the authors demonstrated that selenides are converted directly into the corresponding selenoxides that, after the addition of one equivalent of DTT, restore the selenide, closing the catalytic cycle. When the GPx activity was monitored in a buffered (pH 7.4) aqueous solution using the NADPH/GR/GSH-coupled assay, it was demonstrated that the activity of cyclic selenides was higher than that of the corresponding open-chain analogues. Besides, the optimal ring size is the five-membered ring, with the following order of reactivity: **140** = **141** > **139** > **142** > **138**. From these results, it is also clear that the relative configuration on the carbon atom bearing the polar substituent has no influence in the redox ability of the selenide. Similarly, the investigation of the contribute of the polar substituents confirmed this latter observation, showing a reactivity order of **144** = **145** < **140** = **141** < **143** < **146**. The high reactivity of **146**, in the aforementioned conditions, was explained as a consequence of the deprotonation of the carboxylic unit at pH 7.4, which enhances the reductive ability of the selenide. It is worth highlighting that the results obtained by changing the assay and using the NMR method (H_2O_2, DTT in MeOD-*d*3) were, in some cases, considerably different, suggesting a strong dependence of the analysis on the reaction conditions, and, more specifically on the solvent. The scale of catalytic activity **141** > **140** > **139** = **138** > **142** indicates that, in methanol, the relative configuration plays a role in the GPx-like activity and that the superiority of the cyclic selenide over the open-chain one is considerably reduced. Furthermore, if in water the –OH group is more effective than the corresponding -NH_3Cl, in methanol an opposite trend was observed, although the authors were not able to provide a definitive explanation for this phenomenon (Figure 2.8).[106]

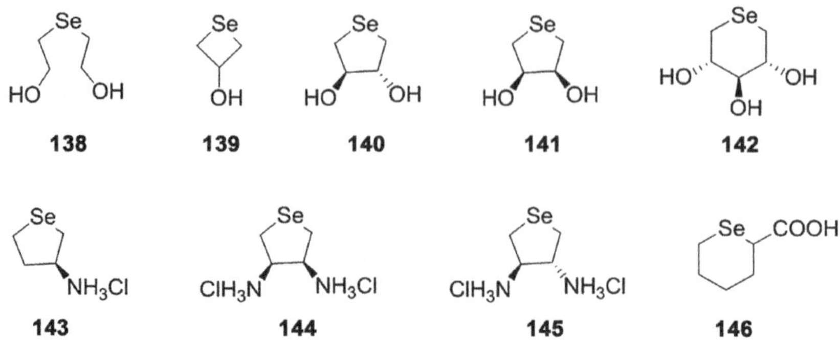

Figure 2.8 Water-soluble cyclic selenides.

During the preparation of this chapter, the same authors reported a further investigation of the mechanism, suggesting that in water, mainly selenide and selenoxide are involved, whereas in methanol becomes prevalent the contribute of a hydroxy perhydroxyselenane intermediated, that suggests a mechanism similar to that depicted in Scheme 2.25.[107]

2.3 Looking Beyond the Simple Concept of GPx Mimetics

2.3.1 Targeting Biologically Relevant Thiols

As illustrated in this chapter, great efforts were devoted to study mimetic strategies to reproduce the catalytic cycle of GPx. It is probably because this selenoenzyme was the first to be discovered that it was thoroughly explored from a mechanistic point of view, demonstrating the crucial role of selenium in the biology of living systems. From one side, the synthesis of different classes of GPx mimics and their detailed mechanistic elucidation in the reaction with peroxides and thiols contributed to refinement and better understanding of the biochemistry of selenium and organoselenium derivatives. From the other, it shows clearly what is meant when selenium is defined as the *Janus* element. One face is the antioxidant mechanism that leads to the beneficial reduction of peroxides and protection from oxidative stress; the other is the pro-oxidant activity, which induces the oxidation of free thiols, stimulating an oxidative offense to the cells. These two pathways are in close competition and the predominance of the beneficial effect or the toxicity depends upon their balance, giving a chemical reason for the very narrow therapeutic window that normally characterizes organoselenium compounds.

The toxic effects of selenium are strictly dependent on its concentration and, obviously, on the nature of organoselenium species in consideration. The pro-oxidant activity can be used as a potent inhibitor of cell growth, in some cases endowed with some tumor specificity. For this reason, some organoselenium compounds are considered good candidates for new chemotherapy treatments, as recently reviewed by Misra *et al.*[108]

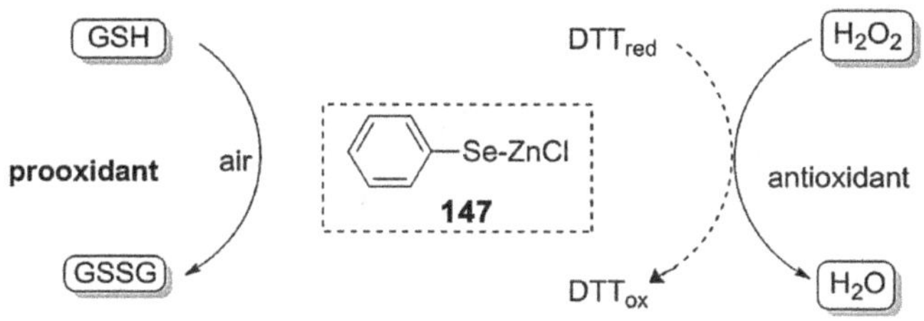

Figure 2.9 The *Janus* behavior of PhSeZnCl **147**.

A clear example of the *Janus* aspect of organoselenium compounds was elucidated during the studies conducted around a novel class of unprecedently stable selenium zinc complexes (Figure 2.9). PhSeZnCl (**147**) was prepared by oxidative insertion of Zn(0) in the commercially available PhSeCl, obtaining through the umpolung of the selenium atom as a stable zinc selenate that can be isolated and stored without any precaution. Some of us demonstrated that, in aerobic conditions, **147** promoted the slow oxidation (47 h–120 h) of several thiols, including PhSH (in THF) and DTT and GSH (in D_2O). In the same paper, the GPx like activity was demonstrated for the reduction of H_2O_2 or ^{t}BuOOH in the presence of GSH, as well as in the reduction of H_2O_2 in the presence of DTT, optimizing the reaction conditions to perform the NMR analysis using D_2O as medium. In this latter test, **147** resulted less efficient than selenocysteine but considerably more efficient than diphenyl diselenide, dispelling the involvement of this latter specie in the catalytic mechanism.[109]

The GPx-like catalysis was confirmed, in a separate study, using two additional methods: the GR-coupled assay and Amplex® Red assay, to effectively demonstrate the peroxide consumption. The results obtained confirmed, in both cases, a strong catalytic activity with a V_{max} higher than those measured for diphenyl diselenide and ebselen. Interestingly, it was observed that upon substituting the selenium with sulfur, the GPx-like activity totally disappeared, highlighting the crucial role of selenium in the peroxide reduction. Considering the kinetic parameters, it was observed that **147** exhibits similar catalytic efficiency for both the co-substrates (GSH and H_2O_2) with a high turnover number (K_{cat}) and lower K_M for the glutathione among a series of tested molecules. For this reason, in some cancer cells, PhSeZnCl exhibits considerable cytotoxicity based on an increased level of ROS production, depletion of intracellular thiols and induction of apoptosis. The perturbation of the redox homeostasis was demonstrated to be a consequence of the reaction between **147** and the free cysteine residue in cellular glutathione *S*-transferase (GSTP) of the GPx. Interestingly, it was also demonstrated that the toxicity can be controlled by microencapsulation of PhSeZnCl in a polylactide microparticle formulation, envisioning a possible application in the chemotherapy treatment of drug-resistant cancers that overexpress this enzyme.[110] The thiol peroxidase activity can be controlled and selectively targeted to specific thiols and it can be obtained or using advanced formulation (like demonstrated in the case of microencapsulation) or by modification of the chemical structure of the active compounds.

GSTP is an enzyme that plays an important role in signaling and in the detoxification from hydrophobic and electrophilic compounds having thiol peroxidase activity. A series of diselenides (**148–150**) was investigated as mild electrophiles to produce a submaximal condition of stress. It was observed that among these compounds, **150**, which according the ^{77}Se NMR seems to be the less electrophilic, produces a moderate amount of hydrogen peroxide and nitroxides, stimulating Nrf2 nuclear translocation and, consequently, the transcription of Nrf2 and a series of phase II genes, including the GSTP one. The global result is a hormetic effect, leading to an increased resistance of the cell against H_2O_2 cytotoxicity. The authors demonstrated that this response is GSTP-dependent, suggesting a redox-sensitive formation of a GSTP/Nrf2 complex (Figure 2.10).[111]

Considering the interaction of organoselenium compounds with free cysteine-containing enzymes, spectrometric evidence demonstrated that ebselen could be correlated to the inhibition of GR. Thiols of Cys58 and Cyst63 attacks the electrophilic selenium of ebselen, affording the corresponding oxidized selenenyl sulfides and the consequent covalent inactivation of the GR's enzymatic functions.[112,113] These results could suggest a toxicity which is not in agreement with several clinical trials that clearly demonstrated that the toxicity of ebselen in human patients is insignificant.

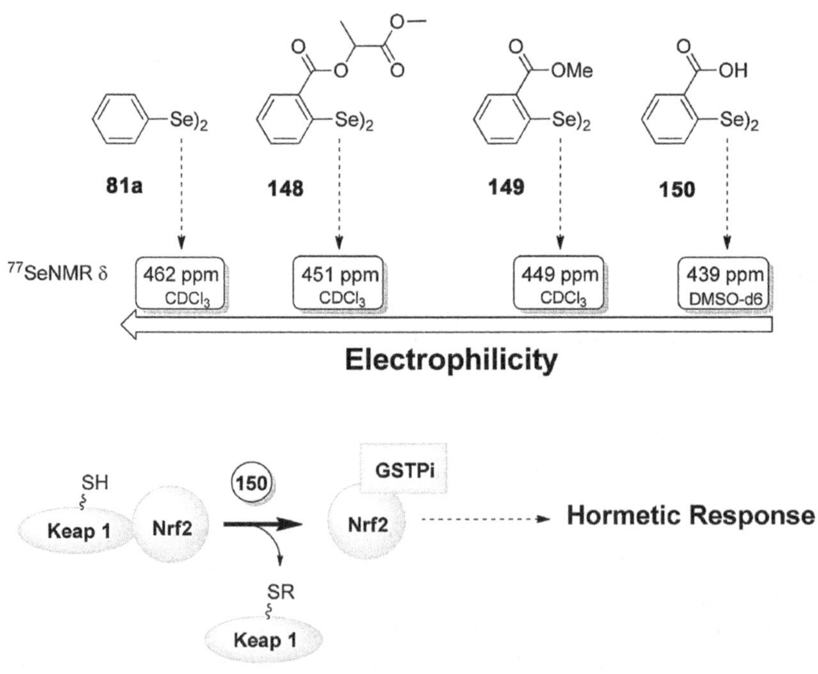

Figure 2.10 The hormetic effect of diselenides and the Nrf2 activation mechanism.

Recently, it was reported that by a similar mechanism proposed for the GR, ebselen could be targeted to the thiol of a bacterial urease. The authors studied many ebselen-like compounds and the corresponding diselenides, observing that the best results were obtained with the original ebselen. In this case, the interaction resulted in a competitive reversible inhibition with a very interesting value of K_i (2.11 nM against *Sporosarcina pasteurii* and 226 nM against *Helicobacter pylori*). The mechanism involves the cysteine 322 in the loop near to the entrance of the catalytic site of the enzyme. An additional advantage of these small-sized selenium molecules is that they have been demonstrated to easily penetrate the cell membrane of Gram-negative bacteria, such as *Escherichia coli*.[114]

2.3.2 Targeting "Zinc-finger" Domains

Even if zinc ions are biologically inert from a redox point of view, when they are inserted in thiolate-zinc(II)-thiolate bridges (like in some zinc-finger domains) or complexed with four thiolates (like in the metallothionein), they become susceptible to the redox status of the cellular environment. They are normally stable in reducing cytoplasmic conditions and assist in the folding of protein, representing a point of steric and conformational constrains for the peptidyl chains. They have been defined as "redox-sensitive molecular switches" and, under this perspective, they control several important cellular processes.[115] As an example, as a consequence of oxidative stress, Zn(II) can be released in the cytoplasm, affecting not only the 3D structure of the involved protein, but also interfering with the cellular signaling cascade. It is reasonable to envision that a selenium-containing redox modulator can interfere with the stability of some zinc-finger domain and this interaction can be explored and used to obtain new classes of therapeutically relevant compounds.

In 2004, Hartwig and co-workers demonstrated that low concentrations of a series of reducible selenium compounds, such as $PhSeO_2H$, PhSeCl, selenocysteine, ebselen and 2-nitrophenylselenocyanate, produced a decrease of the formamide pyrimidine-DNA glycosylase (Fpg), a zinc-finger protein involved in DNA reparation, inactivating this process. Interestingly, this effect was not reproduced by sulfurilated analogues or with other reduced selenium compounds (*e.g.* selenomethionine).[116] In a separate study, Hanas and co-workers observed that, by a similar mechanism, ebselen and sodium selenite were also found to alter the DNA binding of the cysteine with zinc-containing transcription factors.[117]

Recently, a series of 2,2′-diselenobisbenzamides (DISeBAs) were synthesized, proving their anti-HIV activity as retroviral nucleocapsid protein-7 (NCp7) inhibitors. Due to the numerous functions of this protein in both early and late stages of viral life, it represents an innovative and still unexplored target.[118]

Glycine and glutamate DISeBA derivatives **151** and **152** afforded the more interesting results, with broad anti-HIV activity and EC50 values in the low

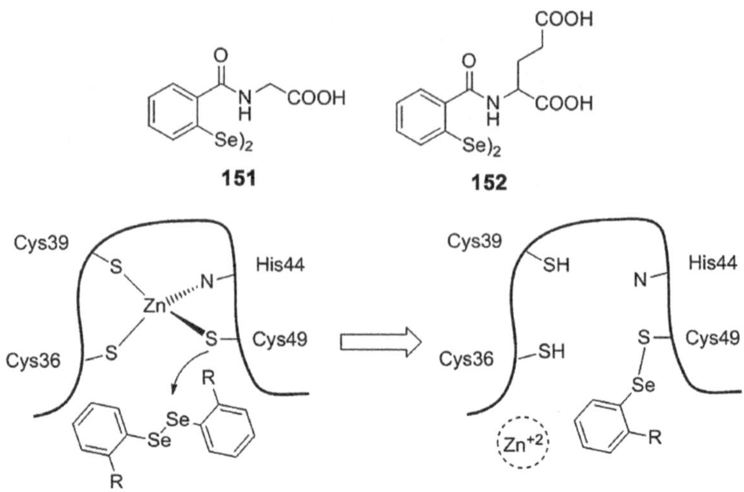

Figure 2.11 Interaction of DISeBas with the zinc-finger domain of protein NCp7.

micromolar range. In addition, **151** and **152** showed virucidal effects, suggesting their potential use in topical anti-HIV treatments. The inhibition of NCp7 has been demonstrated through "time of addition" experiments and it was supposed and confirmed by *in silico* experiments to be based on the interaction of the selenium with the zinc-finger domains of NCp7. It is known that cysteine 49 is the most nucleophilic center in the zinc domain and it is reasonable to suppose that sulfur in Cys-49 attacks the electrophilic selenium atom, allowing the zinc extrusion and the inhibition of the protein functions (Figure 2.11).

Tao, Chen and co-workers have recently reported a similar approach, using ebselen as covalent inhibitor of New Delhi metallo-β-lactamase (NDM-1) for the potential treatment of multidrug-resistant bacterial infections. Ebselen was demonstrated to be able to target the cysteine residue in the zinc domain of the metallo-β-lactamase and this experimental evidence opens the possibility of developing a new class of broad-spectrum inhibitors in consideration of the similar residues contained in the subclass of B1 and B2.

The thiol of a cysteine attacks the selenium of ebselen, affording a covalently bonded selenium-sulfide and promoting the zinc extrusion from the active site of the enzyme. The formation of the Se–S adduct was confirmed by electrospray ionization mass spectrometry analysis.[119]

2.3.3 GPx Mimics as Probes and Biosensors

The ability of organoselenium compounds to respond readily at the redox modification of the chemical environment stimulated researchers to design and realize novel structures to be used as fluorescent probes or to mimic biosensors. These new tools could be used to study redox chemistry in biological systems and to measure peroxides in biologically relevant materials. Han and co-workers recently collected their efforts in a comprehensive account illustrating the progress obtained in the preparation of several fluorescent probes inspired by the GPx and characterized by high sensitivity and selectivity in the region of the near infrared. Another interesting aspect of these probes is the application on confocal fluorescence microscopy to follow some biological redox changes in real time.[120] In these compounds, the active center is the selenium of a diaryl selenide that can switch from selenide to selenoxide, changing the chromophore by conjugation with the aromatic rings.

Vieira and co-workers reported a bio-inspired sensor based on a selenide having GPx mimetic activity immobilized in a film of chitosane-graphite nanoplatelets co-electrodeposited on a gold surface. The authors described the use of this novel biosensor for the detection of hydrogen peroxide in pharmaceutical products using chronoamperometry and the results were in perfect agreement with other standard methods.[121]

Acknowledgements

This manuscript is part of the collaborative and informative projects performed under the umbrella of the International Scientific Network Selenium Sulfur and Redox Catalysis (SeSRedCat). EJL is a recipient of a Research Fellowship from CNPq.

References

1. J. O. Hall, in *Veterinary Toxicology- Basic and Clinical Principles*, ed. R. C. Gupta, Elsevier-Academic Press, Oxford, 2007, pp. 453–460.
2. K. W. Franke and V. R. Potter, *J. Nutr.*, 1934, **8**, 615.
3. H. M. Ohlendorf, A. W. Kilness, J. L. Simmons, R. K. Stroud, D. J. Hoffman and J. F. Moore, *J. Toxicol. Environ. Health*, 1988, **24**, 67.
4. G. Q. Yang, S. Z. Wang, R. H. Zhou and S. Z. Sun, *Am. J. Clin. Nutr.*, 1983, **37**, 872.
5. A. A. Nelson, O. G. Fithugh and H. O. Calvery, *Cancer Res.*, 1943, **3**, 230.
6. L. Flohe, W. A. Günzler and H. H. Schock, *FEBS Lett.*, 1973, **32**, 132.
7. U. Schweizer and N. Fradejas-Villar, *FASEB J.*, 2016, **30**, 3669.
8. V. M. Labunskyy, D. L. Hatfield and V. N. Gladyshev, *Physiol. Rev.*, 2014, **94**, 739.
9. Y. Ge, Z. Qi, Y. Wang, X. Liu, J. Li, J. Xu, J. Liu and J. Shen, *Int. J. Biochem. Cell Biol.*, 2009, **41**, 900.
10. L. Yin, J. Song, P. G. Board, Y. Yu, X. Han and J. Wei, *J. Mol. Recognit.*, 2013, **26**, 38.
11. R. E. Huber and R. S. Criddle, *Arch. Biochem. Biophys.*, 1967, **122**, 164.
12. M. J. Davies, *Chim. Ind.*, 2016, 50.
13. K.-S. Yang, *J. Biol. Chem.*, 2002, **277**, 38029.
14. H. A. Woo, *Science*, 2003, **300**, 653.

15. G. W. Snider, E. Ruggles, N. Khan and R. J. Hondal, *Biochemistry*, 2013, **52**, 5472.
16. R. Brigelius-Flohè, *Free Radical Biol. Med.*, 1999, **27**, 951.
17. B. Ren, W. Huang, B. Akesson and R. Ladenstein, *J. Mol. Biol.*, 1997, **268**, 869.
18. B. Neumeister, I. Besenthal and B. O. Bohm, *Klinikleitfaden Labordiagnostik*, Elsevier Urban & Fisher Verlag, Munchen, 2013, p. 262.
19. V. M. Labunskyy, D. L. Hatfield and V. N. Gladyshev, *Physiol. Rev.*, 2014, **94**, 739.
20. C. Santi, C. Tidei, C. Scalera, M. Piroddi and F. Galli, *Curr. Chem. Biol.*, 2013, **7**, 25.
21. A. J. Pacuła, F. Mangiavacchi, L. Sancineto, E. J. Lenardão, J. Ścianowski and C. Santi, *Curr. Chem. Biol.*, 2016, **9**, 97.
22. A. Barcellos, L. Abenante, M. Sarro, I. Leo, E. J. Lenardão, G. Perin and C. Santi, *Curr. Org. Chem.*, 2016, **20**, DOI: 10.2174/1385272820666161020162113.
23. L. Orian and S. Toppo, *Free Radical Biol. Med.*, 2014, **66**, 65.
24. D. Bhowmick and G. Mugesh, *Org. Biomol. Chem.*, 2015, **13**, 10262.
25. L. A. Wessjohann, A. Schneider, M. Abbas and W. Brandt, *Biol. Chem.*, 2007, **388**, 997.
26. E. E. Alberto, V. do Nascimento and A. L. Braga, *J. Braz. Chem. Soc.*, 2010, **21**, 2032.
27. K. P. Bhabak and G. Mugesh, *Acc. Chem. Res.*, 2010, **43**, 1408.
28. B. K. Sarma and G. Mugesh, *Org. Biomol. Chem.*, 2008, **6**, 965.
29. R. J. Hondal, S. M. Marino and V. N. Gladyshev, *Antioxid. Redox Signaling*, 2013, **18**, 1675.
30. R. J. Hondal and E. L. Ruggles, *Amino Acids*, 2011, **41**, 73.
31. D. Bartolini, S. Ciffolilli, M. Piroddi, G. Murdolo, C. Tortoioli and F. Galli, Biochemistry and Nutrition of Selenium: From Inorganic Forms to Endogenous Proteins, in *Organoselenium Chemistry Between Synthesis and Biochemistry*, ed. C. Santi, Bentham Science Publishers, Sharjah, 2014, pp. 268–327.
32. M. Iwaoka, Antioxidant organoselenium molecules, in *Organoselenium Chemistry Between Synthesis and Biochemistry*, ed. C. Santi, Bentham Science Publishers, Sharjah, 2014, pp. 361–378.
33. R. Lesser and R. Weiss, *Ber. Dtsch. Chem. Ges.*, 1924, **57**, 1077.
34. A. Müller, E. Cadenas, P. Graf and H. Sies, *Biochem. Pharmacol.*, 1984, **33**, 3235.
35. A. Wendel, M. Fausel, H. Safayhi, G. Tiegs and R. Otter, *Biochem. Pharmacol.*, 1984, **33**, 3241.
36. S. Antony and C. A. Bayse, *Inorg. Chem.*, 2011, **50**, 12075.
37. J. K. Pearson and R. J. Boyd, *J. Phys. Chem. A*, 2008, **112**, 1013.
38. K. P. Bhabak and G. Mugesh, *Chem. – Eur. J.*, 2008, **14**, 8640.
39. Z. Wang, W. Li, Y. Wang, X. Li, L. Huang and X. Li, *RSC Adv.*, 2016, **6**, 7139.

40. Z. Wang, Y. Wang, W. Li, Z. Liu, Z. Luo, Y. Sun, R. Wu, L. Huang and X. Li, *Dalton Trans.*, 2015, **44**, 20913.
41. L. Engman and A. Hallberg, *J. Org. Chem.*, 1989, **54**, 2964.
42. R. Cantineau, G. Thiange, A. Plenevaux, L. Christiaens, M. Guillaume, A. Welter and N. Dereu, *J. Labelled Compd. Radiopharm.*, 1985, **23**, 59.
43. A. J. Pacuła, J. Scianowski and K. B. Aleksandrzak, *RSC Adv.*, 2014, **4**, 48959.
44. A. J. Pacuła, K. B. Kaczor, A. Wojtowicz, J. Antosiewicz, A. Janecka, A. Długosz, T. Janecki and J. Scianowski, *Bioorg. Med. Chem.*, 2017, **25**, 126.
45. T. Kálai, G. Mugesh, G. Roy, H. Sies, Z. Berente and K. Hideg, *Org. Biomol. Chem.*, 2005, **3**, 3564.
46. T.-C. Chang, M.-L. Huang, W.-L. Hsu, J.-M. Hwang and L.-Y. Hsu, *Chem. Pharm. Bull.*, 2003, **51**, 1413.
47. S. Shaaban, A. Negm, M. A. Sobh and L. A. Wessjohann, *Eur. J. Med. Chem.*, 2015, **97**, 190.
48. K. P. Bhabak and G. Mugesh, *Chem. - Eur. J.*, 2007, **13**, 4594.
49. S. M. Tan, A. Sharma, D. Y. C. Yuen, N. Stefanovic, G. Krippner, G. Mugesh, Z. Chai and J. B. de Haan, *PLoS One*, 2013, **8**, e69193.
50. K. Satheeshkumar and G. Mugesh, *Chem. - Eur. J.*, 2011, **17**, 4849.
51. B. K. Sarma and G. Mugesh, *Chem. - Eur. J.*, 2008, **14**, 10603.
52. Y. Liu, B. Li, L. Li and H.-Y. Zhang, *Helv. Chim. Acta*, 2002, **85**, 9.
53. X. Yang, Q. Wang and H. Xu, *Carbohydr. Res.*, 2002, **337**, 1309.
54. M. J. Parnham, J. Biedermann, C. Bittner, N. Dereu, S. Leyck and H. Wetzig, *Agents Actions*, 1989, **27**, 306.
55. S. S. Zade, S. Panda, S. K. Tripathi, H. B. Singh and G. Wolmershäuser, *Eur. J. Org. Chem.*, 2004, 3857.
56. S. J. Balkrishna, S. Kumar, G. K. Azad, B. S. Bhakuni, P. Panini, N. Ahalawat, R. S. Tomar, M. R. Dettyc and S. Kumar, *Org. Biomol. Chem.*, 2014, **12**, 1215.
57. S. Kumar, J. Yan, J. Poon, V. P. Singh, X. Lu, M. K. Ott, L. Engman and S. Kumar, *Angew. Chem., Int. Ed.*, 2016, **55**, 3729.
58. K. Selvakumar, P. Shah, H. B. Singh and R. J. Butcher, *Chem. - Eur. J.*, 2011, **17**, 12741.
59. H. J. Reich and C. P. Jasperse, *J. Am. Chem. Soc.*, 1987, **109**, 5549.
60. T. G. Back and B. P. Dyck, *J. Am. Chem. Soc.*, 1997, **119**, 2079.
61. V. P. Singh, H. B. Singh and R. J. Butcher, *Eur. J. Org. Chem.*, 2011, 5485.
62. S. R. Wilson, P. A. Zucker, R.-R. C. Huang and A. Spector, *J. Am. Chem. Soc.*, 1989, **111**, 5936.
63. M. Iwaoka and S. Tomoda, *J. Am. Chem. Soc.*, 1994, **116**, 2557.
64. T. Wirth, *Molecules*, 1998, **3**, 1641.
65. G. Mugesh, A. Panda, H. B. Singh, N. S. Punekar and R. J. Butcher, *J. Am. Chem. Soc.*, 2001, **123**, 839.
66. A. S. Hodage, C. P. Prabhu, P. P. Phadnis, A. Wadawale, K. I. Priyadarsini and V. K. Jain, *J. Organomet. Chem.*, 2012, **720**, 19.
67. K. P. Bhabak and G. Mugesh, *Chem. - Eur. J.*, 2009, **15**, 9846.

68. K. P. Bhabak and G. Mugesh, *Chem. – Asian J.*, 2009, **4**, 974.
69. D. Bhowmick and G. Mugesh, *Tetrahedron*, 2012, **68**, 10550.
70. D. Bhowmick and G. Mugesh, *Org. Biomol. Chem.*, 2015, **13**, 9072.
71. C. A. Collins, F. H. Fry, A. L. Holme, A. A. Yiakouvaki, C. P. Al-Qenaei and C. Jacob, *Org. Biomol. Chem.*, 2005, **3**, 1541.
72. J. Rafique, S. Saba, R. F. S. Canto, T. E. A. Frizon, W. Hassan, E. P. Waczuk, M. Jan, D. F. Back, J. B. T. Rocha and A. L. Braga, *Molecules*, 2015, **20**, 10095.
73. F. Zaccaria, L. P. Wolters, C. F. Guerra and L. Orian, *J. Comput. Chem.*, 2016, **37**, 1672.
74. G. S. Heverly-Coulson and R. J. Boyd, *J. Chem. Theory Comput.*, 2012, **8**, 5052.
75. D. J. Press and T. G. Back, *Org. Lett.*, 2011, **13**, 4104–4107.
76. C. P. Prabhu, P. P. Phadnis, A. Wadawale, K. I. Priyadarsini and V. K. Jain, *J. Organomet. Chem.*, 2012, **713**, 42.
77. P. Prabhu, B. G. Singh, M. Noguchi, P. P. Phadnis, V. K. Jain, M. Iwaoka and K. I. Priyadarsini, *Org. Biomol. Chem.*, 2014, **12**, 2404.
78. V. P. Singh, J. Poon, R. J. Butcher and L. Engman, *Chem. – Eur. J.*, 2014, **20**, 12563.
79. V. P. Singh, J. Poon, R. J. Butcher, X. Lu, G. Mestres, M. K. Ott and L. Engman, *J. Org. Chem.*, 2015, **80**, 7385.
80. L. C. Soares, E. E. Alberto, R. S. Schwab, P. S. Taube, V. Nascimento, O. E. D. Rodrigues and A. L. Braga, *Org. Biomol. Chem.*, 2012, **10**, 6595.
81. E. E. Alberto, L. C. Soares, J. H. Sudati, A. C. A. Borges, J. B. T. Rocha and A. L. Braga, *Eur. J. Org. Chem.*, 2009, 4211.
82. T. Posser, M. B. Moretto, A. L. Dafre, M. Farina, J. B. T. Rocha, C. W. Nogueira, G. Zeni, J. S. Ferreira, R. B. Leal and L. F. Jeferson, *Chem. –Biol. Interact.*, 2006, **164**, 126.
83. V. Nascimento, N. L. Ferreira, R. F. S. Canto, K. L. Schott, E. P. Waczuk, L. Sancineto, C. Santi, J. B. T. Rocha and A. L. Braga, *Eur. J. Med. Chem.*, 2014, **87**, 131.
84. M. R. Straliotto, J. Oliveira, G. Mancini, A. C. D. Bainy, A. Latini, A. M. Deobald, J. B. T. Rocha and A. F. de Bem, *Eur. J. Pharm. Sci.*, 2013, **48**, 717.
85. A. S. Freitas, A. S. Prestes, C. Wagner, J. H. Sudati, D. Alves, L. O. Porciúncula, I. J. Kade and J. B. T. Rocha, *Molecules*, 2010, **15**, 7699.
86. M. Ibrahim, W. Hassan, J. Anwar, A. M. Deobald, J. P. Kamdem, D. O. Souza and J. B. T. Rocha, *Toxicol. In Vitro*, 2014, **28**, 524.
87. T. G. Back and Z. Moussa, *J. Am. Chem. Soc.*, 2002, **124**, 12104.
88. T. G. Back and Z. Moussa, *J. Am. Chem. Soc.*, 2003, **125**, 13455.
89. S. K. Tripathi, S. Sharma, H. B. Singh and R. J. Butche, *Org. Biomol. Chem.*, 2011, **9**, 581.
90. T. G. Back, D. Kuzma and M. Parvez, *J. Org. Chem.*, 2005, **70**, 9230.
91. D. J. Press, E. A. Mercier, D. Kuzma and T. G. Back, *J. Org. Chem.*, 2008, **73**, 4252.

92. N. M. R. McNeil, M. C. Matz and T. G. Back, *J. Org. Chem.*, 2013, **78**, 10369.
93. S. K. Tripathi, U. Patel, D. Roy, R. B. Sunoj, H. B. Singh, G. Wolmershäuser and R. J. Butcher, *J. Org. Chem.*, 2005, **70**, 9237.
94. V. P. Singh, H. B. Singh and R. J. Butcher, *Chem. – Asian J.*, 2011, **6**, 1431.
95. Y. Saito, D. Umemoto, A. Matsunaga, T. Sato and M. Chikuma, *Biomed. Res. Trace Elem.*, 2006, **17**, 423.
96. T. G. Back, Z. Moussa and M. Parvez, *Angew. Chem., Int. Ed.*, 2004, **43**, 1268.
97. S. Braverman, M. Cerkinsky, Y. Kalendar, R. Jana, M. Sprecher and I. Goldberg, *Synthesis*, 2014, **46**, 119.
98. N. M. R. McNeil, D. J. Press, D. M. Mayder, P. Garnica, L. M. Doyle and T. G. Back, *J. Org. Chem.*, 2016, **81**, 7884.
99. D. Kuzma, M. Parvez and T. G. Back, *Org. Biomol. Chem.*, 2007, **5**, 3213.
100. B. K. Sarma, D. Manna, M. Minoura and G. Mugesh, *J. Am. Chem. Soc.*, 2010, **132**, 5364.
101. D. S. Lamani, D. Bhowmick and G. Mugesh, *Org. Biomol. Chem.*, 2012, **10**, 7933.
102. D. Lamani, D. Bhowmick and G. Mugesh, *Molecules*, 2015, **20**, 12959.
103. C. Santi, F. Galli, M. Piroddi and C. Tidei, *Phosphorus, Sulfur Silicon Relat. Elem.*, 2013, **188**, 507.
104. A. S. Hodage, P. P. Phadnis, A. Wadawale, K. I. Priyadarsini and V. K. Jain, *Org. Biomol. Chem.*, 2011, **9**, 2992.
105. V. Nascimento, E. E. Alberto, D. W. Tondo, D. Dambrowski, M. R. Detty, F. Nome and A. L. Braga, *J. Am. Chem. Soc.*, 2012, **134**, 138.
106. K. Arai, F. Kumakura, M. Takahira, N. Sekiyama, N. Kuroda, T. Suzuki and M. Iwaoka, *J. Org. Chem.*, 2015, **80**, 5633.
107. K. Arai, A. Tashiro, Y. Osaka and M. Iwaoka, *Molecules*, 2017, **22**, 354.
108. S. Misra, M. Boylan, A. Selvam, J. Spallholz and M. Björnstedt, *Nutrients*, 2015, **7**, 3536.
109. C. Tidei, M. Piroddi, F. Galli and C. Santi, *Tetrahedron Lett.*, 2012, **53**, 232.
110. D. Bartolini, M. Piroddi, C. Tidei, S. Giovagnoli, D. Pietrella, Y. Manevich, K. D. Tew, D. Giustarini, R. Rossi, D. M. Townsend, C. Santi and F. Galli, *Free Radical Biol. Med.*, 2015, **78**, 56.
111. D. Bartolini, J. Commodi, M. Piroddi, L. Incipini, L. Sancineto, C. Santi and F. Galli, *Free Radical Biol. Med.*, 2015, **88**, 466.
112. D. Bhowmick, S. Srivastava, P. D'Silva and G. Mugesh, *Angew. Chem., Int. Ed.*, 2015, **54**, 844.
113. T. Wirth, *Angew. Chem., Int. Ed.*, 2015, **54**, 10074.
114. K. Macegoniuk, E. Grela, J. Palus, E. Rudzińska-Szostak, A. Grabowiecka, M. Biernat and Ł. Berlicki, *J. Med. Chem.*, 2016, **59**, 8125.
115. K.-D. Kröncke and L.-O. Klotz, *Antioxid. Redox Signaling*, 2009, **11**, 1015.

116. H. Blessing, S. Kraus, P. Heindl, W. Bal and A. Hartwig, *Eur. J. Biochem.*, 2004, **271**, 3190.
117. J. L. Larabee, J. R. Hocker and J. S. Hanas, *J. Inorg. Biochem.*, 2009, **103**, 419.
118. L. Sancineto, A. Mariotti, L. Bagnoli, F. Marini, J. Desantis, N. Iraci, C. Santi, C. Pannecouque and O. Tabarrini, *J. Med. Chem.*, 2015, **58**, 9601.
119. J. Chiou, S. Wan, K.-F. Chan, P.-K. So, D. He, E. W. Chan, T. Chan, K. Wong, J. Tao and S. Chen, *Chem. Commun.*, 2015, **51**, 9543.
120. Z. Lou, P. Li and K. Han, *Acc. Chem. Res.*, 2015, **48**, 1358.
121. E. Zapp, V. Nascimento, D. Dambrowski, A. L. Braga and I. C. Vieira, *Sens. Actuators, B*, 2013, **176**, 782.

CHAPTER 3

Synthesis of Organoselenium Compounds with Potential Biological Activities

FATEH V. SINGH[a] AND THOMAS WIRTH*[b]

[a] Chemistry Division, SAS, VIT University, Chennai Campus, Chennai 600 127, Tamil Nadu, India; [b] School of Chemistry, Cardiff University, Park Place, Main Building, Cardiff CF10 3AT, UK
*Email: wirth@cf.ac.uk

3.1 Introduction

Organoselenium chemistry has been developed as an important tool in organic and medicinal chemistry. Although the first organoselenium compound was reported in 1836, the real development of organoselenium chemistry started after the discovery of the selenoxide elimination in the early 1970s.[1,2] Several organoselenium reagents have been introduced since that time and have been used to achieve various synthetic transformations including selenenylations, selenocyclizations, selenoxide eliminations and 2,3-sigmatropic rearrangements.[3] In the 20th century, several new organic transformations were successfully developed using selenium compounds as catalysts and ligands.[4] The importance of organoselenium chemistry is not only limited to their synthetic chemistry, but also in numerous applications in medicinal chemistry.[5] The role of organoselenium compounds in medicinal chemistry increased dramatically after the discovery of selenocysteine in various mammalian enzymes such as glutathione peroxidase (GPx),[6] iodothyronine deiodinase[7] and thioredoxin reductase.[8] The chemistry and

Organoselenium Compounds in Biology and Medicine: Synthesis, Biological and Therapeutic Treatments
Edited by Vimal Kumar Jain and K. Indira Priyadarsini

Published by the Royal Society of Chemistry, www.rsc.org

biochemistry of organoselenium reagents has become a well-established research area and books,[3c,3d,9] book chapters[10] and review articles[11] have been written to describe their chemical and biological significance. This chapter highlights recent developments in the synthesis and biological activity of organoselenium reagents.

3.2 Synthesis of Organoselenium Compounds as Antioxidants

GPx has been identified as a mammalian selenoenzyme responsible for the destruction of harmful peroxides in various living organisms. This selenoprotein, bearing a selenol (Enz-SeH) moiety in the active site, catalyzes the reduction of harmful peroxides by glutathione (GSH) and protects the biomolecules against oxidative damage. After the discovery of ebselen **1** as a potent antioxidant agent in 1984,[12] various organoselenium compounds have been identified as having GPx activity (Figure 3.1). Organoselenium compounds **2–4** with covalent Se–N bonds are reported to be potent antioxidants.[13] Some research groups investigated selenium compounds **5–8** with weak intramolecular Se···N or Se···O interactions, which exhibit interesting antioxidant properties.[11b,14] Pyridine-based diselenide **9** has been found to be another class of antioxidants.[15]

Furthermore, the synthesis and antioxidant properties of various selenides and diselenides **10–13** bearing OH and COOH functionalities at their terminal positions are also discussed.[16] There are a few other selenium-based compounds, such as cyclic seleninate esters, selenenate esters and

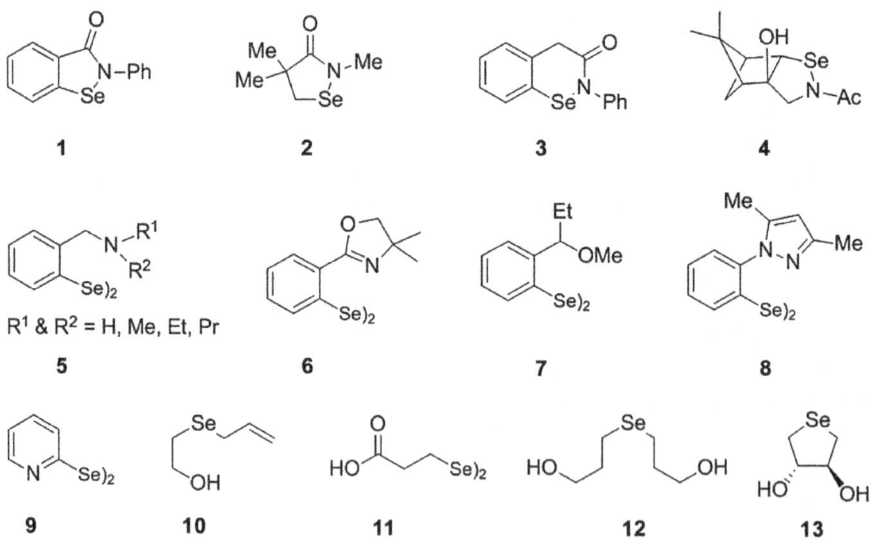

Figure 3.1 Chemical structures of some glutathione peroxidase mimics reported in the literature.

selenenic acid anhydrides which have been investigated as potent antioxidants.[17]

In 2012, Jain and co-workers reported the synthesis and antioxidant activity of a series of nicotinoyl-based selenium compounds (Scheme 3.1). 2-Chloronicotinic acid **14** was used for the synthesis of organoselenium compounds **18–21**. The synthesis of diselenide **20** was achieved in 31% yield by the treatment of Na_2Se_2 with 2-chloronicotinic acid **14** in water. The starting material **14** was reacted with thionyl chloride to form 2-chloronicotinoyl chloride **15**; furthermore, this reacts with different amines **16** to form amides **17**. After that the amide **17** was treated with Na_2Se_2 to form diselenides **18**. Diselenides **18** were converted to corresponding selenides **19** by treatment with molecular chlorine or bromine. Both diselenides **18** and monoselenides **19** have weak intramolecular Se···N interactions. Cyclic diselenide **21** was synthesized by the reaction of 2-chloronicotinoyl chloride **15** with Na_2Se_2 in 21% yield. All the synthesized compounds were characterized with their spectroscopic analysis and examined for their free radical scavenging and GPx mimicking activities. Interestingly, compounds **18a** and **18b** showed significant free radical scavenging activity and exhibit better GPx mimicking activity than ebselen **1**.[18a] In addition, Braga and co-workers reported the synthesis and antioxidant activity of 2-picolylamide-cored diselenides with O···Se non-bonded interactions. In this report, the influence of O···Se interactions on antioxidant activity was also studied and concluded that antioxidant activity may increase with increasing these interactions.[18b]

The same research group reported the synthesis and antioxidant activity of pyridyl and pyrimidyl-substituted selenium compounds (Scheme 3.2). The synthesis of pyridyl and pyrimidyl-based diselenides **23a** and **23b** was achieved by the reaction of Na_2Se_2 in water with heteroaryl bromides **22a** and **22b**, respectively. Additionally, the corresponding monoselenides **24a** and **24b** were obtained in the same reaction, along with diselenides **23a** and **23b**.

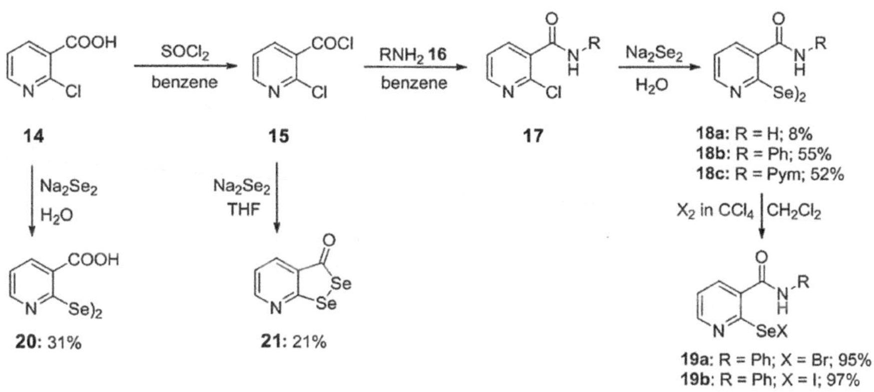

Scheme 3.1 Synthesis of nicotinoyl-based mono- and diselenides **18–21**.

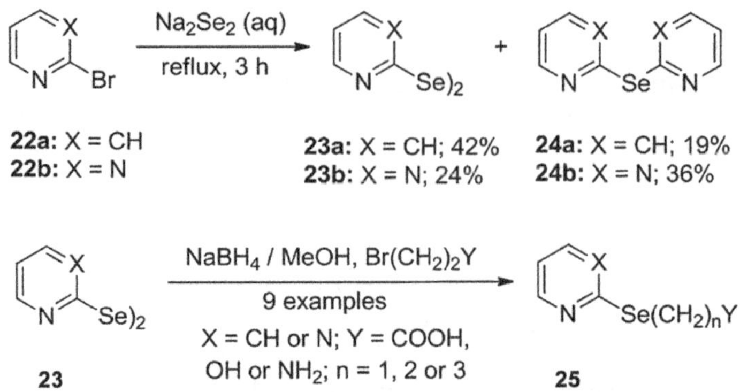

Scheme 3.2 Synthesis of pyridyl and pyrimidyl-substituted mono- and diselenides **23–25**.

Diselenides **23** were reduced with $NaBH_4$ and reacted with alkyl bromides to form various pyridyl/pyrimidyl alkyl selenides **25** in excellent yields (Scheme 3.2). All these compounds were examined with regards to their GPx mimicking activity.

It was observed that symmetrical pyridyl and pyrimidyl monoselenides **24** exhibited higher GPx activity than unsymmetrical monoselenides **25**, except those having amine functionality at their terminal positions (**25**: X = CH, $n = 2$, Y = NH_2 and **25**: X = CH, $n = 3$, Y = NH_2). In addition, the activity profile of pyridyl- and pyrimidyl-substituted selenium compounds was compared with diphenyl mono- and diselenides and showed promising results.[19a] Furthermore, bromo-cored pyridoxine-like diselenides were synthesized and the effect of the bromo moiety on GPx was studied in detail. This study showed that bromo functionality enhanced the GPx mimicking activity of pyridoxine-like diselenides significantly.[19a]

In 2010, Mugesh and co-workers reported the synthesis and GPx mimicking activity of spirodiazaselenurane **29** (Scheme 3.3). The synthesis of spirodiazaselenurane **29** was achieved in two chemical steps. In the first step, diarylselenide **28** was synthesized by the *ortho*-lithiation of benzanilide **26** followed by the addition of $Se(dtc)_2$ **27**. In the second step, diarylselenide **28** was oxidized to spirodiazaselenurane **29** using hydrogen peroxide as the oxidant. The GPx activity of spirodiazaselenurane **29** was evaluated and compared with its tellurium analogue (spirodiazatellurane). It was found that the tellurium compounds are more effective as GPx mimics than their selenium counterparts, due to the fast oxidation of the tellurium centre in the presence of peroxide.[20]

Mugesh and co-workers also reported the synthesis of spirodiazaselenurane **33** bearing electron-withdrawing and donating functionalities on the aromatic ring (Scheme 3.4). The synthesis of spirodiazaselenurane **33** was achieved by reaction of 2,2-selenobis-(benzoyl chloride) **30** and anilines **31** functionalized with electron-withdrawing and electron-donating substituents followed by the oxidation using hydrogen peroxide.

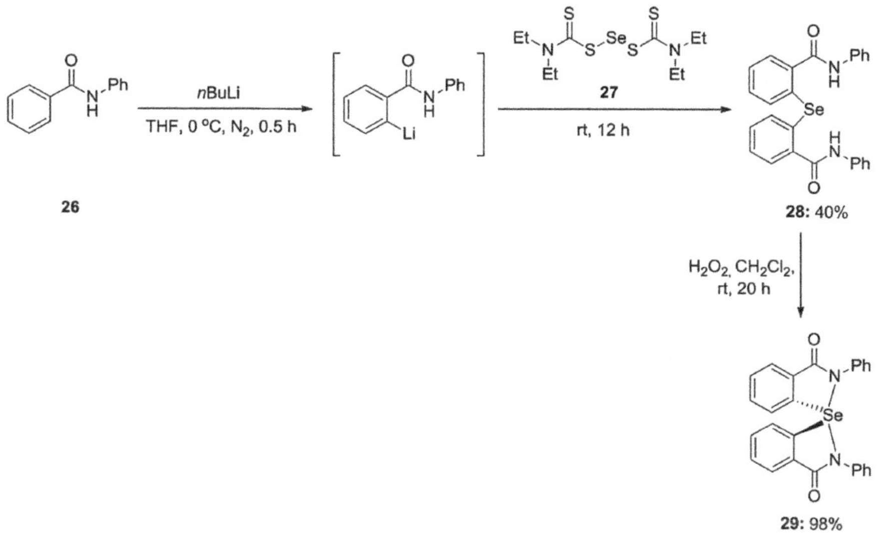

Scheme 3.3 Synthesis of spirodiazaselenurane **29** *via* the formation of diarylselenide **28**.

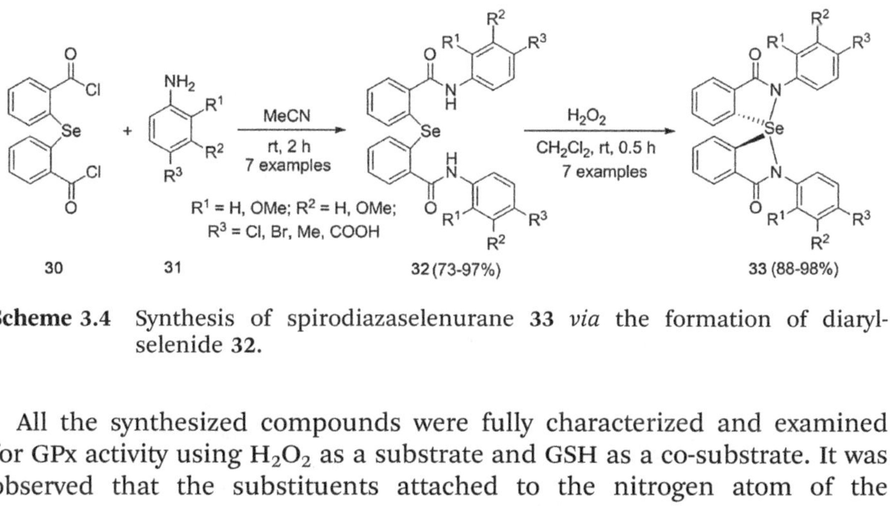

Scheme 3.4 Synthesis of spirodiazaselenurane **33** *via* the formation of diarylselenide **32**.

All the synthesized compounds were fully characterized and examined for GPx activity using H_2O_2 as a substrate and GSH as a co-substrate. It was observed that the substituents attached to the nitrogen atom of the selenazole ring play a vital role for the GPx activity. The introduction of electron-donating functionalities on the phenyl ring enhances the GPx activity significantly of both diarylselenides **32** and spirodiazaselenuranes **33**. The similar enhancement of the GPx activity of compounds **32** and **33** was not observed in the case of electron-withdrawing functionalities on the phenyl ring.[21a] Recently, Back and co-workers developed water-soluble selenuranes with enhanced GPx mimicking activity.[21b]

In 2011, Bhuyan and Mugesh disclosed the synthesis of selenium analogues **35** and **36** of catropril **34** and the related selenium compounds **37–40** (Figure 3.2). The synthesis of the selenium analogues **35** and **36** was achieved in

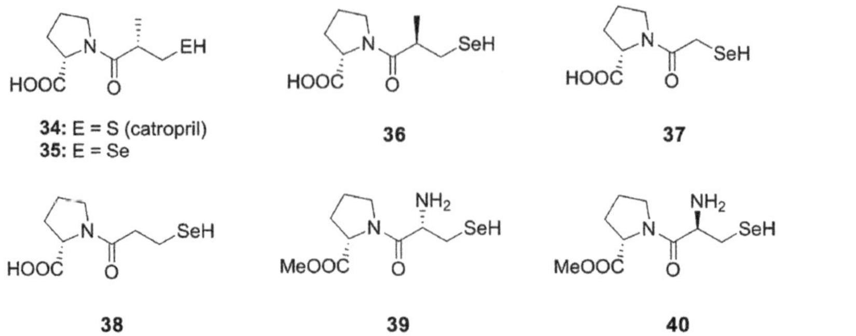

Figure 3.2 Chemical structures of catropril **34** and the selenium analogues **35–40**.

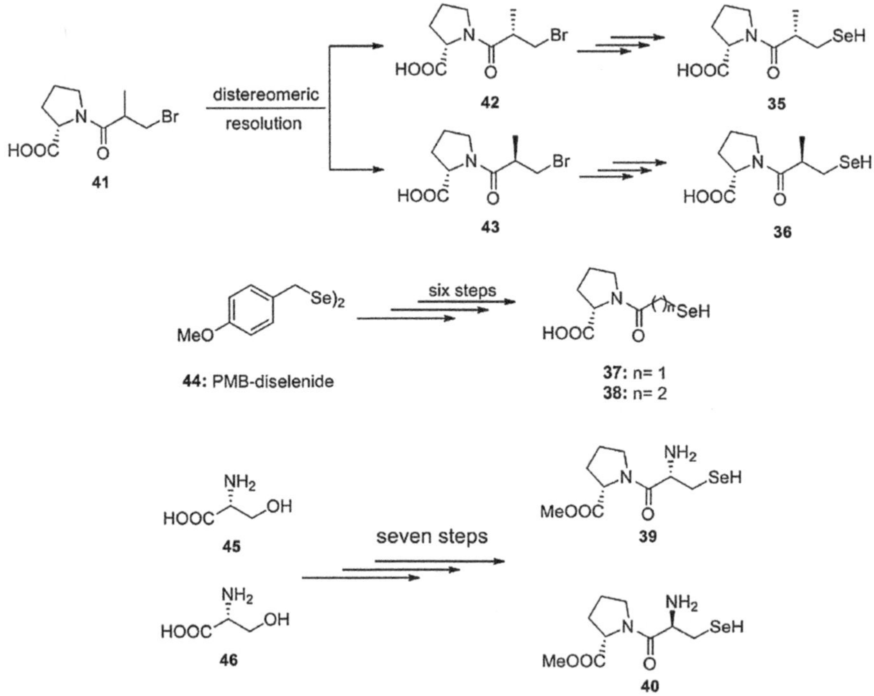

Scheme 3.5 Synthesis of selenium derivatives **35–40**.

three chemical steps after the diastereomeric resolution of the L-proline derivative **41** (Scheme 3.5). Furthermore, selenols **37** and **38** were synthesized in six steps overall starting from PMB-diselenide **44**. The synthesis of compounds **39** and **40** was achieved by starting (*R*)-2-amino-3-hydroxypropanoic acid **45** and (*S*)-2-amino-3-hydroxypropanoic acid **46** in seven chemical steps (Scheme 3.5).[22]

The angiotensin-converting enzyme (ACE) catalyzes the conversion of angiotensin I (Ang I) to angiotensin II (Ang II).[23] ACE also cleaves the

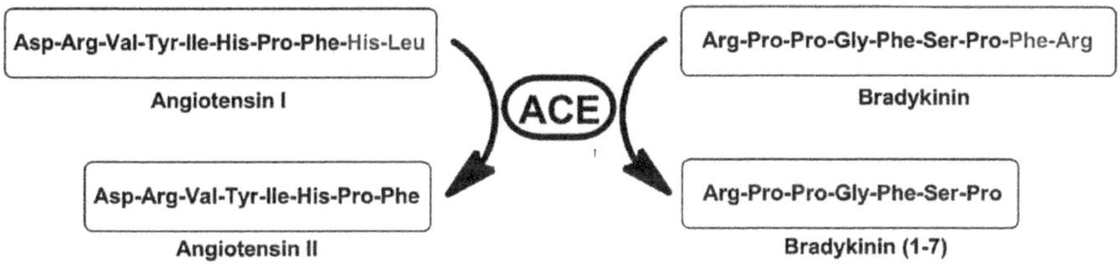

Figure 3.3 Angiotensin-converting enzyme (ACE)-catalyzed conversion of **Ang I** to **Ang II** and bradykinin to bradykinin (1–7).

terminal dipeptide of bradykinin, a vasodilating nonapeptide, to inactivate this hormone (Figure 3.3).[24] The inhibition of ACE is generally used as a method for the treatment of hypertension. 'Oxidative stress' is another disease state caused by an imbalance in the production of oxidants and antioxidants. A number of studies suggest that hypertension and oxidative stress are interdependent. Therefore, ACE inhibitors, having antioxidant properties are considered beneficial in the treatment of hypertension.

All the synthesized compounds were characterized and examined for ACE inhibition activity. Interestingly, the selenium analogues of captopril **35** and **36** showed good inhibition against ACE. In addition, the same compounds exhibit significant protection against peroxynitrite-mediated nitration of tyrosyl residues in peptides and proteins.[22]

The same research group extended the work and reported the synthesis of Sec-containing tripeptides **51**. The tripeptides **51** bearing Sec-residues at the N-terminal were synthesized by solution phase peptide synthesis. Initially, deprotection of the Boc group in compound **47** was performed using trifluoroacetic acid (TFA) to form amino derivative **48**. Compound **48** was coupled with p-methoxybenzyl ether (PMB)-protected Sec **35** to form tripeptides **49** using *N,N'*-dicyclohexylcarbodiimide (DCC) as coupling reagent. On cleavage of the PMB group by iodine, the corresponding diselenides **50** were produced. Finally, diselenides **50** were converted into Sec-containing tripeptides **51** by removal of the Boc group followed by reduction using $NaBH_4$ as reducing agent (Scheme 3.6).[25]

Additionally, selenium containing dipeptide **55** was synthesized for the same purpose. Initially, the starting material L-piperidine-2-carboxylic acid methyl ester **52a** was coupled with PMB-protected Sec **35** to form dipeptide **52b** using DCC as a coupling reagent. On cleavage of the PMB group by iodine the corresponding diselenide **53** was formed. Removal of the Boc group using TFA generated diselenide **54**. Finally, diselenide **54** was converted into the Sec-containing dipeptide **55** by reduction with $NaBH_4$ (Scheme 3.7).[25]

All these compounds were examined with respect to ACE inhibition activity. It is noteworthy that Sec-containing tripeptides **51** showed better ACE inhibition activity than Sec-containing dipeptide **55**. In addition, peroxynitrite-scavenging and GPx-like activity of the Sec-containing peptides **51** and **55** were examined and compared with Cys-containing peptides. It was

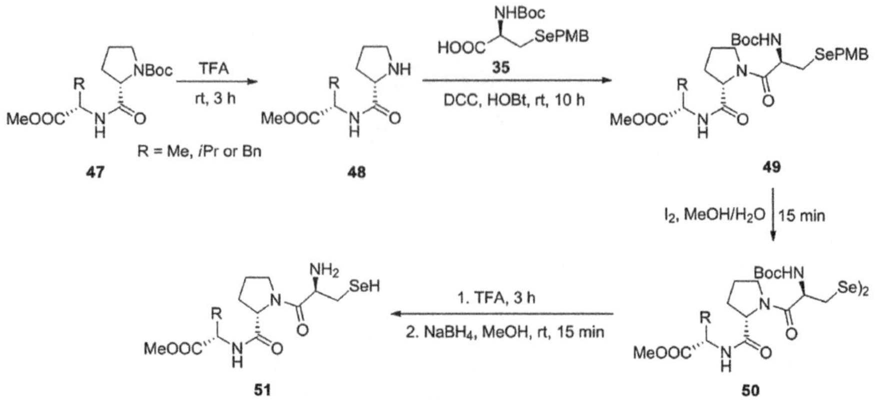

Scheme 3.6 Synthesis of Sec-containing tripeptides **51** from the proline-based compound **47**.

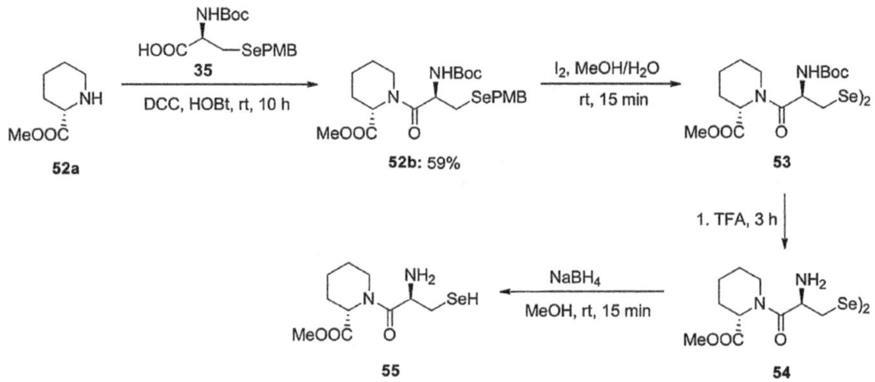

Scheme 3.7 Synthesis of Sec-containing dipeptide **55** starting from L-piperidine-2-carboxylic acid methyl ester **52a**.

concluded that the selenocysteine-containing peptides may provide better protection against oxidative damage compared to the cysteine-containing peptides.[25]

In 2012, Wang *et al.* reported the synthesis of selenium-containing polysaccharides and their antioxidant activity. In this report, the selenium-containing derivatives of *Artemisia sphaerocephala* polysaccharide with the highest Se content of 1703 $\mu g\,g^{-1}$ were synthesized using H_2SeO_3/HNO_3 and $BaCl_2$ as a catalyst. The antioxidant activity of these compounds was evaluated and the results suggest that selenium-containing polysaccharide could be applied as a novel selenium source in dietary supplements, with potent antioxidant properties.[26]

Furthermore, Lopez and co-workers reported the synthesis and antioxidant activity of various L-fuco-configured selenosugars **57–61**.[27] Compound **56** was used as key intermediate and synthesized in eight steps from

L-arabinose.[28] Next, the intermediate **56** was converted to *O*-protected selenane **57** *via* a double intramolecular nucleophilic substitution with sodium selenide generated *in situ* by reducing elemental selenium with sodium borohydride in ethanol. *O*-Protectedselenane **57** was further converted into the unprotected isofucoselenofagomine **58** in 67% yield (Scheme 3.8) by treatment with TFA. Compound **57** was treated with methyl triflate in anhydrous nitromethane to form unprotected methyl selenonium triflate **61**. Finally, the synthesis of compound **60** was achieved by oxidation of selenane **57** followed by reaction with TFA (Scheme 3.8). All compounds were tested for their GPx-like activity and selenane derivative **57** showed the best antioxidant properties.[27]

Furthermore, Lopez and co-workers reported the synthesis of *O*-alkyl selenocarbamates **63** and **65** from the *N*-aryl isoselenocyanates **62** and **64**, respectively.[29] This reaction was performed in ethanol at reflux temperature and reaction products **63** and **65** were isolated in excellent yields (Scheme 3.9). The starting materials *N*-aryl isoselenocyanates **62** and **64** were

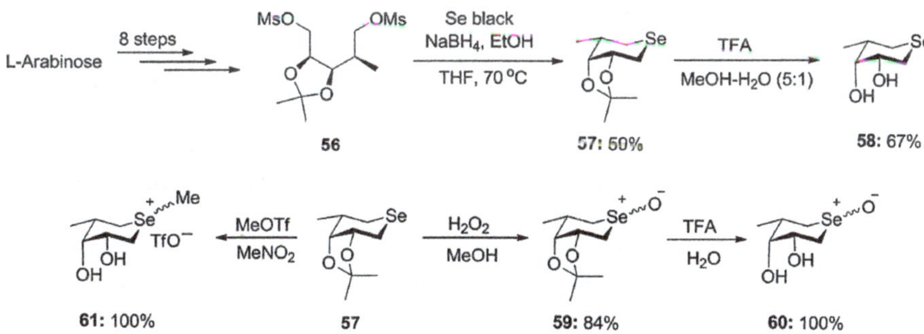

Scheme 3.8 Synthesis of selenosugars such as selenane **57**, selenoxide **60** and selenonium triflate **61** starting from intermediate **56**.

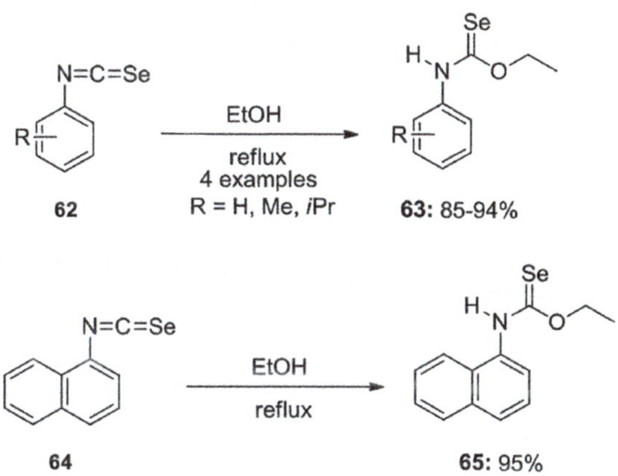

Scheme 3.9 Synthesis of *O*-alkyl selenocarbamates **63** and **65** from *N*-aryl isoselenocyanates **62** and **64**, respectively.

synthesized from the corresponding *N*-arylformamides using procedures from the literature.[30]

Additionally, a reaction of phenyl isoselenocyanate **62** was performed with L-valine ethyl esterhydrochloride **66** in the presence of an excess of triethylamine and *N*-phenyl selenohydantoin **67** was isolated in a quantitative yield (Scheme 3.10). The same reaction was undertaken with 1-naphthyl isoselenocyanate **62** under similar reaction conditions and *N*-naphthyl selenohydantoin **67** was isolated in 25% yield along with the formation of corresponding selenourea derivative **68** in 53% yield (Scheme 3.10).

Selenourea derivative **68** showed excellent GPx activity while selenocarbomate derivatives **63** and **65** were found to be the least efficient GPx mimics. In addition, all the compounds were tested for other biological activities, such as H_2O_2 scavengers, nitric acid scavengers and inhibition of lipid peroxidation. Selenohydantoin **67** was found to be most effective as an H_2O_2 scavenger, while selenocarbomates **63** were found more efficient as nitric acid scavengers. Interestingly, selenocarbomate **65** also showed a significant inhibition against lipid peroxydation.[30]

In 2014, the Braga research group reported the synthesis and GPx activities of nitrogen-containing aryl and alkyl diselenides.[31] Anthranilic acid **70** was used as starting material for the synthesis of the targeted diselenides. The aromatic diselenide **72** was synthesized through diazotation of anthranilic acid **70** followed by nucleophilic substitution with Na_2Se_2 generated *in situ* from elemental selenium and sodium borohydride.[32] Diselenide **72** was converted into nitrogen-containing aryl diselenides **73** with reaction of 1-phenylpropan-2-amine in the presence of 4-dimethylaminopyridine (DMAP) and an excess of triethylamine using existing procedures (Scheme 3.11).[32]

Bromocarboxylic acids **74** were used as starting material for the synthesis of nitrogen-containing aliphatic diselenides **76.** Initially, aliphatic acid diselenides **75** were synthesized by nucleophilic substitution with Na_2Se_2 generated *in situ* by reducing elemental selenium with sodium borohydride.[32] After that, diselenides **75** were directly converted into nitrogen-containing aryl diselenides **76** with the reaction of 1-phenylpropan-2-amine in the presence of DMAP and an excess of triethylamine (Et_3N) using existing procedures (Scheme 3.12).

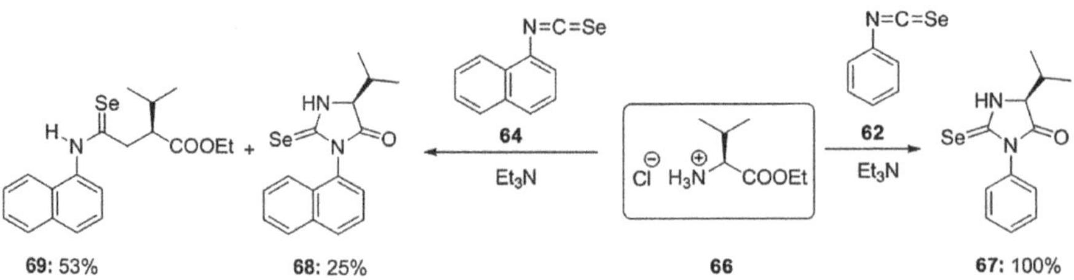

Scheme 3.10 Synthesis of *N*-aryl selenohydantoin **67** and **68** and *N*-naphthyl selenourea **69** by the reaction of *N*-aryl isoselenocyanates **62** and **64** with **66**.

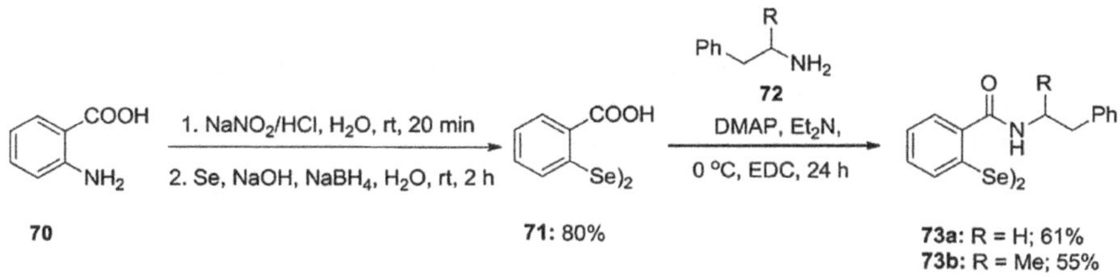

Scheme 3.11 Synthesis of nitrogen-containing diselenides **73** from anthranilic acid **70** *via* the formation of diselenide **71**.

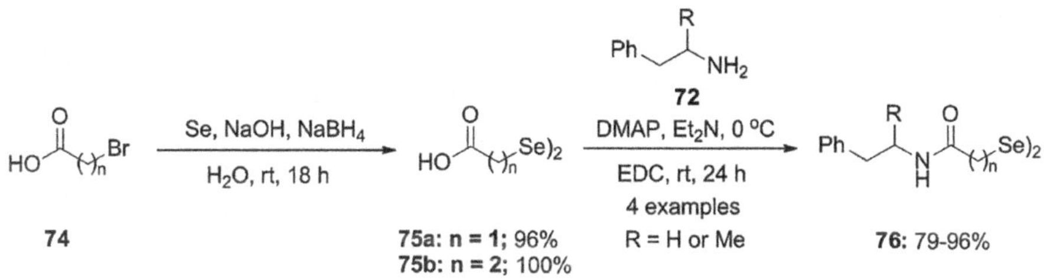

Scheme 3.12 Synthesis of nitrogen-containing aliphatic diselenides **76** from bromo-carboxylic acids **74** *via* diselenides **75**.

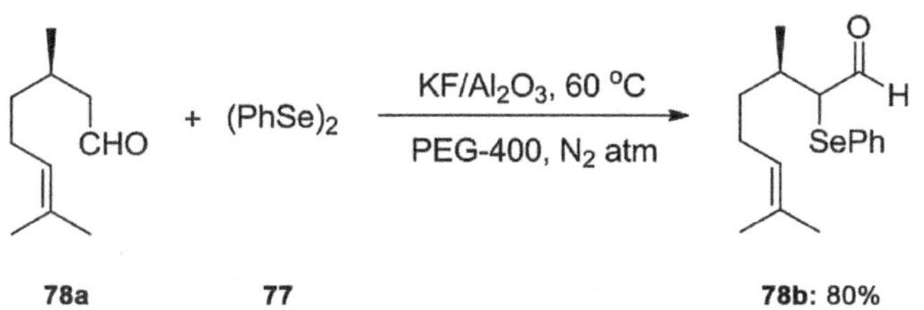

Scheme 3.13 Synthesis of α-phenylselenocitronellal (PhSeCIT) **78b** by the reaction of citronellal **78a** with diphenyl diselenide **77**.

All nitrogen-containing diselenides **73** and **76** were characterized using spectroscopic analysis and examined for their GPx activity. It was observed that nitrogen-containing aryl diselenides **73** showed better GPx activity than aliphatic diselenides **76**. It was also concluded that aryl diselenide **73** (R = Me) exhibits nine times more GPx activity than the standard organoselenium drug ebselen **1**.

Furthermore, Savegnago and co-workers reported the synthesis of α-phenylselenocitronellal (PhSeCIT) **78b** by the reaction of citronellal **78a** with diphenyl diselenide **77** using a solid-supported catalyst (KF/Al_2O_3) and PEG-400 as a clean and recyclable reaction medium (Scheme 3.13).[33] The use of this catalytic system is advantageous because can be re-used up to four times.[34] The compound PhSeCIT **78b** was evaluated using *in vitro* and *ex vivo* assays. PhSeCIT **78b** showed promising activity in both *in vitro* and *ex vivo* assays without any toxicological effect.[33]

Furthermore, the synthesis of various chiral ebselen analogues **81** was reported along with their antioxidant activity.[35] During the synthesis of these analogues, diselenide **71** was used as a key intermediate which was prepared by the diazotation of anthranilic acid **70** followed by reaction with disodium diselenide. Diselenide **71** was refluxed in thionyl chloride to form 2-(chloroseleno)benzoyl chloride **79**. Finally, dichloride **79** was treated with chiral amines **80** in the presence of triethylamine in dry diethylether and ebselen derivatives **81** were obtained in good-to-excellent yields (Scheme 3.14). The GPx activities of the chiral ebselen analogues **81** were examined and compared to ebselen **1**. The biological studies of these analogues clearly indicates that the chiral substituent on the nitrogen atom did not provide a substantial increase in activities and the newly synthesized compounds **81** showed similar activities to ebselen **1**.[35]

Break *et al.* developed the synthesis of selenonucleosides by the ribosylation of selenols such as pyridineselenol and pyridazineselenol. These compounds showed interesting antioxidant properties and this research was the first to report selenonucleosides acting as antioxidants.[36]

In addition, the synthesis of nitrogen-containing diselenide **82** was reported and compared its antioxidant properties with diphenyldiselenide **77** (Figure 3.4). The aim of this study was to investigate the effects of weak intramolecular Se···N interactions. GPx-like antioxidant activities of both compounds were evaluated using H_2O_2 or Bu^tOOH as substrates and compound **82** was found to be two times more active than diphenyldiselenide **77**. This study clearly demonstrates that weak intramolecular Se···N interactions play a vital role in antioxidant activity.[37]

In 2005, the same research group synthesized the nitrogen-containing diselenide **83** following a reported procedure[38] and compared its GPx-like antioxidant activity to diphenyldiselenide **77**. The nitrogen-containing diselenide **83** showed two-fold GPx-like antioxidant activity compared to **77**.

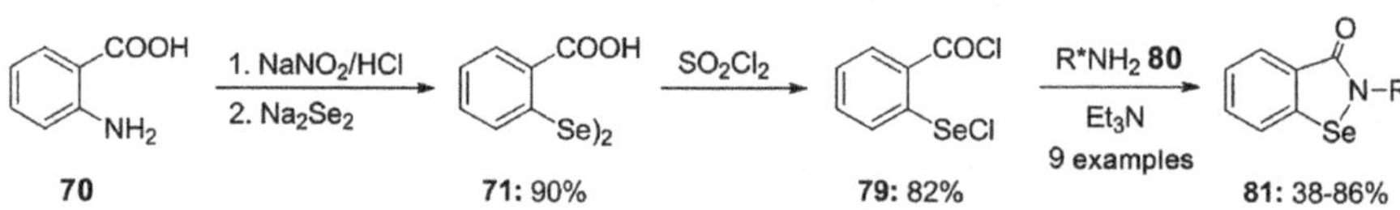

Scheme 3.14 Synthesis of chiral ebselen analogues **81** from anthranilic acid **70** *via* formation of dichloride **79**.

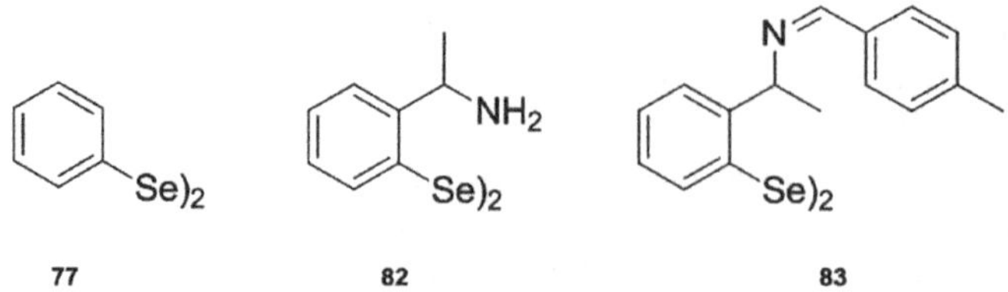

Figure 3.4 The structures of diphenyldiselenide 77 and nitrogen-containing diselenides **82** and **83**.

In addition, diselenide **83** showed a better activity profile than **77** against lipid peroxidation. This study was further evidence of the importance of weak intramolecular Se···N interactions in the pharmacological properties of diselenides.[39]

Mugesh and co-workers also reported the synthesis and antioxidant activity of cyclic isoselenazoles **87**.[40] Diselenide **84** was used as starting material and synthesized by lithiation of 3-methoxybenzaldehyde with *n*BuLi in the presence of *N*,*N*,*N*′-trimethylethylenediamine, followed by selenium insertion and oxidative workup.[41] Compound **84** was converted to the corresponding Schiff bases **85** on treatment with primary amines **16**. Schiff bases **85** were reduced to diselenides **86** using $NaBH_4$ as reducing agent. Interestingly, diselenides **86** were found to be quite unstable and cyclized to isoselenazoles **87** during purification (Scheme 3.15).[41]

All these compounds were evaluated with regards to their GPx-like activity. Cyclic isoselenazoles **87** showed excellent GPx activity and provide remarkable cytoprotection to human cells, mainly by exhibiting antioxidant activities in the presence of cellular thiols. The cytotoxicity of the isoselenazoles was examined and found to be significantly lower than that of ebselen.[40]

Furthermore, Singh and co-workers reported the synthesis and GPx-like activity of cyclohexyl-based selenide **89** and diselenide **90**.[42] In order to synthesize diselenide **90**, β-chlorocyclohexenal **88** was treated with disodium diselenide (Na_2Se_2) in oxolane/tetrahydrofuran (THF) and a mixture of both selenide **89** and diselenide **90** was obtained (Scheme 3.16). Both compounds were separated using column chromatography using 2% EtOAc in petroleum ether as eluent. The starting material β-chlorocyclohexenal **88** was synthesized by reacting cyclohexanone with phosphoryl chloride in

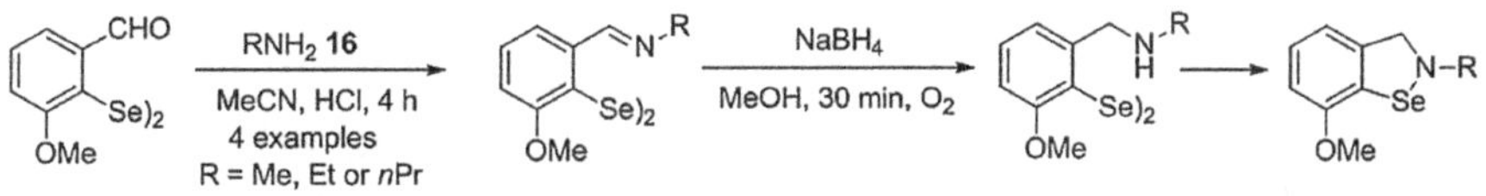

84 85 86 87

Scheme 3.15 Synthesis of isoselenazoles **87** from diselenide **84** *via* intermediate **85**.

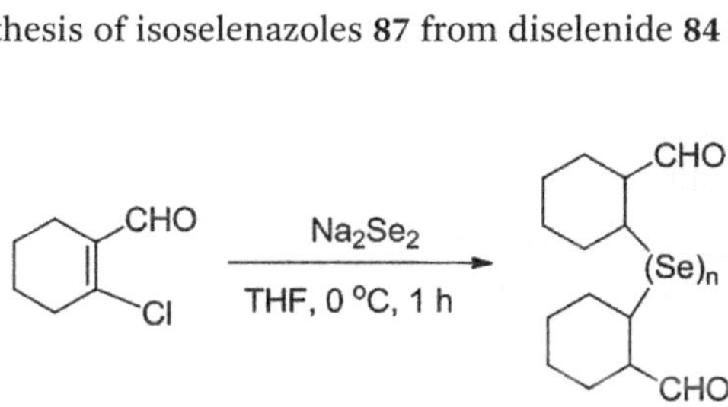

88

89: n = 1
90: n = 2

Scheme 3.16 Synthesis of cyclohexyl-based selenide **89** and diselenides **90** from β-chlorocyclohexenal **88**.

N,*N*-dimethylformamide following literature procedures.[43] After synthesis and characterization, both compounds **89** and **90** were examined for their GPx-like activity and compared with standard drug ebselen **1**. Diselenide **90** showed good activity, even better than ebselen, while selenide **89** did not show any significant GPx activity.[42]

The same research group used compound **89** for the synthesis of cyclohexyl-substituted oxygen-containing selenide **91** and cyclohexene-fused spiroselenuranes **92** and **93**.[44] Compound **89** was reduced to oxygen-containing selenide **91** using $NaBH_4$ as reducing agent in ethanol. Compound **91** was further oxidized to cyclohexene-fused spiroselenuranes **93** using H_2O_2 as oxidizing agent in dichloromethane. In addition, compound **89** was also treated with H_2O_2 to form another spiroselenurane derivative **92** (Scheme 3.17). All compounds **91–93** were examined for their GPx-like activity and compared with diselenide **90** and ebselen **1**. None of the tested compounds **91–93** showed any significant GPx-like antioxidant activity.[44]

In 2013, Savegnago *et al.* reported the synthesis of 4-arylseleno-7-chloroquinolines **96** by the reaction of 4,7-dichloroquinoline **94** with diaryl diselenides **95** using KOH as base, DMSO as solvent at 100 °C under air atmosphere. Various electron withdrawing and donating substituents on the diaryldiselenide were successfully tolerated and reaction products **96** were isolated in high yields (Scheme 3.18). The antioxidant activity of **96** (R = H) was examined and compared with its tellurium analogue. The biological results revealed that the tellurium analogue showed better antioxidant activity than **96** (R = H).[45]

The same research group reported the synthesis of selenium-containing chrysin derivatives. Chrysin is a naturally occurring flavone. The synthesis of

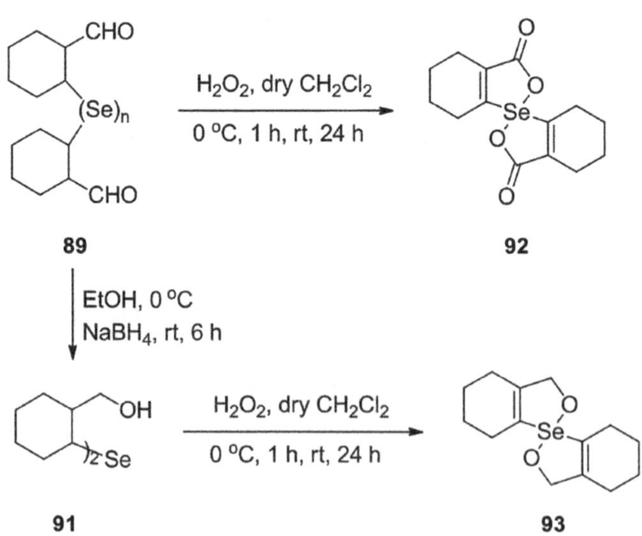

Scheme 3.17 Synthesis of oxygen-containing selenide **91** and cyclohexene-fused spiroselenuranes **92** and **93**.

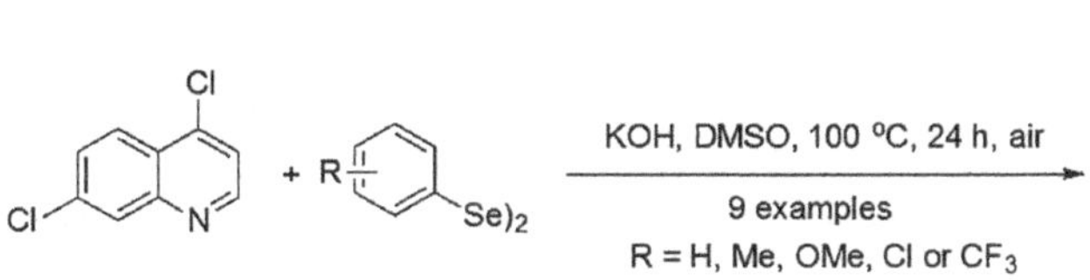

94 **95** **96**: 78-90%

Scheme 3.18 Synthesis of 4-arylseleno-7-chloroquinolines **96** by the reaction of 4,7-dichloroquinoline **94** with diaryl diselenides **95**.

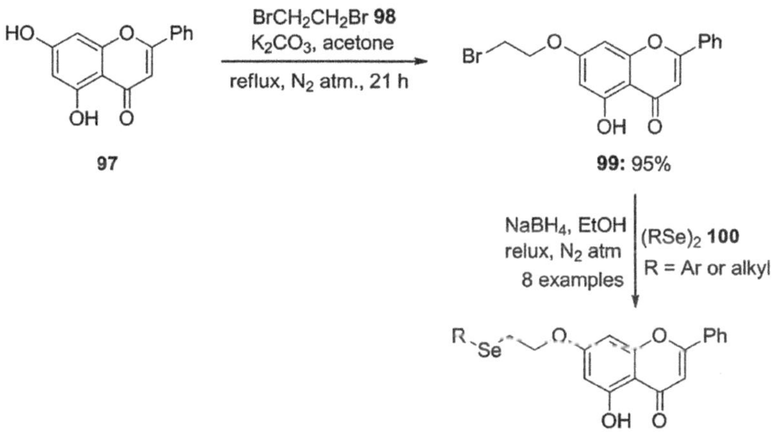

97 **101**: 57-96%

Scheme 3.19 Synthesis of Se-containing chrysin derivatives **101** from chrysin **97** *via* intermediate **99**.

selenium-containing chrysin derivatives **101** was achieved in high yields by the reaction of 7-(2-bromoethoxy)-chrysin **99** with diselenide **100** in the presence of sodium borohydride in ethanol at reflux temperature under inert atmosphere (Scheme 3.19). 7-(2-Bromoethoxy)-chrysin **99** was obtained by refluxing chrysin **97** with 1,2-dibromoethane **98** in dry acetone using K_2CO_3 as base in inert atmosphere. The antioxidant properties of Se-containing chrysin derivatives **101** were examined using three different *in vitro* assays and compared with similar Te-containing chrysin derivatives. Se-containing chrysin derivatives **101** exhibited lower antioxidant activities than the Te-containing chrysin derivatives.[46]

Recently, Dornelles and co-workers reported the synthesis and antioxidant activity of organoselenium compounds **106** and **107**. Both organoselenium compounds **106** and **107** were synthesized in good yields from 5-aryl-1,3,4-oxadiazoles **102** and thiadiazole-2-thiols **103**, respectively, by treatment with α-chloromethyl arylselenides in acetonitrile under reflux (Scheme 3.20). Most of these compounds showed significant antioxidant properties.[47]

The catalytic cycle of the GPx-mimicking activity is depicted in Scheme 3.21. The selenol is oxidized by the peroxide to selenenic acid

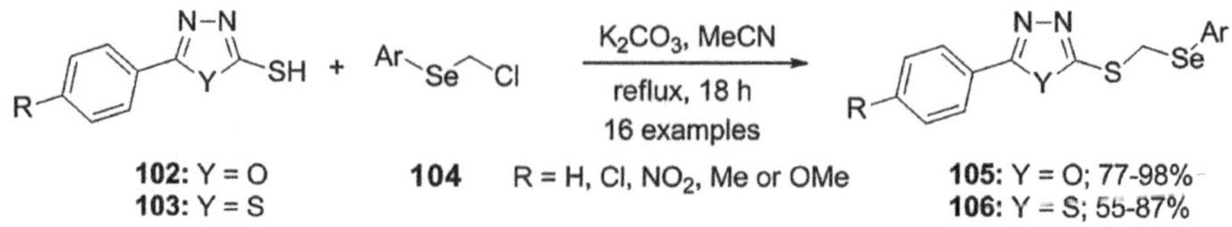

Scheme 3.20 Synthesis of organoselenium compounds **105** and **106** from **102** and **103**, respectively.

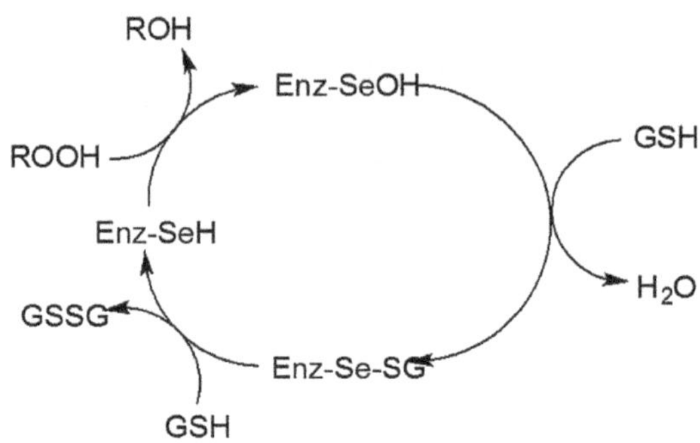

Scheme 3.21 Catalytic cycle of glutathione peroxidase (GPx) mimics.

(Enz-SeOH), which then reacts with GSH to form a selenenyl sulfide adduct (Enz-Se-SG). A second glutathione molecule then regenerates the active form of the enzyme by reacting with Enz-Se-SG to form oxidized glutathione (GSSG) as the byproduct (Scheme 3.21).

3.3 Organoselenium Compounds as Antithyroid Agents

The medicinal applications of organoselenium compounds is not limited to GPx-like antioxidant activity; there also investigations in which organoselenium reagents have been used for the iodination of thyroid hormones. Iodothyronine deiodinases are mammalian selenoenzymes that catalyze the regioselective deiodination of thyroid hormones.[7]

In 2005, Roy and Mugesh reported a series of organoselenium compounds **109** and **110** bearing the *N*-methylimidazole pharmacophore.[48] The synthesis of compound **109** was achieved from intermediate **108** by alkylation with methyl iodide or benzyl bromide. Compound **108** was synthesized by the reaction of 1-methyl-1*H*-imidazole **107** with *n*BuLi and elemental selenium. In addition, compound **108** was converted into **110** when it was reacted with α,α,-dibromo-*m*-xylene (Scheme 3.22). The antithyroid activity of the synthesized compounds was examined and all the compounds showed a significant protection of thyroid gland against oxidative damage.[48]

There was a report in 2007 in which synthesis and the antithyroid activities of selones with *N*,*N*-disubstituted imidazoles were discussed. This

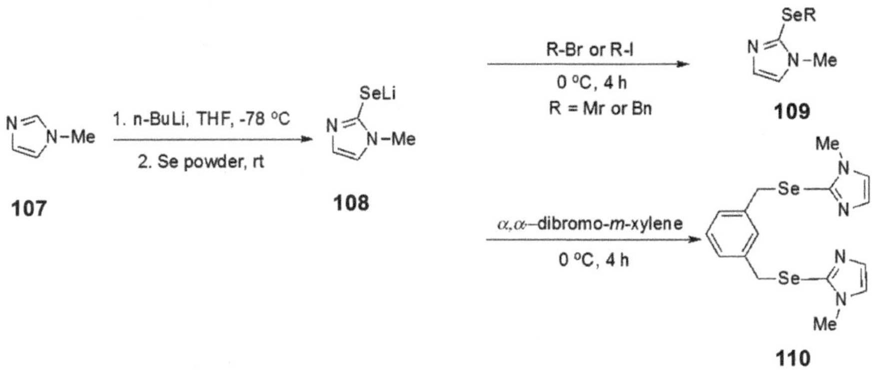

Scheme 3.22 Synthesis of organoselenium compounds **109** and **110** bearing the *N*-methylimidazole pharmacophore from 1-methyl-1*H*-imidazole **107**.

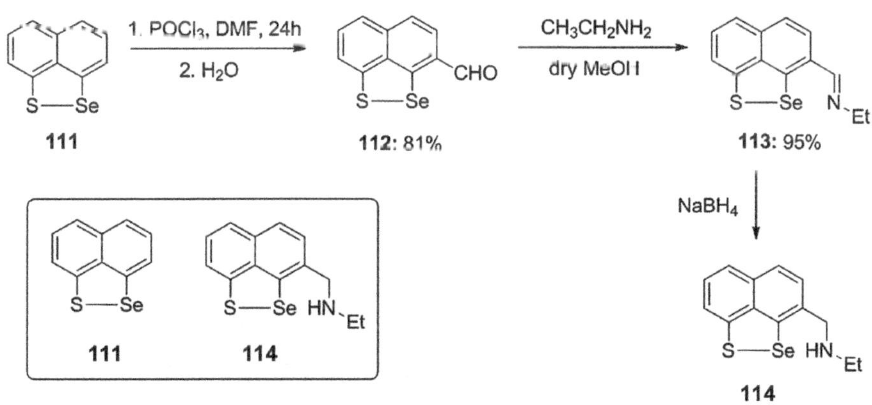

Scheme 3.23 Synthesis of nitrogen-containing naphthyl-based selenosulfide **114** from selenosulfide **111**.

research article mainly focused on bioinorganic aspects of inhibition of thyroid hormone biosynthesis, so all the details are not included in this section.[49]

Furthermore, Manna and Mugesh reported the synthesis of naphthyl-based selenium compounds **111** and **114** having a selenenyl sulfide (–Se–S–) bond.[50] The synthesis of compound **111** was achieved by a known procedure[51] and was used as starting material for the synthesis of nitrogen-containing selenosulfide **114**. Compound **111** was converted to naphthaldehyde **112** by the reaction with $POCl_3$ in dimethylformamide (DMF) followed by hydrolysis. After that, compound **112** was treated with ethyl amine to form imine **113**, which was further reduced to nitrogen-containing selenosulfide **114** using $NaBH_4$ as reducing agent (Scheme 3.23). In addition, both compounds **111** and **114** were successfully used for the deiodination of thyroxine by activating iodothyronine deiodinase enzyme.[50]

Furthermore, the same research group synthesized other naphthyl-based compounds **115–120** (Figure 3.5). These compounds were synthesized using a similar procedure, discussed earlier.[51] All the compounds were examined for the deiodination of thyroxine. Compounds having two selenol groups **118** and **119** showed much better deiodination activity than those having only one selenol group (**115–117**). It was noted that compound **120** (having a basic amino group in close proximity to one of the selenol moieties) showed 100 times better deiodination activity than compound **118**.[52]

After the discovery of promising activities of compound **120** (with the basic amino group in close proximity to one of the selenol moieties), similar compounds **122–125** were synthesized by the same authors (Figure 3.6). The synthesis of **122–125** was achieved by the condensation of compound **121** with different amines followed by the reduction of imine intermediates.

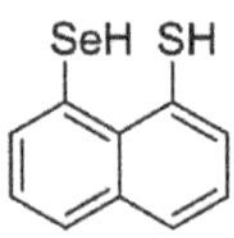

115

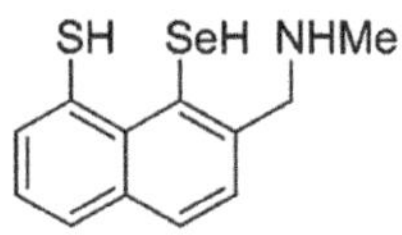

116

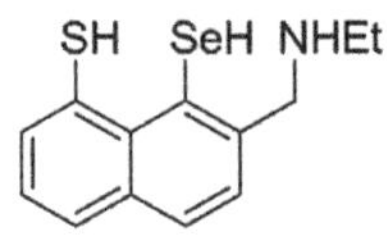

117

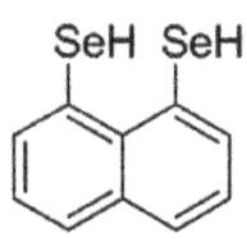

118

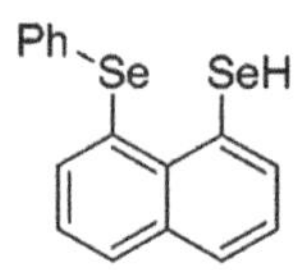

119

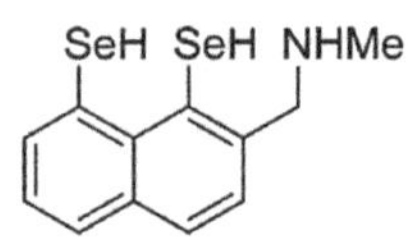

120

Figure 3.5 The structures of naphthyl-based compounds **115–120** having one or two selenol groups.

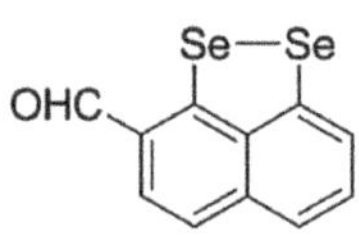

121

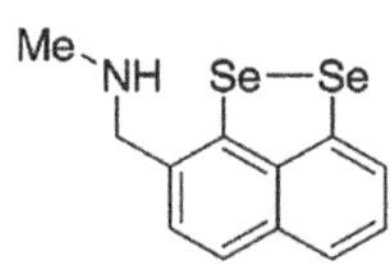

122

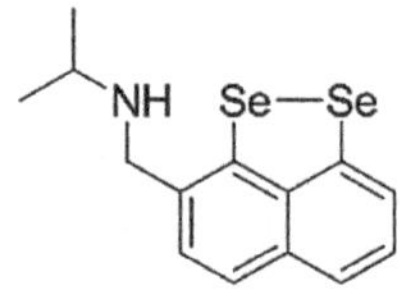

123

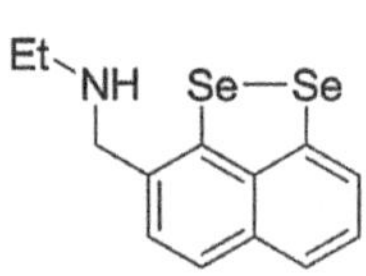

124

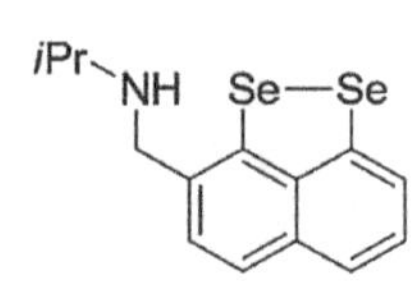

125

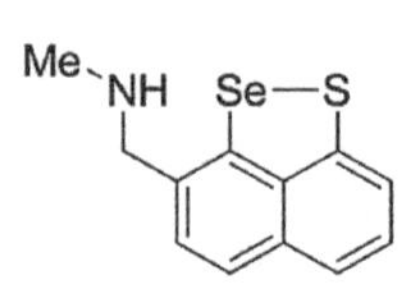

126

Figure 3.6 The structures of naphthyl-based compounds **122–126** having basic amino group in close proximity to the selenol moiety.

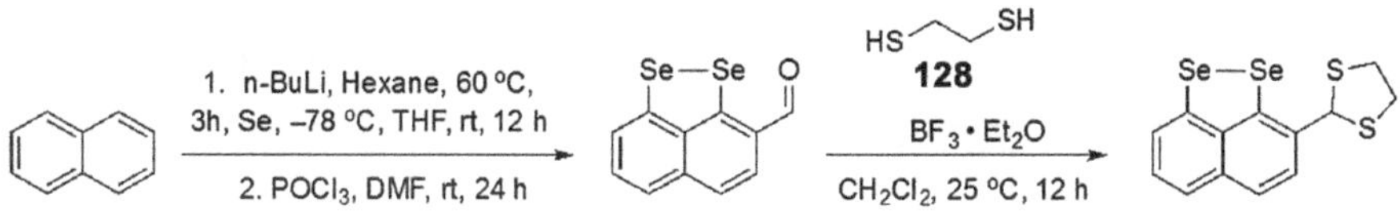

Scheme 3.24 Synthesis of cyclic diselenide **129** from naphthalene **127** in two steps.

Compound **121** was synthesized from the cyclic diselenide derived from **118** using a previously reported procedure.[51] Compounds **122–125** showed excellent deiodination activity, but compound **126** had a lower activity profile.[53]

Raja and Mugesh substituted selenium for tellurium and examined related compounds for their deiodinase activity. These compounds showed remarkable effects as enzyme mimetics for the deiodination of thyroid hormone. These tellurium compounds have been identified as the first tellurium-containing deiodinase mimetics.[54]

In 2014, Mondal and Mugesh reported synthesis and deiodination activity of cyclic diselenides **129** with a 1,3-dithiolane at the *ortho*-position of one of the selenium moieties. The synthesis of compound **129** was completed in two chemical steps. In the first step, diselenide **121** was synthesized from naphthalene by the insertion of selenium followed by the reaction with $POCl_3$. Finally, compound **121** was treated with ethane-1,2-dithiol **128** to form product **129** in good yields (Scheme 3.24). Compound **129** was examined for its deiodinase activity and found to be a deiodinase mimetic for the deiodination of iodothyroamines and endogenous thyroid hormones.[55]

3.4 Organoselenium Compounds as Anticancer Agents

In 1985, the first report on the application of organoselenium compounds as anticancer agents was published.[56] Since that report various organoselenium compounds have been developed for the treatment of cancer.[57–62] In 2012, Li, Wirth and co-workers reported the synthesis and anticancer activity of selenium-containing naphthalimides **133** and **136**.[63] The intermediate 2-aryl-selanyl-ethylamines **132** were synthesized from 2-chloroethylamine with organylselenolate anions which were generated *in situ* by treatment of the corresponding diselenides with $NaBH_4$ in ethnaol.[64] The synthesis of selenonaphthalimides **133** was achieved from 6-nitrobenzo[de]isochromene-1,3-dione **130**. In the first step, compound **130** was refluxed with different amines **16** under argon atmosphere to form 6-nitro-naphthalimides **131**. Compound **131** was then reacted with amine **132** in DMF at room temperature and final products **133** were isolated in good yields. In addition, compound **136** was prepared in 35% yield by the reaction of the bromo compound **134** with the known selenomorpholine **135**[65] in 2-methoxyethanol (Scheme 3.25).

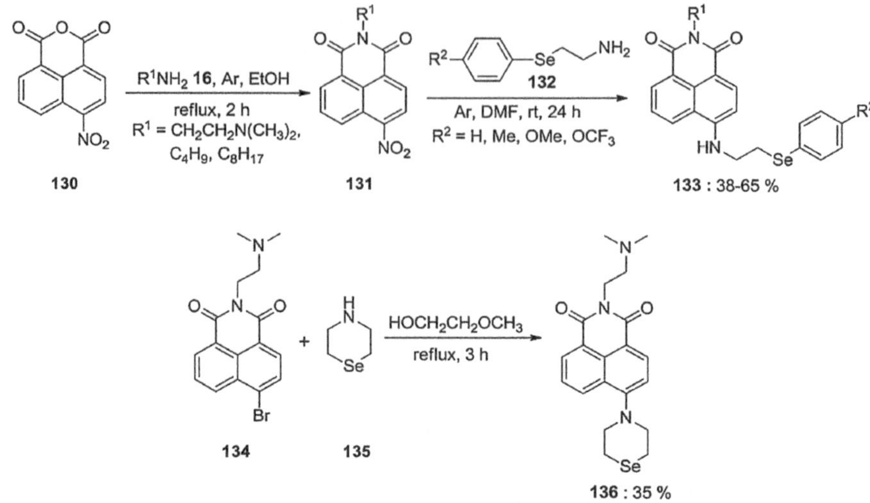

Scheme 3.25 Synthesis of selenium-containing naphthalimides **133** and **136** from **130** and **134**, respectively.

All these compounds were tested on their anticancer activity against K562 and MCF-7 cell lines. It was observed that most of the tested compounds showed moderate anticancer activities with IC_{50} values between 10^{-6} and 10^{-5} M. Compound **136** showed the highest activity, with IC_{50} values ~5.3 μM against K562 and MCF-7 cell lines.[63]

Furthermore, Bhattacharya and his research group reported the synthesis of 2-[5-selenocyanato-pentyl]-7-amino-benzo[de]isoquinoline-1,3-dione (ANOS) **141**.[66] The synthesis of target compound **141** was achieved in two steps starting from 4-amino naphthalimide **137**. In first step, the alkyl bromo compound **139** was prepared by reaction of 4-amino naphthalimide **137** with 1,5-dibromopentane **138** in the presence of MeONa in methanol at room temperature. Compound **139** was then treated with potassium selenocyanate **140** in acetone to get the desired selenocyanate **141** (Scheme 3.26). The synthesis of compound **141** was also reported in 2010 by same research group.[67] The biological studies of this compound illustrated that the administration of ANOS **141** significantly potentiated the therapeutic efficacy of cyclophosphamide by reducing the tumor burden and chemotherapy-induced toxicity in the host.[66]

Park and co-workers reported the synthesis and anticancer activity of 3-allylseleno-6-alkylthiopyridazines **146** in MCF-7 cells.[68] The synthesis of thiopyridazines **146** was achieved by starting 3,6-dichloropyridazine **142** by two different routes. In the first route, 3,6-dichloropyridazine **142** was reacted with disodium diselenide to form the diselenide **143** which was converted into **144** by the reaction with hydrazine, allyl bromide and tetrabutylammonium bromide (TBAB) in the presence of a base in THF. After that, compound **144** was treated with different thiols **145** in a mixture of water and methanol to form 3-allylseleno-6-alkylthiopyridazines **146**

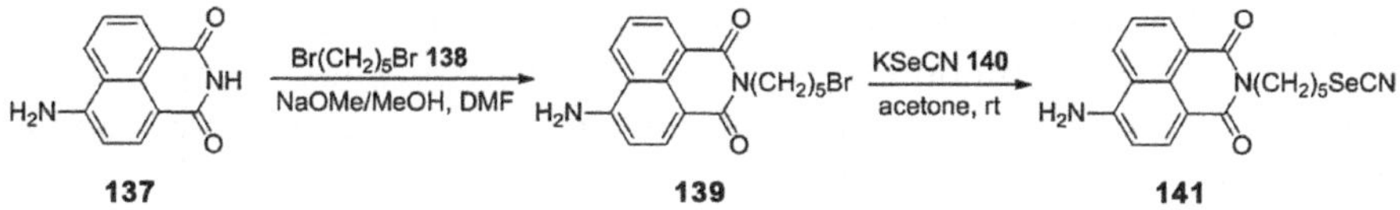

Scheme 3.26 Synthesis of ANOS **141** from 4-amino naphthalimide **137** *via* intermediate **139**.

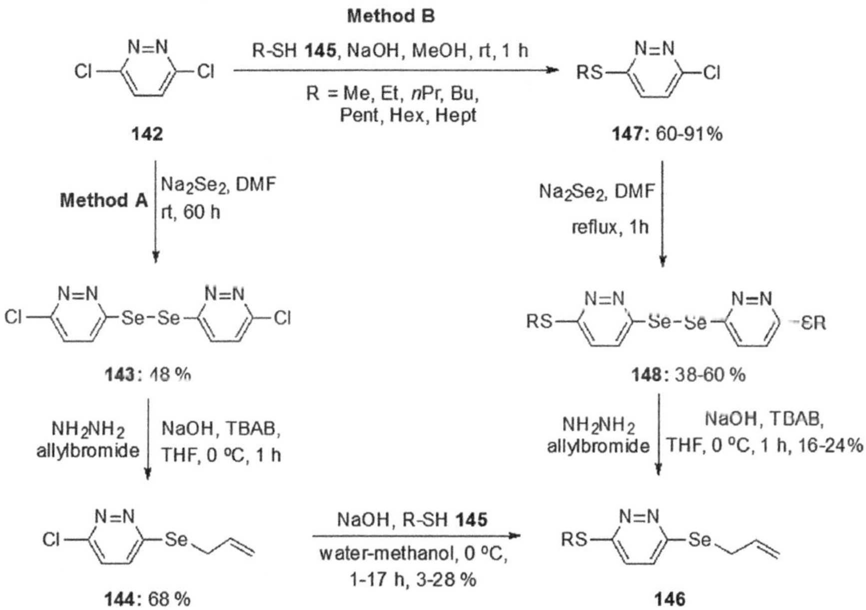

Scheme 3.27 Synthesis of 3-allylseleno-6-alkylthiopyridazines **143** from 3,6-dichloropyridazine **137** *via* methods A and B.

(Scheme 3.27, method A). In the second route, 3,6-dichloropyridazine **142** was treated with different thiols in methanol to form thioethers **147** in good yields. Furthermore, thioethers **147** were treated with disodium diselenide to form corresponding diselenides **148** which were transformed into 3-allylseleno-6-alkylthiopyridazines **146** by the reaction with hydrazine, allyl bromide and TBAB (Scheme 3.27, method B). The evaluation of antiproliferative activities against breast cancer (MCF-7) cells in CCK-8 assays was conducted. All compounds showed significant inhibition in the growth of these cell lines and compound **139** showed even higher potency than the standard anticancer drug fluorouracil.[68]

The synthesis and antitumor activity of functionalized organoselenocyanates was reported by Hamama and co-workers.[69] 4-Aminophenylselenocyanate **149** was used as starting material prepared by the reaction of aniline with triselenium dicyanide.[70] Compound **149** was hydrolyzed to diselenide **150** under alkaline conditions using a reported procedure.[71] Compounds **152a** and **152b** were synthesized by diazotization of 4-aminophenylselenocyanate

149 followed by the reaction with malononitrile **151a** and ethyl cyanoacetate **152b**, respectively. Compounds **155a** and **155b** were prepared using a similar strategy. Furthermore, compound **149** was used to make selenium-containing thalidomide analogues **154a** and **154b** by reaction with **153a** and **153b**, respectively. A similar approach was used for the synthesis of another selenium-containing thalidomide analogue **156** (Scheme 3.28).

In addition, diazo compounds **152a,b** and **155b** were employed as material for the preparation of selenium containing diazopyrazoles **158**, **159**, **161** and **162**. Compound **152a** was treated with hydrazines **157a,b** to form 3,5-diamino-pyrazoles **158a,b**, respectively. Similarly, compound **152b** was reacted with hydrazines **157a,b** and **160** to form 3-aminopyrazol-5-one derivatives **159a,b** and **161**. Finally, compound **155b** was refluxed with hydrazine in ethanol to furnish diselenide-3,5-diaminopyrazole **162** (Scheme 3.29).

All the compounds were examined for *in vitro* antitumor screening using Ehrlich ascites cells. It was observed that compound **159b** has the highest

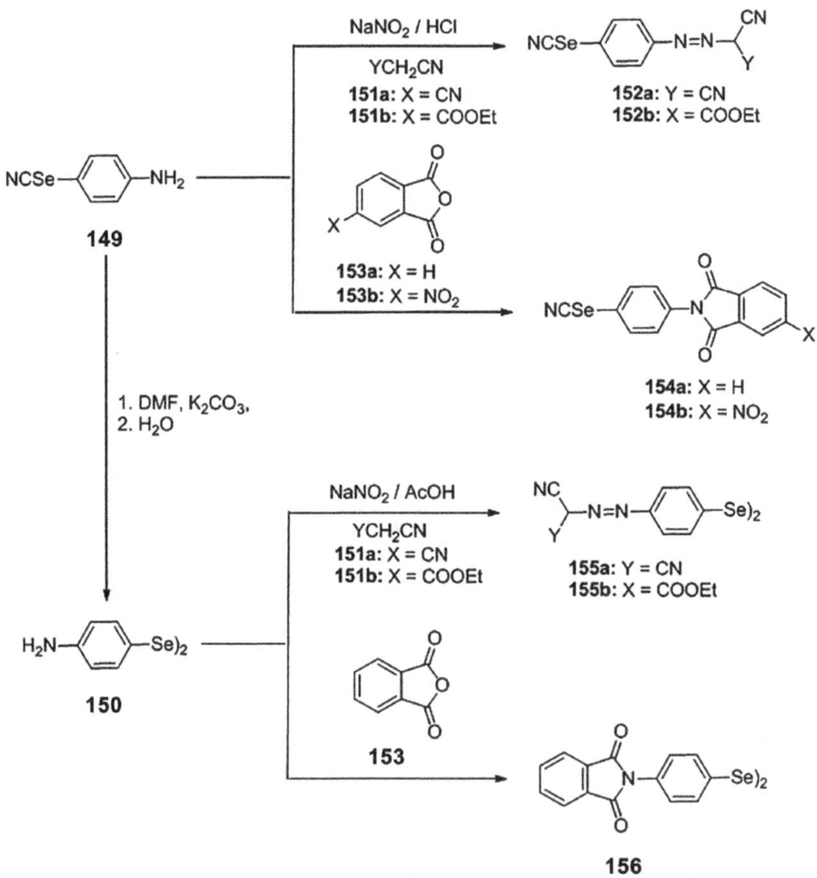

Scheme 3.28 Synthesis of selenium-containing compounds **152**, **154**, **155** and **156** starting from 4-aminophenylselenocyanate **149**.

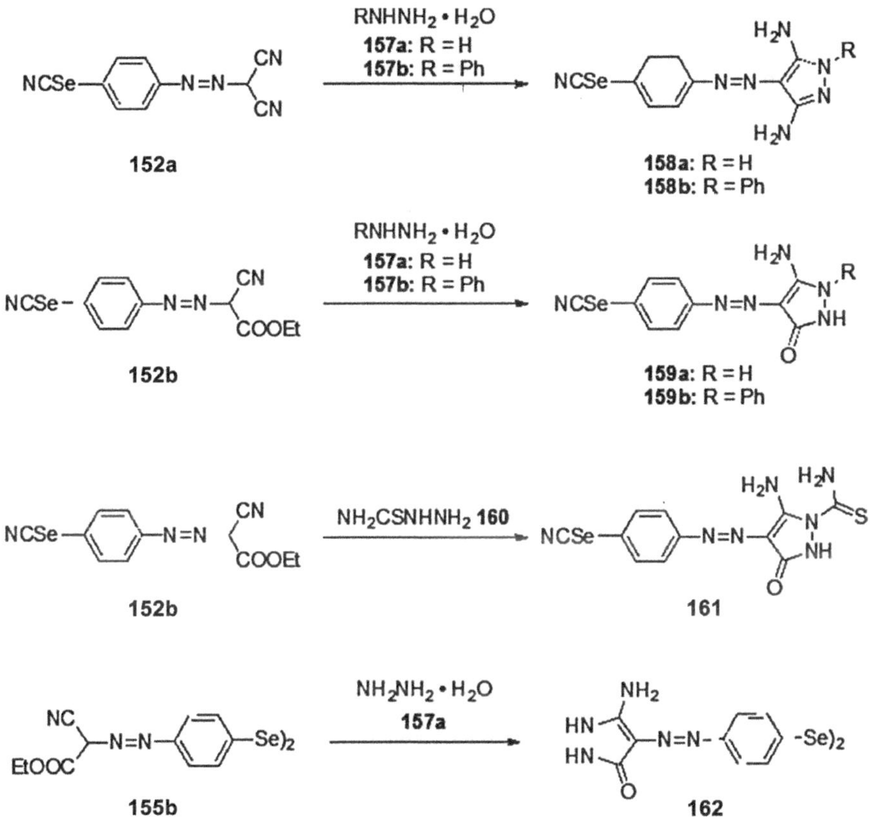

Scheme 3.29 Synthesis of diazo pyrazoles **158**, **159**, **161** and **162** from compounds **152a**, **152b**, **152b** and **155b**, respectively.

cytotoxic activity followed by compounds **162**, **158b**, **155a** and **159a**. Selenium-containing thalidomide analogues **154a** and **156** showed the lowest cytotoxicity. The rest of the tested compounds (**152a,b** and **155b**) exhibited moderate activities.

Zhang and co-workers performed the synthesis and investigated the antiproliferative activity of 3,4-diaryl-1,2,5-selenadiazols **166**.[72] Functionalized benzaldehydes **163** were used to synthesize 3,4-diaryl-1,2,5-selenadiazols **166**. The synthesis of functionalized benzil intermediates **164** was achieved in four chemical steps following reported procedures.[73,74] Benzil derivatives **164** were heated with hydroxyl amine at 90 °C in ethanol in the presence of pyridine to form the imine derivatives **165**. Finally, 3,4-diaryl-1,2,5-selenadiazols **166** were prepared by reaction of imines **165** with selenium dioxide (SeO_2) at reflux temperature in DMF (Scheme 3.30).

Overall, a series of 19 compounds was synthesized and characterized as 3,4-diaryl-1,2,5-selenadiazols **166**. The compounds were tested for their *in vitro* antiproliferative activities. Among these, compounds **166a** ($R^1 = R^2 = R^3 = R^5 = OMe$, $R^4 = H$), **166i** ($R^1 = R^2 = R^3 = OMe$, $R^4 = H$,

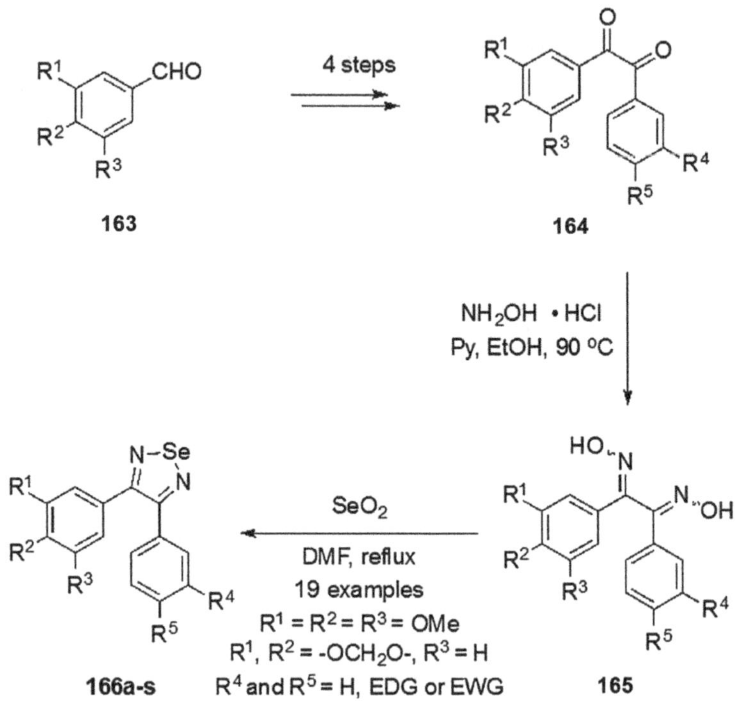

Scheme 3.30 Synthesis of 3,4-diaryl-1,2,5-selenadiazols **166** from benzaldehydes **163**.

$R^5 = Br$), **166n** (R^1, R^2 = –OCH_2O–, $R^3 = H$, $R^4 = OH$, $R^5 = OMe$) and **166s** ($R^1 = R^2 = R^3 = R^5 = OMe$, $R^4 = NHMe$) showed high potency against different tumor cell lines.

Recently, the synthesis and antitumor activity of selenium-containing triazoles has been reported.[75] Selenium-containing triazoles such as **171, 177, 179, 183, 185** and **186–194** have been synthesized in good yields. Initially, selenium-containing dihydropyran naphthoquinones-based triazoles **171, 177** and **179** were prepared from lapachol **167** as staring material (Scheme 3.31).

α-Lapachone **172** was synthesized by acid-catalyzed cyclization of **167**; later it was converted into compound **173** in 80% yield in a reaction with *N*-bromosuccinimide (NBS) in CCl_4. Bromo compound **173** was treated with sodium azide to form **178**. Selenium-containing *p*-naphthoquinone-based triazole analogue **179** was prepared in good yield by copper(I)-catalyzed click reaction between **178** and alkyne **170** (Scheme 3.31).[76]

Monobromo compound **173** was reacted with 3-ethynylaniline **174** in dichloromethane to form intermediate **175** using conditions mentioned in a published protocol.[77] Furthermore, intermediate **175** was transformed to another *p*-naphthoquinone-based triazole derivative **179** by copper(I)-catalyzed click reaction with compound **176**. Finally, intermediate **167** was transformed to dibromo compound **168** by the reaction with H_2SO_4 followed

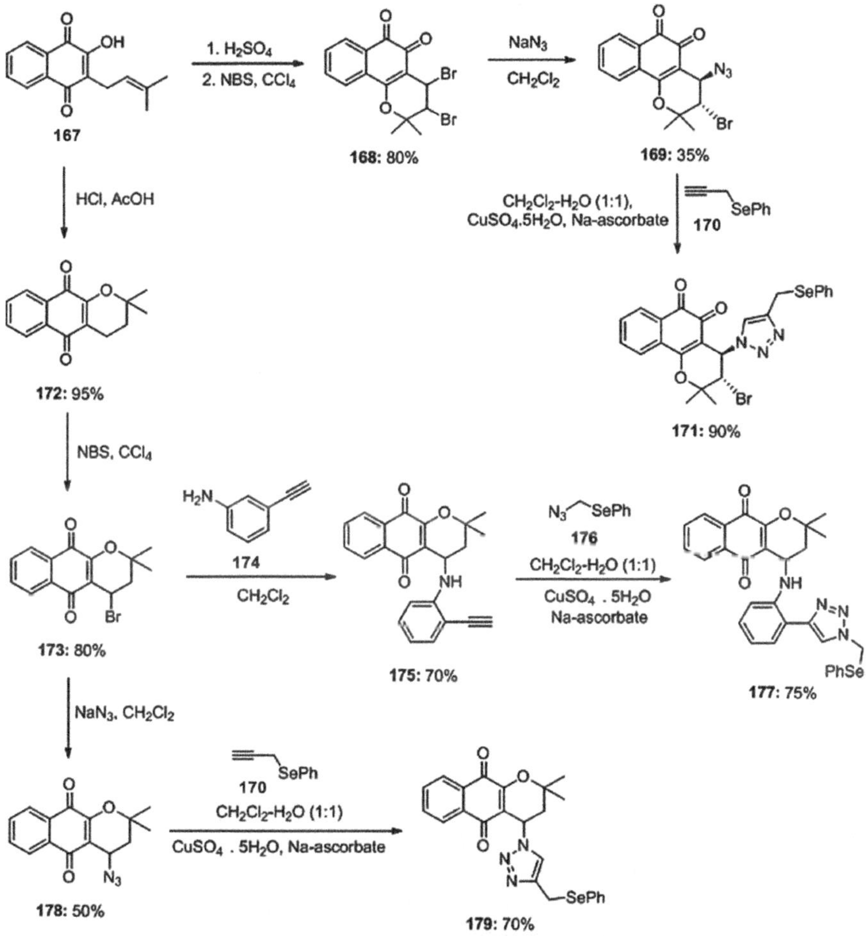

Scheme 3.31 Synthesis of selenium-containing dihydropyran naphthoquinone-based triazoles **171**, **177** and **179** from lapachol **167**.

by reaction with NBS in CCl_4. Dibromo intermediate **168** was treated with sodium azide in dichloromethane to form **169**, which was further converted to selenium-containing *o*-naphthoquinone based triazole analogue **171** by copper(I)-catalyzed click reaction between **169** and alkyne **170** (Scheme 3.31).

After achieving the synthesis of selenium-containing dihydropyran naphthoquinone-based triazoles **171**, **177** and **179**, efforts were directed towards the synthesis of selenium-containing dihydrofuran naphthoquinone-based triazoles **183** and **185**. The synthesis of both targeted compounds **183** and **185** (Scheme 3.32) was achieved by a similar approach described during the discussion of Scheme 3.30. In addition, some other selenium-containing naphthoquinone-triazole compounds **186–194** were synthesized (Figure 3.7).

These compounds were then evaluated for their antitumor activity against six types of cancer cell lines: HL-60 (human promyelocytic leukemia cells),

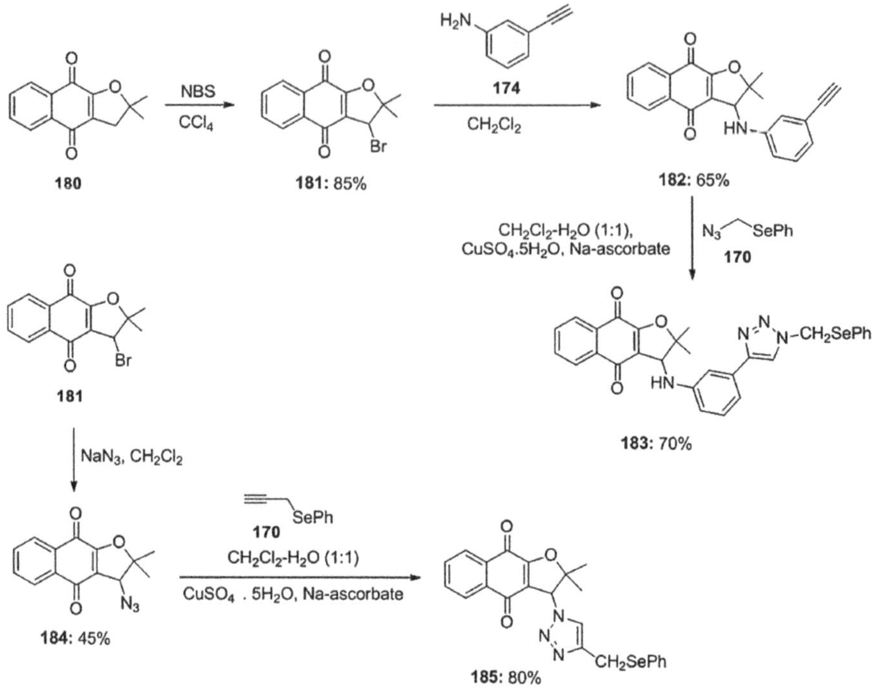

Scheme 3.32 Synthesis of selenium-containing dihydrofuran *p*-naphthoquinone-based triazoles **183** and **185**.

HCT-116 (human colon carcinoma cells), PC3 (human prostatecells), SF295 (human glioblastoma cells), MDA-MB-435 (melanoma cells) and OVCAR-8 (human ovarian carcinoma cells). Compounds **186** and **187** showed promising anticancer activity, while the other selenium-containing triazoles only exhibited moderate anticancer activity.[75]

In 2013, Qi and co-workers developed the synthesis of sucrose selenious ester **196** (sucrose-Se) by the reaction of sucrose **195** and selenium oxychloride (Figure 3.8). Sucrose selenious ester **196** was examined for its anticancer effect in the cervical cancer cell line HeLa through a MTT assay. It was found to be quite effective and inhibited the proliferation of the HeLa cell line in a dose-dependent manner. Additionally, sucrose-Se **196** was effective to inhibit proliferation of bladder carcinoma cell line 5637, human malignant melanoma cell line A375 and gastric carcinoma cell line MGC-803.[78]

Antunes and co-workers developed the synthesis of 6-selenocaffeine **198** from **197** (Figure 3.8). 6-Selenocaffeine **198** was tested for its cytotoxic potential and showed low cytotoxic potential in both MCF-10A and MCF-7 breast cells. Interestingly, 6-selenocaffeine is used to modulate the activity of the anticancer agents doxorubicin and oxaliplatin in MCF-7 breast cancer cells when used in a combination therapy.[79]

Furthermore, Zhong and co-workers developed the synthesis of selenadiazole derivative (BSBD) **199** (Figure 3.8), which showed improved

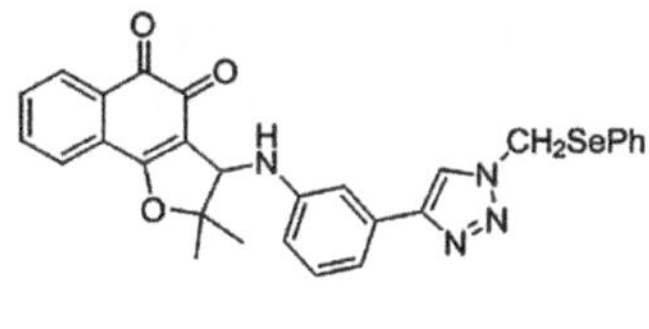

186: 50%

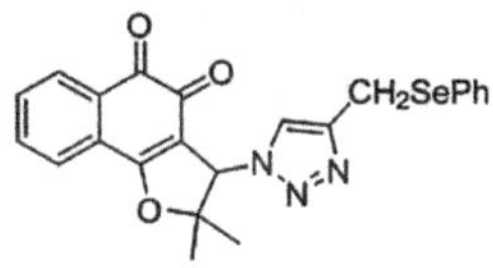

187: 80%

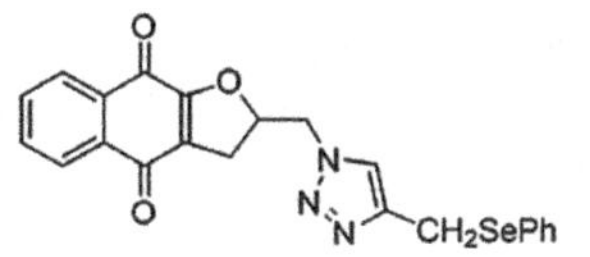

188: 90%

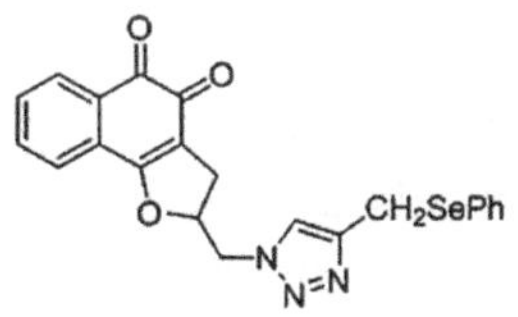

189: 85%

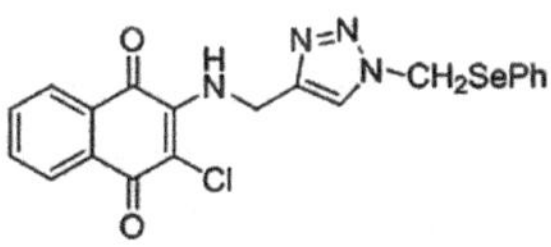

190: 65%

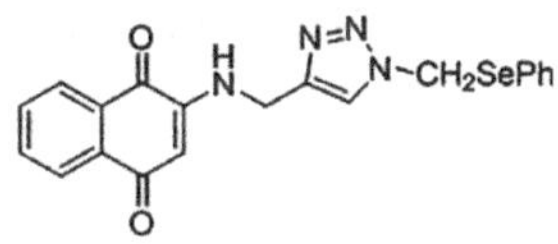

191: 60%

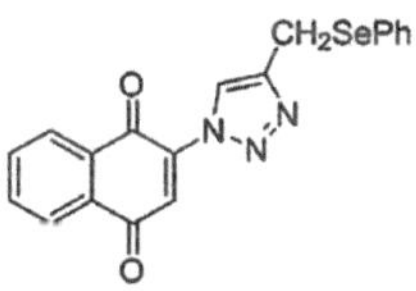

192: 60%

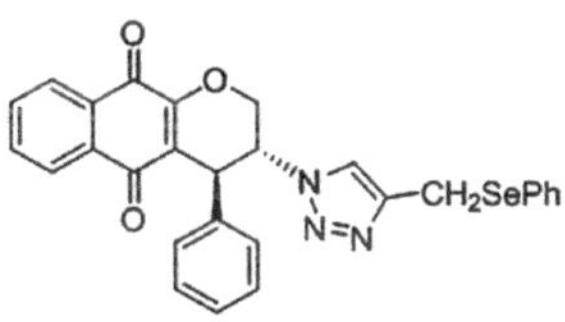

193: 70%

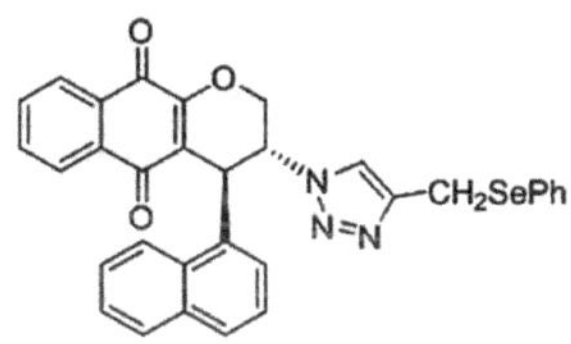

194: 65%

Figure 3.7 The structures of selenium-containing naphthoquinone-triazole compounds **186–194**.

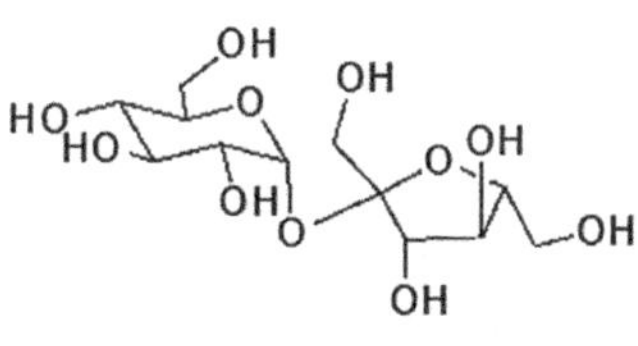

195

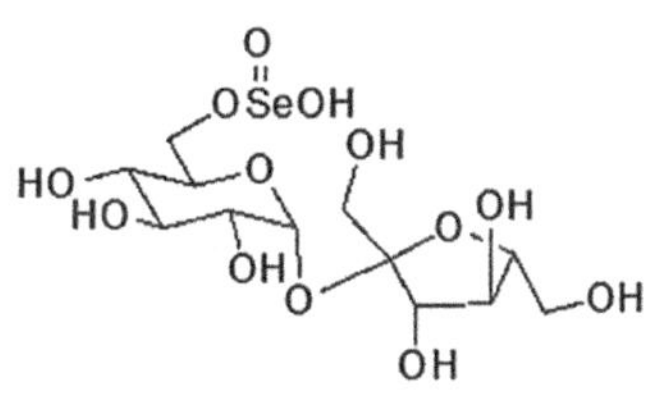

196

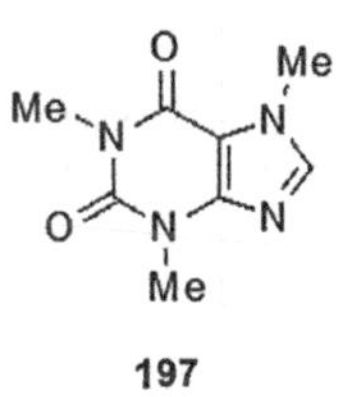

197

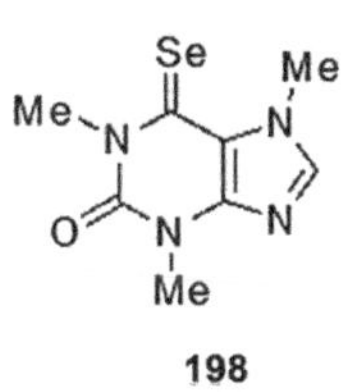

198

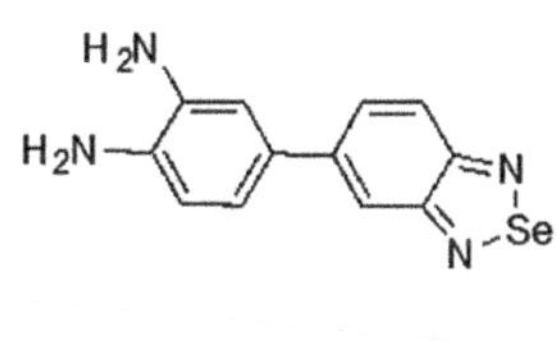

199

Figure 3.8 The structures of sucrose selenious ester **196**, 6-selenocaffeine **198** and selenadiazole derivative **199**.

anticancer activity and greater selectivity than inorganic selenium. It was concluded from mechanistic studies that BSBD induces apoptosis in human glioma cells by dephosphorylation of AKT.[80]

A significant percentage of patients develop lymphedema disease after treatment of cancer.[81] Recently, Pfister *et al.* compiled a review article in which they highlight that sodium selenite showed good results for the reduction of cancer-related lymphedema.[82]

3.5 Selenium Compounds as Anti-Alzheimer's Disease Agents

In recent years, studies have been published in which organoselenium compounds were used for the treatment of Alzheimer's disease. In 2013, Li and others reported the synthesis of tacrine-based ebselen analogues **205** starting from anthranilic acid **70**.[83] Initially, tacrine-based compound **202** was prepared by reaction with cyclohexanone **200** in $POCl_3$ at reflux temperature, followed by the reaction of diamine **201** (Scheme 3.33).

Diselenide **203** was prepared by diazotization of **70** followed by reaction with sodium diselenide and refluxed in thionyl chloride using a catalytic amount of DMF to yield dichloride **204**, which was further treated with the

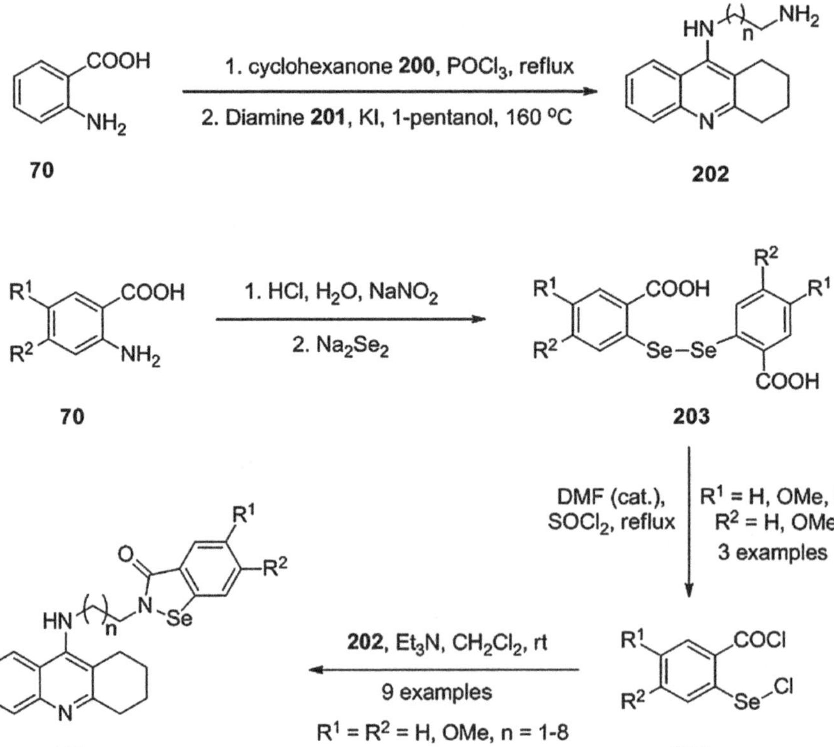

Scheme 3.33 Synthesis of tacrine-based ebselen analogues **205** from anthranilic acid **70**.

amino compound **202** to form tacrine-based ebselen analogues **205** in good yield (Scheme 3.33). All the synthesized compounds were tested for their acetylcholinesterase (AChE)- and butylcholinesterase (BuChE)-inhibiting activity. Interestingly, tacrine-based ebselen analogue **205** ($R^1 = R^2 = OMe$, $n = 5$) was found to be the most promising agent to inhibit AChE and BuChE, with IC_{50} values of 2.55 and 2.80 nM, respectively. In addition, all the synthesized compounds **205** were found to be effective antioxidants.[83]

Furthermore, Li and co-workers prepared a series of new ebselen analogues **208** and evaluated their role as cholinesterase inhibitors.[84] The dichloro intermediate **204** was used as starting material which was prepared from anthranilic acid **70** in two chemical steps shown in Scheme 3.32. The dichloro intermediate **204** was reacted with functionalized anilines in dichloromethane in the presence of Et_3N to form ebselen analogues **208** in good yield (Scheme 3.34).

Newly synthesized ebselen analogues **208** were evaluated for their role as AChE and BuChE inhibitors. It was observed that most compounds showed good inhibition against AChEs and BuChE, while two compounds **208** ($R = OMe$, $n = 0$, $X = CH_2$, $Y =$ piperidine and $R = OMe$, $n = 3$, $X = O$, $Y =$ piperidine) showed the most promising results. Simultaneously, compounds **208** also showed GPx-like antioxidant activity.[84]

The same research group reported synthesis and multifunctional anti-Alzheimer's disease activity of selenium-containing clioquinol derivatives **217**.[85] Quinolin-8-ol **209** was used as the starting material for the preparation of selenium-containing clioquinol derivatives **217**. Intermediates **210** and **211** were synthesized by reported procedures using reaction conditions mentioned in Scheme 3.34.[86] Intermediate **211** was diazotized and then reacted with KSeCN **212** to form selenocyanate derivative **213**. Compound **213** was further stirred in methanol in the presence of base to form clioquinol-diselenide **214**, which was reduced with $NaBH_4$ and reacted with alkyl halide

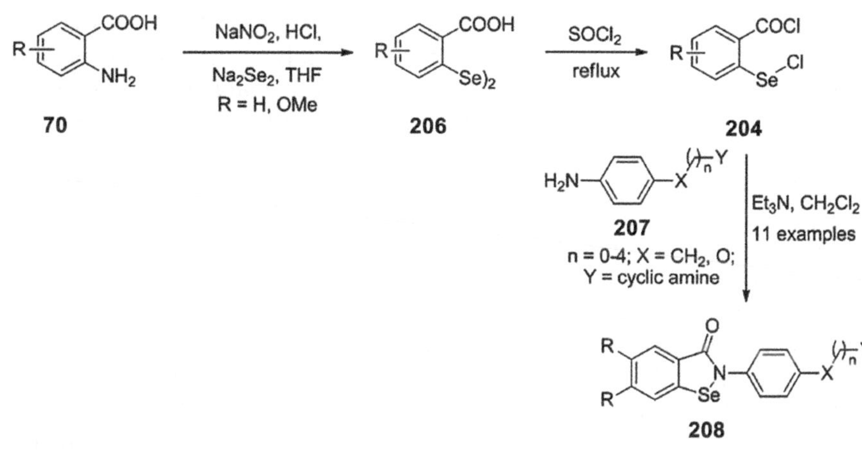

Scheme 3.34 Synthesis of various new ebselen analogues **208**.

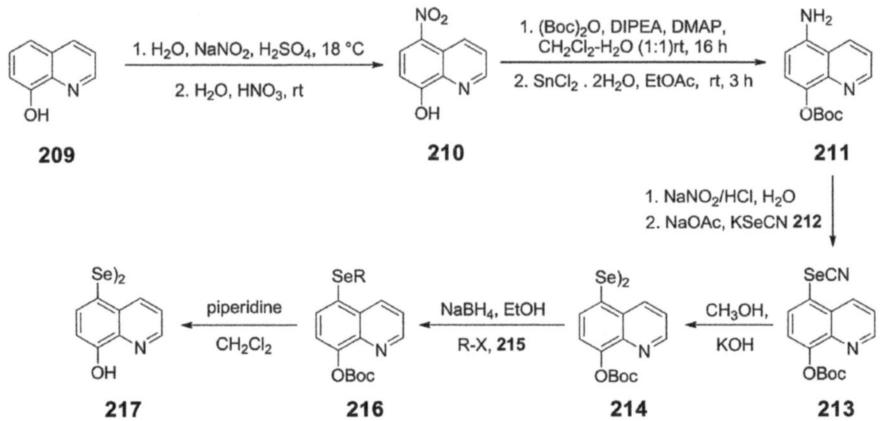

Scheme 3.35 The synthesis of selenium-containing clioquinol derivatives **217** starting from quinolin-8-ol **209**.

to produce alkyl aryl selenides **216**. Finally, selenides **216** were transformed into target compounds **217** by removal of the Boc group by reaction with piperidine (Scheme 3.35).

Clioquinol derivatives **217** were evaluated for the inhibition of Cu(II)-induced Aβ1–42 aggregation. Most of the compounds showed significant inhibition while compound **217** (R = propargyl) showed the best result. This compound is a promising candidate for the development of a novel drug for Alzheimer's disease.[85a] Braga and his research group developed dihydropyrimidinone-cored selenium compounds as AChE inhibitors.[85b,c]

3.6 Selenium Compounds as Antimicrobial Agents

In 2010, Radhakrishna *et al.* developed the synthesis and investigated the antibiotic activity of different selenium-containing carboxylic acids and their esters. These selenium compounds were found to be quite effective against certain bacteria such as *Staphylococcus aureus*, *Salmonella typhimurium*, *Escherichia coli* and *Bacillus subtilis.*[87] In 2015, Khanna and co-workers demonstrated the antimicrobial activity of cycloalkeno-1,2,3-selenadiazoles **221**.[88] The synthesis of cycloalkeno-1,2,3-selenadiazoles **221** was achieved by selenium dioxide-mediated cyclization of semicarbazones **220**. Semicarbazones **220** were prepared by the reaction of cycloalkanones **218** and semicarbazide hydrochloride **219** using a reported procedure (Scheme 3.36).[89]

In addition, 1,2,3-selenadiazoles **221** were evaluated for their antimicrobial activity against various pathogenic bacteria using an agar disc diffusion method. All compounds showed significant activity against *E. coli*, *S. aureus*, *Salmonella typhi* and *Pseudomonas aeruginosa*. Cyclopenteno-1,2,3-selenadiazole **221** ($n = 1$) was found to be the best antibacterial agent against the highly resistant *P. aeruginosa.*[89]

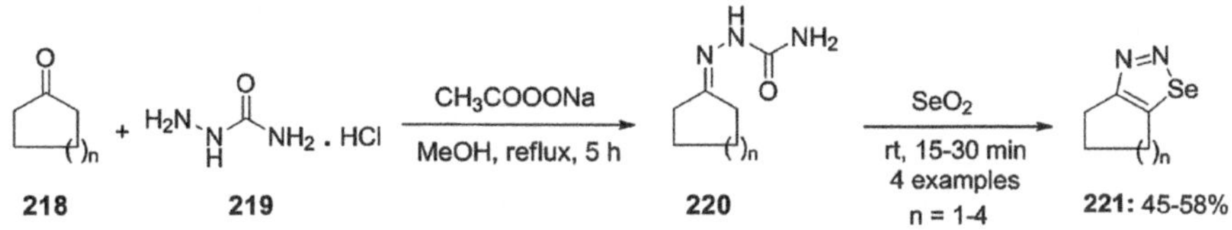

Scheme 3.36 The synthesis of cycloalkeno-1,2,3-selenadiazoles **221** from cycloalkanones **218** *via* semicarbazones **220**.

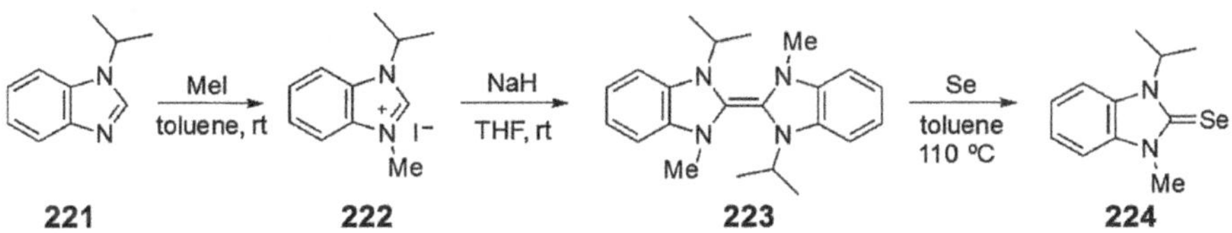

Scheme 3.37 Synthesis of 1-isopropyl-3-methylbenzimidazole-2-selenone **224** starting from 1-isopropyl-1*H*-benzo[*d*]imidazole **221**.

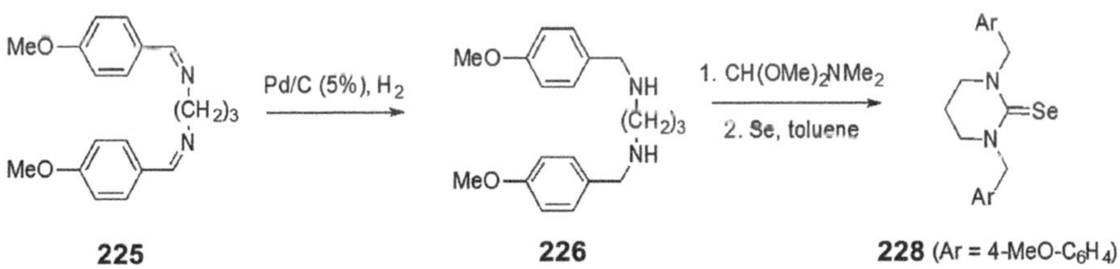

Scheme 3.38 Synthesis of 1,3-di-*p*-methoxybenzylpyrimidine-2-selenone **228** from compound **225** in three chemical steps.

Talas *et al.* developed the synthesis of 1-isopropyl-3-methylbenzimidazole-2-selenone **224** starting from 1-isopropyl-1*H*-benzo[*d*]imidazole **221**. The key intermediate **223** was prepared by methylation of compound **221** followed by base-catalyzed condensation. Finally, tetra substituted alkene intermediate **223** was heated with selenium in toluene at 110 °C to form product **224** in good yield (Scheme 3.37).[90]

Additionally, 1,3-di-*p*-methoxybenzylpyrimidine-2-selenone **228** was synthesized from compound **225** in three chemical steps. Product **228** was prepared by the reaction of **226** with *N*,*N*-dimethylformamide dimethylacetal ($CH(OMe)_2NMe_2$) followed by heating with selenium in toluene at 110 °C (Scheme 3.38).[90] Compounds **224** and **228** were tested for their antimicrobial activity using Gram-positive and Gram-negative bacteria. Both selenium compounds showed antimicrobial activity, but 1-isopropyl-3-methylbenzimidazole-2-selenone **224** showed better antimicrobial activity than 1,3-di-*p*-methoxybenzylpyrimidine-2-selenone **228**.[90]

Recently, Bhasin and co-workers investigated the antimicrobial activity of imidazo[1,2-*a*]pyridine-based organoselenium compounds.[91] The synthesis of derivative **234** was achieved in high yields by reaction of 2-chloroimidazo[1,2-*a*]pyridines **232** with aryl/heteroaryl selenols, generated *in situ* by reduction of diselenides **233** with hypophosphorous acid. The key intermediate **232** were prepared from 2-aminopyridines **229** in three chemical steps. Initially, 2-aminopyridine **229** was converted to 2-iminopyridine derivative **230** with chloroacetic acid in the presence of Et_3N. Intermediate **230** was cyclized to 2-chloroimidazo[1,2-*a*]pyridines **232** by refluxing with $POCl_3$ in toluene. On nitration, compound **232** provided the key intermediate **233** in good yields (Scheme 3.39). Many compounds **234** showed significant antimicrobial activity.[91]

Recently, Pietrella and co-workers synthesized diselenides **20, 72, 77, 235** and **236** (Figure 3.9) and studied their effectiveness in inhibiting microbial biofilm formation, which is involved in wound infection. Diselenides

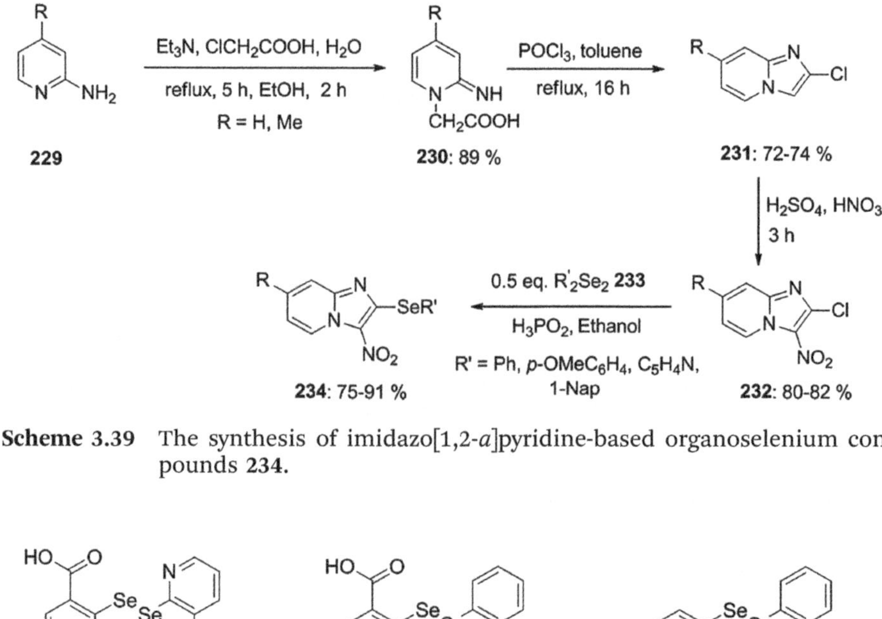

Scheme 3.39 The synthesis of imidazo[1,2-*a*]pyridine-based organoselenium compounds **234**.

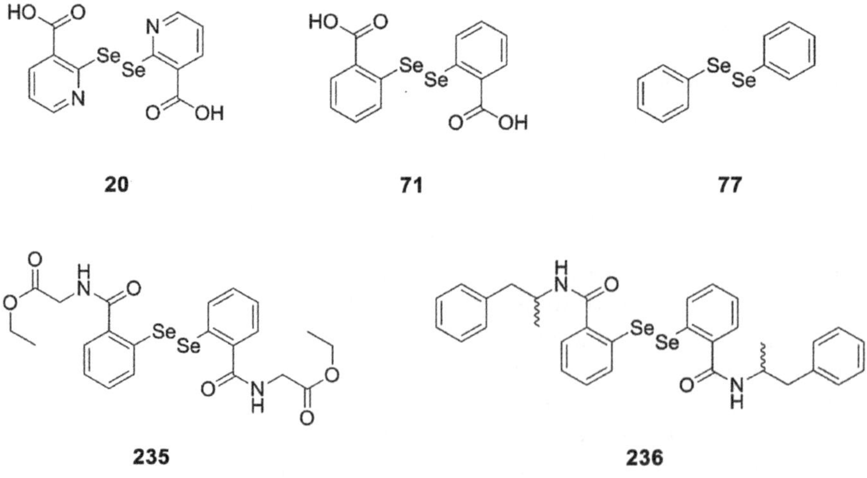

Figure 3.9 The structures of various diselenides **20, 71, 77, 235** and **236**.

bis[ethyl *N*-(2′-selenobenzoyl)glycinate] **235** and bis[2′-seleno-*N*-(1-methyl-2-phenylethyl) benzamide] **236** were found to be promising compounds for preventing and treating microbial wound infections.[92]

3.7 Organoselenium Compounds as Antiviral Agents

In 1990, Sidwell and co-workers investigated the utility of selenazofurin to inhibit the infections of phlebovirus during *in vivo* studies.[93] The organoselenium drug ebselen **1** also possesses antiviral properties, although it is better known for the problems associated with inflammatory and oxidative stress.[94] Jeong and co-workers reported the synthesis and antiviral activity of 4′-selenonucleosides.[95] The cyclic selenium compound **239** was synthesized in four steps using alcoholic starting material **235**. The synthesis of uracil-based 4′-selenonucleoside was achieved by the reaction of compound **239** with *m*-CPBA and uracil followed by deprotection of the protected alcohol using TFA.[96] The selenonucleosides **241–243** were synthesized using a similar procedure (Scheme 3.40). All four selenonucleosides **241–243** were evaluated for their antiviral activities and **241** and **243** showed potent antiviral activities with EC_{50} 2.3 μM and 2.9 μM, respectively.[97]

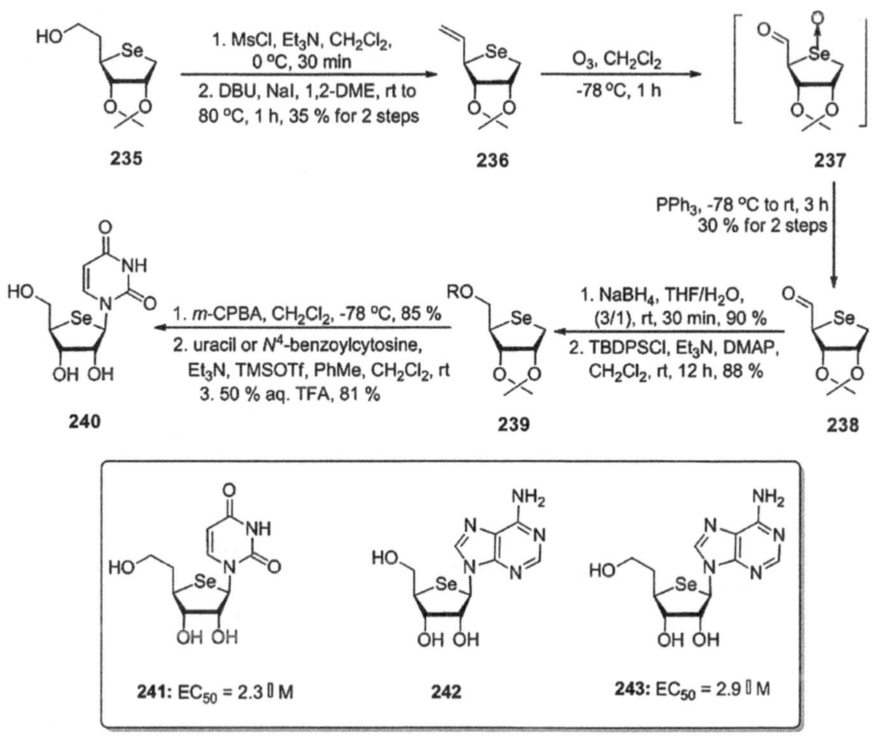

Scheme 3.40 Synthesis of uracil- and adenine-based selenonucleosides **240–243** from starting material **235**.

The same research group synthesized two other organoselenium compounds, selenoacyclovir **251** and selenoganciclovir **259**.[98] The synthesis of selenoacyclovir **251** was achieved in 94% yield by the reaction of purine intermediate **248** with TBAF followed by $SHCH_2CH_2OH$ **250** in methanol in the presence of sodium methoxide. The purine intermediate **248** was synthesized from **244** in four chemical steps (Scheme 3.41).

Selenoganciclovir **257** was prepared in 70% yield by the reaction of selenide **258** with $SHCH_2CH_2OH$ **250** in methanol in the presence of sodium methoxide. The purine-based selenide **258** was synthesized from **252** in six chemical steps (Scheme 3.42). Both compounds **252** and **259** were tested for

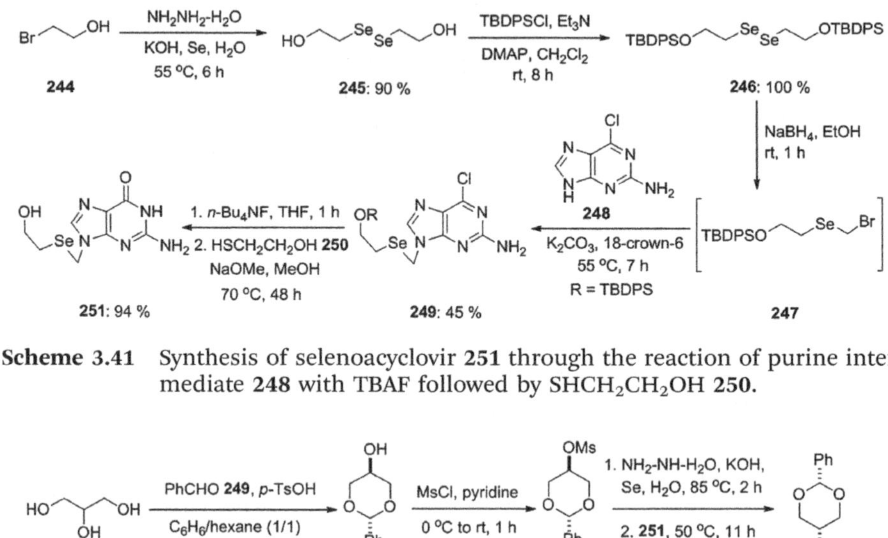

Scheme 3.41 Synthesis of selenoacyclovir **251** through the reaction of purine intermediate **248** with TBAF followed by $SHCH_2CH_2OH$ **250**.

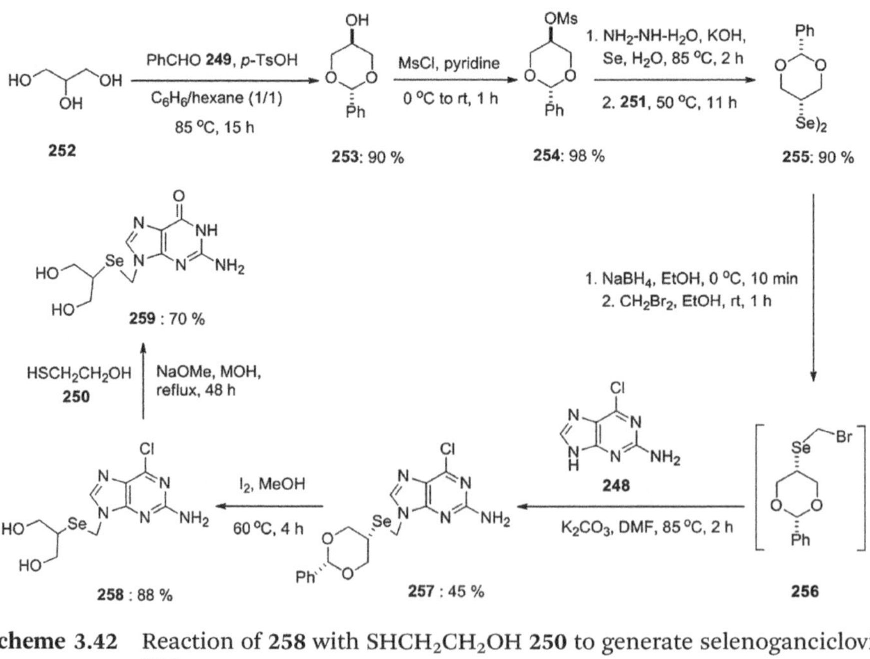

Scheme 3.42 Reaction of **258** with $SHCH_2CH_2OH$ **250** to generate selenoganciclovir **259**.

their antiviral activity. Compound **251** showed promising anti-herpes simplex virus-1 and -2 activities while **259** exhibited moderate anti-human cytomegalovirus activity.[98a] In addition, diselenobisbenzamides (DISeBAs) were developed as nucleocapsid protein 7 (NCp7) inhibitors.[98b] Nucleocapsid protein 7 (CCp7) plays a vital role in the replication of RNA during the early stage of viral infections.[98c] This protein represents an interesting target for the development of antiviral drugs.

3.8 Organoselenium Compounds as Xanthine Oxidase Inhibitors

Xanthine oxidase is an important enzyme and plays a vital role in the biosynthesis of uric acid.[99] The overproduction of uric acid can lead to hyperuricemia which is main cause of gout.[100] Xanthine oxidase is an important target for the treatment of gout. In 2014, Zhang and coworkers reported the synthesis of 2-aryl-4-methyl-1,3-selenazole-5-carboxylic acids **265** as potent xanthine oxidase inhibitors. 4-Cyanophenol **260** was used as starting point for the synthesis of compounds **260**. 1,3-Selenazole intermediate **262** was prepared from 4-cyanophenol **260** *via* formation of benzoselenoamide intermediate **261**. 1,3-Selenazole intermediate was further converted into **264a** directly and into **264b** *via* formation of another intermediate **263**. Finally, 2-aryl-4-methyl-1,3-selenazole-5-carboxylic acids **265** were synthesized in moderate-to-excellent yields by reaction of alkyl halides followed by ester deprotection (Scheme 3.43).[101]

The new compounds **265** were evaluated for their xanthine oxidase inhibitory activity and showed significant inhibition at nanomolar

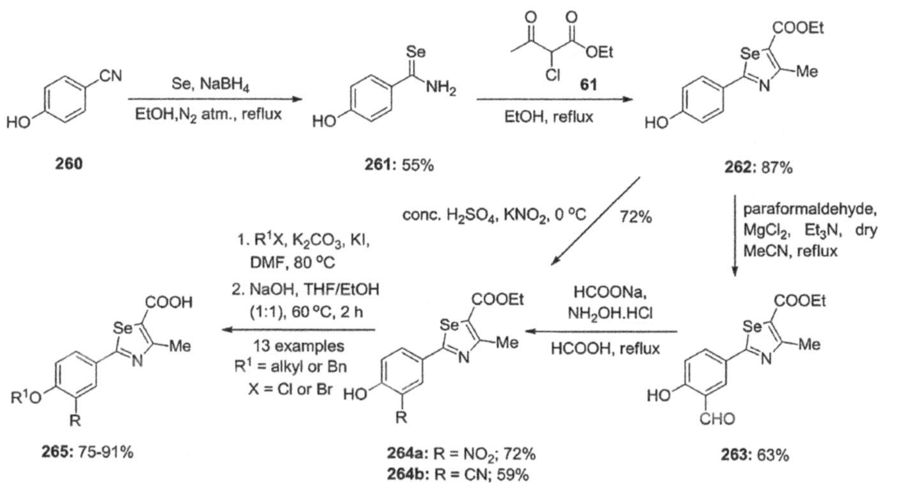

Scheme 3.43 Synthesis of 2-aryl-4-methyl-1,3-selenazole-5-carboxylic acids **265** from 4-cyanophenol **260**.

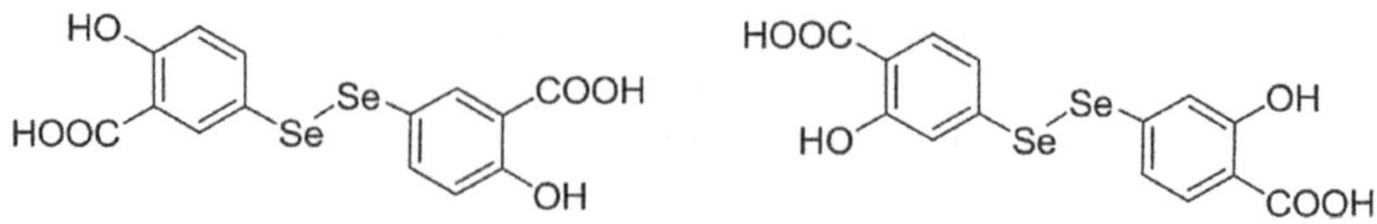

266 **267**

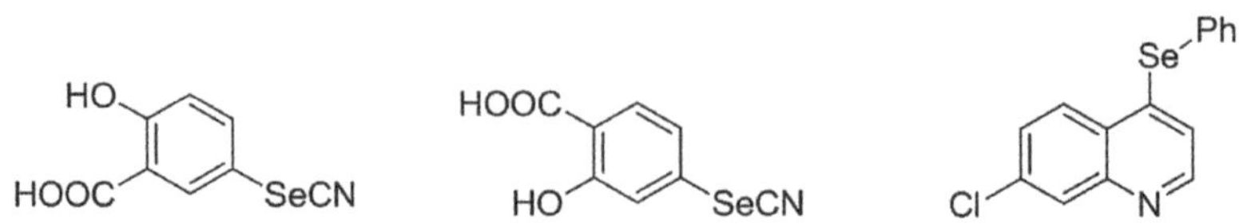

268 **269** **270**

Figure 3.10 The structures of selenium-based anti-inflammatory agents **266–270**.

concentrations. Compound **265** (R = CN and R^1 = allyl) was found to be the most potent xanthine oxidase inhibitor (IC_{50} = 5.5 nM) in comparison to the drug febuxostat (IC_{50} = 18.6 nM).[101]

3.9 Organoselenium Compounds as Anti-inflammatory Agents

Ebselen is one of the selenium-containing anti-inflammatory drugs.[102] In 2003, Rocha and co-workers identified diselenides as a new class of anti-inflammatory agents.[103] In 2014, Nogueira and co-workers reported the synthesis and anti-inflammatory activity of salicyclic-based diselenides **266**, **267** and selenocynates **268**, **269** (Figure 3.10).[104] Recently, Wilhelm and co-workers identified 4-phenylselenyl-7-chloroquinoline **270** as a potent anti-inflammatory compound (Figure 3.10).[105]

3.10 Organoselenium Compounds as Antidiabetic Agents

There are only a few reports in the literature in which organoselenium compounds have been shown to possess significant utility in the treatment of diabetes. In 2012, Lopez and co-workers synthesized L-fuco-selenosulars **57–61** (Scheme 3.7).[27] The synthesis of these compounds is shown in Scheme 3.8. Interestingly, selenoxide **60** was found to be a good inhibitor of α-L-fucosidase. Compound **60** was the first example of a selenosugar inhibiting a glycosidase enzyme.[28] The same research group also investigated the glycosidase inhibitory activity of *O*-alkyl selenocarbamates **63**, *N*-phenyl selenohydantoin **67** and *N*-naphthyl selenohydantoin **67**.[29] Most of the synthesized compounds showed significant inhibition against

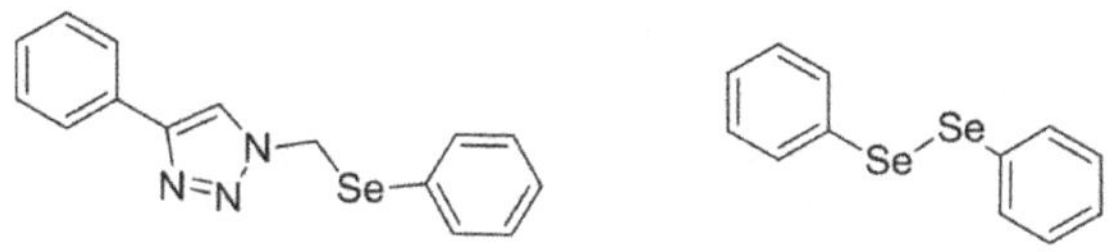

271 **77**

Figure 3.11 The structures of antidepressant agents 4-phenyl-1-(phenylselanylmethyl)-1,2,3-triazole **271** and diphenyl diselenide **77**.

glycosidase enzymes. The synthesis of these compounds is described in Schemes 3.9 and 3.10.[29]

3.11 Organoselenium Compounds as Antidepressants

There is also some evidence of organoselenium compounds acting as antidepressants. In 2014, Savegnago and co-workers reported the synthesis and antidepressant activity of α-phenylselenocitronellal (PhSeCIT) **78b**.[33] The synthesis of compound **78b** is discussed in Scheme 3.13. PhSeCIT **78b** was evaluated for its antidepressant activity and promising results were found.[34] The same research group synthesized 4-phenyl-1-(phenylselanylmethyl)-1,2,3-triazole **271** which was identified as an antidepressant agent (Figure 3.11).[106] Barbosa and co-workers reported a detailed biological study to investigate the effect of diphenyl diselenide **77** supplementation on depressive-like behavior triggered by methimazole exposure in female rats (Figure 3.11).[107]

References

1. C. J. Lowig, *Poggendorff's Ann. Phys.*, 1836, **37**, 552.
2. (a) J. L. Huguet, *Adv. Chem. Ser.*, 1967, **76**, 345; (b) D. N. Jones, D. Mundy and R. D. Whitehouse, *J. Chem. Soc., Chem. Commun.*, 1970, 86; (c) R. Walter and J. Roy, *J. Org. Chem.*, 1971, **36**, 2561.
3. (a) L. C. Wilkins, B. A. R. Günther, M. Walther, J. R. Lawson, T. Wirth and R. L. Melen, *Angew. Chem., Int. Ed.*, 2016, **55**, 11292; (b) J. Tancock and T. Wirth, *Molecules*, 2015, **20**, 10866; (c) S. A. Shahzad, C. Vivant and T. Wirth, *Org. Lett.*, 2010, **12**, 1364; (d) S. A. Shahzad and T. Wirth, *Angew. Chem., Int. Ed.*, 2009, **48**, 2588; (e) D. M. Freudendahl, S. A. Shahzad and T. Wirth, *Eur. J. Org. Chem.*, 2009, 1649; (f) D. M. Freudendahl, S. Santoro, S. A. Shahzad, C. Santi and T. Wirth, *Angew. Chem., Int. Ed.*, 2009, **48**, 8409; (g) D. M. Freudendahl, S. Santoro, S. A. Shahzad, C. Santi and T. Wirth, *Angew. Chem., Int. Ed.*, 2009, **48**, 8409; (h) D. M. Browne and T. Wirth, *Curr. Org. Chem.*, 2006, **10**, 1893; (i) T. Wirth, *Angew. Chem., Int. Ed.*, 2000, **39**, 3742.

4. (a) C. Santi, R. Di Lorenzo, C. Tidei, L. Bagnoli and T. Wirth, *Tetrahedron*, 2012, **68**, 10530; (b) F. V. Singh and T. Wirth, *Org. Lett.*, 2011, **13**, 6504; (c) F. V. Singh and T. Wirth, in *Organoselenium Chemistry*, ed. T. Wirth, Wiley-VCH, 2011, p. 321; (d) D. M. Browne, O. Niyomura and T. Wirth, *Org. Lett.*, 2007, **9**, 3169; (e) A. L. Braga, D. S. Ludtke, F. Vargas and R. C. Braga, *Synlett*, 2006, 1453.
5. (a) S. P. Thomas, K. Satheeshkumar, G. Mugesh and T. N. G. Row, *Chem. – Eur. J.*, 2015, **21**, 679; (b) T. Wirth, *Angew. Chem., Int. Ed.*, 2015, **54**, 10074; (c) S. A. Akhoon, T. Naqvi, S. Nisar and M. A. Rizvi, *Asian J. Chem.*, 2015, **27**, 2745; (d) V. P. Singh, J.-F. Poon, R. J. Butcher and L. Engman, *Chem. – Eur. J.*, 2014, **20**, 12563; (e) V. Nascimento, E. E. Alberto, D. W. Tondo, D. Dambrowski, M. R. Detty, F. Nome and A. L. Braga, *J. Am. Chem. Soc.*, 2012, **134**, 138.
6. L. Flohe, E. A. Gunzler and H. H. Schock, *FEBS Lett.*, 1973, **32**, 132; (b) J. T. Rotruck, A. L. Pope, H. E. Ganther, A. B. Swanson, D. G. Hafeman and W. G. Hoekstra, *Science*, 1973, **179**, 588.
7. (a) D. Behne, A. Kyriakopoulos, H. Meinhold and J. Kohrle, *Biochem. Biophys. Res. Commun.*, 1990, **173**, 1143; (b) J. Kohrle, *Methods Enzymol.*, 2002, **347**, 125.
8. T. Tamura and T. C. Stadtman, *Proc. Natl. Acad. Sci. U. S. A.*, 1996, **93**, 1006.
9. (a) T. Wirth, *Organoselenium Chemistry: Synthesis and Reactions*, Wiley-VCH, Germany, 2011; (b) S. Patai and Z. Rappoport, *The Chemistry of Organic Selenium and Tellurium Compounds*, Wiley, vol. 3, 2012; (c) T. G. Back, *Organoselenium Chemistry*, Oxford University Press, Oxford, 1999; (d) T. Wirth, *Organoselenium Chemistry: Modern Developments in Organic Synthesis*, Topics in Current Chemistry, Springer, Berlin, vol. 208, 2000; (e) D. Liotta, *Organoselenium Chemistry*, John Wiley & Sons, New York, 1987; (f) C. Paulmier, *Selenium Reagents and Intermediates in Organic Synthesis*, Pergamon Press, Oxford, 1986; (g) S. Patai and Z. Rappoport, *The Chemistry of Organic Selenium and Tellurium Compounds*, Wiley, vol. 1, 1986; (h) S. Patai and Z. Rappoport, *The Chemistry of Organic Selenium and Tellurium Compounds*, Wiley, vol. 2, 1987; (i) K. C. Nicolaou and N. A. Petasis, *Selenium in Natural Products Synthesis*, CIS, Philadelphia, 1984; (j) A. Krief and L. Hevesi, *Organoselenium Chemistry I*, Springer, Berlin, 1988; (k) D. L. Klayman and W. H. H. Günther, *Organic Selenium Compounds: Their Chemistry and Biology*, Wiley, New York, 1973.
10. (a) B. J. Bhuyan, D. S. Lamani, G. Mugesh and T. Wirth, in *Handbook of Chalcogen Chemistry*, ed. F. A. Devillanova and W. W. du Mont, RSC, 2013, vol. 2, p. 25; (b) D. Bhowmick and G. Mugesh, in *Patai Series: Organic Selenium and Tellurium Compounds*, ed. Z. Rappoport, John Wiley & Sons, 2013, vol. 4, p. 1175; (c) A. L. Braga and J. Rafique, in *Patai Series: Organic Selenium and Tellurium Compounds*, ed. Z. Rappoport, John Wiley & Sons, 2014, vol. 4, p. 1175; (d) F. V. Singh and T. Wirth, in

Patai Series: Organic Selenium and Tellurium Compounds, ed. Z. Rappoport, John Wiley & Sons, 2012, vol. 3, p. 303; (e) C. W. Nogueira and J. B. T. Rocha, in *Patai Series: Organic Selenium and Tellurium Compounds*, ed. Z. Rappoport, John Wiley & Sons, 2012, vol. 3, p. 1277; (f) D. M. Freudendahl and T. Wirth, in *Selenium and Tellurium Chemistry*, ed. J. D. Woolins and R. S. Laitinen, Springer, 2011, p. 41; (g) B. J. Bhujan and G. Mugesh, in *Organoselenium Chemistry*, ed. T. Wirth, Wiley-VCH, 2011, p. 361; (h) P. L. Beaulieu and R. Déziel, in *Organoselenium Chemistry: A Practical Approach*, ed. T. G. Back, Oxford University Press, Oxford, 1999, p. 35; (i) Y. Nishibayashi and S. Uemura, in *Topics in Current Chemistry*, ed. T. Wirth, 2000, vol. 208, p. 201.

11. (a) H. J. Reich and R. J. Hondal, *ACS Chem. Biol.*, 2016, **11**, 821; (b) A. L. Braga, F. A. R. Barbosa, S. Saba, R. F. S. Canto and J. Rafique, *Curr. Org. Chem.*, 2016, **20**, 166–188; (c) M. Ninomiya, D. R. Garud and M. Koketsu, *J. Coord. Chem.*, 2011, **255**, 2968; (d) A. J. Mukherjee, S. S. Zade, H. B. Singh and R. B. Sunoj, *Chem. Rev.*, 2010, **110**, 4357; (e) E. E. Alberto, V. do Nascimentob and A. L. Braga, *J. Braz. Chem. Soc.*, 2010, **21**, 2032; (f) K. P. Bhabak and G. Mugesh, *Acc. Chem. Res.*, 2010, **43**, 1408; (g) B. K. Sarma and G. Mugesh, *Org. Biomol. Chem.*, 2008, **6**, 965; (h) J. Mlochowski, K. Kloc, R. Lisiak, P. Potaczek and H. Wojtowicz, *ARKIVOC*, 2007, **6**, 14; (i) C. Narajji, M. D. Karvekar and A. K. Das, *Indian J. Pharm. Sci.*, 2007, **69**, 344; (j) G. Guillena and D. J. Ramon, *Tetrahedron: Asymmetry*, 2006, **17**, 1465; (k) M. Tiecco, L. Testaferri, F. Marini, L. Bagnoli, C. Santi, A. Temperini, S. Sternativo and C. Tomassini, *Phosphorus, Sulfur Silicon Relat. Elem.*, 2005, **180**, 729; (l) G. Mugesh, W.-W. du Mont and H. Sies, *Chem. Rev.*, 2001, **101**, 2125; (m) N. Petragnani, H. A. Stefani and C. J. Valduga, *Tetrahedron*, 2001, **57**, 1411; (n) X. Ren, L. Yang, J. Liu, D. Su, D. You, C. Liu, K. Zhang, G. Luo, Y. Mu, G. Yan and J. Shen, *Arch. Biochem. Biophys.*, 2001, **387**, 250; (o) M. Tiecco, *Top. Curr. Chem.*, 2000, **208**, 7; (p) G. Mugesh and H. B. Singh, *Chem. Soc. Rev.*, 2000, **29**, 347; (q) T. Wirth, *Tetrahedron*, 1999, **55**, 1; (r) S. Uemura, *Phosphorus, Sulfur Silicon Relat. Elem.*, 1998, **136–138**, 219; (s) L. A. Wessjohann and U. Sinks, *J. Prakt. Chem.*, 1998, **340**, 189; (t) Y. Nishibayashi and S. Uemura, *Rev. Heteroat. Chem.*, 1996, **14**, 83; (u) M. Tiecco, L. Testaferri, M. Tingoli, L. Bagnoli, F. Marini, C. Santi and A. Temperini, *Gazz. Chim. Ital.*, 1996, **126**, 635; (v) H. J. Reich and S. Wollowitz, *Org. React.*, 1993, **44**, 1; (w) H. J. Reich, *Acc. Chem. Res.*, 1979, **12**, 22.
12. (a) A. Müller, E. Cadenas, P. Graf and H. Sies, *Biochem. Pharmacol.*, 1984, **33**, 3235; (b) A. Wendel, M. Fausel, H. Safayhi, G. Tiegs and R. Otter, *Biochem. Pharmacol.*, 1984, **33**, 3241.
13. (a) B. K. Sharma and G. Mugesh, *Chem. – Eur. J.*, 2008, **14**, 10603; (b) K. P. Bhabak and G. Mugesh, *Chem. – Eur. J.*, 2007, **13**, 4594; (c) B. K. Sharma and G. Mugesh, *J. Am. Chem. Soc.*, 2005, **127**, 11477; (d) T. G. Back and D. P. Dyck, *J. Am. Chem. Soc.*, 1997, **119**, 2079;

(e) P. V. Jacquemin, L. E. Christiaens, M. J. Renson, M. J. Evers and N. Dereu, *Tetrahedron Lett.*, 1992, **33**, 3863; (f) S. R. Wilson, P. A. Zucker, R. R. C. Huang and P. A. Spector, *J. Am. Chem. Soc.*, 1989, **111**, 5936; (g) H. J. Reich and C. P. Jasperse, *J. Am. Chem. Soc.*, 1987, **109**, 5549.

14. (a) S. K. Tripathi, S. Sharma, H. B. Singh and R. J. Butcher, *Org. Biomol. Chem.*, 2011, **9**, 581; (b) W. Hassan, S. Pinton, J. T. da Rocha, A. M. Deobald, A. L. Braga, C. W. Nogueira, A. S. Latini and J. B. T. Rocha, *Chem. Biol. Interact.*, 2011, **190**, 35; (c) K. P. Bhabak and G. Mugesh, *Chem. – Eur. J.*, 2009, **15**, 9846; (d) K. P. Bhabak and G. Mugesh, *Chem. – Eur. J.*, 2008, **14**, 8640; (e) G. Mugesh, A. Panda, H. B. Singh, N. S. Punekar and R. J. Butcher, *J. Am. Chem. Soc.*, 2001, **123**, 839; (f) G. Mugesh, A. Panda, H. B. Singh and R. J. Butcher, *Chem. – Eur. J.*, 1999, **5**, 1411; (g) M. Iwaoka and S. Tomoda, *J. Am. Chem. Soc.*, 1994, **116**, 2557; (h) T. Wirth, *Molecules*, 1998, **3**, 164.
15. C. A. Collins, F. H. Fry, A. L. Holme, A. Yiakouvaki, A. Al-Qenaei, C. Pourzand and C. Jacob, *Org. Biomol. Chem.*, 2005, **3**, 1541.
16. (a) V. K. Jain and K. I. Priyadarsini, *Proc. Natl. Acad. Sci.*, 2010, **80A**, 269; (b) F. Fumakura, B. Mishra, K. I. Priyadarsini and M. Iwaoka, *Eur. J. Org. Chem.*, 2010, 440; (c) H. Johansson, O. Svartström, P. Phadnis, L. Engman and M. K. Ott, *Bioorg. Med. Chem.*, 2010, **18**, 1783; (d) D. Plano, Y. Baquedano, E. Ibanez, I. Jimenez, J. A. Palop, J. E. Spallholz and C. Sanmartin, *Molecules*, 2010, **15**, 7292; (e) P. Belenky, K. L. Bogan and C. Brenner, *Trends Biochem. Sci.*, 2007, **32**, 12; (f) T. G. Back, Z. Moussa and M. Parvez, *Angew. Chem., Int. Ed.*, 2004, **43**, 1268; (g) T. G. Back and Z. Moussa, *J. Am. Chem. Soc.*, 2002, **124**, 12104; (h) Y. Rojanasakul, J. Ye, F. Chen, L. Wang, N. Cheng, V. Castranova, V. Vallyathanand and X. Shi, *Mol. Cell. Biochem.*, 1999, 119.
17. (a) S.-C. Yu, A. Borchert, H. Kuhn and I. Ivanov, *Chem. – Eur. J.*, 2008, **14**, 7066; (b) T. G. Back, D. Kuzma and M. Parvez, *J. Org. Chem.*, 2005, **70**, 9230; (c) S. S. Zade, H. B. Singh and R. J. Butcher, *Angew. Chem., Int. Ed.*, 2004, **43**, 4513; (d) T. G. Back and Z. Moussa, *J. Am. Chem. Soc.*, 2003, **125**, 13455; (e) T. G. Back and Z. Moussa, *J. Am. Chem. Soc.*, 2002, **124**, 12104.
18. (a) C. P. Prabhu, P. P. Phadnis, A. P. Wadawale, K. I. Priyadarsini and V. K. Jain, *J. Organomet. Chem.*, 2012, **713**, 42; (b) J. Rafique, S. Saba, R. F. S. Canto, T. E. A. Frizon, W. Hassan, E. P. Waczuk, M. Jan, D. F. Back, J. B. T. Da Rocha and A. L. Braga, *Molecules*, 2015, **20**, 10095.
19. (a) A. S. Hodage, C. P. Prabhu, P. P. Phadnis, A. Wadawale, K. I. Priyadarsini and V. K. Jain, *J. Organomet. Chem.*, 2012, **720**, 19; (b) V. P. Singh, J.-F. Poon, R. J. Butcher, X. Lu, G. Mestres, M. K. Ott and L. Engman, *J. Org. Chem.*, 2015, **80**, 7385.
20. B. K. Sarma, D. Manna, M. Minoura and G. Mugesh, *J. Am. Chem. Soc.*, 2010, **132**, 5364.

21. (a) D. S. Lamani, D. Bhowmick and G. Mugesh, *Org. Biomol. Chem.*, 2012, **10**, 7933; (b) N. M. R. McNeil, D. J. Press, D. M. Mayder, P. Garnica, L. M. Doyle and T. G. Back, *J. Org. Chem.*, 2016, **81**, 7884.
22. B. J. Bhuyan and G. Mugesh, *Org. Biomol. Chem.*, 2011, **9**, 1356.
23. (a) H. Y. T. Yang, E. G. Erdos and Y. J. Levin, *Pharmacol. Exp. Ther.*, 1971, **177**, 291; (b) E. G. Erdosand and H. Y. T. Yang, *Life Sci.*, 1967, **6**, 569; (c) H. Y. T. Yang and E. G. Erdos, *Nature*, 1967, **215**, 1402.
24. (a) E. D. Sturrock, R. Natesh, J. M. V. Rooyen and K. R. Acharya, *Cell. Mol. Life Sci.*, 2004, **61**, 2677; (b) R. Natesh, S. L. U. Schwager, E. D. Sturrock and K. R. Acharya, *Nature*, 2003, **421**, 551; (c) M. R. W. Ehlers and J. F. Riordan, *Biochemistry*, 1991, **30**, 7118; (d) M. R. W. Ehlers, E. A. Fox, D. J. Strydom and J. F. Riordan, *Proc. Natl. Acad. Sci. U. S. A.*, 1989, **86**, 7741.
25. B. J. Bhuyan and G. Mugesh, *Org. Biomol. Chem.*, 2012, **10**, 2237.
26. J. Wang, B. Zhao, X. Wang, J. Yao and J. Zhang, *Int. J. Biol. Macromol.*, 2012, **51**, 987.
27. P. Merino-Montiel, O. Lopez and J. G. Fernandez-Bolanos, *Tetrahedron*, 2012, **68**, 3591.
28. V. Ulgar, J. G. Fernandez-Bolanos and M. J. Bols, *J. Chem. Soc., Perkin Trans. 1*, 2002, 1242.
29. P. Merino-Montiel, S. Maza, S. Martos, O. Lopez, I. Maya and J. G. Fernandez-Bolanos, *Eur. J. Pharm. Sci.*, 2013, **48**, 582.
30. O. Lopez, S. Maza, V. Ulgar, I. Maya and J. G. Fernandez-Bolanos, *Tetrahedron*, 2009, **65**, 2556.
31. V. Nascimento, N. L. Ferreira, R. F. S. Canto, K. L. Schott, E. P. Waczuk, L. Sancineto, C. Santi, J. B. T. Rocha and A. L. Braga, *Eur. J. Med. Chem.*, 2014, **87**, 131.
32. X. Yang, Q. Wang, Y. Tao and H. Xu, *J. Chem. Res., Synop.*, 2002, **4**, 160.
33. F. N. Victoria, R. Anversa, F. Penteado, M. Castro, E. J. Lenardão and L. Savegnago, *Eur. J. Pharmacol.*, 2014, **742**, 131.
34. F. N. Victoria, C. S. Radatz, M. Sachini, R. G. Jacob, G. Perin, W. P. da Silva and E. J. Lenardao, *Tetrahedron Lett.*, 2009, **50**, 6761.
35. M. Elsherbini, W. S. Hamama, H. H. Zoorob, D. Bhowmick, G. Mugesh and T. Wirth, *Heteroat. Chem.*, 2014, **25**, 320.
36. L. M. Break, M. A. Mohamed and S. H. Abdel-Hafez, *Orient. J. Chem.*, 2014, **30**, 163.
37. M. Ibrahim, W. Hassan, J. Anwar, A. M. Deobald, J. P. Kamdem, D. O. Souza and J. B. T. Rocha, *Toxicology*, 2014, **28**, 524.
38. (a) A. L. Braga, M. W. Paixao and G. Marin, *Synlett*, 2005, 111675; (b) D. Liu, Q. Dai and X. Zhang, *Tetrahedron*, 2005, **61**, 6460.
39. M. Ibrahim, N. Muhammad, M. Naeem, A. M. Deobald, J. P. Kamdem and J. B. T. Rocha, *Toxicology*, 2015, **29**, 947.
40. D. Bhowmick, S. Srivastava, P. D'Silva and G. Mugesh, *Angew. Chem., Int. Ed.*, 2015, **54**, 8449.

41. (a) D. L. Comins and J. D. Brown, *J. Org. Chem.*, 1984, **49**, 1078; (b) D. J. Press, N. M. R. McNeil, M. Hambrook and T. G. Back, *J. Org. Chem.*, 2014, **79**, 9394.
42. P. R. Prasad, H. B. Singh and R. J. Butcher, *J. Organomet. Chem.*, 2016, **814**, 42.
43. J. M. F. Gagan, A. G. Lane and D. Lloyd, *J. Chem. Soc. C*, 1970, 2484.
44. P. R. Prasad, H. B. Singh and R. J. Butcher, *Molecules*, 2015, **20**, 12670.
45. L. Savegnago, A. I. Vieira, N. Seus, B. S. Goldani, M. R. Castro, E. J. Lenardão and D. Alves, *Tetrahedron Lett.*, 2013, **54**, 40.
46. S. F. Fonseca, D. B. Lima, D. Alves, R. G. Jacob, G. Perin, E. J. Lenardao and L. Savegnago, *New J. Chem.*, 2015, **39**, 3043.
47. A. C. Sauer, J. G. Leal, S. T. Stefanello, M. T. B. Leite, M. B. Souza, F. A. A. Soares, O. E. D. Rodrigues and L. Dornelles, *Tetrahedron Lett.*, 2017, **58**, 87.
48. G. Roy and G. Mugesh, *J. Am. Chem. Soc.*, 2005, **127**, 15207.
49. G. Roy, D. Das and G. Mugesh, *Inorg. Chem. Acta*, 2007, **360**, 303.
50. D. Manna and G. Mugesh, *Angew. Chem., Int. Ed.*, 2010, **49**, 9246.
51. E. Block, V. Eswarakrishnan, M. Gernon, G. Ofori-Okai, C. Saha, K. Tang and J. Zubieta, *J. Am. Chem. Soc.*, 1989, **111**, 658.
52. D. Manna and G. Mugesh, *J. Am. Chem. Soc.*, 2011, **113**, 9980.
53. D. Manna and G. Mugesh, *J. Am. Chem. Soc.*, 2012, **134**, 4269.
54. K. Raja and G. Mugesh, *Angew. Chem., Int. Ed.*, 2015, **54**, 7674.
55. S. Mondal and G. Mugesh, *Chem. – Eur. J.*, 2014, **20**, 11120.
56. K. EI-Bayoumy, *Cancer Res.*, 1985, **45**, 3631.
57. B. S. Reddy, S. Sugie, H. Maruyama, K. El-Bayoumy and P. Marra, *Cancer Res.*, 1987, **47**, 5901.
58. E. S. Fiala, M. E. Staretz, G. A. Pandya, K. El-Bayoumy and S. R. Hamilton, *Carcinogenesis*, 1998, **19**, 557.
59. B. S. Reddy, A. Rivenson, N. Kulkarni, P. Upadhyaya and K. EI-Bayoumy, *Cancer Res.*, 1992, **52**, 5635.
60. T. Tanaka, H. Kohno, M. Murakami, S. Kagami and K. El-Bayoumy, *Cancer Res.*, 2000, **60**, 3713.
61. T. Tanaka, H. Makita, K. Kawabata, H. Mori and K. E1-Bayoumy, *Cancer Res.*, 1997, **57**, 3644.
62. M. D. Short, Y. Xie, L. Li, P. B. Cassidy and J. C. Roberts, *J. Med. Chem.*, 2003, **46**, 3308.
63. L. Zhao, J. Li, Y. Li, J. Liu, T. Wirth and Z. Li, *Bioorg. Med. Chem.*, 2012, **20**, 2558.
64. S. V. Amosova, N. A. Makhaeva, A. V. Martynov, V. A. Potapov, B. R. Steele and I. D. Kostas, *Synthesis*, 2005, **10**, 1641.
65. L. Hu, Z. Chen, S. Lu, X. Li, Z. Liu and H. Xu, *Phosphorus, Sulfur Silicon Relat. Elem.*, 2004, **179**, 1065.
66. S. S. Roy, P. Chakraborty, J. Biswas and S. Bhattacharya, *Biochimie*, 2014, **105**, 137.

67. S. S. Roy, P. Ghosh, S. U. Hossain, P. Chakraborty, J. Biswas, S. Mandal, A. Bhattacharya and S. Bhattacharya, *Bioorg. Med. Chem. Lett.*, 2010, **20**, 6951.
68. C. Kim, S.-B. Kim and M.-S. Park, *Arch. Pharmacal Res.*, 2014, **37**, 452.
69. S. Shaabana, M. A. Arafata, H. E. Gafferb and W. S. Hamama, *Pharma Chem.*, 2014, **6**, 186.
70. D. R. Garud, M. Koketsu and H. Ishihara, *Molecules*, 2007, **12**, 504.
71. A. Krief, W. Dumont and C. Delmotte, *Angew. Chem., Int. Ed.*, 2000, **39**, 1669.
72. Q. Guan, F. Yang, D. Guo, J. Xu, M. Jiang, C. Liu, K. Bao, Y. Wu and W. Zhang, *Eur. J. Med. Chem.*, 2014, **87**, 1.
73. M. Wu, Q. Sun, C. Yang, D. Chen, J. Ding, Y. Chen, L. Lin and Y. Xie, *Bioorg. Med. Chem. Lett.*, 2007, **17**, 869.
74. E. Napolitano, R. Fiaschi and Marsili, *Gazz. Chim. Ital.*, 1988, **118**, 415.
75. E. H. G. da Cruz, M. A. Silvers, G. A. M. Jardim, J. M. Resende, B. C. Cavalcanti, I. S. Bomfim, C. Pessoa, C. A. de Simone, G. V. Botteselle, A. L. Braga, D. K. Nair, I. N. N. Namboothiri, D. A. Boothman and E. N. da Silva, *Eur. J. Med. Chem.*, 2016, **122**, 1.
76. (a) V. V. Rostovtsev, G. L. Green, V. V. Fokin and K. B. Sharpless, *Angew. Chem., Int. Ed.*, 2002, **41**, 2596; (b) H. C. Kolb, M. G. Finn and K. B. Sharpless, *Angew. Chem. Int. Ed.*, 2001, **40**, 2004.
77. T. T. Guimaraes, M. C. F. R. Pinto, J. S. Lanza, M. N. Melo, R. L. Monte-Neto, I. M. M. de Melo, E. B. T. Diogo, V. F. Ferreira, C. A. Camara, W. O. Valença, R. N. de Oliveira, F. Frezard and E. N. da Silva, *Eur. J. Med. Chem.*, 2013, **63**, 523.
78. P. Guo, P. Zhao, J. Liu, H. Ma, J. Bai, Y. Cao, Y. Liu, H. He and C. Qi, *Biol. Trace Elem. Res.*, 2013, **151**, 301.
79. I. L. Martins, J. P. Miranda, N. G. Oliveira, A. S. Fernandes, S. Gonçalves and A. M. M. Antunes, *Molecules*, 2013, **18**, 5251.
80. Y. Zhang, S. Zheng, S.-M. Ngai, W. Zheng, J. Li, T. Chen and X. Zhong, *Chem. Pharm. Bull.*, 2014, **62**, 994.
81. E. D. Paskett, J. A. Dean, J. M. Oliveri and J. P. Harrop, *J. Am. Soc. Clin. Oncol.*, 2012, **30**, 3726.
82. C. Pfister, H. Dawzcynski and F.-J. Schingale, *J. Trace Elem. Med. Biol.*, 2016, **37**, 111.
83. F. Mao, J. Chen, Q. Zhou, Z. Luo, L. Huang and X. Li, *Bioorg. Med. Chem. Lett.*, 2013, **23**, 6737.
84. Z. Luo, L. Liang, J. Sheng, Y. Pang, J. Li, L. Huang and X. Li, *Bioorg. Med. Chem.*, 2014, **22**, 1355.
85. (a) Z. Wang, Y. Wang, W. Li, F. Mao, Y. Sun, L. Huang, X. Li, *ACS Chem. Neurosci.*, 2014, **5**, 952; (b) R. F. S. Canto, F. A. R. Barbosa, V. Nascimento, A. S. de Oliveira, I. M. C. Brighente and A. L. E. Braga, *Org. Biomol. Chem.*, 2014, **12**, 3470; (c) F. A. R. Barbosa, R. F. S. Canto, S. Saba, J. Rafique and A. E. Braga, *Bioorg. Med. Chem.*, 2016, **24**, 5762.

86. (a) N. B. Patel, J. C. Patel and S. H. Modi, *J. Saudi Chem. Soc.*, 2011, **15**, 167; (b) U. K. Mazumder, M. Gupta, S. Bhattacharya, S. S. Karki, S. Rathinasamy and S. S. J. Thangavel, *J. Enzyme Inhib. Med. Chem.*, 2004, **19**, 185.
87. P. M. Radhakrishna, K. C. Sharadamma, H. M. Vagdevi, P. M. Abhilekha, S. R. Mubeen and K. Nischal, *Int. J. Chem.*, 2010, **2**, 149.
88. A. A. Jadhav, V. P. Dhanwe, P. G. Joshi and P. K. Khanna, *Chem. Heterocycl. Compd.*, 2015, **51**, 102.
89. M. Al-Smadi and S. Ratrout, *Molecules*, 2004, **9**, 957.
90. Z. S. Talas, Y. Gok, I. Ozdemir, B. Ates, S. Gunal and I. Yilmaz, *Pak. J. Pharm. Sci.*, 2015, **28**, 611.
91. S. Kumar, N. Sharma, I. K. Maurya, A. K. K. Bhasin, N. Wangao, P. Brandao, V. Felix, K. K. Bhasin and R. K. Sharma, *Eur. J. Med. Chem.*, 2016, **123**, 916.
92. L. Sancineto, M. Piccioni, S. D. Marco, R. Pagiotti, V. Nascimento, A. L. Braga, C. Santi and D. Pietrella, *BMC Microbiol.*, 2016, **16**, 220.
93. D. F. Smee, J. H. Hutfman, L. L. Hall, J. W. Huggins and R. W. Sidwell, *Antiviral Chem. Chemother.*, 1990, **1**, 211.
94. S. Mukherjee, W. S. Weiner, C. E. Schroeder, D. S. Simpson, A. M. Hanson, N. L. Sweeney, R. K. Marvin, J. Ndjomou, R. Kolli, D. Isailovic, F. J. Schoenen and D. N. Frick, *ACS Chem. Biol.*, 2014, **9**, 2393.
95. P. K. Sahu, G. Kim, J. Yu, J. Y. Ahn, J. Song, Y. Choi, X. Jin, J.-H. Kim, S. K. Lee, S. Park and L. S. Jeong, *Org. Lett.*, 2014, **16**, 5796.
96. L. S. Jeong, D. K. Tosh, H. O. Kim, T. Wang, X. Hou, H. S. Yun, Y. Kwon, S. K. Lee, J. Choi and L. X. Zhao, *Org. Lett.*, 2008, **10**, 209.
97. J. Yu, J.-H. Kim, H. W. Lee, V. Alexander, H.-C. Ahn, W. J. Choi and L. S. Jeong, *Chem. – Eur. J.*, 2013, **19**, 5528.
98. (a) P. K. Sahu, T. Umme, J. Yu, A. Nayak, G. Kim, M. Noh, J.-Y. Lee, D.-D. Kim and L. S. Jeong, *J. Med. Chem.*, 2015, **58**, 8734; (b) L. Sancineto, A. Mariotti, L. Bagnoli, F. Marini, J. Desantis, N. Iraci, C. Santi, C. Pannecouque and O. Tabarrini, *J. Med. Chem.*, 2015, **58**, 9601; (c) J. A. Thomas and R. J. Gorelick, *Virus Res.*, 2008, **134**, 39.
99. (a) F. Borges, E. Fernandes and F. Roleira, *Curr. Med. Chem.*, 2002, **9**, 195; (b) M. Ferrari, M. Sgobba, M. C. Gamberini and G. Rastelli, *Eur. J. Med. Chem.*, 2007, **42**, 1028.
100. P. Pacher, A. Nivorozhkin and C. Szabo, *Pharmacol. Rev.*, 2006, **58**, 87.
101. Q. Guan, Z. Cheng, X. Ma, L. Wang, D. Feng, Y. Cui, K. Bao, L. Wu and W. Zhang, *Eur. J. Med. Chem.*, 2014, **85**, 508.
102. B. K. Sharma and G. Mugesh, *Chem. – Eur. J.*, 2008, **14**, 10603.
103. C. W. Nogueira, E. B. Quinhones, E. A. C. Jung, G. Zeni and J. B. T. Rocha, *Inflammation Res.*, 2003, **52**, 56.
104. P. M. Chagas, S. G. Rosa, M. H. M. Sari, C. E. S. Oliveira, R. F. S. Canto, S. C. A. da Luz, A. L. Braga and C. W. Nogueira, *Pharmacol., Biochem. Behav.*, 2014, **118**, 87.

105. M. Pinz, A. S. Reis, V. Duarte, M. J. da Rocha, B. S. Goldani, D. Alves, L. Savegnago, C. Luchese and E. A. Wilhelm, *Eur. J. Pharmacol.*, 2016, **780**, 122.
106. F. Donato, M. G. D. Gomes, A. T. R. Goes, N. Seus, D. Alves, C. R. Jesse and L. Savegnago, *Life Sci.*, 2013, **93**, 393.
107. G. R. M. Dias, T. M. D. Almeida, J. H. Sudati, F. Dobrachinski, S. Pavin, F. A. A. Soares, C. W. Nogueira and N. B. V. Barbosa, *Physiol. Behav.*, 2014, **124**, 116.

CHAPTER 4

The Relevant Chemistry of Imidazoline-2-selone Donors with Potential Biological Applications

V. LIPPOLIS

Università degli Studi di Cagliari, Dipartimento di Scienze Chimiche e Geologiche, S.S. 554 Bivio per Sestu, Monserrato (CA) 09042, Italy
Email: lippolis@unica.it

4.1 Introduction

Over the past two decades, thio-amide and seleno-amide compounds, in particular imidazoline-2-chalcogenone derivatives, have received considerable attention, either as potential anti-thyroid drugs in the treatment of hyperthyroidism[1–4] or as potential anti-oxidants for preventing metal-mediated oxidative damage of DNA.[5–7]

The chemical structure of the most commonly used anti-thyroid drugs are shown in Figure 4.1, together with the thyroid hormones **T4** (L-thyroxine) and **T3** (3,3′,5-triiodo-L-thyronine).

1-Methyl-2-mercaptoimidazole (**1**, MMI) and 6-*n*-propyl-2-thiouracil (**5**, PTU) are used clinically in the United States in the treatment of hyperthyroidism (overproduction of thyroid hormones **T4** and **T3**), while **1** and its analogue carbimazole (**7**, CBZ) are used in most European and Asian countries. The mechanism of action of these drugs is not completely

Organoselenium Compounds in Biology and Medicine: Synthesis, Biological and Therapeutic Treatments
Edited by Vimal Kumar Jain and K. Indira Priyadarsini

Published by the Royal Society of Chemistry, www.rsc.org

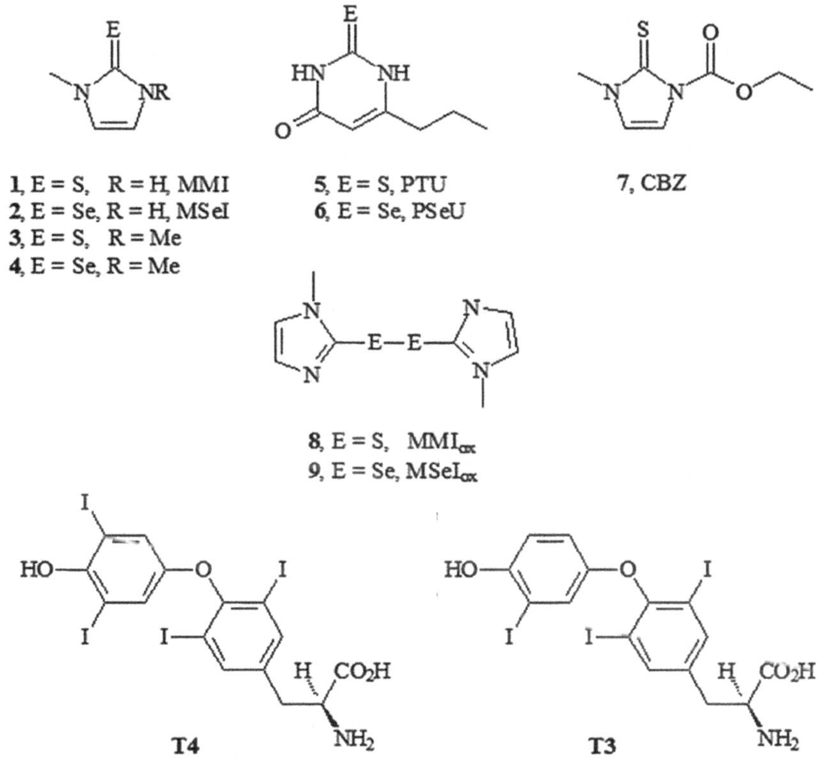

Figure 4.1 Thiourea-based anti-thyroid drugs, their selenium analogoues and thyroid hormones **T4** and **T3**.

understood. **T4**, the main secretory product of the thyroid gland, is produced *via* iodination of tyrosyl residues on thyroglobulin by the thyroid peroxidase (TPO)/H_2O_2/I^- system. Mono deiodination of **T4** by iodothyronine deiodinase affords the biologically active hormone **T3**. In the initially proposed general mode of action, **1** and **5** would reduce the concentration of oxidized iodide produced by the TPO/H_2O_2 system with the formation of hypothetical sulfenyl iodide derivatives evolving towards other metabolites (disulfides, sulfenic acid and desulfurated molecules), and thus diverting the oxidized iodide from the natural substrate (thyroglobulin-linked tyrosyl residues).[8–11] As most investigations confirm that **1** and **5** irreversibly inhibit TPO, even in the presence of excess iodide, another proposed mechanism that would block the biosynthesis of thyroid hormones involves the binding of these anti-thyroid drugs or metabolites derived from the reaction of them with oxidized iodide, to the heme centre of TPO.[2,12,13] Other experimental evidence and model studies suggest that **5**, in contrast to **1** (showing irreversible effects only on the TPO-catalyzed iodination), can also inhibit one of the iodothyronine deiodinase enzymes present in the thyroid gland (in particular ID-I), thus reducing the conversion of **T4** to **T3** by forming a stable selenylsulfide with the seleno-cysteine residue of the enzyme.[14,15]

Interestingly, the selenium analogues of thiourea-based anti-thyroid drugs appear to have a different mechanism of action in inhibiting the thyroid activity. In iron-containing lactoperoxidase (LPO) inhibition experiments *in vitro* (phosphate buffer, pH = 7.4), it has been found that compound **2** (MSeI, Figure 4.1),[16] and also *N,N′*-disubstituted imidazoline-2-selone derivatives[3,17] inhibit the catalytic activity of the enzyme by reducing the H_2O_2 required for the oxidation of the iron centre in LPO. In contrast to **1**, in the case of **2**, the LPO activity can be recovered by increasing the concentration of H_2O_2, and in the presence of glutathione (GSH), **2** can constitute a redox cycle involving the catalytic reduction of H_2O_2, thus mimicking the glutathione peroxidase (GPx) activity *in vitro* (with 80 times higher activity than ebselen). These results, justified by the fact that selenium compounds are more susceptible to oxidation than their sulfur analogues, indicate that selenium-based anti-thyroid drugs may also act better as anti-oxidants and protect thyroid cells from oxidative damage by controlling the H_2O_2 level and scavenging reactive oxygen species (ROS).

It is relevant to observe that the treatment of compound **10** with the LPO/H_2O_2/I^- system (LPO has been shown to behave very similarly to TPO toward iodination of thyroglobulin and other iodide acceptors) produces the diselenide dication **11** as confirmed by X-ray analysis, while the reaction with I_2 in CH_2Cl_2 produces the T-shaped hypervalent adduct **12** (Figure 4.2).[3] Therefore, *in vivo*, beside the possibility that thiourea-based anti-thyroid drugs inhibit the biosynthesis of thyroid hormones through the diversion of oxidized iodide, there is the possibility that their selenium analogues react with either I_2 or activated iodine (I^+ produced by the TPO/H_2O_2/I^- system) to form different metabolites. The necessity of identifying the species that are produced in the reaction of anti-thyroid drugs with the TPO/H_2O_2/I^- system within the thyroid gland has stimulated the study of the reactivity of thioamides and selenoamide compounds, in particular imidazoline-2-chalcogenone derivatives, towards dihalogens and inter-halogens (see later).

Interest in thione (C=S) and selone (C=Se) compounds, in particular imidazoline-2-chalcogenone derivatives as anti-oxidants comes from the properties of naturally occurring thiol- and seleno-histidines such as ergothioneine (**17**) and selenoneine (**18**) (Figure 4.3), which are structural analogues of the more common selenium-containing amino acids cysteine (**13**), selenocysteine (**14**), methionine (**15**) and selenomethionine (**16**).

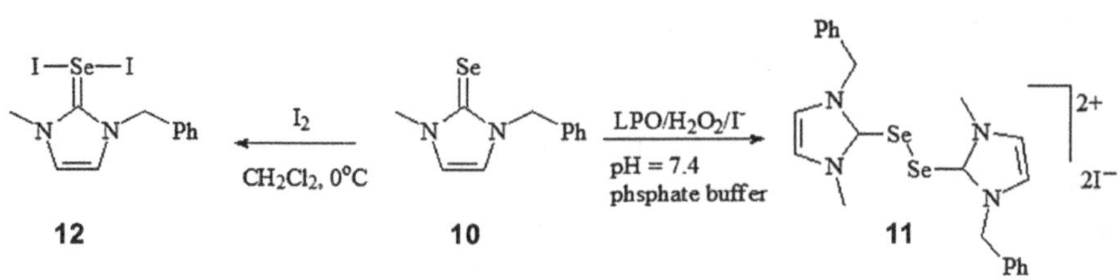

Figure 4.2 Reactivity of *N,N′*-disubstituted imidazoline-2-selone **10** with I_2 and the LPO/H_2O_2/I^- system.[3]

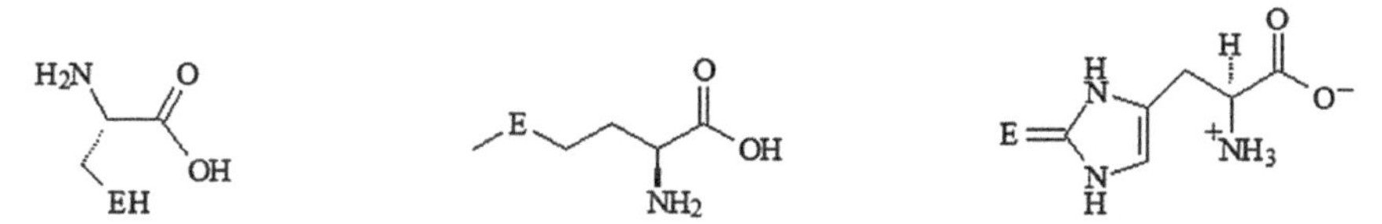

13, E = S, Cysteine
14, E = Se, Selenocysteine

15, E = S, Methionine
16, E = Se, Selenomethionine

17, E = S, Ergothioneine
18, E = Se, Selenoneine

Figure 4.3 Ergothioneine and selenoneine compared to other common sulfur and selenium compounds in biological systems.

In fact, these compounds seem to have multifunctional anti-oxidant abilities and mechanisms of action,[17–20] including reduction of copper-mediated oxidative damage of DNA caused *via* metal binding.[6,21,22] For this reason, compounds **3** and **4** and their analogues have been extensively studied in their coordination to transition metal ions,[23] and in particular to Cu^{2+}.

4.2 General Reactivity of Heterocyclic Pentatomic thio- and Seleno-amides Towards Dihalogens and Inter-halogens

As mentioned in the Introduction, the molecular interactions of anti-thyroid drugs with I_2 have been the subject of many investigations, because of their possible implications in the mechanism of action of these drugs *in vivo*.

It is well known that S/Se molecular donors generally form in solution 1 : 1 charge-transfer (C.T.) adducts with I_2, whose formation constants can be easily measured spectrophotometrically due to the red-shift of the absorption band of free I_2 upon interaction with a donor.[24]

Data reported in Figure 4.4[25] clearly indicate that, in general, (C=Se) selone compounds are stronger donors towards I_2 than the corresponding (C=S) thione analogues. The presence of two nitrogen atoms in the pentatomic ring makes the donor stronger, compared to the presence of only one nitrogen atom, and an increase in the formation constant for the corresponding 1 : 1 I_2 C.T. adducts is observed going from benzoimidazole systems to imidazolidine analogues, and finally to imidazoline derivatives.

The methylated analogues of MMI (**1**) and MSeI (**2**), **3** and **4**, respectively, are the strongest donors among the compounds considered. This also follows an increase of the calculated negative charge on the chalcogen atom going from **23**, **24** to **25**, **26** and to **3**, **4**.

The equilibrium reaction between MMI (**1**) and I_2 in CH_2Cl_2 has recently been reinvestigated by Laurence *et al.*,[26] who calculated that the resulting 1 : 1 I_2 C.T. adduct of the thione tautomer of **1** is favored by 13.2 kJ mol^{-1} compared to that of the thiol tautomer. The formation constant at 25 °C for the $\mathbf{1}\cdot\mathbf{I_2}$ C.T. adduct in CH_2Cl_2 is reported to be 84 700 or 92 400 L mol^{-1} when the influence of **1** self-association is also taken into consideration.[26]

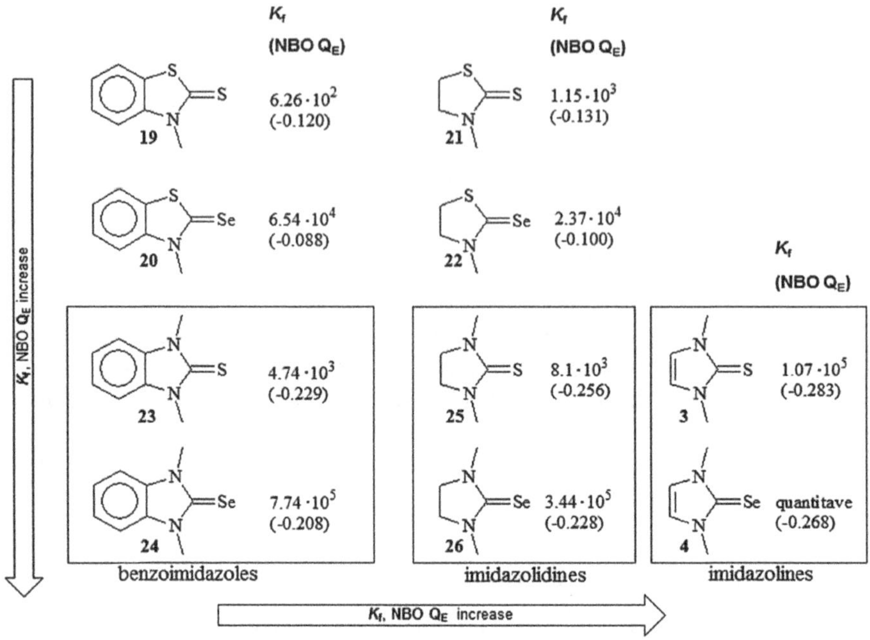

Figure 4.4 Measured formation constants (K_f, [L mol^{-1}]) of 1 : 1 I_2 charge-transfer (C.T.) adducts (CH_2Cl_2, 25 °C) for the pentatomic heterocyclic thio- and seleno-amides donors: *N*-methylbenzothiazole-2-chalcogenones (**19**, **20**), *N*-methylthioimidazolidine-2-chalcogenones (**21**, **22**), *N,N′*-dimethylbenzoimidazole-2-chalcogenones (**23**, **24**), *N,N′*-dimethylimidazolidine-2-chalcogenones (**25**, **26**), *N,N′*-dimethylimidazoline-2-chalcogenones (**3**, **4**)[25] and calculated natural bond orbital (NBO) charges on the chalcogen atoms (Q_E, [e]).[25]

Both values locate **1** among the strongest S-donors toward I_2 (Figure 4.4) and this can explain why **1** could act as an efficient "I_2 sponge" in the thyroid gland. Analogously, the value of the formation constant for the 1 : 1 I_2 C.T. adduct of the Se analogue of MMI, **2**, is estimated to be of the order of magnitude of 10^6.[27] All this is in agreement with the higher polarizability of Se as compared with S, and with a better match between the energies of the Se lone pair and the σ^* orbital of I_2. Imidazoline-2-selones give more stable adducts with I_2 than the corresponding sulfur isologues. The very good donor properties of imidazoline-2-chalcogenone ligands towards I_2 can also be deduced from the observed linear relationship between the lnK_f and the ionization energy of the lone pair of the donor atom calculated from UV photoelectron spectra.[28] The lowest ionization energy among the considered compounds is measured for **4** followed by **2** and their S-isologues.

The take-home message from solution studies on the interaction between chalcogen donor molecules and I_2 is that in general Se-donors bind I_2 more strongly than the corresponding S-isologues; thiones and selones are stronger donors than sulfides and selenides, but more importantly the values of the formation constants of the 1 : 1 I_2 C.T. adducts can spread over a wide range, depending strongly on the chemical surroundings of the donor atom. Small variations on the chemical environment of the chalcogen donor atom can have huge effects on the formation constant of the corresponding I_2 C.T. adducts. Considering thione and selone ligands, the best donors are thio- and seleno-amides, in particular imidazoline-2-chalcogenone derivatives.[24]

The nature of the products obtainable at the solid state strongly depend on the identity of the chalcogen atom (S or Se) and on its chemical surroundings.[29] Diorganosulfur and selenium compounds (R_2E, E = S, Se) generally give 1 : 1 C.T. spoke adducts and T-shaped hypervalent adducts from the reaction with dihalogens and inter-halogens.

In Figure 4.5 the X-ray the crystal structure of the products isolated from the reaction of representative R_2E (E = S, Se) donor molecules with with I_2, IBr and Br_2 are shown, together with some relevant structural features. The greater lengthening of the halogen–halogen bond observed in the case of selenium donors demonstrates once again the stronger donor (Lewis base) properties of these compounds towards dihalogens and inter-halogens as compared to sulfur donors. Furthermore, the shortening of the chalcogen–halogen bond observed on going from I_2 to IBr and to Br_2 indicates an increase of the acceptor (Lewis acid) properties of the dihalogens in the order $I_2 < IBr < Br_2$. Selenium donors with Br_2 always give T-shaped adducts (see later).

The maximum structural diversity for the isolated products at the solid state is observed with thioamides and selenoamides, in particular heterocyclic pentatomicchalcogenoamides.[29,35] The nature of the products depends on the acid/base properties of the reagents and on the experimental conditions used and the reaction molar ratios. Just considering I_2 as reagent

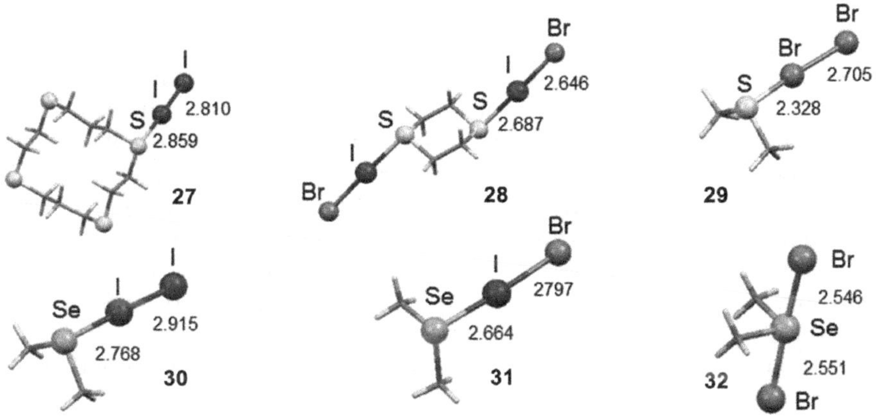

Figure 4.5 Structural features of the products isolated from the reaction of representative diorganosulfur or selenium compounds with I_2, IBr and Br_2.[30–34] All reported bond distances are in Å.

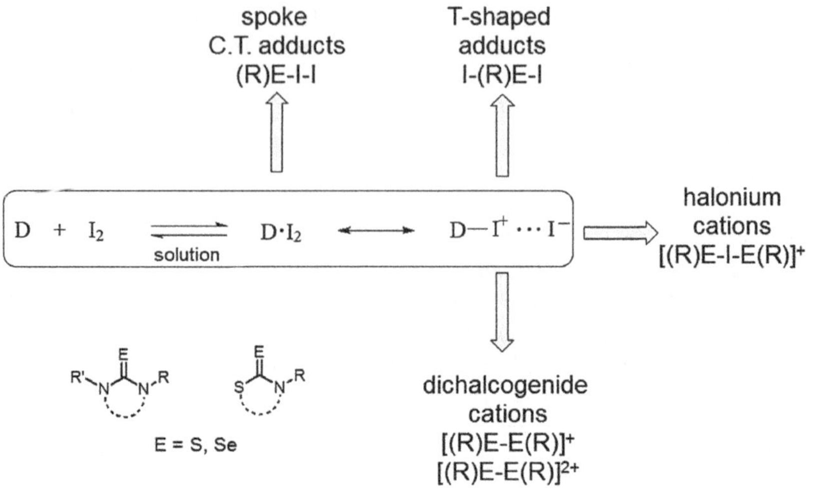

Figure 4.6 Schematic representation of the main type of products obtainable at the solid state from the reaction of I_2 with heterocyclic thioamides and selenoamides.

(Figure 4.6) beside C.T. and T-shaped adducts, oxidation products of the donors can form at the solid state such as dichalcogenide cations $[(R)E\text{-}E(R)]^{n+}$ (R = organic framework; E = S, Se; $n = 1, 2$) featuring a chalcogen–chalcogen bond or halonium complex cations $[(R)E\text{-}I\text{-}E(R)]^{+}$ (E = S, Se) featuring an almost linear E-I-E framework, balanced by polyhalides of different complexity. The polarization of the halogen–halogen bond in the initially formed C.T. adducts in solution is often considered at the base of this reactivity.

As an example of the richness of this chemistry, all products isolated and structurally characterized starting from *N*-methylbenzothiazole-2-thione (**19**) and *N*-methylbenzothiazole-2-selone (**20**) are schematically shown in Figure 4.7.[36–42] Apart from a few exceptions, the products isolated from the reaction of **19** with dihalogens and inter-halogens are different from those isolated from the reactions of **20**, despite using the same dihalogen and experimental conditions. Interestingly, no dichalcogenide cations are obtained with these donors; the reaction with Br_2 affords T-shaped adducts in both cases. With the other dihalogens, the formation of 1 : 1 spoke adducts and extended spoke adducts is observed. Halonium complexes are formed with both ligands by reaction with I_2, and a rare case of desulfuration is observed from the reaction of **19** with excess of ICl.[39]

A similar dechalcogenation is observed in the reaction of *N*-methylthiazolidine-2-selone (**22**) with Br_2 in a 1 : 2 molar ratio (Figure 4.8).[41]

The reaction of **22** with I_2 and IBr affords the corresponding 1 : 1 C.T. adducts at the solid state,[42] similar to that observed in the case of the S-donor isologue.[44]

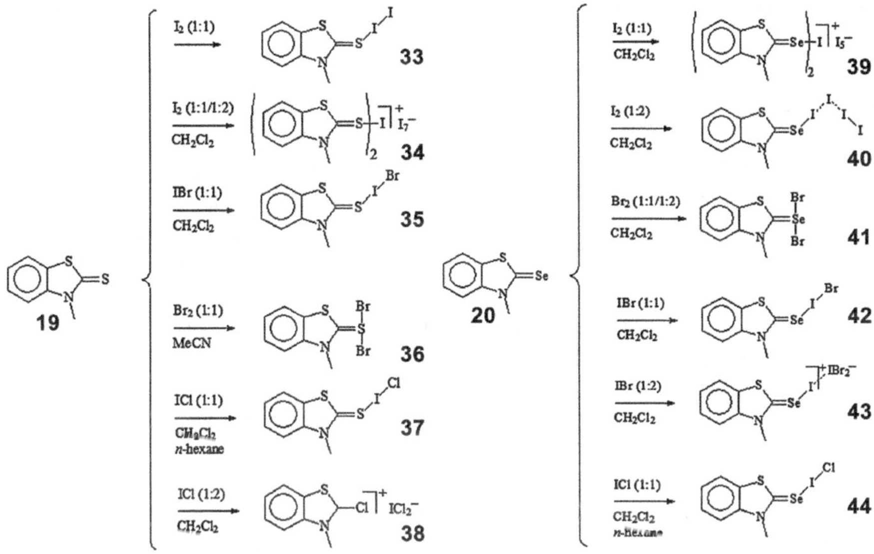

Figure 4.7 Schematic representation of the compounds obtained at the solid state from the reactions of **19** and **20** with dihalogens and inter-halogens.[36–42] This scheme, like the others of the same kind reported in this chapter are intended purely as an illustration of the various compounds, with no implications of the nature of the chemical bonds involved.[43]

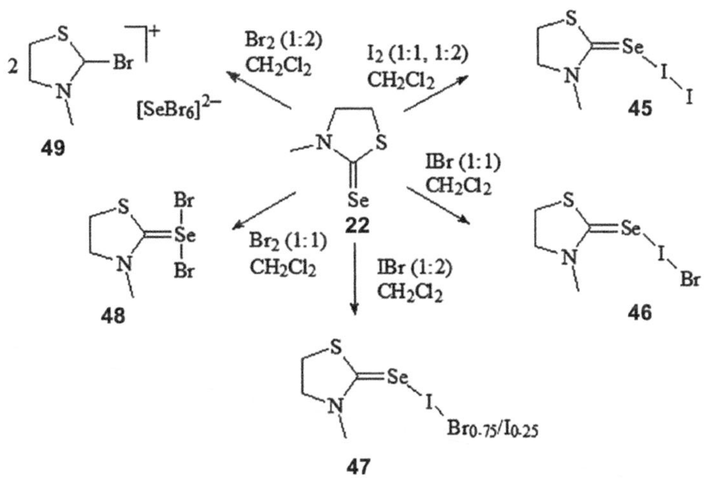

Figure 4.8 Schematic representation of the compounds obtained at the solid state from the reactions of **22** with dihalogens and inter-halogens.[41,42]

The structural features of the 1 : 1 C.T. spoke adducts of donors such as **19**, **20** and **22** are similar to those observed in the case of diorgano R_2E ($E = S$, Se) donors, with selenoamides being stronger donors than the sulfur analogues, thus causing a greater lengthening of the halogen–halogen bond in the corresponding 1 : 1 C.T. spoke adducts.

These kinds of heterocyclic thioamides and selenoamides can also afford at the solid state extended spoke adducts in which a second dihalogen molecule interacts with the first molecule bound to the chalcogen atom (see Figure 4.9a for the 1 : 2 extended I_2 spoke adduct **51** of benzothiazole-2-thione (**50**)).[45] In the case of **20**, following the sequence of the halogen–halogen bond distances in the isolated 1 : 2 extended I_2 spoke adduct **40** (Figure 4.9b), we can consider this compound to be formed by an iodide ion, I^-, interacting with a selenyl iodide cation and an I_2 molecule. The corresponding 1 : 2 extended IBr spoke adduct **43** (Figure 4.9c) can be seen as a selenyl iodide cation interacting with a trihalide species IBr_2^-.[42] These type of halogen-rich compounds are also formed at the solid state with other strong acyclic chalcogen donors such as thiourea **52** (Figure 4.9d, compound **53**)[46] or trialkylphosphane selenides and sulfides.[47]

Beside C.T. adducts, the reaction of dihalogens, in particular I_2 with benzothiazole-2-thione and benzothiazole-2-selenone derivatives can also afford iodonium complexes (**34**, **39** in Figure 4.7) featuring a linear E-I-E framework (E = S, Se). This kind of reactivity is more common among thione (C=S) donors such as **19**,[37] thiourea **52**,[46,48] benzimidazole-2-thione[36] and thiazolidine-2-thione[45] than among selone (CSe) donors. Compound **39** (Figure 4.7) was the first reported iodonium complex of a selenium donor.[37]

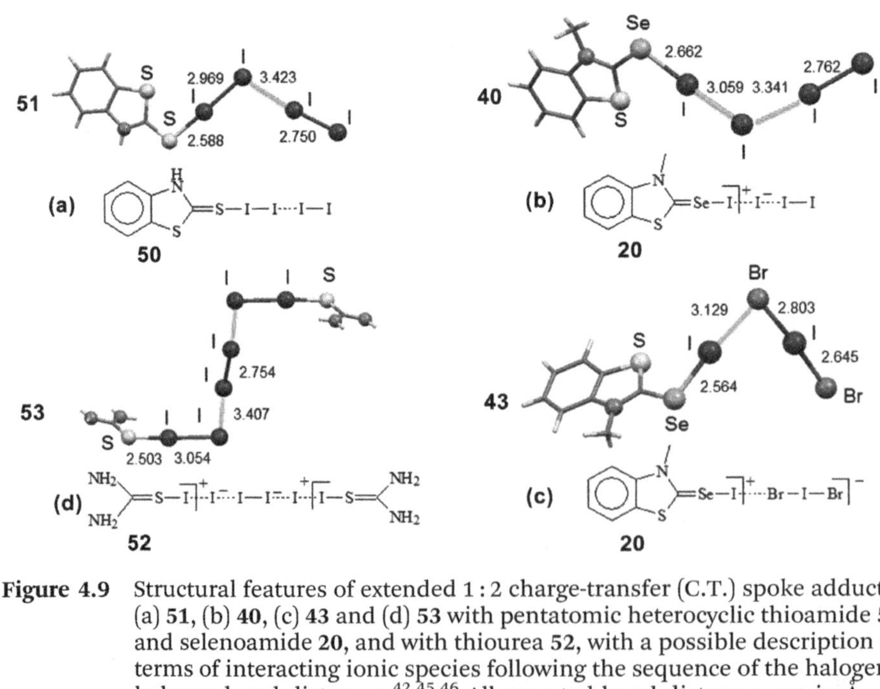

Figure 4.9 Structural features of extended 1 : 2 charge-transfer (C.T.) spoke adducts: (a) **51**, (b) **40**, (c) **43** and (d) **53** with pentatomic heterocyclic thioamide **50** and selenoamide **20**, and with thiourea **52**, with a possible description in terms of interacting ionic species following the sequence of the halogen–halogen bond distances.[42,45,46] All reported bond distances are in Å.

Singh and co-workers have reported the synthesis of iodonium complexes of dialkylbenzoimidazole-2-selone derivatives **54** and **55** starting from their T-shaped I_2 adducts **56** and **57**[49] and oxidizing them with $AgBF_4$ (Figure 4.10).[50]

Considering chalcogen donors **19–24** as the most representative of their respective classes of pentatomic heterocyclic thioamides and selenoamides, we can say that on the basis of the data so far available the reaction with Br_2 always brings T-shaped adducts,[29,35,51,52] while, to a lesser extent, reactions with I_2 and inter-halogens generally afford C.T. spoke adducts and iodonium complexes (Figure 4.11).[29,35]

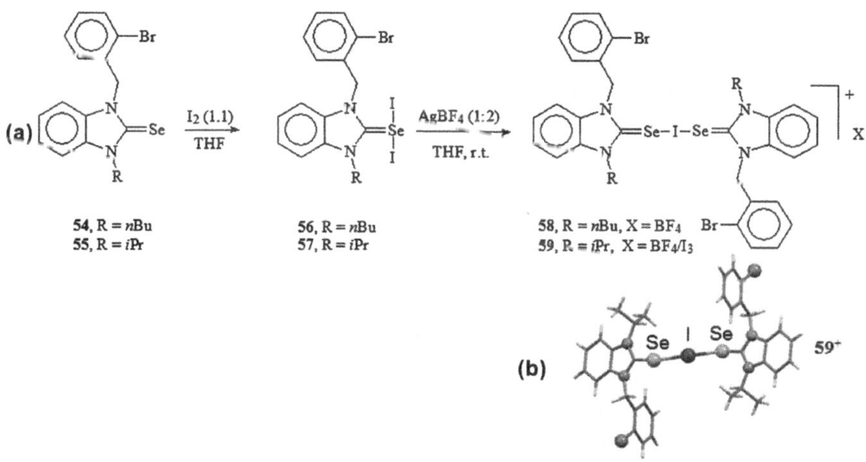

Figure 4.10 (a) Reactivity of dialkylbenzoimidazole-2-selone derivatives **54** and **55** with I_2 and formation of iodonium complexes **58** and **59** following the oxidation of T-shaped I_2 adducts **56** and **57**, respectively, with $AgBF_4$; (b) X-ray crystal structure of the iodonium complex cation $\mathbf{59}^+$.[50]

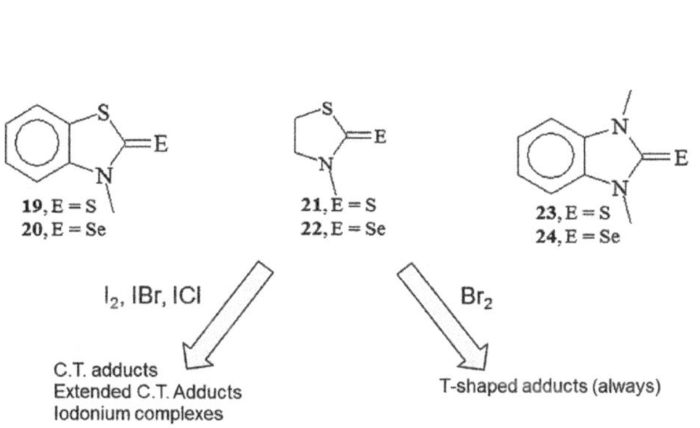

Figure 4.11 Schematic representation of the main type of products obtainable at the solid state from the reaction of dihalogens and inter-halogens with benzothiazole-2-chalcogenones **19**, **20**, thiazolidine-2-chalcogenones **21**, **22** and benzoimidazole-2-chalcogenones **23**, **24**.

Only structural analogues of **24** bearing different alkyl substituents have been proven to afford T-shaped adducts from the reaction with I_2.[49] Interestingly, halonium complexes featuring a central Br or Cl atom are still unknown, as are T-shaped adducts of S-donors with iodine or inter-halogens, which therefore still represent a synthetic challenge.

The formation at the solid state of disulfide or diselenide cations $[(R)E\text{-}E(R)]^{n+}$ (E = S, Se; $n = 1, 2$) from the direct reaction of these donors with dihalogens and inter-halogens has not yet been observed. Recently, Singh and co-workers obtained the first examples of diselenide dications from the classes of pentatomic heterocyclic donors reported in Figure 4.11, by reacting transition metals, in particular Fe, as powder with T-shaped dihalogen adducts of dialkylbenzoimidazole-2-selone derivatives.[53]

Moving to the right on the scale of donor strength towards I_2, defined in Figure 4.4, and considering *N,N′*-dimethylimidazolidine-2-selone **26** as a donor, we start observing the prevailing formation of diselenide dications following the reaction with dihalogens and inter-halogens (Figure 4.12).

Interestingly, the reaction of **26** with I_2 in CCl_4 affords the unique mixed valence compound **62** in which the C.T. I_2 adduct of **26** and its diselenide dication balanced by I_3^- ions co-exist in the same crystal and are connected *via* soft–soft Se···I interactions.[55]

A clear preference to form T-shaped adducts and diselenide dications from the reaction with dihalogens and inter-halogens is shown by *N,N′*-dimethyl-4-imidazoline-2-selene (**4**) (Figure 4.13). As expected, **4** reacts with Br_2 to give the T-shaped adduct **64**. However, the T-shaped adduct **66** is obtained from the reaction with I_2. In contrast, the reactions with IBr and ICl afford in both cases diselenide dications balanced by halogenide anions which strongly interact with the selenium atoms to give 3c,4e (**65**) or 4c,6e (**63**) chalcogen–halogen frameworks.

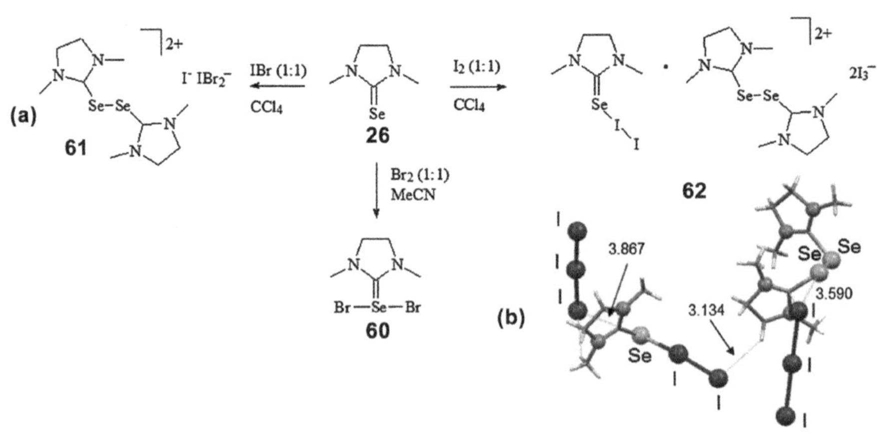

Figure 4.12 (a) Schematic representation of the compounds obtained at the solid state from the reactions of **26** with dihalogens and inter-halogens;[25,54,55] (b) structural features of the compound $(\mathbf{26}\cdot I_2)(\mathbf{26}_2)^{2+}2I_3^-$ (**62**).[55] All reported bond distances are in Å.

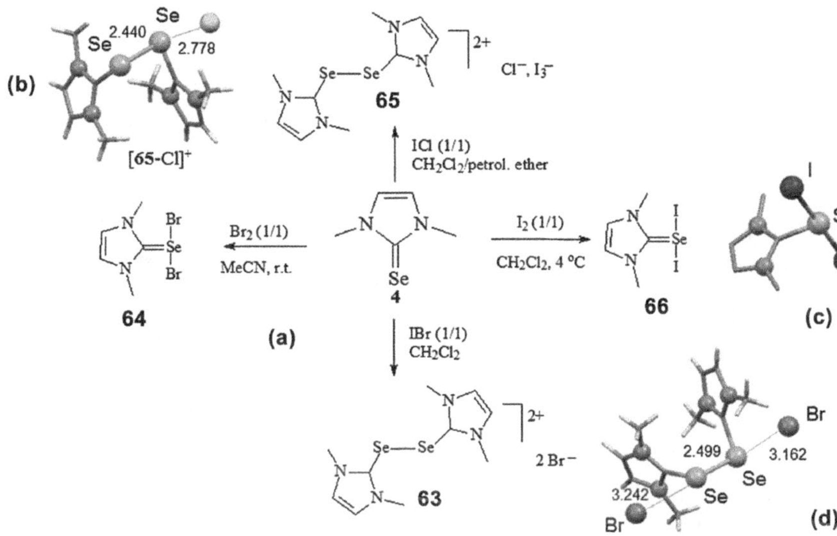

Figure 4.13 (a) Schematic representation of the compounds obtained at the solid state from the reactions of **4** with dihalogens and inter-halogens;[54,56,57] structural features of (b) cation $[65\text{-Cl}]^+$ in compound **65** (I_3^- not shown);[54] (c) compound **66**;[56] (d) compound **63**.[54] All reported bond distances are in Å.

Interestingly, the reaction of **4** with the LPO/H_2O_2/I^- system in phosphate buffer at pH 7.4 produces exclusively the diselenide dication analogous to compounds **63** and **65**, but having I^- as counter-anions.[58] Although this is in contrast to the reaction of **4** with I_2 in CH_2Cl_2 (Figure 4.13) which produces the T-shaped adduct **66**,[56] it indicates the importance of the experimental conditions in determining at the solid state the products of the reactions between chalcogen donors and dihalogens and inter-halogens.

The compounds obtained by Kuhn *et al.* from the reaction of imidazoline-2-selone derivatives **67** and **68** (Figure 4.14) with I_2 are significant both from a structural and a synthetic point of view.[59] In **69**, the I_2 T-shaped adduct of **67** behaves as a donor towards a I_2 molecule *via* one of the terminal iodine atoms; in **70**, the I_2 T-shaped adduct of **68** is strongly polarized and the compound can be seen as a selenyl iodide cation interacting head-to-tail with triodide ions (Figure 4.14).

In any case, compared to benzothiazole-2-chalcogenone, thiazolidine-2-chalcogenone and benzoimidazole-2-chalcogenone derivatives (Figure 4.11), compound **4** shows a clearly higher tendency to undergo oxidation after reaction with dihalogens and inter-halogens with a prevalence of the formation of diselenide dications and hypervalent T-shaped adducts.

In addition, bis-4-imidazoline-2-selone derivatives such as **71** and **72** generally afford T-shaped adducts from the reaction with dihalogens and inter-halogens, with some peculiar differences that are generally dependent

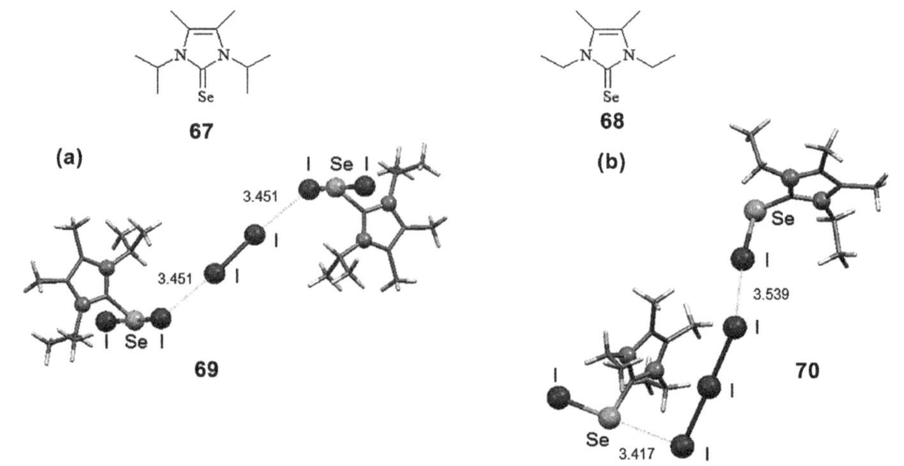

Figure 4.14 Structural features of the compounds **69** and **70** obtained at the solid state from the reactions of (a) **67** and (b) **68**, respectively, with I_2.[59] All reported bond distances are in Å.

on the reaction conditions used or strictly related to the different lengths of the bridge between the two imidazoline rings.

Both **71** and **72** give the corresponding 1:2 T-shaped adducts following the reaction with Br_2 in MeCN (not shown in Figure 4.15).[60] In the case of **71**, T-shaped adducts are also formed at the solid state from the reaction with inter-halogens (compounds **73**, **75**, **76**). However, in all cases structure refinement procedures suggest mixed occupied halogen sites, which, according to vibrational analysis corresponds to the presence in the crystals of mixed halogen T-shaped adducts.[61] These findings clearly indicate the tendency of inter-halogens (IX, X = Br, Cl) to disproportionate in solution, giving rise to I_2 and X_2 molecules. The reaction of **71** with I_2 instead affords the 1:2 C.T. adduct **74**.[56] In contrast, the reaction of **72** with I_2 in the same solvent (CH_2Cl_2) affords the corresponding 1:2 T-shaped adduct **80**,[56] which is similar to **66** (Figure 4.13) obtained from the reaction of **4** with I_2. The reactivity of **72** with IBr is quite interesting; the reaction in MeCN using a 1:2 molar ratio affords compound **82**, which still is the only known example of an asymmetric T-shaped adduct featuring two different halogen atoms bound to the chalcogen centre, and also demonstrates the possibility of performing the oxidative addition of inter-halogens to chalcogenone donors.[62] On the changing the **72**/IBr reaction molar ratio to 1:4, the cyclic diselenide dication **81** is isolated instead.

Interestingly, hypervalent T-shaped I_2 adducts of imidazoline-2-selone donors can be used as starting materials to obtain diselenide dications at the solid state that are not formed directly from the reaction with I_2, but they can generally be isolated from the reaction with inter-halogens (Figures 4.13 and 4.15). This can be achieved following the reaction with elemental Te or Se (Figure 4.16a and 4.16b).

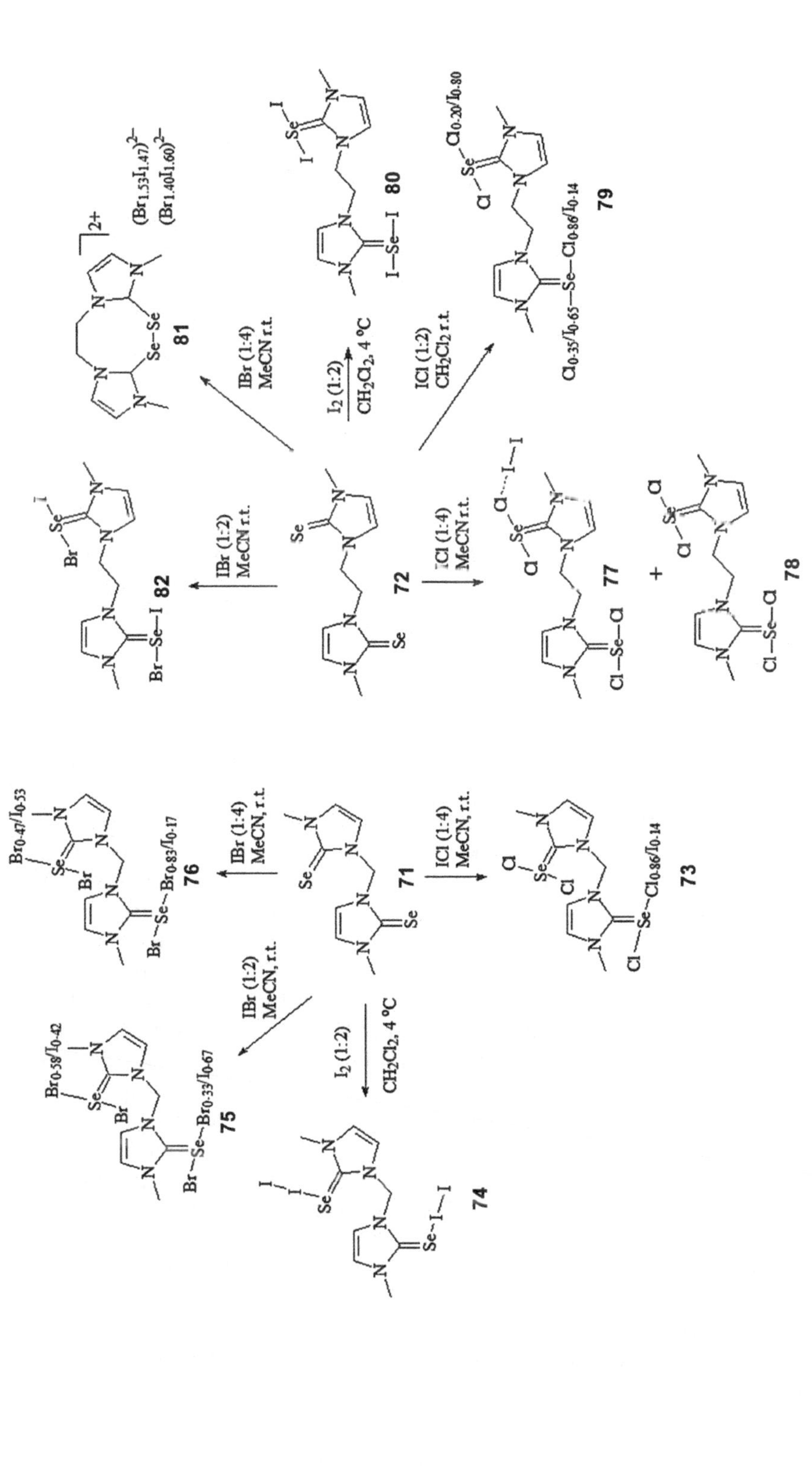

Figure 4.15 Schematic representation of the compounds obtained at the solid state from the reactions of **71** and **72** with dihalogens and inter-halogens.[56,60,61]

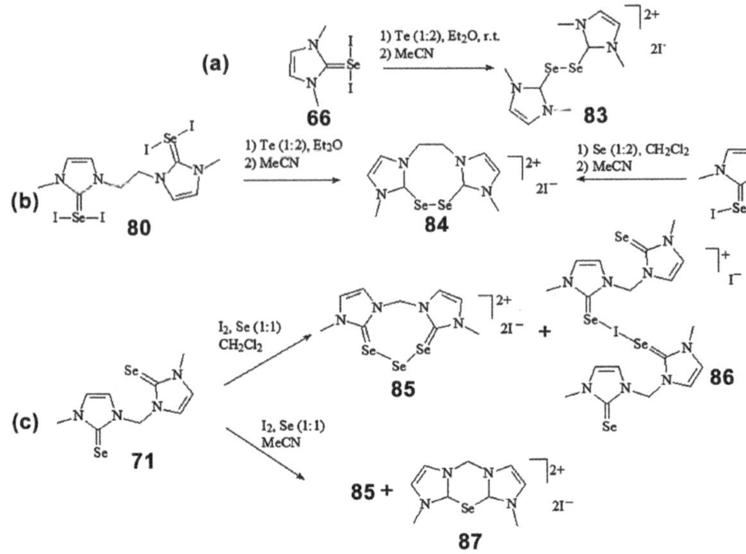

Figure 4.16 Schematic representation of the compounds obtained at the solid state from (a) the reaction of T-shaped I_2 adduct **66** with Te,[54] (b) the reactions of **80** with Te or Se[54] and (c) the reaction of **71** with I_2/Se in different solvents.[63]

This strategy was used by Khun *et al.* in the reaction of **66** with $AgBF_4$ to obtain the same diselenide dication as in **63** and **65**, but having BF_4^- counteranions.[59] Analogously, Singh and co-workers were able to synthesize iodonium complexes and diselenide cations of dialkylbenzoimidazole-2-selone derivatives starting from their T-shaped dihalogen adducts and reacting them with $AgBF_4$ (Figure 4.10)[50] and transition metal powders,[53] respectively. Interestingly, the one-pot reaction of **71** with I_2 and Se (1 : 1 : 1 molar ratio) affords, depending on the nature of the solvent, two different mixtures of products at the solid state.[63] Using CH_2Cl_2 as the solvent the products obtained are the compound **85** featuring a mixed-valence dication in which a formal Se^{2+} is coordinated by a neutral bidentate unit of **71**, and the polymeric iodonium complex **86**. Using MeCN as the solvent, beside **85**, the compound **87** is formed featuring a different dication whose formation can formally be derived from a deselenation process of the starting donor **71** driven by the formation of an intramolecular six-membered ring.[63]

A deselenation process in the reaction of selenoamides with oxidizing agents is rare, but not unknown. The reaction of *N*-methylthiazolidine-2-selone **22** with Br_2 to afford **49**, in which deselenation of the organic molecule is accompanied by oxidation of the Se atom and formation of the $SeBr_6^{2-}$ anion, has already been discussed (Figure 4.8). The electrochemical oxidation of *N*-methyl benzothiazole-2-selone **20** affords the compound $[Se(\mathbf{20})_4]^{2+}(BF_4)_2$ featuring a noncyclic dication in which a central Se^{2+} is coordinated by four units of **20** forming a square planar Se_5 framework.[64]

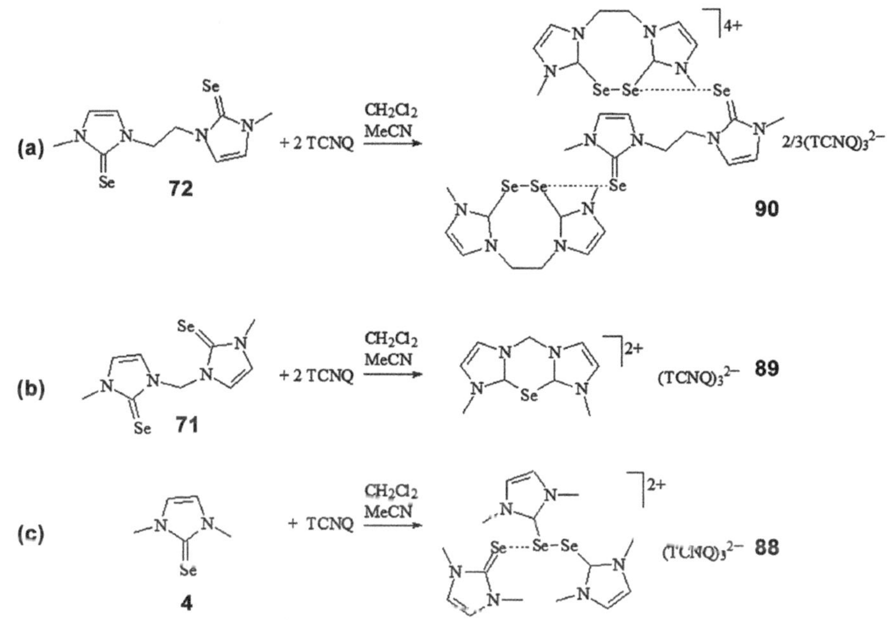

Figure 4.17 Schematic representation of the compounds obtained at the solid state from the reaction of (a) **72**, (b) **71**, (c) **4** with tetracyanoquinodimethane (TCNQ).[65–67]

Deselenation and formation of mixed valence compounds have been observed in the reactions of **4**, **71** and **72** with tetracyanoquinodimethane (TCNQ) to afford semiconductor systems (Figure 4.17).[65–67]

The dication in **89** is similar to that present in **87** (Figure 4.16), while the mixed-valence products **88** and **90** from the reactions of **4** and **72**, respectively with TCNQ feature diselenide dications analogous to those present in **83** and **81** and **84**, but interacting with the selenium atom of a neutral donor molecule to afford Se–Se–Se three-body systems (Figure 4.17).

It is important to recall at this point that thiourea-based antithyroid drugs and their Se analogues show additional beneficial effects under oxidative stress conditions and may, therefore, protect thyroid cells from oxidative damage. In particular, it has been shown that these molecules can effectively inhibit protein tyrosine nitration mediated by peroxynitrite ($ONOO^-$, PN). The significant aspect in connection with the reactivity discussed here is that experimental evidence suggests that inhibition of PN-mediated tyrosine nitration occurs *via* the elimination of sulfur and selenium moieties from thione/selone drugs and the formation of H_2EO_3 species ($E = S$, Se) and the corresponding imidazole derivatives (Figure 4.18 for **4**).[68,69]

Considering the reactivity of the most common pentatomic heterocyclic chalcogenoamides with dihalogens and inter-halogens, it is mandatory to ask whether there is a common origin to all products observed. In 1997, Husebye and co-workers proposed a reaction scheme in which the cation

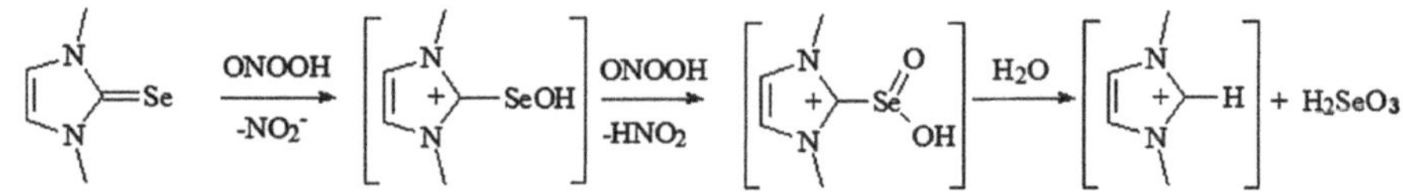

4 **91** **92** **93**

Figure 4.18 Possible mechanism for the elimination of selenium from **4** by peroxynitrite.[68]

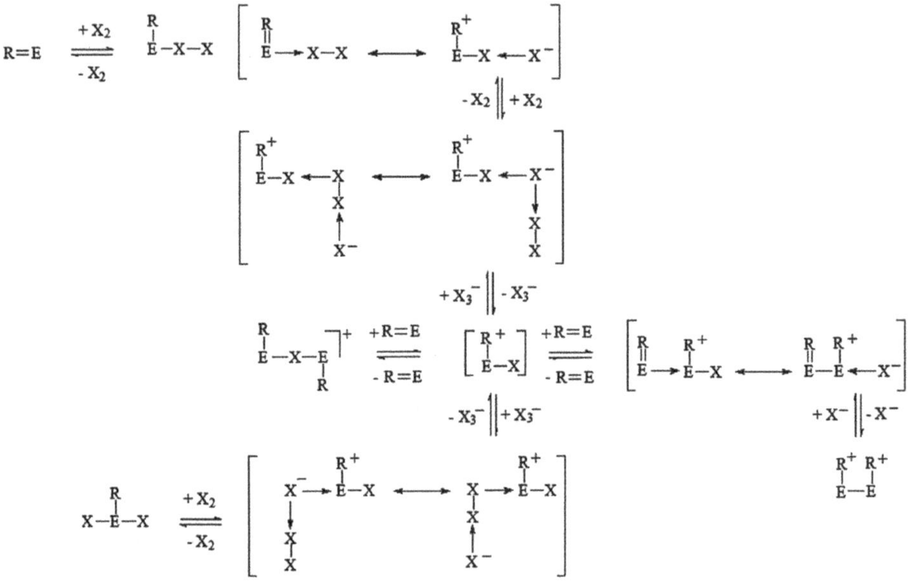

Figure 4.19 Possible reaction pathways to the formation of the known isolated products at the solid state from R=E/X_2 systems in solution (R=E = chalcogenone donor; X_2=dihalogen).[70]

$[RE\text{-}X]^+$ (Husebye's cation; E = S, Se) would represent the key intermediate in the formation of all final products (Figure 4.19).[70]

This cation would originate from the initially formed C.T. adduct by heterolytic cleavage of the polarized halogen–halogen bond thanks to the intervention of a second dihalogen molecule and removal of a trihalide. Then, dicahogenide dications would form from the interaction of a second donor molecule at the chalcogen site of $[RE\text{-}X]^+$ and the removal of a halogenide ion. Halonium complexes would generate from the interaction of a second donor molecule at the halogen site of $[RE\text{-}X]^+$. T-shaped adducts would form from the interaction of a trihalide at the chalcogen site of $[RE\text{-}X]^+$ and the removal of a dihalogen molecule. Interestingly, highly polarized C.T. adducts, extended C.T. adducts with different bond polarization pathways, dichalcogenide dications interacting with halide ion and T-shaped adducts interacting with dihalogen molecules are all structural archetypes observed at the solid state. Therefore, this chemical scheme is very well supported by experimental evidence in terms of isolation and structural characterization of all hypothesized intermediates.

Furthermore, the calculated frontier molecular orbitals for optimized Husebye's cations of different pentatomic heterocyclic chalcogenoamides show that the lowest-energy unoccupied molecular orbital (LUMO) is σ* in nature in all cases and it is mainly localized along the chalcogen–halogen bond.[26] Therefore, a nucleophilic attack at both the chalcogen and the halogen site of Husebye's cation, along the chalcogen–halogen direction, is possible, and this further supports the reaction scheme in Figure 4.19 and the central role that the hypothetical $[RE\text{-}X]^+$cation might have in the formation of the products isolated at the solid state.

In Table 4.1, the calculated natural bond orbital (NBO) charges on chalcogen and halogen atoms are reported from the optimized geometries of the donors **19–26**, **3** and **4**, and for the corresponding Husebye's cation with bromine and iodine.[26]

If we consider the calculated NBO charge on the chalcogen atom for the cations $[RE\text{-}Br]^+$, we see that it is always positive and higher than that calculated on the Br atom. Consequently, a nucleophilic attack on this cation is likely to occur at the chalcogen site in agreement with the experimental observation that C.T. adducts between these kind of donors and Br_2 are not known. On comparing the calculated NBO charges for the cations $[RE\text{-}I]^+$, the observation is that the charge on the terminal halogen is always positive, and in the case of S-donors, it is also higher than that on the chalcogen atom. This supports the preferential formation of C.T. adducts between thioamides and I_2 and inter-halogens. In contrast, the positive charge calculated on the Se atom is higher than that on the terminal iodine in $[RE\text{-}I]^+$, which is in agreement with the preferential tendency of pentatomic heterocyclic selenoamides to form T-shaped hypervalent rather than C.T. adducts. Density functional theory (DFT) calculations indicate that T-shaped dihalogen adducts of **4** are more stable than the corresponding C.T. adducts.[61] However, both type of adducts have been isolated from the reaction of some considered

Table 4.1 Natural bond orbital charges Q [e] on E (E = S, Se), X (X = Br, I) atoms in RE (**19–26**, **3** and **4**) and in their corresponding $[LE\text{-}X]^+$ ions.

RE	E	Q_E RE	Q_E $[RE\text{-}Br]^+$	Q_E $[RE\text{-}I]^+$	Q_{Br} $[RE\text{-}Br]^+$	Q_I $[RE\text{-}I]^+$	$\Delta Q_E{}^a$ $[RE\text{-}Br]^+$	$\Delta Q_E{}^a$ $[RE\text{-}I]^+$
19	S	−0.120	0.232	0.116	0.017	0.151	0.352	0.236
20	Se	−0.088	0.396	0.280	−0.059	0.069	0.484	0.368
21	S	−0.131	0.281	0.156	0.048	0.183	0.412	0.287
22	Se	−0.100	0.438	0.314	−0.037	0.094	0.538	0.414
23	S	−0.229	0.191	0.083	0.000	0.131	0.419	0.312
24	Se	−0.208	0.346	0.236	−0.070	0.055	0.554	0.444
25	S	−0.256	0.196	−0.080	0.046	0.179	0.452	0.176
26	Se	−0.228	0.356	0.233	−0.038	0.091	0.584	0.461
3	S	−0.283	0.196	0.085	0.004	0.135	0.479	0.368
4	Se	−0.268	0.355	0.244	−0.074	0.051	0.623	0.512

$^a\Delta Q_E[e] = Q_E([REX]+) - Q_E(RE)$.

selenoamides with I_2 and inter-halogens (see earlier). In terms of a general preference to afford C.T. or T-shaped adducts, if we consider the total number of crystal structures of both structural archetypes present in the Cambridge Crystallographic Data Centre (CCDC) database for any kind of chalcogen donor, we can observe that on decreasing the electronegativity difference between the halogen and the chalcogen, the number of C.T. adduct structures increases and that of T-shaped hypervalent adduct structures decreases.[71] Indeed, no T-shaped hypervalent adducts with I_2 of S-donors are known, whereas only two I_2 C.T. adducts of Te-donors are known.[72,73]

In order to evaluate the tendency to form dichalcogenide dications $[(RE)_2]^{2+}$, the difference between the charge on the chalcogen atom in the Husebye's cations and the charge on the chalcogen atom on the respective neutral donors can be considered.[26] Interestingly, the greater differences are calculated in the order **4** > **26** > **24** ≈ **22** > **20** ≈ **3** > **25** > **21** ≈ **23** > **2**. In fact, imidazoline-2-selone derivatives very frequently afford diselenide dications following the reaction with dihalogens and inter-halogens, in contrast to all the other donors considered (see earlier). Therefore, the charge distribution calculated on the hypothetical relevant Husebye's cation can account for the preference observed in the formation of the products at the solid state from reactions between pentatomic heterocyclic thioamides/selenoamides and dihalogens/inter-halogens. Interestingly, among the thioamides considered, the methylated analogue of MMI (**3**) should have the highest tendency to be oxidized at least to the corresponding disulfide dication.

The high tendency of **3** and **4** to be oxidized in the reactions with dihalogens and inter-halogens is also supported by the calculated energy of their frontier molecular orbitals; the highest occupied molecular orbital (HOMO) of **4** is higher in energy than the LUMO of each acceptor X_2 (X = I, Br, Cl). This is also true for **3**, except for the comparison with the LUMO of I_2.[74] In fact, in the reaction with I_2 we observe the formation of a T-shaped adduct in the case of **4**[56] and a C.T. adduct in the case of **3**.[75]

4.2.1 MMI (2) and MSeI (3) and Their Reactivity Towards I_2

As discussed, the mechanisms of action of thiourea-based anti-thyroid drugs and their selenium analogues in the treatment of hyperthyroidism appear to be different and are still not clear. Many synthetic efforts have been devoted to identifying the metabolites produced in their reaction with the TPO/H_2O_2/I^- system, putting at the basis for the explanation of the observed different inhibitory activities the controversial keto-enol tautomerism in thiourea-based anti-thyroid drugs and their selenium analogues.[1,3,16,76,77]

The observed reactivity of **1** and **2** with I_2 is in agreement with the highest tendency of imidazoline-chalcogenone derivatives among their pentatomic heterocyclic thioamide and selenoamide congeners to be oxidized by dihalogens and inter-halogens (see earlier).

Although it is true that, contrary to **1**, **2** undergoes a spontaneous aerial oxidation to produce the corresponding diselenide **9** (it is necessary to

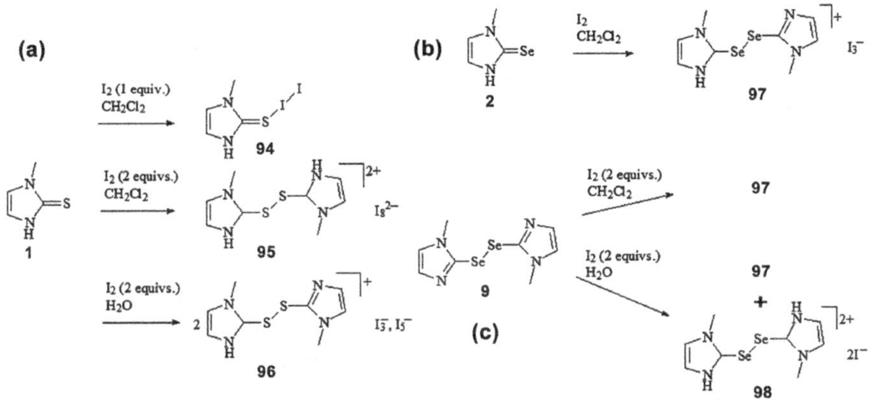

Figure 4.20 Schematic representation of the compounds obtained at the solid state from the reaction of (a) **1**,[79,80] (b) **2**, (c) **9**[81] with I_2 in different experimental conditions.

operate in controlled conditions to isolate **2**),[77,78] both compounds show a very similar chemical reactivity towards I_2 (Figure 4.20).

The reaction of **1** with I_2 in a 1 : 1 molar ratio in CH_2Cl_2 affords the C.T. adducts **94**,[79] conversely **1** can be oxidized to the disulfide dication **95** or the disulfide cation **96** by using an excess of I_2 in CH_2Cl_2 and H_2O as the solvent media, respectively.[79,80] Similar diselenide cations are obtained starting from either **9** or the less air-stable **2** by changing the experimental conditions (compounds **97** and **98** in Figure 4.20)[81] (a compound similar to **98**, but having Br^- as counter anions can be obtained from the reaction in H_2O of **2**, generated *in situ* from **9** and $NaBH_4$, and Br_2).[58]

Interestingly, **1** and **2** both inhibit LPO-catalysed oxidation of ABTS and LPO-catalysed iodination of tyrosine, and both compounds forms ionic dichalcogenides which might therefore be effective intermediates in the inhibition of thyroid hormone synthesis. The different behaviour as anti-oxidants can be re-conducted to the so-called selenium paradox, which is the ability of selenium compared to sulfur to become rapidly oxidized and then be rapidly reduced.[82]

4.3 Anti-oxidant Action of Imidazoline-2-chalcogenone Derivatives *via* Metal Binding

ROS $O_2^{\bullet-}$, H_2O_2, $HO^\bullet$, $ONOO^-$ and $^1O_2$ are natural bioproducts of cellular respiration that can cause oxidative damage to cellular components such as lipids, proteins and DNA. The hydroxyl radical is the most dangerous ROS and it is typically generated catalytically by the Fe^{2+}- and Cu^{2+}- mediated reduction of H_2O_2 (Fenton-like reaction: $Cu^+ + H_2O_2 \rightarrow Cu^{2+} + HO^- + HO^\bullet$). Normal cells can control the level of labile metal ions, but under oxidative stress conditions, the increase in labile iron and copper concentrations can bring an increase in $HO^\bullet$ generation.[6]

Interest in the coordination chemistry of imidazoline-2-chalcogenone derivatives towards copper(II) and iron(II) arises from several studies indicating metal-binding-mediated anti-oxidant properties for the naturally occurring thione (C=S)- and selone (C=Se)-containing compounds ergothioneine (**17**) and selenoneine (**18**) (Figure 4.3). In fact, compounds **3** and **4** (Figure 4.1) represent good mimics to understand how Cu^{2+} and Fe^{2+} binding by **17** and **18** might prevent metal-mediated oxidative damage.

As reported by Brumaghim and co-workers,[83] the reactions of **3** and **4** with Cu^{2+} as triflate (TfO^-) in a 1 : 1 molar ratio in MeCN result in the reduction of the metal centre and formation of the complex cation $[Cu(MeCN)_4]^+$ with the concomitant oxidation of the imidazoline-2-chalcogenone compounds and formation of the corresponding dichalcogenide dications. Treatment of $Cu(OTf)_2$ with excess of **3** and **4** in a mixed-solvent system affords the complexes [Cu(**3**)$_3$](OTf) (**99**) and [Cu(**4**)$_4$](OTf) (**101**), respectively, accompanied by the disulfide dication triflate **100** and the triselenide dication triflate **102** (Figure 4.21), similar to that formed from the reaction of **4** with TCNQ (Figure 4.17). Interestingly, no copper coordination compounds featuring seleno-hystidine or MSeI (**2**) are known.

From a kinetic point of view, **4** reduces Cu^{2+} three times faster than **3** in the presence of oxygen (under O_2-free conditions the reaction is significantly slower with both ligands) despite electrochemical data that suggest a higher stabilization of Cu^{2+} by coordination of **4** ($Cu^{2+/+}$ reduction potential is −217 and −373 V for **99** and **101**, respectively).[83] Both Cu^{2+} and Cu^+ stabilization might be important in preventing copper redox cycling under biological conditions, thus leading to a decreasing generation of damaging $HO^•$.[6,83]

To compare the reactivity of analogous Cu^+ complexes in establishing a possible antioxidant mechanism linked to metal binding of selone and

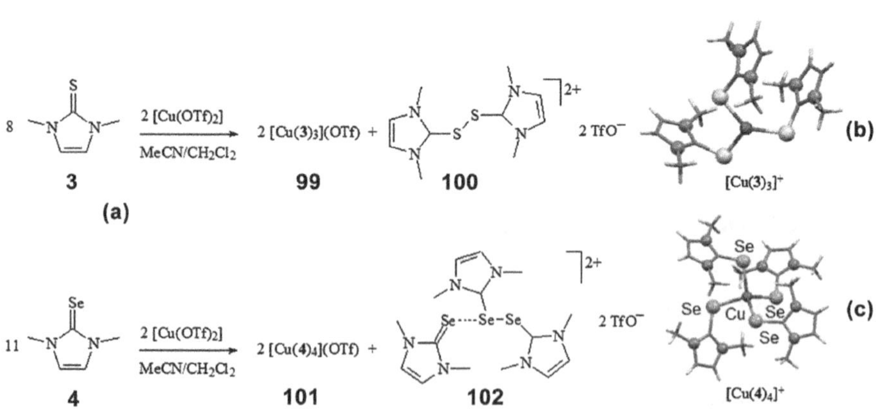

Figure 4.21 (a) Schematic representation of the compounds obtained at the solid state from the reactions of excess of **3** and **4** with Cu(TfO)$_2$,[83] structural features of (b) [Cu(**3**)$_3$]$^+$ in **99**[83] and (c) [Cu(**4**)$_4$]$^+$ in **101**.[83]

thione compounds, Brumaghim and co-workers have synthesised and fully characterised a series of tris(pyrazolyl)methane Cu^+ complexes of general formula $[Tpm^RCu(L)]BF_4$ (L = **3**, **4**, MeCN; R = H, Me, *i*Pr).[84] Oxidation of $[Tpm^{Me}Cu(MeCN)]BF_4$ with H_2O_2 or O_2 results in the immediate oxidation of Cu^+ to Cu^{2+}. Oxidation of $[Tpm^{Me}Cu(\mathbf{3/4})]BF_4$ with H_2O_2 in MeCN affords $[Tpm^{Me}Cu(MeCN)]^+$, dimethylimidazolium cation and oxidized sulfur and selenium species.[84] Therefore, both coordinated **3** and **4** (the latter more efficiently than the former) prevent Cu^+ oxidation by H_2O_2, thus providing an example of targeted ROS sacrificial scavenging antioxidant activity determined by metal binding. DFT calculations indicate that the HOMOs of the complexes $[Tpm^{Me}Cu(\mathbf{3/4})]^+$ have a significant S/Se p character, in contrast to the HOMO of the $[Tpm^{Me}Cu(MeCN)]^+$ complex, which is mainly located on the metal centre. Therefore, chalcogenone ligands in $[Tpm^{Me}Cu(\mathbf{3/4})]^+$ are preferentially oxidized, protecting Cu^+ from being oxidized and confirming the importance of metal binding in antioxidant activity. Interestingly, the reaction of $[Tpm^{iPr}Cu](OTf)_2$ with **9** (Figure 4.1) in a 1 : 1 molar ratio in a $MeCN/CH_2Cl_2$ solvent mixture affords the Cu^{2+} complex $[Tpm^{iPr}Cu(\mathbf{9})](OTf)_2$ in which the diselenide ligand binds the metal centre in a bidentate fashion through the imidazolyl nitrogens.[83] Conversely, the reaction of **9** with $Cu(ClO4)_2 \cdot 6H_2O$ (1 : 1 molar ratio) in MeOH leads to an unexpected deselenation of **9** and formation of the mononuclear complex **103** in which two bis(imidazoline)monoselenide units chelate the metal centre in the equatorial plane of a distorted coordination geometry (Figure 4.22).[85] Complex **103** reduces H_2O_2 by using GSH as co-substrate, thus mimicking GPx *in vitro*.[85]

Interestingly, on changing the denticity of the chalcogenone ligands it is possible to fine-tune the $Cu^{2+/+}$ reduction potential in the resulting coordination compounds over a 470 mV range, with the copper selone complexes having more negative reduction potentials than their thione analogues. Furthermore, the use of mono- and polydentate chalcogenone ligands and a Cu^+ starting salt in the synthetic procedures also allows diversification of the coordination mode of the metal centre with formation of polynuclear complexes.

In fact, the reaction of $[Cu(MeCN)_4]BF_4$ with **4** in a 1 : 2.5 molar ratio in MeCN affords the binuclear complex **104** in which the two metal centres are bridged by the Se atom of a unit of **4**, forming a bent CuSeCu core. Each copper(I) centre is further bonded to other two units of **4** reaching a distorted

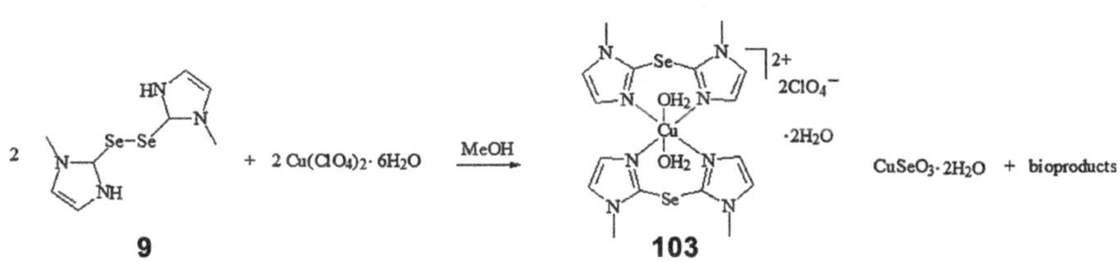

Figure 4.22 Deselenation of **9** upon reaction with copper(II) in MeOH.[85]

trigonal planar geometry (Figure 4.23). A similar complex is formed using **3** in the same experimental conditions.[86]

From a structural point of view, the complexes obtained from the reaction of $[Cu(MeCN)_4]BF_4$ with 1.5 equiv. of compound **71**, or 1 equiv. of **71** followed by the addition of 1 equiv. of **4** are quite interesting (Figure 4.24).[86]

In **105** only one ligand unit bridges the two metal centres, while the other two units are coordinated to only one Cu^+ ion to afford an overall Cu_2Se_2 core. A similar complex is formed using the sulfurated analogues of **71**.[86] In **106**, the dinuclear complex also exhibits a Cu_2Se_2 core with **71** featuring an unusual bridging monodentate/bidentate (μ-k^1 : k^2) coordination mode.

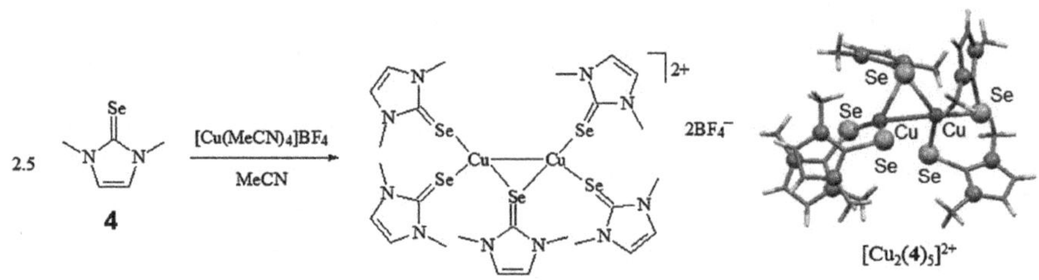

(a) 104 (b)

Figure 4.23 (a) Schematic representation of the compound obtained at the solid state from the reactions of **4** and $[Cu(MeCN)_4]BF_4$;[86] (b) structural features of $[Cu_2(4)_5]^+$ in **104**.[86]

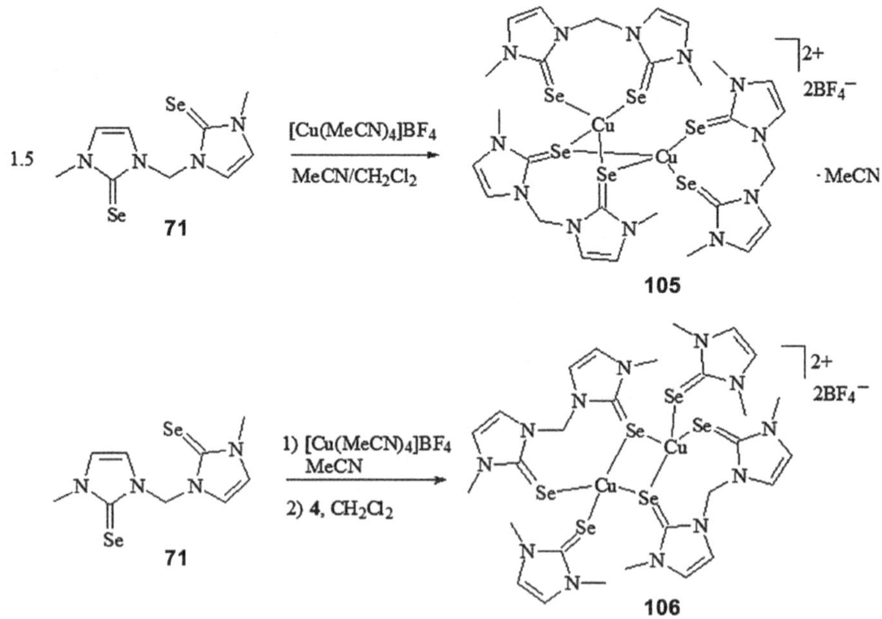

Figure 4.24 Schematic representation of the compounds obtained at the solid state from the reactions of **71** with $[Cu(MeCN)_4]BF_4$.[86]

The distorted tetrahedral coordination sphere around each metal ion is reached with the terminal coordination of a unit of **4**. A similar complex is obtained by replacing **4** with **3**.[86] These binuclear complexes have more negative $Cu^{2+/+}$ reduction potentials than their sulfur analogues, which become more negative on increasing the length of the methylene linker in the bidentate chalcogenone ligands. Only in a few cases is the measured reduction potential lower than −324 mV (*vs.* normal hydrogen electrode), which would imply the possibility of the complexes, if formed *in vivo*, resisting cellular reductants such as NADH, thus preventing copper redox cycling and copper-mediated DNA damage. In any case, the complex redox behaviour of the copper complexes of imidazoline-2-chalcogenone derivatives is the basis of the antioxidant abilities of these ligands in preventing oxidative DNA damage *via* metal binding.

4.4 Conclusion

Although imidazoline-2-chalcogenone derivatives ligands have been known for a long time, they have been the object of a recent renewed interest due to their potential biological applications. This has stimulated many chemists to study their reactivity towards dihalogens, inter-halogens and specific metal ions such as Cu^{2+} and Fe^{2+} to understand the possible mechanisms at the basis of their activities as anti-thyroid drugs and anti-oxidants by metal binding. Experimental evidence suggests that the peculiar redox and donor properties of the selenium derivatives forms the basis of the differences in the chemical reactivity and biological properties observed, compared to those of the sulfur analogues.

Acknowledgements

The Università degli Studi di Cagliari and the Fondazione di Sardegna are kindly are acknowledged for financial support of projects related to the topic developed in this chapter.

References

1. G. Roy and G. Mugesh, *Phosphorus, Sulfur, Silicon Relat. Elem.*, 2008, **183**, 908.
2. D. Manna, G. Roy and G. Mugesh, *Acc. Chem. Res.*, 2013, **46**, 2706.
3. G. Roy, P. N. Jayaram and G. Mugesh, *Chem. - Asian. J.*, 2013, **8**, 1910.
4. E. Rijntjes, P. M. Scholz, G. Mugesh and J. Köhrle, *Eur. Thyroid J.*, 2013, **2**, 252.
5. E. E. Battin, M. T. Zimmerman, R. R. Ramoutar, C. E. Quarles and J. L. Brumaghim, *Metallomics*, 2011, **3**, 503.
6. M. T. Zimmerman, C. A. Bayse, R. R. Ramoutar and J. L. Brumaghim, *J. Inorg. Biochem.*, 2015, **145**, 30.
7. E. E. Battin and J. L. Brumaghim, *Cell Biochem. Biophys.*, 2009, **55**, 1.

8. D. S. Cooper, *N. Engl. J. Med.*, 2005, **352**, 905.
9. B. Davidson, M. Soodak, J. T. Neary, H. V. Strout, J. D. Kieffer, H. Mover and F. Maloof, *Endocrinology*, 1978, **103**, 871.
10. A. Taurog, M. L. Dorris and F. S. Guziec Jr., *Endocrynology*, 1989, **124**, 30.
11. D. R. Doerge, N. M. Cooray and M. E. Brewster, *Biochemistry*, 1991, **30**, 8960.
12. R. Basosi, N. Niccolai and C. Rossi, *Biophys. Chem.*, 1978, **8**, 61.
13. D. R. Doerge, *Biochemistry*, 1988, **27**, 3697.
14. G. G. J. M. Kuiper, M. H. A. Kester, R. P. Peeters and T. J. Visser, *Thyroid*, 2005, **15**, 787.
15. W.-W. du Mont, G. Mugesh, C. Wismatch and P. G. Jones, *Angew. Chem., Int. Ed.*, 2001, **40**, 2486.
16. G. Roy and G. Mugesh, *J. Am. Chem. Soc.*, 2005, **127**, 15207.
17. K. P. Bhabak and G. Mugesh, *Inorg. Chim. Acta*, 2010, **363**, 2812.
18. O. I. Aruoma, M. Whiteman, T. G. England and B. Halliwell, *Biochem. Biophys. Res. Commun.*, 1997, **231**, 389.
19. O. I. Aruoma, J. P. E. Spencer and N. Mahmood, *Food Chem. Toxicol.*, 1999, **37**, 1043.
20. C. E. Hand and J. F. Honek, *J. Nat. Prod.*, 2005, **68**, 293.
21. D. Akanmu, R. Cecchini, O. I. Aruoma and B. Halliwell, *Arch. Biochem. Biophys.*, 1991, **288**, 10.
22. B. Z. Zhu, L. Mao, R. M. Fan, J. G. Zhu, Y. N. Zhang, J. Wang, B. Kalyanaraman and B. Frei, *Chem. Res. Toxicol.*, 2011, **24**, 30.
23. B. S. Stadelman and J. L. Brumaghim, *ACS Symp. Ser.*, 2013, **1152**, 33.
24. M. C. Aragoni, M. Arca, F. A. Devillanova, A. Garau, F. Isaia, V. Lippolis and G. Verani, *Coord. Chem. Rev.*, 1999, **184**, 271.
25. M. C. Aragoni, M. Arca, F. Demartin, F. A. Devillanova, A. Garau, F. Isaia, F. Lelj, V. Lippolis and G. Verani, *Chem. – Eur. J.*, 2001, **7**, 3122.
26. C. Laurence, M. J. ElGhomari, J.-Y. Le Questel, M. Berthelot and R. Mokhlisse, *J. Chem. Soc., Perkin Trans. 1*, 1998, **2**, 1545.
27. F. A. Devillanova and G. Verani, *Tetrahedron*, 1981, **37**, 1803.
28. F. A. Devillanova, F. Isaia, V. Lippolis and G. Verani, *Spectrochim. Acta, Part A*, 1996, **52**, 1569.
29. P. D. Boyle and S. M. Godfrey, *Coord. Chem. Rev.*, 2001, **223**, 265.
30. A. J. Blake, F. Cristiani, F. A. Devillanova, A. Garau, L. M. Gilby, R. O. Gould, F. Isaia, V. Lippolis, S. Parsons, C. Radek and M. Schröder, *J. Chem. Soc., Dalton Trans.*, 1997, 1337.
31. C. Knobler, C. Baker, H. Hope and J. D. McCullough, *Inorg. Chem.*, 1971, **10**, 697.
32. G. B. M. Vaugham, A. J. Mora, A. N. Fitch, P. N. Gates and A. S. Muir, *J. Chem. Soc., Dalton Trans.*, 1999, 79.
33. S. M. Godfrey, C. A. McAuliffe, R. G. Pritchard and S. Sarwar, *J. Chem. Soc., Dalton Trans.*, 1997, 1031.
34. S. M. Godfrey, C. A. McAuliffe, R. G. Pritchard and S. Sarwar, *J. Chem. Soc., Dalton Trans.*, 1997, 3501.

35. M. Arca, F. Demartin, F. A. Devillanova, A. Garau, F. Isaia, V. Lippolis and G. Verani, *Trends Inorg. Chem.*, 1999, **6**, 1.
36. G. J. Corban, S. K. Hadjikakou, N. Hadjiliadis, M. Kubicki, E. R. T. Tiekink, I. S. Butler, E. Drougas and A. M. Kosmas, *Inorg. Chem.*, 2005, **44**, 8617.
37. F. Demartin, P. Deplano, F. A. Devillanova, F. Isaia, V. Lippolis and G. Verani, *Inorg. Chem.*, 1993, **32**, 3694.
38. L. Koskinen, P. Hirva, A. Hasu, S. Hääskeläinen, J. Koivistoinen, M. Pettersson and M. Haukka, *CrystEngComm*, 2015, **17**, 2718.
39. P. D. Boyle, W. I. Cross, S. M. Godfrey, C. A. McAuliffe, R. G. Pritchard and S. Teat, *J. Chem. Soc., Dalton Trans.*, 1999, 2219.
40. P. D. Boyle, W. I. Cross, S. M. Godfrey, C. A. McAuliffe, R. G. Pritchard and S. J. Teat, *J. Chem. Soc., Dalton Trans.*, 1999, 2845.
41. F. A. Devillanova, P. Deplano, F. Isaia, V. Lippolis, M. L. Mercuri, S. Piludu, G. Verani and F. Demartin, *Polyhedron*, 1998, **17**, 305.
42. F. Cristiani, F. Demartin, F. A. Devillanova, F. Isaia, V. Lippolis and G. Verani, *Inorg. Chem.*, 1994, **33**, 6315.
43. F. Demartin, F. A. Devillanova, F. Isaia, V. Lippolis, P. D. Boyle and S. M. Godfrey, *J. Chem. Soc., Dalton Trans.*, 2000, 1959.
44. M. C. Aragoni, M. Arca, F. Demartin, F. A. Devillanova, T. Gelbrich, A. Garau, M. B. Hursthouse, F. Isaia and V. Lippolis, *Cryst. Growth Des.*, 2007, **7**, 1284.
45. V. Daga, S. K. Hadjikakou, N. Hadjiliadis, M. Kubicki, J. H. Z. dos Santos and I. S. Butler, *Eur. J. Inorg. Chem.*, 2002, **7**, 1718.
46. P. D. Boyle, J. Christie, T. Dyer, S. M. Godfrey, I. R. Howson, C. McArthur, B. Omar, R. G. Pritchard and G. Rh. Williams, *J. Chem. Soc., Dalton Trans.*, 2000, 3106.
47. W.-W. du Mont, M. Bätcher, C. Daniliuc, F. A. Devillanova, C. Druckenbrodt, J. Jeske, P. G. Jones, V. Lippolis, F. Ruthe and E. Seppälä, *Eur. J. Inorg. Chem.*, 2008, 4562.
48. G. H.-Y. Lin and H. Hope, *Acta Crystallogr., Sect. B: Struct. Crystallogr. Cryst.*, 1972, **28**, 643.
49. S. T. Manjare, S. Sharma, H. B. Singh and R. J. Butcher, *J. Organomet. Chem.*, 2012, **717**, 61.
50. S. T. Manjare, H. B. Singh and R. J. Butcher, *Eur. J. Inorg. Chem.*, 2013, 2161.
51. M. C. Aragoni, M. Arca, F. Demartin, F. A. Devillanova, A. Garau, F. Isaia, V. Lippolis and G. Verani, *Dalton Trans.*, 2005, 2252.
52. S. T. Manjare, S. Yadav, H. B. Singh and R. J. Butcher, *Eur. J. Inorg. Chem.*, 2013, 5344.
53. S. Yadav, S. T. Manjare, H. B. Singh and R. J. Butcher, *Dalton Trans.*, 2016, **45**, 12015.
54. F. Bigoli, F. Demartin, P. Deplano, F. A. Devillanova, F. Isaia, V. Lippolis, M. L. Mercuri, M. A. Pellinghelli and E. F. Trogu, *Inorg. Chem.*, 1996, **35**, 3194.
55. F. Demartin, F. A. Devillanova, F. Isaia, V. Lippolis and G. Verani, *Inorg. Chim. Acta*, 1997, **255**, 203.

56. F. Bigoli, M. A. Pellinghelli, P. Deplano, F. A. Devillanova, V. Lippolis, M. L. Mercuri and E. F. Trogu, *Gazz. Chim. Ital.*, 1994, **124**, 445.
57. D. J. Williams, D. VanDerveer, B. R. Crouse, R. R. Raye, T. Carter, K. S. Hagen and M. Brewer, *Main Group Chem.*, 1997, **2**, 61.
58. G. Roy, K. P. Bhabak and G. Mugesh, *Cryst. Growth Des.*, 2011, **11**, 2279.
59. N. K. R. Fawzi, T. Kratz, M. Steiman and G. G. Henkel, *Phosphorous, Sulfur, Silicon Related Elem.*, 1996, **112**, 225.
60. F. Bigoli, P. Deplano, F. A. Devillanova, V. Lippolis, M. L. Mercuri, M. A. Pellinghelli and E. F. Trogu, *Eur. J. Inorg. Chem.*, 1998, **1**, 137.
61. E. J. Juárez-Pérez, M. C. Aragoni, M. Arca, A. J. Blake, F. A. Devillanova, A. Garau, F. Isaia, V. Lippolis, R. Núñez, A. Pintus and C. Wilson, *Chem. – A Eur. J.*, 2011, **17**, 11497.
62. M. C. Aragoni, M. Arca, A. J. Blake, F. A. Devillanova, W.-W. Du Mont, A. Garau, F. Isaia, V. Lippolis, G. Verani and C. Wilson, *Angew. Chem. Int. Ed.*, 2001, **40**, 4229.
63. P. J. Jones and V. Lippolis, Unpublished results.
64. C. Adamo, F. Demartin, P. Deplano, F. A. Devillanova, F. Isaia, F. Lelj, V. Lippolis, P. Lukes and M. L. Mercuri, *J. Chem. Soc., Chem. Commun.*, 1996, **7**, 873.
65. F. Bigoli, P. Deplano, F. A. Devillanova, A. Girlando, V. Lippolis, M. L. Mercuri, M. A. Pellinghelli and E. F. Trogu, *Inorg. Chem.*, 1996, **35**, 5403.
66. F. Bigoli, P. Deplano, F. A. Devillanova, A. Girlando, V. Lippolis, M. L. Mercuri, M. A. Pellinghelli and E. F. Trogu, *J. Mater. Chem.*, 1998, **8**, 1145.
67. F. A. Devillanova, A. Garau, F. Isaia, V. Lippolis, G. Verani, A. Cornia, A. C. Fabretti and G. Girlando, *J. Mat. Chem.*, 2000, **10**, 1281.
68. K. P. Bhabak and G. Mugesh, *Chem. – Eur. J.*, 2010, **16**, 1175.
69. K. P. Bhabak, K. Satheeshkumar, S. Jayavelu and G. Mugesh, *Org. Biomol. Chem.*, 2011, **9**, 7343.
70. M. D. Rudd, S. V. Lindeman and S. Husebye, *Acta Chem. Scand.*, 1997, **51**, 689.
71. M. C. Aragoni, M. Arca, F. A. Devillanova, F. Isaia and V. Lippolis, *Cryst. Growth Design*, 2012, **12**, 2769.
72. E. Faoro, G. Manzoni de Oliveira, S. Schultz Lang and C. Bicca Pereira, *J. Organomet. Chem.*, 2010, **695**, 1480.
73. M. Tretiakov, Y. G. Shermolovich, A. P. Singh, P. P. Samuel, H. W. Roesky, B. Niepotter, A. Visscher and D. Stalke, *Dalton Trans.*, 2013, **42**, 12940.
74. M. C. Aragoni, M. Arca, F. A. Devillanova, P. Grimaldi, F. Isaia, F. Lelj and V. Lippolis, *Eur. J. Inorg. Chem.*, 2006, **11**, 2166.
75. F. Freeman, J. W. Ziller, H. N. Po and M. C. Keindl, *J. Am. Chem. Soc.*, 1988, **110**, 2586.
76. G. Roy, D. Das and G. Mugesh, *Inorg. Chim. Acta*, 2007, **360**, 303.
77. V. K. Landry, M. Minoura, K. Pang, D. Buccella, B. V. Kelly and G. Parkin, *J. Am. Chem. Soc.*, 2006, **128**, 12490.

78. G. Roy, M. Nethaji and G. Mugesh, *J. Am. Chem. Soc.*, 2004, **126**, 2712.
79. F. Isaia, M. C. Aragoni, M. Arca, F. Demartin, F. A. Devillanova, G. Floris, A. Garau, M. B. Hursthouse, V. Lippolis, R. Medda, F. Oppo, M. Pira and G. Verani, *J. Med. Chem.*, 2008, **51**, 4050.
80. M. C. Aragoni, M. Arca, F. Demartin, F. A. Devillanova, A. Garau, F. Isaia, V. Lippolis and G. Verani, *J. Am. Chem. Soc.*, 2002, **124**, 4538.
81. G. Roy, M. Nethaji and G. Mugesh, *Org. Biomol. Chem.*, 2006, **4**, 2883.
82. H. J. Reich and R. J. Hondal, *ACS Chem. Biol.*, 2016, **11**, 821.
83. M. M. Kimani, H. C. Wang and J. L. Brumaghim, *Dalton Trans.*, 2012, **41**, 5248.
84. M. M. Kimani, J. L. Brumaghim and D. VanDerveer, *Inorg. Chem.*, 2010, **49**, 9200.
85. G. Roy, M. Nethaji and G. Mugesh, *Inorg. Chem. Commun.*, 2006, **9**, 571.
86. M. M. Kimani, D. Watts, L. A. Graham, D. Rabinovich, G. P. A. Yap and J. Brumaghim, *Dalton Trans.*, 2015, **44**, 16313.

CHAPTER 5

Organoselenium Cations: Structures, Reactivity and Applications

SANGIT KUMAR*[a] AND HARKESH B. SINGH*[b]

[a] Indian Institute of Science Education and Research Bhopal, Department of Chemistry, Bhopal By-Pass Road Bhauri, 462 066, Bhopal, India;
[b] Indian Institute of Technology Bombay, Department of Chemistry, Powai, 400 076, Mumbai, India
*Email: sangitkumar@iiserb.ac.in; chhbsia@chem.iitb.ac.in

5.1 Introduction

The past two decades have witnessed a steady progress in the field of organochalcogen chemistry. A variety of organochalcogen compounds are being used as biological models,[1] reagents in synthetic organic chemistry,[2] re-generable sensors,[3] precursors for various smart materials to address the challenges society faces in meeting the energy requirements through sustainable and green sources such as water splitting and solar cells,[4] and remedies for various diseases.[5] In this context, diverse classes of organochalcogen compounds are being discovered and isolated; in particular, organoselenium cations are of current interest. The instability and high reactivity of organoselenium cations due to their incomplete valence shell (six valence electrons) has attracted considerable attention from chemists around the world. In particular, the strong electrophilic and/or oxidizing nature of these cations in various synthetic reactions has been explored. Some of the noteworthy reactions include the addition of organoselenenium

Organoselenium Compounds in Biology and Medicine: Synthesis, Biological and Therapeutic Treatments
Edited by Vimal Kumar Jain and K. Indira Priyadarsini

Published by the Royal Society of Chemistry, www.rsc.org

ions across a carbon–carbon double or triple bond for the functionalization of alkenes or alkynes,[6] their catalytic role in intramolecular cyclization of alkenoic acids,[7] and electrophilic addition of arylselenenium cations to electron-rich arenes for the synthesis of unsymmetrical diaryl chalcogenides.[8] The enhancement of the Lewis acidic character of organoselenenium ions by several bases has been utilized by Denmark and Chi for the stereoselective functionalization of alkenes (see later).[9]

The nomenclature of organoselenium cations is shown in Chart 5.1. The classical approach is based on the nomenclature of nitrogen analogues, *i.e.* nitrenium is equivalent to selenenium, nitronium to selenonium and cyclic aziridinium to seleniranium cations.[10] Organoselenium(II) cations are thus named as selenenium, organoselenium(III) and/or (IV) cations as selenonium, and organoselenium(VI) cations as hypervalent selenonium. A variety of organoselenium cations have been isolated with diverse electronic environments around the selenium atom and naming organoselenium

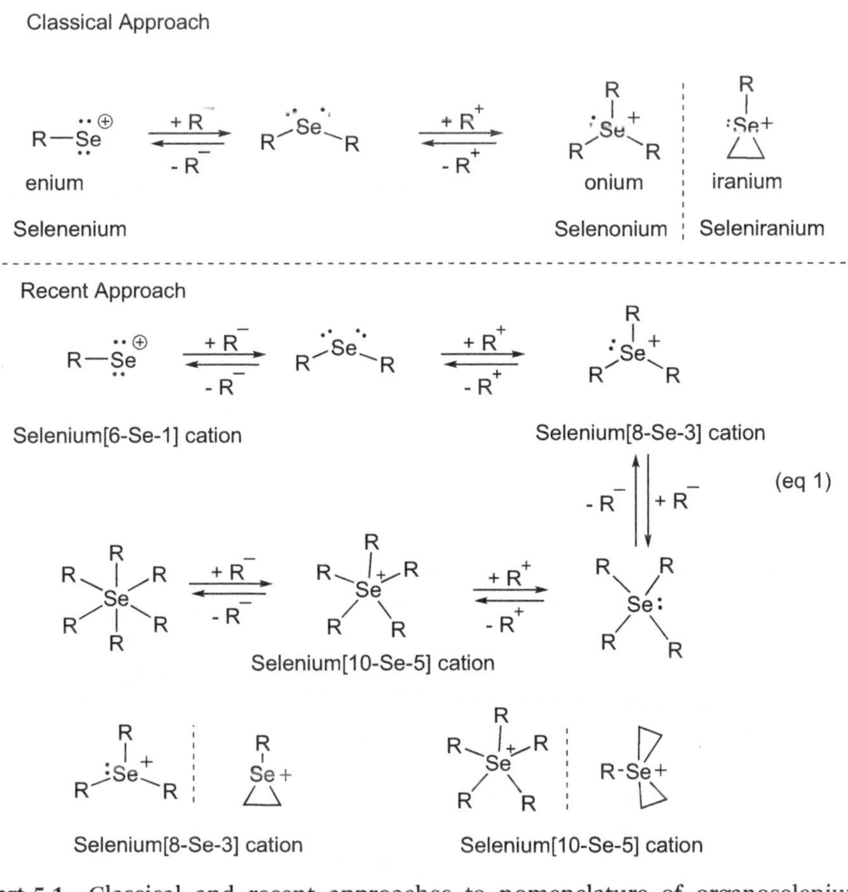

Chart 5.1 Classical and recent approaches to nomenclature of organoselenium cations.

cations using the classical approach may not be sufficient to depict their electronic structure. Therefore, organoselenium cations are best named along with the electronic arrangement around the selenium atom as [number of valence electron around selenium-Se-number of covalent bonds] cations, *i.e.* an organoselenenium cation is named as a selenium[6-Se-1] cation, an organoselenonium cation as a selenium[8-Se-3] cation, and a hypervalent organoselenonium cation as a selenium[10-Se-5] cation. This nomenclature is advantageous as it provides the number of covalent bonds and the lone pairs around the central selenium atom in the cations. In addition, it is possible to calculate the charge on the selenium using the formal charge concept.

5.2 Organoselenenium Cations

In the past few decades, synthetic chemists have focused their attention on increasingly challenging inorganic compounds. This has been facilitated in part by the availability of various reactive and versatile reagents, specially designed ligands with desired steric or electronic features and state-of-the-art facilities to characterize moisture- and air-sensitive compounds. The role of ligand design and reaction conditions in the isolation of selenium[6-Se-1] cation **1** are discussed shortly.[11] Readers, particularly beginners in organoselenium chemistry might ask why one should attempt to isolate the notoriously reactive organoselenium cations. Moreover, the isolated organoselenium cations (if one succeeds in isolating them!) may be of little or no significant use in terms of applications, because the reactive species isolated by applying intramolecular coordination or bulky ligands are generally found to be so stable that they do not react. Apart from academic interest in isolating reactive species that have not been isolated before, the isolation of organoselenium cations with fewer than eight valence electrons could provide (a) a new avenue in catalysis where organoselenium cations catalyse a reaction in a sustainable manner; (b) a better rationale for the design of new catalysts to accomplish formidable transformations in an enantioselective manner, namely enantioselective bromolactonization and lactamization reactions; (c) synthesis of existing and novel organoselenium compounds by benign and affordable synthetic routes; and (d) the preparation of smart materials for energy storage to meet the energy requirements of society.

Although phenylselenenium cations with triflate and sulfate anions were synthesized by Murata and Suzuki[12] in 1987 and Bartoli and co-workers[13] in 1989, respectively, the phenylselenenium cation could not be characterized structurally. Its existence has been established in solution *via* its electrophilic addition to carbon–carbon double bonds. Further organoselenium compounds, in which selenium bears a positive charge ($ArSe^+$ X^- where X^- = coordinating anions F^-, Cl^-, Br^- or non-coordinating anions BF_4^-, OTf^-, SbF_6^-, PF_6^-, and OSO_3H^-), have been proposed as electrophilic selenium reagents.[14] Similarly, an intermediate selenenyl sulfide (EnzSeSR, Enz = enzyme) where selenium bears a partial positive charge has been proposed in the catalytic cycle of the glutathione peroxidase selenoenzyme for the reduction of various peroxides. Despite the establishment of the presence of organoselenium cations in various areas of chemistry several years ago, the very first organoselenenium cation **1** was isolated and structurally characterized by Fujihara *et al.* in 1995 (Scheme 5.1).[11]

While attempting to prepare the respective selenoxide from the oxidation of 2,6-bis[(dimethylamino)methyl]phenyl methyl selenide, having a *N*,*C*,*N*-pincer ligand, with *t*-BuOCl, the Se-demethylated product, *i.e.* selenenium cation **1**, was isolated. Phenyl methyl selenide and 2-(*N*,*N*-dimethylaminobenzyl)methyl selenide, upon oxidation with *t*-BuOCl are expected to give the corresponding selenoxides, namely $PhSe(O)CH_3$ and $(2\text{-}C_6H_4CH_2NMe_2)\text{-}Se(O)CH_3$, respectively.[1] In the present case, the Se-demethylation was induced by the neighbouring group participation of the two nitrogen atoms. It seems that the pincer ligand having two nitrogen donor atoms is responsible for the isolation of the selenenium cation **1**, as the substitution on the two nitrogen atoms provides steric protection to the selenium cation, and most importantly the electron deficiency of the selenium atom is largely compensated by electron donation from the nitrogen atoms. Anion exchange by bulky PF_6^- enhances the crystallinity of the selenenium cation **1**. Spectroscopic characterization of **1** by ^{77}Se nuclear magnetic resonance (NMR) spectroscopy shows a chemical shift at δ 1208 ppm, which is significantly downfield shifted compared to the parent selenide ($\delta = 90$ ppm). The significant downfield chemical shift clearly established the positive charge on the selenium atom. The intramolecular Se···N distances (2.154 Å and 2.180 Å) are considerably shorter than the $\Sigma\, r_{vdw}$ (Se, N = 3.5 Å). Poleschner's[15] and Silvestru's[16] groups have isolated selenenium cations (**2–5**) with HF_2^- and other halide anions (Scheme 5.2).

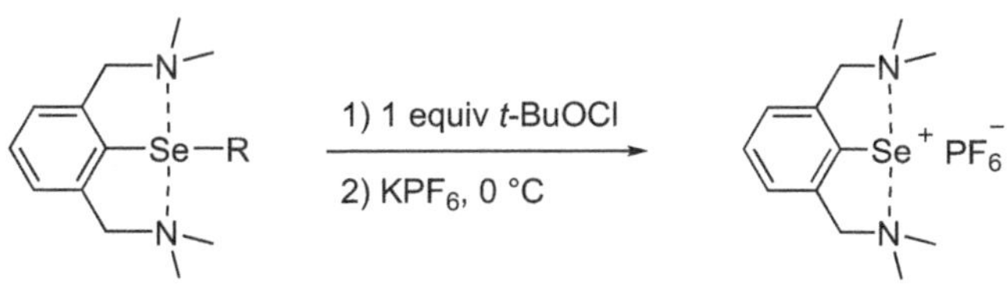

1

Scheme 5.1 Synthesis of the first organoselenium cation **1**.

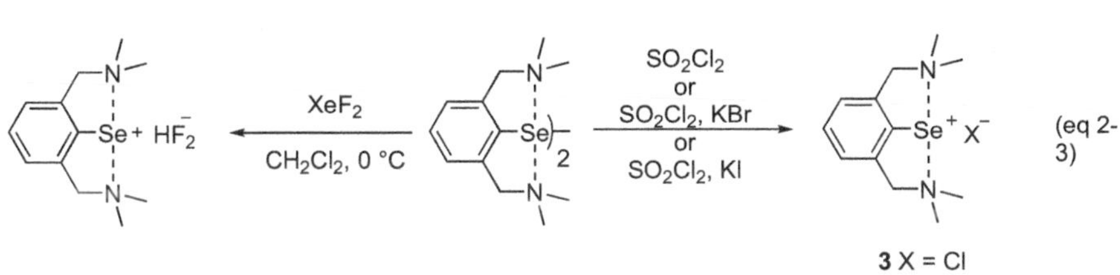

2

3 X = Cl
4 X = Br
5 X = I

Scheme 5.2 Isolation of selenenium cations **2–5**.

Structural characterization of the tellurium analogue of the selenenium cation **1** took another 16 years. A slightly modified bulky pincer ligand, 2,6-$[O(CH_2CH_2)_2NCH_2]_2C_6H_3$ has been successfully applied by Silvestru and co-workers[17] in the structural characterization of an organotellurenium cation with tellurium–nitrogen bond lengths of Te–N 2.373 Å. Since this, several reports have appeared in the literature on the stabilization of RTe^+.[18] The chemistry of tellurium cations is beyond the scope of this chapter and is not discussed further.

Later, Du Mont and co-workers[18a] isolated a pentafluorinated [6-Se-1] cation **6** by the intermolecular coordination of a dimesityl telluride ligand and use of a hexafluroroantimonate anion while conducting the reaction between dimesityl telluride, bis(pentafluorophenyl)diselenide, Br_2 and $AgSbF_6$ (Chart 5.2). The structure shows a strong intermolecular Te···Se interaction, as the distance between the two [Te–Se 2.575(5) Å] is

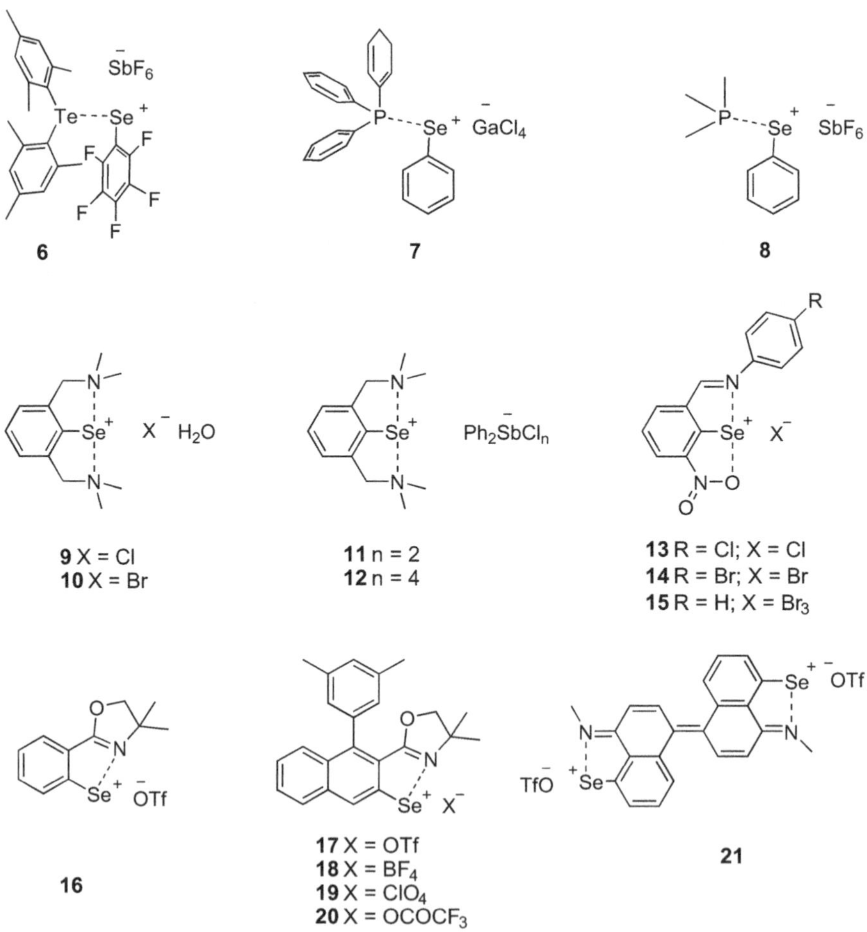

Chart 5.2 Organochalcogenium cations.

significantly shorter than the sum of their van der Waals radii (3.96 Å) and close to the sum of their covalent radii (2.51 Å). Solution studies on cation **6** could not be performed satisfactorily as ^{77}Se NMR spectra showed peaks only at the chemical shifts of the diselenide [$(C_6F_5Se)_2$ $\delta = 370$ ppm] and selenide [$(C_6F_5)_2Se$, $\delta = 176$ ppm]. Similarly, mass spectrometry peaks corresponding to the cation could not be observed. It seems that the intermolecular coordination, although very strong in nature, is not enough to stabilize cation **6** in solution, and as a consequence it converts to its diselenide and monoselenide. Russell and co-workers[19] have isolated phenylselenenium cations **7** and **8** using the intermolecularly coordinating ligands triphenyl- and trimethyl phosphine with $GaCl_4^-$ and SbF_6^- anions, respectively. Both the selenenium cations have been completely characterized both in solution and the solid state using intermolecular donor ligands for the first time. Nonetheless, these cations may be viewed as ionic quasi-phosphonium salts and may not be truly considered as selenenium cations. This can be partly confirmed by their ^{77}Se NMR spectra in which selenium resonates at $\delta = \sim 300$ ppm, suggesting significant electron density around selenium, contrary to that expected for selenenium cations. It seems that pincer ligands having two intramolecular nitrogen donor atoms are very efficient in stabilizing selenenium cations, as is evident from the structures of **9** and **10** in which a water molecule is incorporated. Nonetheless, selenenium cations could be isolated successfully.[20] Moreover, selenenium cations **11** and **12** having $[Ph_2SbCl_2]^-$ and $[Ph_2SbCl_4]^-$ counter anions have been isolated and structurally characterized.[20b]

The syntheses of peculiar isoselenazolone-derived selenenium cations **13–15** stabilized by nitrogen and oxygen donor atoms have been reported by our group.[21] The precursor for selenenium cations **13–15** was synthesized by treating *N*-(2-bromo-3-nitrobenzylimino)benzene, an *N*,*O*-pincer type of ligand, with *n*BuSeNa generated *in situ*. The selenenium cation **13** with chloride as counter anion was isolated when ^{n}butyl[*N*-(3-nitrobenzylimino)]phenyl selenide **22** was treated with sulfuryl chloride (SO_2Cl_2). The selenenium cations with bromide **14** and tribromide **15** as counter anions were synthesized by treating ^{n}butyl[*N*-(3-nitrobenzylimino)]phenyl selenide with Br_2 (Scheme 5.3). The presence of secondary Se···O/Se···N intramolecular interactions in **14** and **15** has been confirmed by single-crystal X-ray studies. Again, a pincer type electronic arrangement consisting of nitrogen and oxygen atoms seems responsible for the isolation of selenenium cations **13–15**.

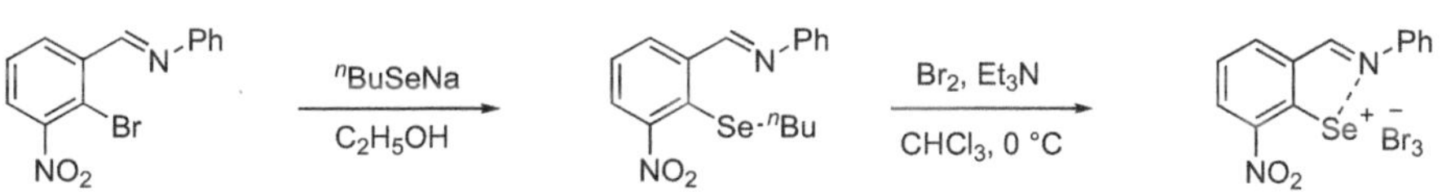

22 **15**

Scheme 5.3 Synthesis of selenenium salts.

Substituents incorporating a single donor atom have also been successfully employed by our group to obtain organoselenenium cations **16–21** (Chart 5.2 and Scheme 5.4).[22] The addition of bromine to the respective diorgano diselenide followed by quenching with bulky non-coordinating triflate anion yielded [2-(4,4-dimethyl-2-oxazolinyl)-phenyl]selenenium triflate **16** and [2-(1-(3,5-dimethylphenyl)-2-naphthyl)-4,5-dihydro-4,4-dimethyl-1,3-oxazole]selenenyl cations **17–20.** The selenenium cation **16** is stable under ambient conditions only for 10 h, but in contrast, selenenium cations **17–20,** derived from bulky ligands and having an extended aromatic conjugated ring system are stable under ambient conditions for several months. A similar reaction of silver triflate with [8-(dimethylamino)-1-naphthyl]selenenyl(II) bromide afforded 4,4′-bis(methylimino)-1,1′-binaphthylene)-5-diselenenium triflate **21**, instead of the expected monocation [8-(dimethylamino)-1-naphthyl]selenenium(II) triflate **23.** Diselenenium

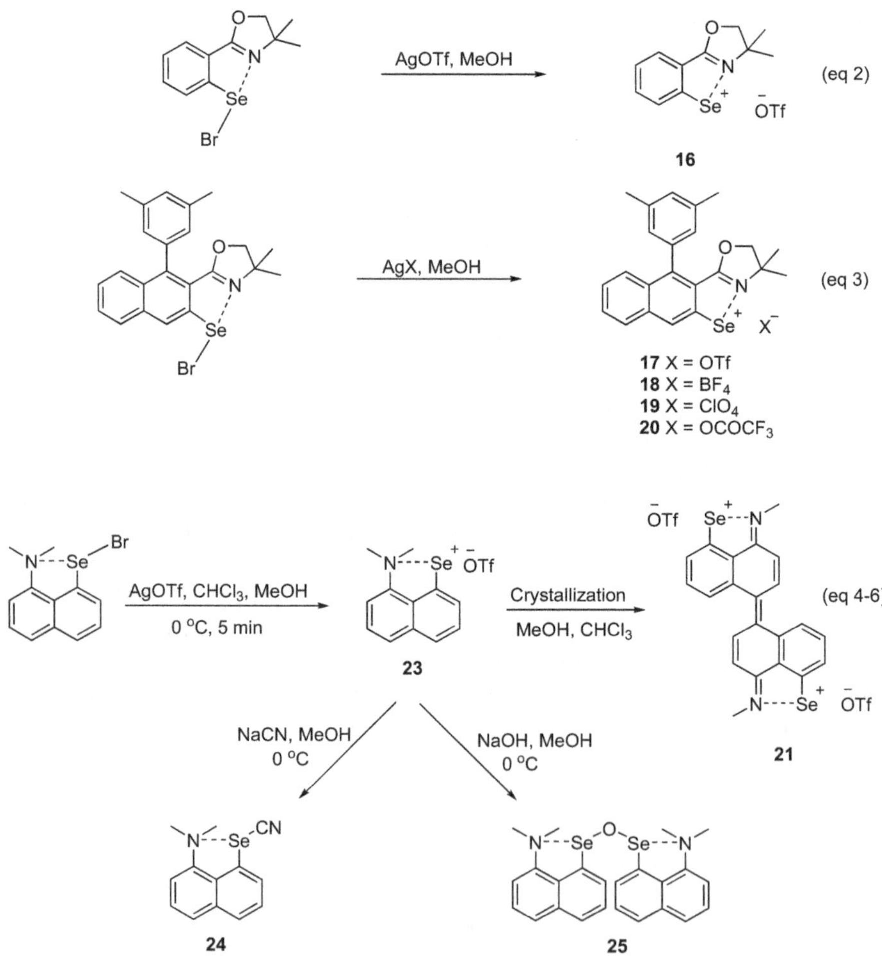

Scheme 5.4 Synthesis of diselenenium ions.

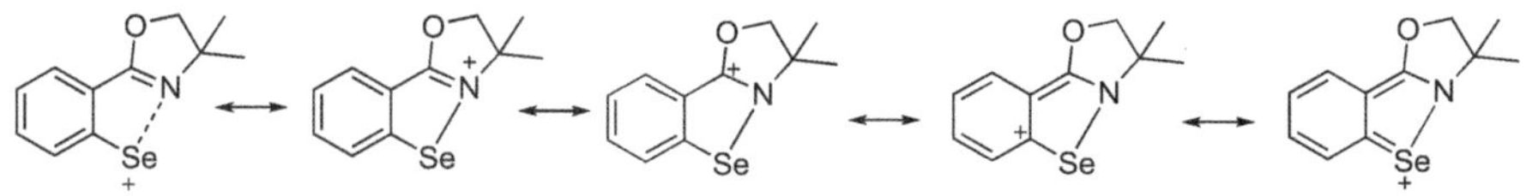

Chart 5.3 Pseudo-aromatic five-membered Se–N ring.

ion **21** is quite stable under ambient conditions for 2 months, whereas the expected selenenium cation **23** turned out to be less stable and its existence could only be indirectly confirmed by its reaction with NaCN and NaOH. The reactions lead to the formation of selenium cyanide **24** and selenium oxide **25**, respectively. The ^{77}Se NMR chemical shifts (950–985 ppm) of **16–20** are significantly upfield shifted compared to the reported chemical shifts (1185–1215 ppm) for the related selenenium cations derived from pincer ligands having two sp^3 nitrogen atoms.[11,16] This is indicative of less positive character on selenium in cations **16–20**. Cation **23** is the exception and shows a significantly downfield shift at 1462 ppm. Furthermore, the Se···N1 (sp^2) bond distance [1.8563(15) Å] in **16** is shorter than that in **1** [2.052(3) Å]. This is the shortest distance reported for any selenenium ion [2.180–1.862 Å],[11,20,22] and is even shorter than the sum of the covalent radii of Se and N atoms (1.91 Å).[22] Further correlation of experimental findings with theoretical investigations by calculating natural bonding orbital charges and Wiberg bond indices indicates that the positive charge is delocalized over the molecule in compounds **16–20**. This suggests that the strong Se···N secondary interaction leads to the formation of a pseudo-aromatic five membered ring (Chart 5.3) and this allows the positive charge of the selenium cation to drift onto the ring and consequently to the aromatic ring. Extended conjugation by an addition of a dimethyl substituted phenyl aromatic ring in **17–20** further improves the stability of selenenium cations. The results further indicate that the isolation of stable selenenium salts is facilitated not only by steric bulk and intramolecular coordination, but also by extended conjugation.

5.3 Preparation of Selenium Dications

The synthesis and characterization of selenium polycations is of current interest. Selenium polycations have an incomplete valence shell and are therefore very reactive and difficult to isolate. Another difficulty in the isolation of dicationic selenium species is the unavailability of suitable precursors, as selenium commonly exists as the tetrahalides (SeX_4; X = Cl, Br, I).[23] Unfortunately, these Lewis acid starting precursors readily undergo redox reactions in the presence of strong Lewis bases and often produce selenium in its elemental form, thus precluding the possibility of isolating selenium dications. For the first time, highly charged (dicationic) Se species were generated by exploiting low-valent $SeCl_2$.[23b] The dications have been made isolable by the reaction of a strong Lewis base with a Lewis acidic selenium precursor, which results in delocalization of the polycationic charge. Early

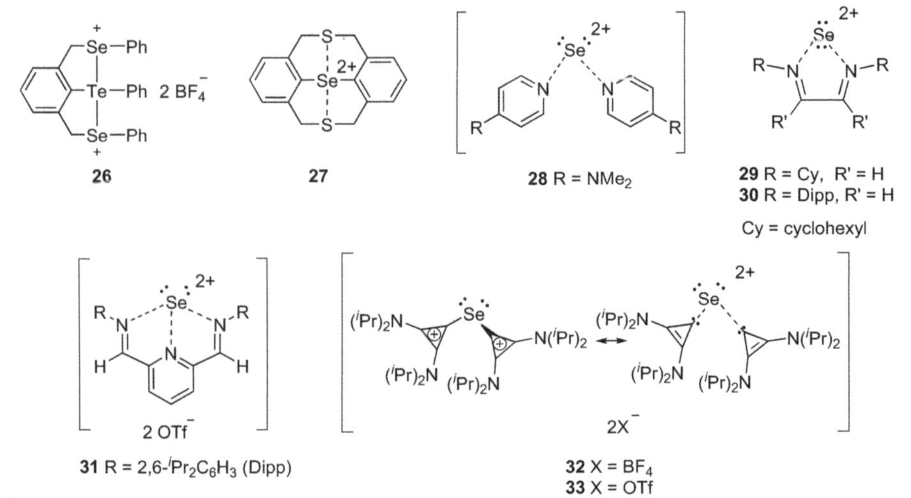

Chart 5.4 Stable selenium(II) dications.

developments in the dicationic chemistry of selenium are highlighted by the work of Furukawa and co-workers.[24] However, these are restricted to rare examples of annulated derivatives with covalent linkages to the main-group element of interest. They reported a new type of dicationic σ-tellurane (λ^4-tellane), $[10\text{-Te-4(C2Se2)}]^{2+}\cdot 2\ BF_4^-$ **26** and dicationic σ-selenurane (λ^4-selane) $[10\text{-Se-4(C2S2)}]^{2+}\cdot$ **27** (Chart 5.4). Ragogna's group has put in a lot of effort to stabilise and understand the reactivity of selenium(II) dications.[23,25–27] For the first time, in 2006, they introduced a heterocyclic selenium carbenoid and synthesised a dicationic N-heterocyclic selenium carbene analogue.[26b] Subsequently, Ragogna's and other groups reported a different type of selenenium dication, where the dicationic derivatives were stabilized by monodentate (**28**) and bidentate (**29–30**) ligands.[26] The problem with both these systems **28–30** is that they are highly air- and moisture-sensitive and start decomposing within few a hours of their synthesis.

Later, Ragogna *et al.*[27] utilized a tridentate pyridine-based pincer ligand (diiminopyridine, DIP) for the isolation of selenium dication **31**. This cation showed remarkable stability in the open air for a period of 3 weeks. Selenium dication **31** adopts a trigonal bipyramidal geometry based on the AX_3E_2 electron pair configuration around the selenium centre. The cation shows a T-shaped structure in which both imine nitrogen atoms occupy axial positions and the pyridine nitrogen and two lone pairs occupy the equatorial positions. The selenium–nitrogen distances are slightly shorter than typical selenium–nitrogen bonds and slightly longer than those in other selenium cations. This could be due to the increase in coordination number from 2 to 3. Alcarazo and co-workers[28] reported the synthesis, structure, and reactivity of cyclopropenyl-1-ylidene stabilized Se(II) dications **32** and **33**.

Evans, Block and Glass and co-workers[29] prepared organoselenium [8-Se-3] cations **34–36** by electrochemical and chemical oxidation of diselena-, selenathia-, and tellurathiamesocycles (Chart 5.5). The X-ray crystal structures of the Se–Se dimer of the product of one-electron oxidation of 1,5-selenathiocane (**35**), and the monomeric two-electron oxidized product **36** are reported. Electrochemical study demonstrates that the chemical stabilities follow the order; Te^{+}–Te >Se^{+}–Se >S^{+}–S, as well as Se^{+}–Se >Se^{+}–S and Te^{+}–Te >Te^{+}–S. Furthermore, the stability of these cationic species is attributed to the transannular chalcogen bond.

Dutton and co-workers[30] reported the reactions of aromatic group 16 thiophene, selenophene, and tellurophene rings with the I(III) oxidants, PhI(OAc)(OTf) and $[PhI(Pyr)_2][OTf]_2$ (Pyr = pyridine) (Scheme 5.5). In the reaction, oxidative processes take place with generation of PhI as the reduction product. Reaction of the substituted selenophene using two equivalents of $[PhI(Pyr)_2][OTf]_2$ gave **37**. This was established using ^{77}Se NMR spectroscopy, which showed one peak at 628 ppm. Reaction of selenophene with PhI(OAc)(OTf) gave **38**. X-ray diffraction studies confirmed the product as iodonium salt **38**, an unreported derivative of selenophene. Although **37** and **38** are a di-cation and mono-cation, respectively, in both cations, none of the selenium atoms bears a positive charge. This is also supported by their ^{77}Se NMR chemical shifts, 627 ppm for **37** *vs.* 620 ppm for the precursor and 726 ppm for **38** *vs.* 644 ppm for selenophene.

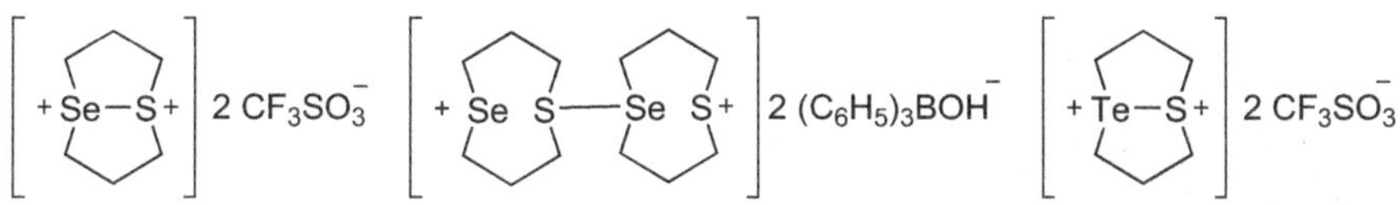

34 **35** **36**

Chart 5.5 Organoselenium [8-Se-3] dications.

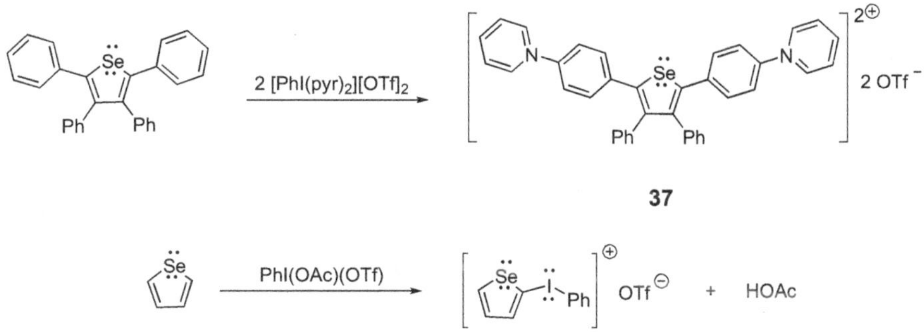

37

38

Scheme 5.5 Synthesis of [8-Se-2] cations **37** and **38**.

5.4 Organoseleniranium and Selenirenium Cations

The organoselenium(IV) cations (seleniranium ions) **39–40** were proposed in 1974 as intermediates in the seleno functionalization of alkenes by the addition of organoselenenium cations to alkenes (Scheme 5.6). The functionalization of alkenes is discussed in detail under applications (Section 5.6).[2,12,13]

The proposed seleniranium or [8-Se-3] cation plays a decisive role in determining the stereoselectivity in the reaction.[2b] Wirth and co-workers[31] have reported the α-methoxyselenenylation reaction of styrene with a series of chiral selenium electrophiles. Later on, Denmark and co-workers provided many insights into the structure and selectivity of the organoselenium(IV) cations and proposed that the selenium(IV) cation can be activated by the addition of a Lewis base. This study provided insight to circumvent the "olefin-to-olefin" transfer process and characterized the cations **41–48** by multinuclear NMR spectroscopy (Chart 5.6).[10,32]

Structural characterization of selenirenium cations was provided by Poleschner and Seppelt by using sterically bulky alkyne substrates (Scheme 5.7).[33] Selenirenium cations were obtained either by the reaction of an alkyne with the phenylselenenium cation or by the reaction of 1-halo-2-phenylseleno alkenes with silver hexafluoroantimonate (Scheme 5.7).

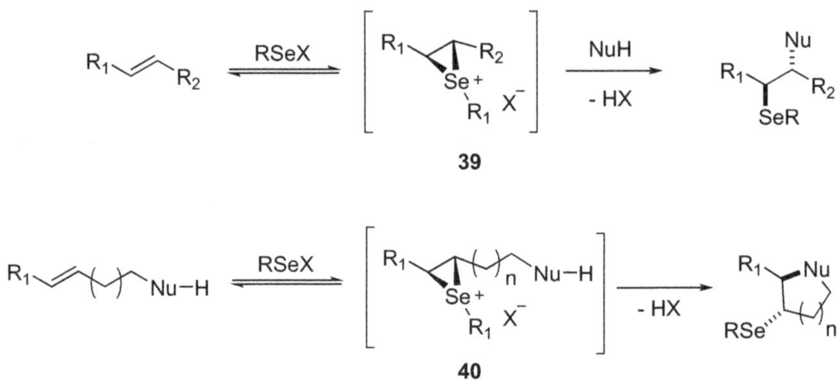

Scheme 5.6 Stereoselective functionalization of non-activated C–C double bonds.

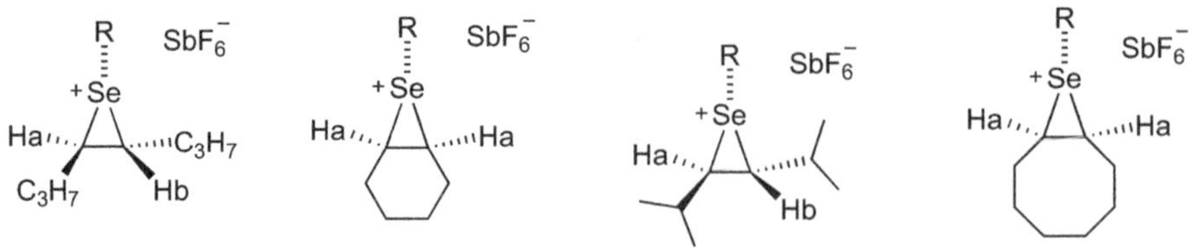

41 R = C_6H_5 **43** R = C_6H_5 **45** R = C_6H_5 **47** R = C_6H_5
42 R = C_4H_9 **44** R = C_4H_9 **46** R = C_4H_9 **48** R = C_4H_9

Chart 5.6 Organoselenium(IV) cations.

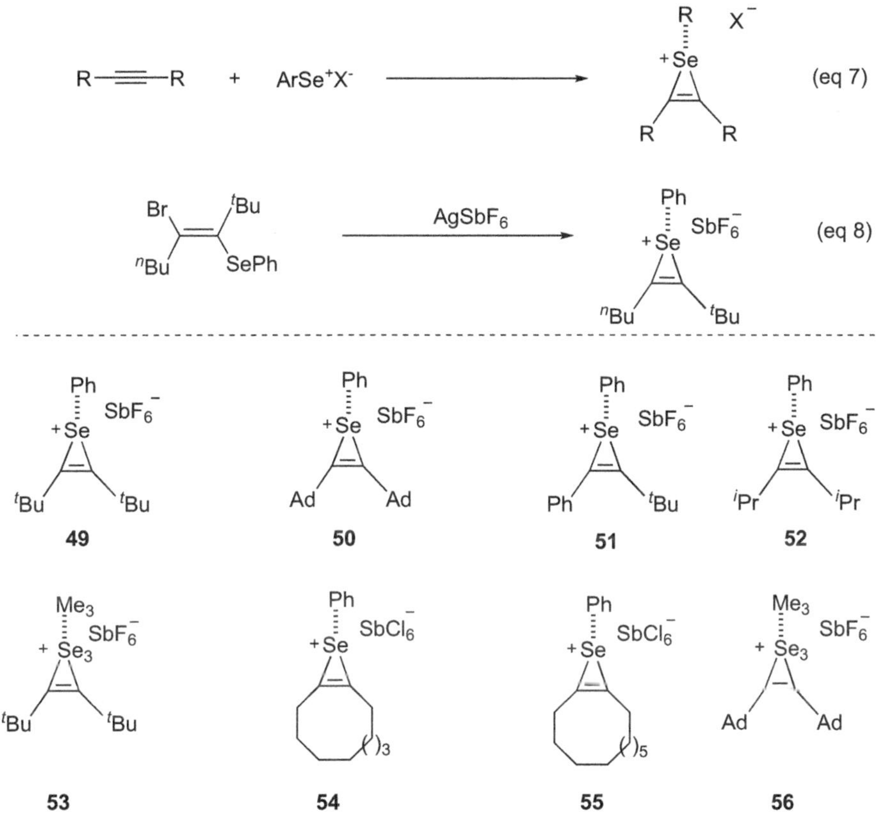

Scheme 5.7 Preparation of selenirenium cations.

The ^{77}Se NMR study of a series of selenirenium cations **49–56** shows a chemical shift range from −200 to −50 ppm which is upfield shifted compared to selenenium cations. Similarly, upfield shifts in ^{13}C NMR resonances were observed for the cyclic carbons. The isolated selenirenium cations we noticed to be stable only under an argon atmosphere and considerable decomposition was observed after a week.

Knight *et al.* isolated a series of organochalcogen cations **57–62** and the dication **63** derived from acenaphthene and naphthalene ligands and stabilized by *peri*-chalcogen intramolecular coordination (Chart 5.7).[34] The monocationic chalcogenium salts **57–63** were synthesised by a methylation reaction performed by treating MeOTf in a 1 : 1 molar ratio with the corresponding naphthalene and acenaphthene chalcogenide derivative. The ^{77}Se NMR chemical shifts are in the range 320–425 ppm for these [8-Se-3] cations and are indicative of a smaller positive charge on the selenium nucleus as compared to [6-Se-1] cations.

Huang and colleagues[35] have described the synthesis of several chiral [8-Se-3] cations. These were derived from camphor methyl selenide and

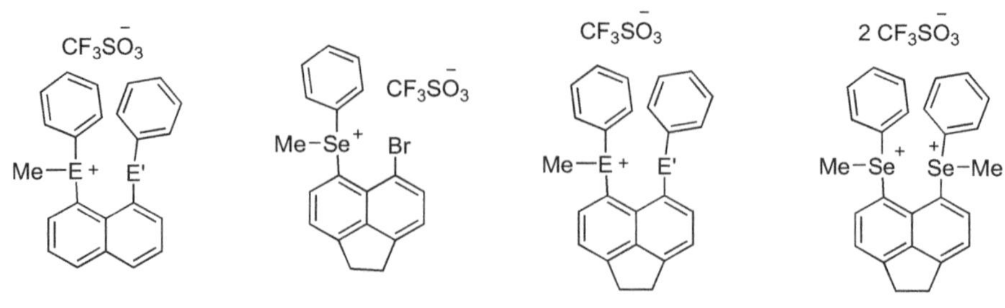

57 E/E' = Se
58 E = Te E' = Se

59

60 E/E' = Se
61 E = Se E' = S
62 E = Te E' = Se

63

Chart 5.7 Examples of monocationic and dicationic selenonium salts.

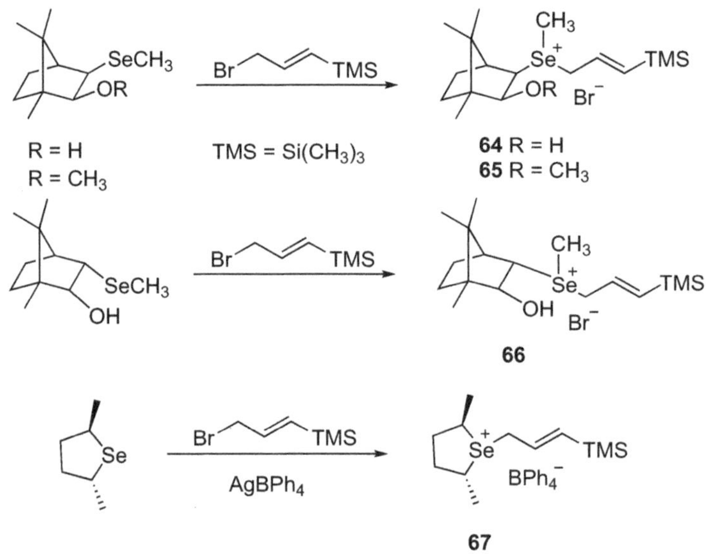

Scheme 5.8 Preparation of [8-Se-3] cations.

chiral tetrahydroselenophene (Scheme 5.8). Three cations **64, 66** and **67** have been structurally characterized. The selenium [8-Se-3] cations were applied in the preparation of chiral 1,2,3-trisubstituted cyclopropanes (see later, Scheme 5.27).

Midura *et al.*[36] synthesized chiral organoselenonium [8-Se-3] cations **68–70** derived from (−)-menthol and (+)-limonene ligands (Scheme 5.9) and used them as precursors for ylide generation for the asymmetric cyclopropanation of vinylphosphonates (see later, Scheme 4.28). Organoselenium cations were obtained by the reaction of diorgano selenides with benzyl bromide or ethyl 2-chloroacetate, followed by anion exchange with silver triflate or silver tetrafluoroborate. The synthesized cations are stable and have been characterized by multinuclear (^{1}H, ^{13}C, and ^{77}Se) NMR spectroscopy.

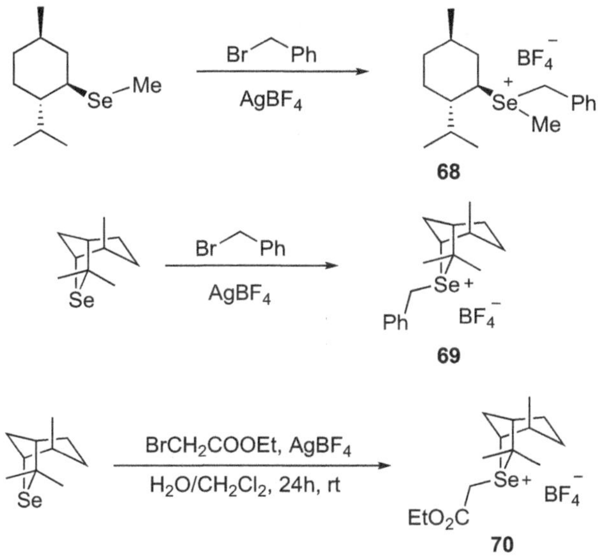

Scheme 5.9 Preparation of selenium[8-Se-3] cations.

5.5 Diselenium Cations

Interestingly, Poleschner and Seppelt[37] used the most important class of organoselenium compounds, the diorgano diselenides, for the preparation of diselenenium cations using $XeF_2/BF_3 \cdot OEt_2$ as a one-electron oxidant (Scheme 5.10). In addition, $R_4Se_4^{2+}$ $(TfO^-)_2$ have been prepared by oxidation of the dialkyl dichalcogenides, R_2E_2, with the one-electron oxidant NO^+ TfO^- ($Tf = CF_3SO_2$). When iodine was used as an oxidizing reagent along with the one-electron oxidant NO^+ TfO^-, a dicationic four-membered ring (72) and neutral charge transfer adduct of dimethyl diselenide and diiodine 73 were also obtained.[38]

Our group has prepared dicationic diselenides *via* oxidation of selones with different oxidizing agents (Scheme 5.11).[39] Alternative systematic investigations suggested that the reduction of dihaloselones to dicationic diselenides required the presence of strong Lewis acid metal halides and selones for the redox reactions. Alternatively, the selone adducts of metal halides, when treated with halogens, provide the desired methodology to synthesize dicationic diselenides.

Devillanova and co-workers[40] have synthesized a compound of formula $[SeL_4][BF_4]_2$ by electrochemical oxidation of *N*-methylbenzothiazole-2(3*H*)-selone, where the $[SeL_4]^{2+}$ cation can be formally described by analogy with transition-metal complexes, being formed from a central Se^{2+} cation coordinated by four neutral molecules of L to achieve an almost square-planar arrangement of five selenium atoms. The crystal structure consisting of bicyclic Se_{10}^{2+} cations and of layered chloroidobismuthate anions with the cations located between the anionic planes has been reported by Beck and Eck.[41]

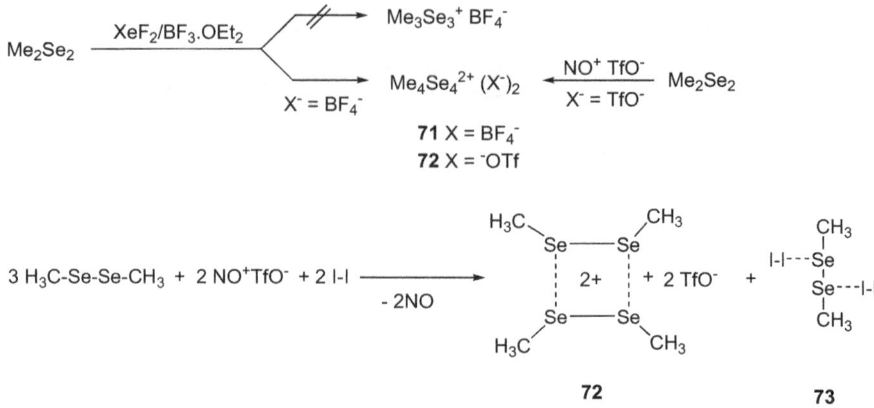

Scheme 5.10 Oxidation of Me_2Se_2 to dimethyl diselenide mono cation using a one-electron oxidant.

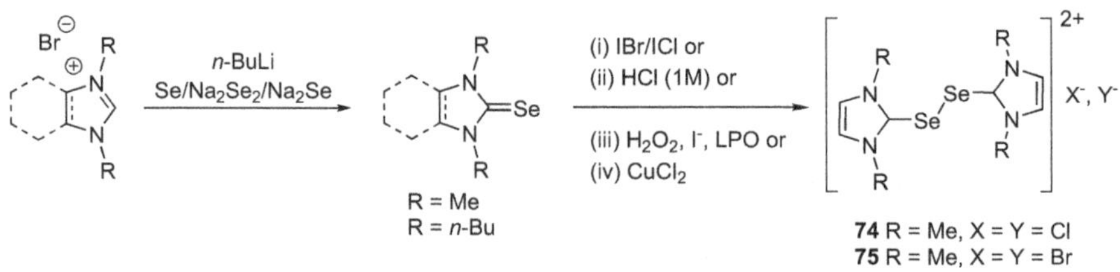

Scheme 5.11 Oxidation of selones to diselenide dications.

This dication contains the greatest number of selenium atoms (10) reported so far.

5.6 Applications of Organoselenium Cations

Organoselenenium cations, prepared *in situ* from diorgano diselenides were used for the first time in the functionalization of alkenes (Scheme 5.6). At present, there are not many reports of the applications of isolated organoselenium cations. The very first isolated selenenium cation **1** was used for the oxidation of benzenethiol by atmospheric oxygen (Scheme 5.12).[11]

The progress of the reaction was monitored using ^{77}Se NMR spectroscopy, which indicated the formation of selenenyl sulfide **76** ($\delta = 1214$ ppm). The reaction with an additional equivalent of PhSH yielded diphenyl disulfide and intermediate **77**, which regenerates the selenenium cation **1** by atmospheric oxidation. This suggests that oxidation of benzenethiol is possible by air in the presence of selenenium cation **1**.

Selenenium cations have also been studied as mimics of glutathione peroxidase selenoenzyme (Scheme 5.13).[21] The selenenium cation **12** oxidizes to the corresponding selenoxide and related analogues **78–83** by the action of an excess of hydrogen peroxide followed by reaction with benzenethiol.

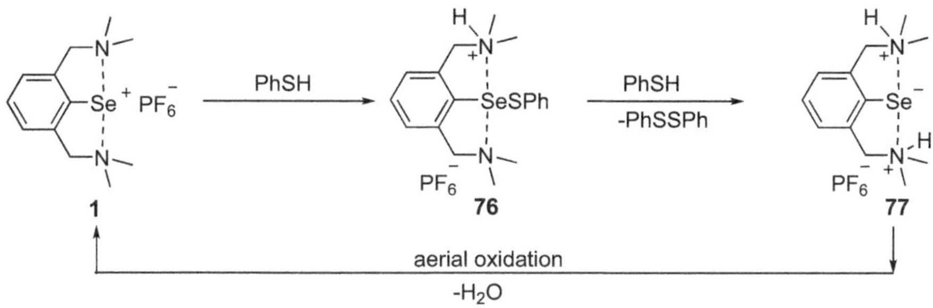

Scheme 5.12 Oxidation of PhSH to diphenyl disulphide.

The selenenium cations showed better thiol peroxidase activity in the oxidation of benzenethiol to diphenyl disulfide than the organoselenium drug ebselen. Selenenium cation **18** derived from naphthalene shows a visible color change in acidic solution.[22] In basic conditions, the selenenium cation exists in the non-cationic form **84**, but produces a dark colored solution in an acidic medium (Scheme 5.14).

It seems that conjugation between the two naphthalene rings is broken down by hydrolysis of the C=C double bond, leading to a colorless species (**84**). Under acidic conditions, the species occurs as the highly conjugated blue-colored selenenium cation **18**. Apart from these reports, there is no study of applications of structurally characterized organoselenium cations, but there are several reports in which organoselenium, particularly selenenium cations, were prepared *in situ* and used in many reactions. Most of the examples presented here delineate the role of isolated selenenium cations. Bartoli *et al.* have prepared phenylselenenium cation (**85**) using a persulfate oxidant (Scheme 5.15).[13] It is worth mentioning that only the isolated phenylselenenium sulfate is successful in the functionalization of alkenes. Otherwise, the reaction mixture of diphenyl diselenide, persulfate and alkene provides a polymeric undesired product. Persulfate is an economical one-electron oxidizing agent and easily oxidizes diphenyl diselenide phenylselenenium cation. However, this oxidant polymerizes alkenes. Therefore, it seems practical to isolate phenylselenenium cations and then use them in the functionalization of alkenes (Scheme 5.15).

Another important application of arylselenenium cations was realized in the electrophilic substitution of aromatic rings for the synthesis of unsymmetrical diaryl and aryl alkyl selenides by our group.[8] Synthesis of unsymmetrical diaryl selenides is difficult either by conventional metalation using organolithium, lithium–halogen exchange and Grignard routes, or by transition metal-catalyzed coupling of aryl halides with suitable aryl selenium precursors. Conventional routes require a strictly inert atmosphere and a precursor with a halogen leaving group or an excess of diaryl diselenide. Similarly, transition metal-catalyzed reactions require harsh reaction conditions along with high catalyst loading, as chalcogen containing precursors poison the catalyst. The generation of the phenylselenenium

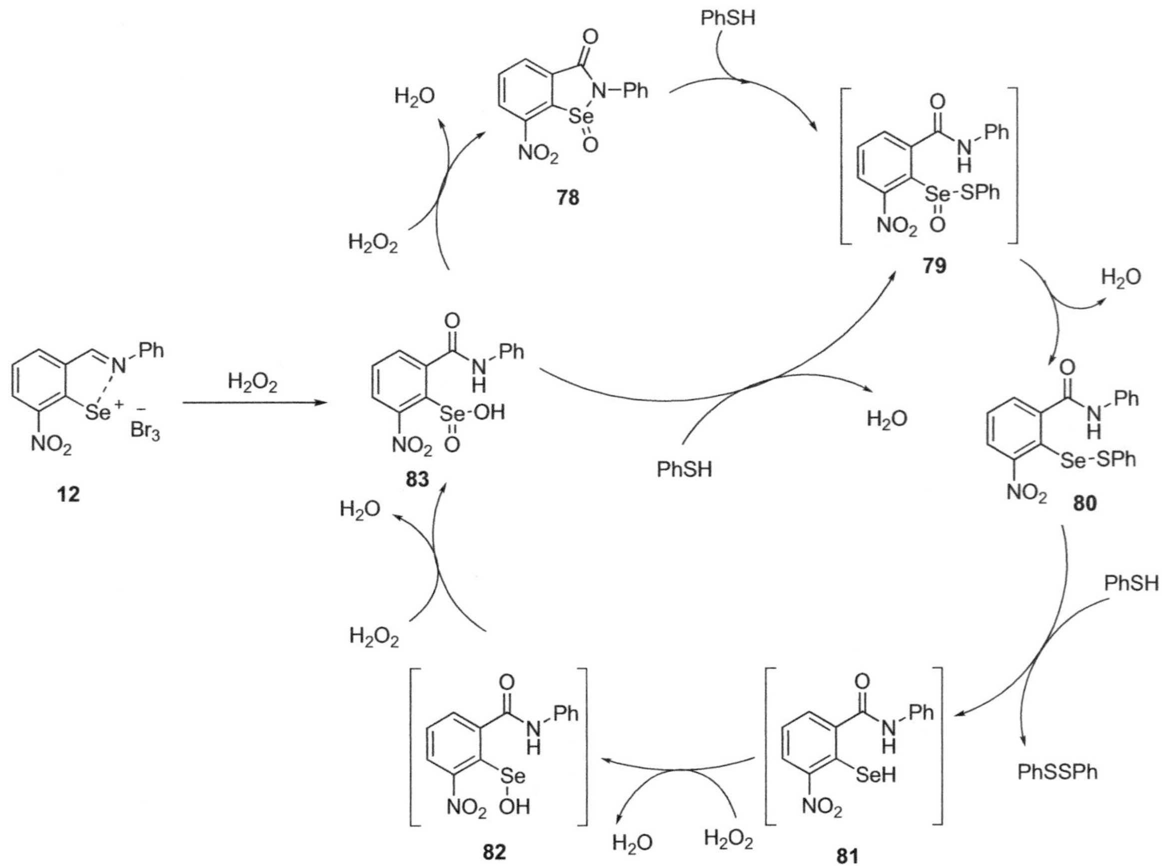

Scheme 5.13 Selenenium cations as glutathione peroxidase (GPx) models.

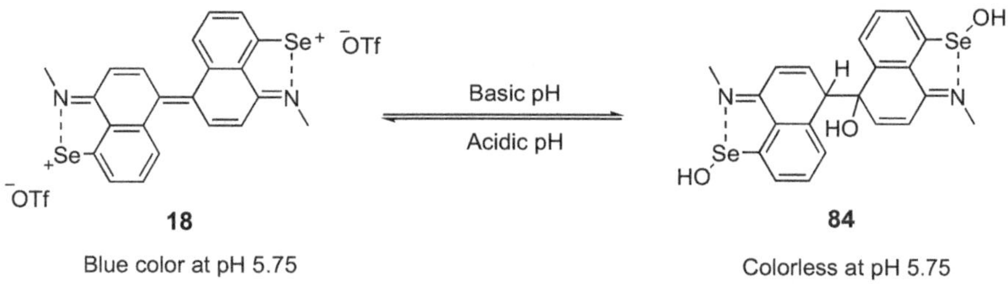

Scheme 5.14 Colorimetric pH-responsive selenenium cations.

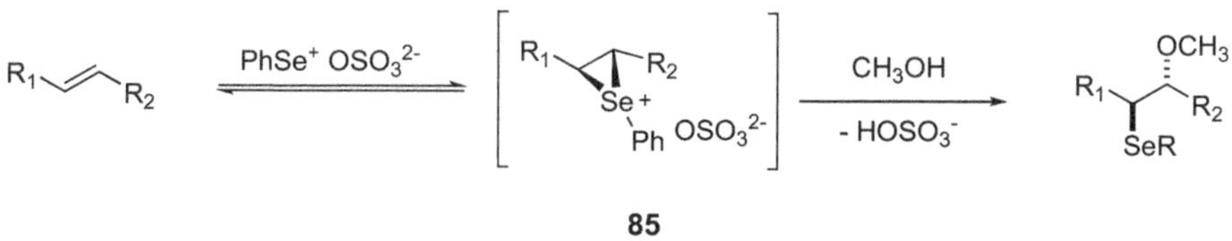

Scheme 5.15 Phenylselenomethoxylation of alkenes.

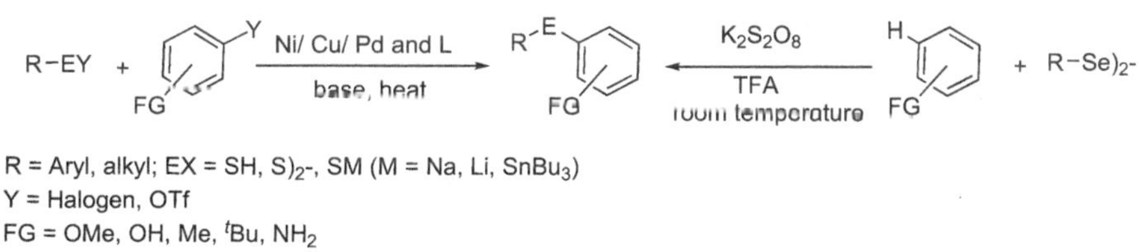

Scheme 5.16 Synthesis of diorganoselenides.

cation allows the synthesis of unsymmetrical diaryl selenides at room temperature without involving strict manipulations and expensive chemicals. Most importantly, the use of the phenylselenenium cation enables the synthesis of usymmetrical diaryl selenides with sensitive OH, COOH and NH_2 functional groups. These functionalities are important in further diversification of diaryl selenides and cannot be easily accessed either by transition metal or metalation methods (Scheme 5.16).

Most importantly, the aryl selenide moiety can be incorporated into biological molecules such as vitamin E, δ-tocopherol and other related phenolic antioxidants which are highly electron rich, which undergo electron transfer chemistry under conventional methods and do not provide the desired arylselenated product (Scheme 5.17).

It is well established that incorporation of an organoselenium moiety in the close proximity of an –OH functionality in phenols enhances their radical chain-breaking antioxidant activity. The treatment of phenylselenenium cation generated *in situ* by the reaction of persualfate with diphenyl diselenide provided the desired organoselenated tocopherols in good yields (Scheme 5.18).[42] Although the arylselenenium cation has been used

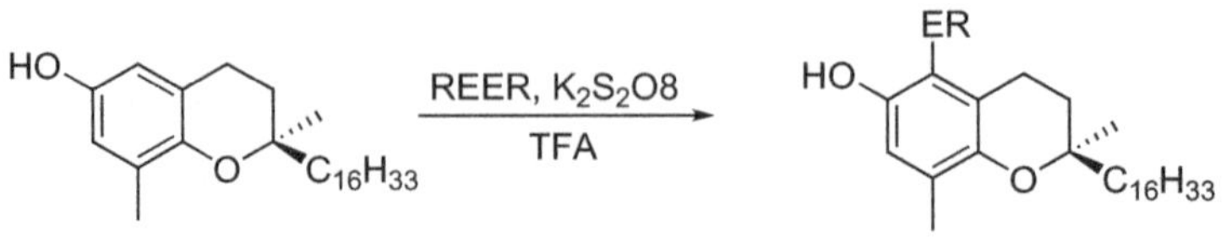

Scheme 5.17 Synthesis of alkylseleno-δ-tocopherols **86–87**.

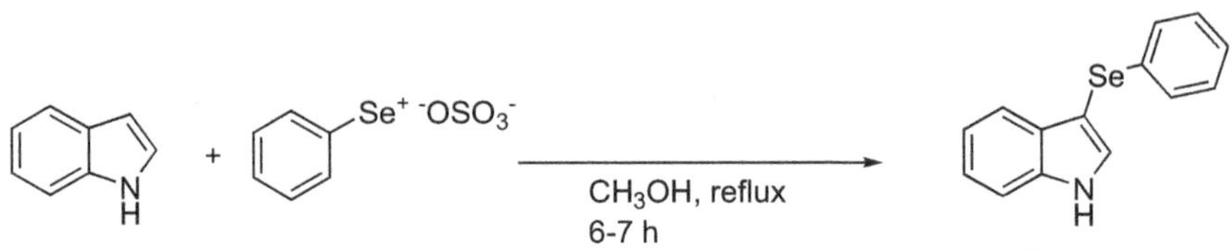

88

Scheme 5.18 Synthesis of **88**.

successfully for the aryl selenation of arenes, a highly acidic medium is required to accomplish the transformation. Substrates such as indoles or pyridines are sensitive to acidic conditions and not compatible with the reaction conditions.

Our group has also studied arylselenation and thiolation of an indole moiety in a benign methanol solvent system (Scheme 5.18).[8b] In methanol, phenylselenenium sulfate was formed under reflux conditions and subsequently treated with indole in the same pot to obtain the desired arylselenide **88**. The catalytic applications of the *in situ* generated phenylselenenium cation have been studied by Singh and Wirth[43] for the synthesis of butenolides from butenoic acids (Scheme 5.19). In the reaction, the phenylselenenium cation was generated by the reaction of diphenyl diselenide with [bis(trifluoroacetoxy)iodo]benzene as a stoichiometric oxidant in acetonitrile to give the highest yields (Scheme 5.19). Using 5 mol% of the catalyst is critical, since lower catalytic loadings resulted in a side product.

Wirth and co-workers[31] have synthesized a series of new chiral diselenides as precursor molecules for the generation of powerfully electrophilic selenenium cations **89** and **91**, which add with high stereoselectivities to alkenes (Schemes 5.20 and 5.21).

Several conclusions regarding the conditions required to achieve very high selectivity in the functionalization of alkenes can be drawn from the studies of Wirth and co-workers,[31] particularly in the methoxyselenenylation of alkenes. Although electrophilic selenium reagents can be easily prepared by the addition of chlorine or bromine to diorgano diselenides, bromide is not a good nucleophile for alkenes such as styrene and cyclohexene as it does not add to the carbon–carbon double bond in these substrates. Most importantly, the reversibility of the addition of bromide to the alkene leads to a

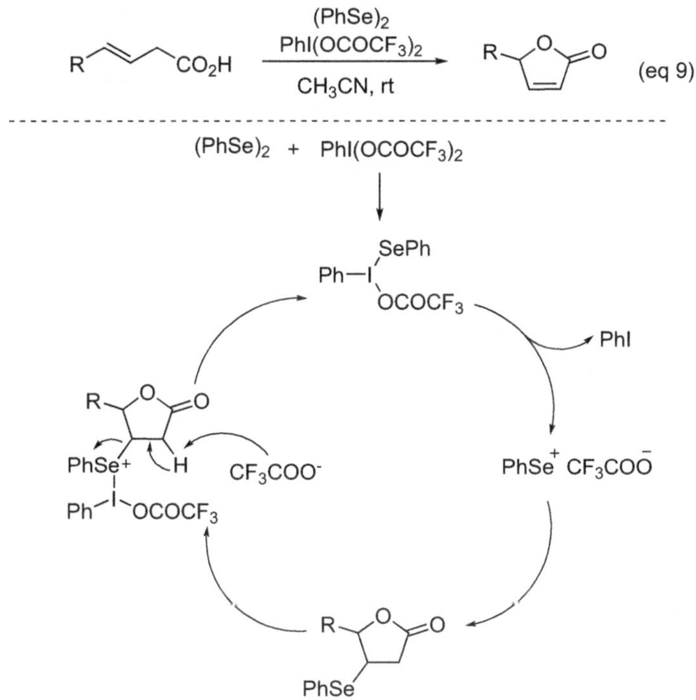

Scheme 5.19 Electrophilic selenium reagents.

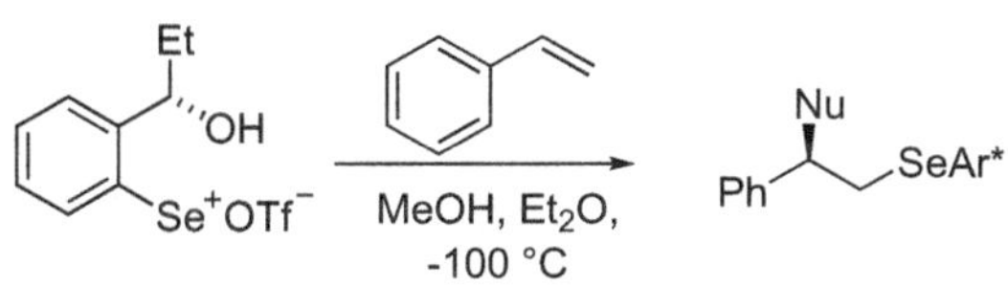

89

Scheme 5.20 Chiral selenenium cation addition to alkene.

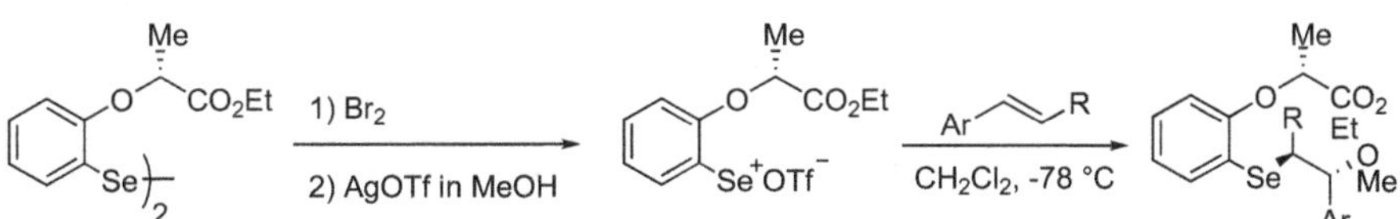

90 **91**

Scheme 5.21 A selenenium cation containing two heteroatoms.

potential decrease in stereoselectivity. Denmark *et al.* have shown that formation of seleniranium ion destroys the selectivity in these reactions. Therefore, it is better to exchange the halide ion *in situ* with a non-coordinating anion such as triflate or sulfate.

The stereochemistry-determining step is the formation of the seleniranium intermediate **93** by the attack on the alkene of the selenium electrophile **92** (Scheme 5.22). It has been suggested that the stability of seleniranium intermediate **93** decides the enantioselectivity in the selenomethoxylation of alkenes. The more stable the seleniranium ion, the less is the chance for it to dissociate to the alkene and the phenylselenenium cation, *i.e.* the backward reaction, and hence a good selectivity is expected. The stability of the seleniranium ion is enhanced by the interaction of a heteroatom with the cationic selenium. Back and co-workers[44] have reported a series of chiral selenenium cations **96–101** which act as electrophiles in various selenomethoxylation reactions (Chart 5.8).

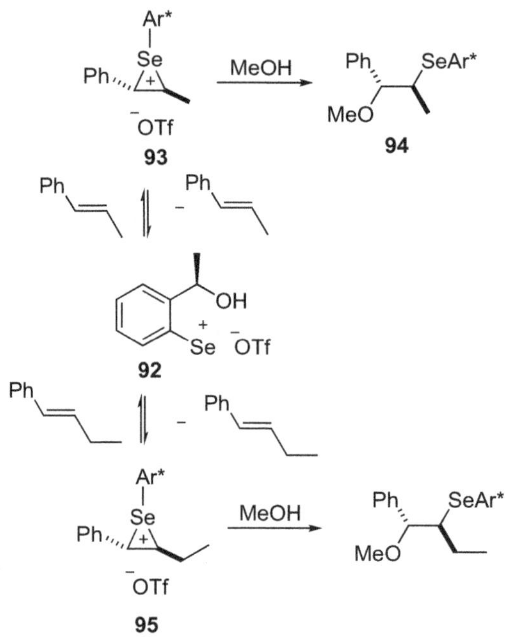

Scheme 5.22 Stability of seleniranium intermediates **93** and **95**.

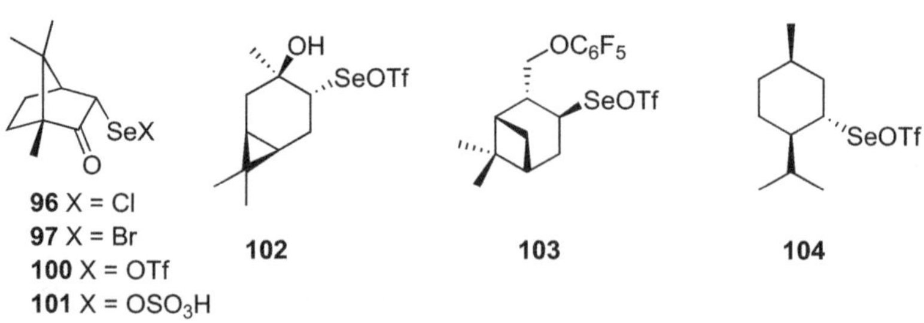

Chart 5.8 Terpene-based electrophilic selenium reagents.

Recent advances in selenenylation have been made by Tiecco *et al.*[45] in an asymmetric azidoselenenylation reaction (Scheme 5.23). This allows further transformations into aziridines and triazoles. It is remarkable that this reaction occurs with a very high level of facial selectivity and with Markovnikov orientation. The sulfur atom in the chiral side chain in electrophile **105** seems to play an important role in this reaction.

Denmark *et al.*[46] attempted a diastereoselective selenoetherification using achiral phenylselenium succinimidate in the presence of a chiral Lewis base catalyst (Scheme 5.24). It was envisioned that as the *in situ*-formed phenylselenenium cation is a Lewis acid, it could be stabilized by a chiral Lewis base and thus interaction of Lewis acid and chiral base would provide a

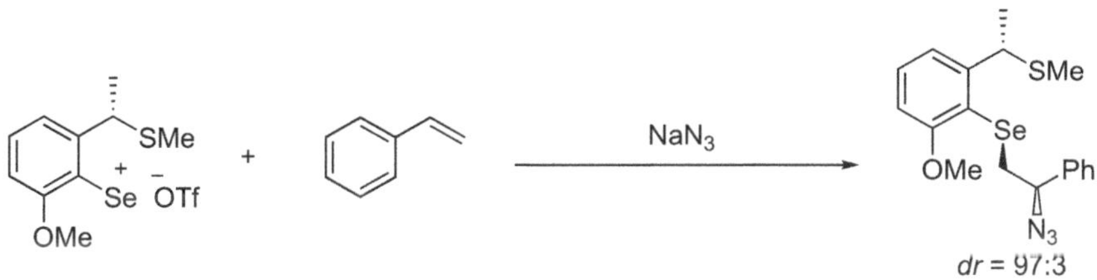

105

Scheme 5.23 Sulfur-coordinated selenenium cation for asymmetric azidoselenenylation.

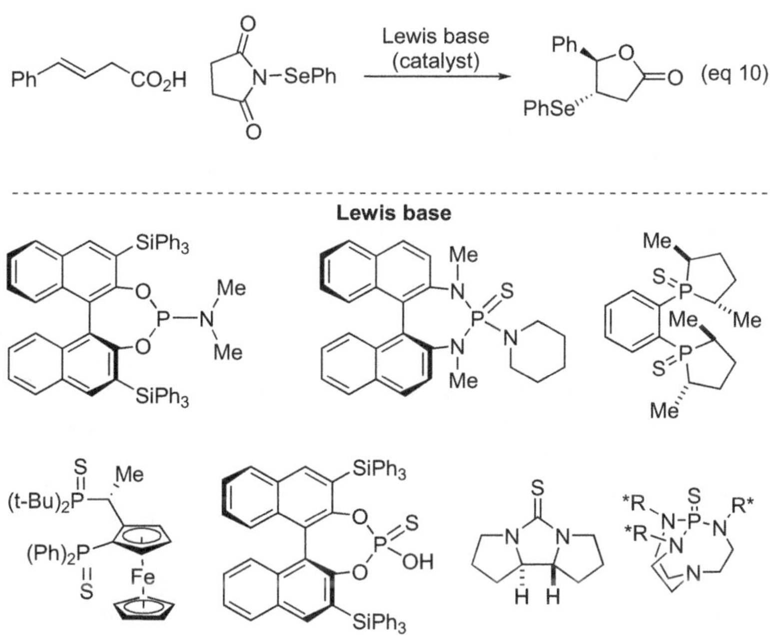

Scheme 5.24 PhSe$^+$ stabilized by intermolecular coordination of chiral Lewis base.

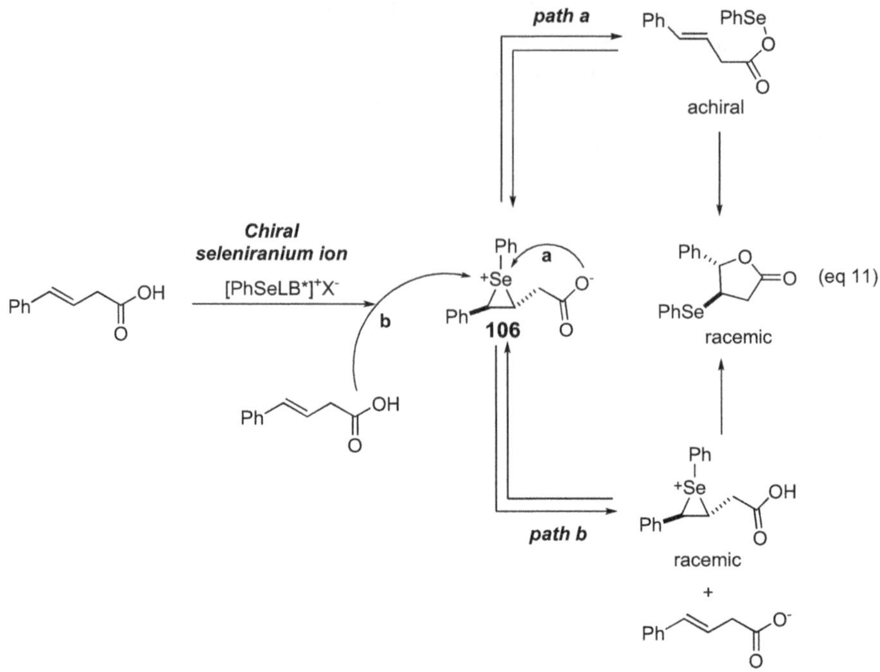

Scheme 5.25 Possible pathways for racemization of phenyseleniranium cation.

stable seleniranium intermediate leading to a distereoselective seleno-etherification of alkenes (Schemes 5.24 and 5.25). Disappointingly, the diastereoselectivity of the reaction was very low. As discussed earlier, selenenium cations **7** and **8** stabilized by intermolecular coordination of phosphorus were characterized using single-crystal X-ray crystal structure determination, but were unstable in solution. Presumably the coordination of the selenenium cation by the chiral Lewis base is also weak, and as a consequence the resulting seleniranium cation **106** racemized by paths a and b due to its poor stability.

Recently, Yeung *et al.*[47] accomplished the enantioselective selenolactonization of phenyl substituted alkenoic acids using chiral Lewis bases derived from quinine or diazines and phenylselenium succinimidate (Scheme 5.26). Here, high enantioselectivity has been achieved by the interaction of the substrate with the chiral Lewis base and not by its interaction with the phenylselenenium cation.

Huang and co-workers[35] provided a rationale for using selenonium ylides or selenium[8-Se-3] cations in the synthesis of chiral cyclopropanes rather than sulfur and tellurium ylides (Scheme 5.27). Although chiral sulfonium ylides provided high enantioselecitivity in the cyclopropanation reaction, they have the tendency to initiate rearrangement reactions, and as a result the yields of the asymmetric cyclopropanation are sometimes not satisfactory. Chiral telluronium ylides provided chiral cyclopropanes in yields of

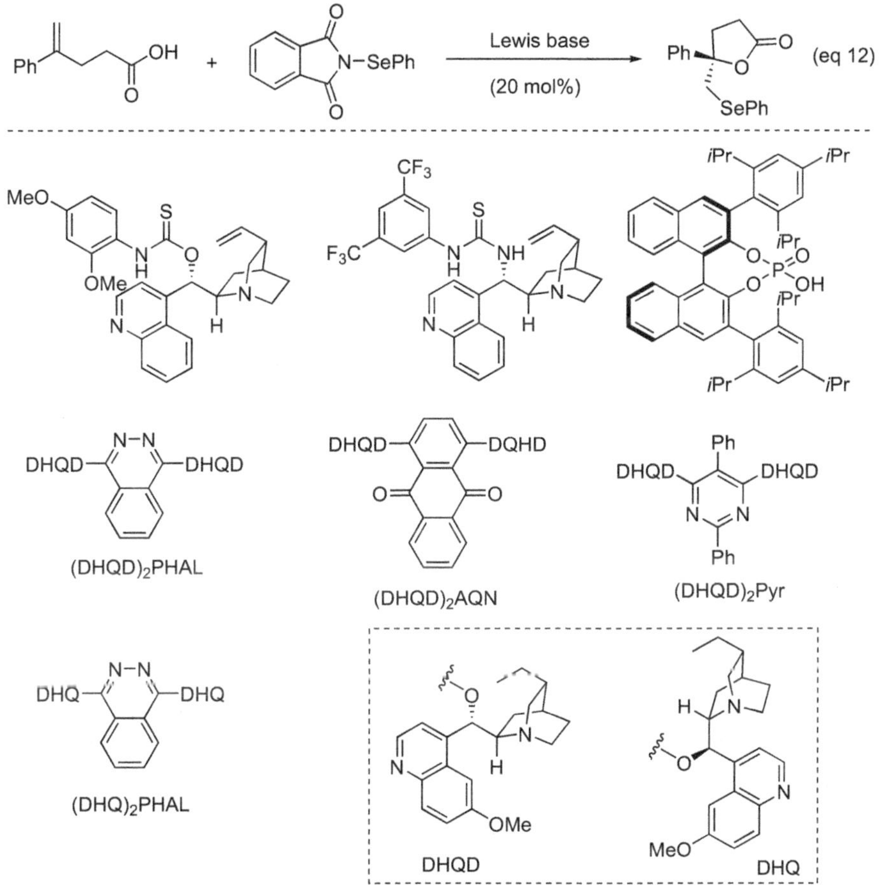

Scheme 5.26 Enantioselective delivery of achiral phenyselenenium cation.

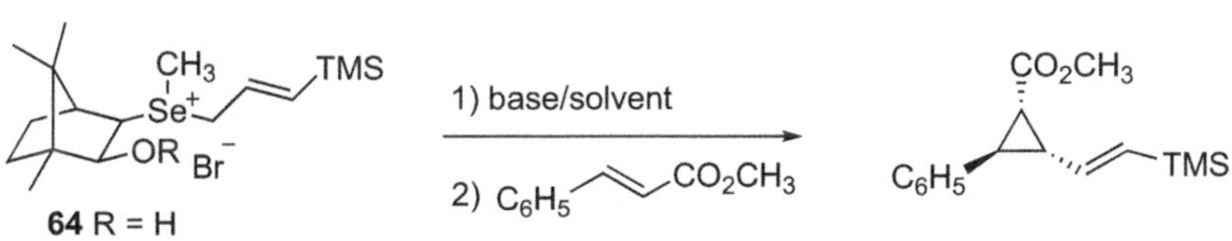

64 R = H
65 R = CH$_3$

Scheme 5.27 Enantioselective cyclopropanation by chiral [8-Se-3] cations.

up to 99% with excellent diastereoselectivities and enantioselectivities, but the chiral telluride, as the precursor of the telluronium ylide is unstable and this makes it difficult to recover and reuse. The selenonium ylides **64–67** lead to facile synthesis and recyclability of the chiral selenides along with high generality, controllable selectivities and high enantioselectivities in cyclopropanation reactions.

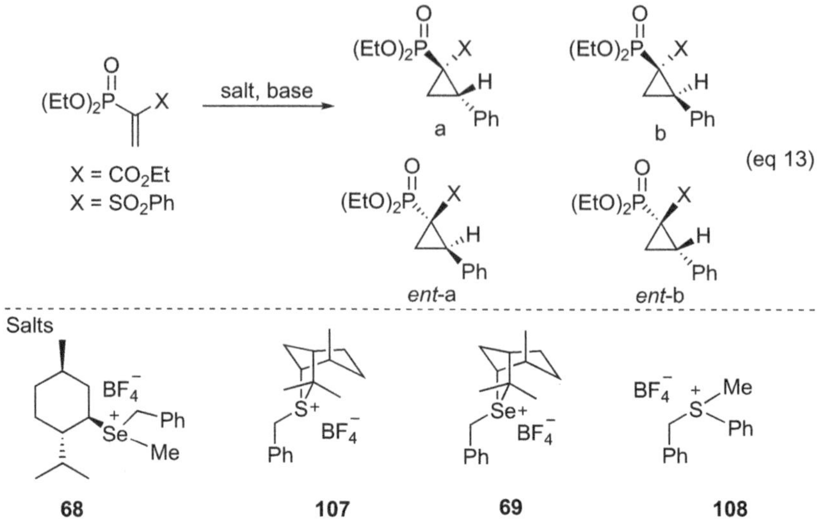

Scheme 5.28 Enantioselective cyclopropanation by chiral [8-Se-3] cations.

The asymmetric cyclopropanation of a vinylphosphonate using optically active selenonium ylides or [8-Se-3] cations **68–70** derived from (−)-menthol and (+)-limonene has been developed by Midura *et al.* (Scheme 5.28).[36] The ylides were generated *in situ* by the reaction of the selenonium salts with a potassium carbonate or DBU base. The transfer of the CHPh and $CHCO_2Et$ groups to the cyclopropane ring showed moderate diastereoselectivity and excellent enantioselectivity. It was observed that the selenium cations obtained from cyclic isoselenocineole provided better enantioselectivities than the cations derived from methyl menthyl selenides. The sulfur derived cations **107** and **108** showed similar diastereoselectivity and enantioselectivity.

Acknowledgements

SK thanks DST-SERB New Delhi (EMR/2015/000061) and IISER Bhopal for generous funding and heartily acknowledges and thanks Dr Shailesh Kumar and Dr Rahul Kadu for their valuable suggestions. HBS is grateful to DST, New Delhi for the award of J C Bose National Fellowship.

References

1. (a) B. J. Bhuyan and G. Mugesh, in *Biological and Biochemical Aspects of Selenium Compounds, in Organoselenium Chemistry* ed. T. Wirth, 2012, p. 361; (b) G. Mugesh and H. B. Singh, *Chem. Soc. Rev.*, 2000, **29**, 347.
2. (a) F. V. Singh and T. Wirth, *Selenium Compounds as Ligands and Catalysts, in Organoselenium Chemistry*, ed. T. Wirth, 2012, p. 321; (b) C. Santi, R. D. Lorenzo, C. Tidei, L. Bagnoli and T. Wirth, *Tetrahedron*, 2012, **68**, 10530.

3. (a) S. T. Manjare, Y. Kim and D. G. Churchill, *Acc. Chem. Res.*, 2014, **47**, 2985; (b) S. J. Balkrishna, A. S. Hodage, S. Kumar, P. Panini and S. Kumar, *RSC Adv.*, 2014, **4**, 11535.
4. A. Kremer, C. Aurisicchio, F. D. Leo, B. Ventura, J. Wouters, N. Armaroli, A. Barbieri and D. Bonifazi, *Chem. – Eur. J.*, 2015, **21**, 15377.
5. L. Zhao, J. Li, Y. Li, J. Liu, T. Wirth and Z. Li, *Bioorg. Med. Chem.*, 2012, **20**, 2558.
6. F. V. Singh and T. Wirth, in *Stereoselective Reactions of Organoselenium Reagents Including Catalysis, Chemistry of Organic Selenium and Tellurium Compounds*, ed. S. Patai and Z. Rappoport, Wiley, 1986–1987, Chichester, UK, 2012, vol. 3, p. 303.
7. (a) D. M. Browne, O. Niyomura and T. Wirth, *Org. Lett.*, 2007, **9**, 3169; (b) S. J. Balkrishna, C. D. Prasad, P. Panini, M. R. Detty, D. Chopra and S. Kumar, *J. Org. Chem.*, 2012, **77**, 9541.
8. (a) C. D. Prasad, S. J. Balkrishna, A. Kumar, B. S. Bhakuni, K. Shrimali, S. Biswas and S. Kumar, *J. Org. Chem.*, 2013, **78**, 1434; (b) C. D. Prasad, S. Kumar, M. Sattar, A. Adhikary and S. Kumar, *Org. Biomol. Chem.*, 2013, **11**, 8036.
9. S. E. Denmark and H. M. Chi, *J. Am. Chem. Soc.*, 2014, **136**, 8915.
10. S. E. Denmark, W. E. Kuester and M. T. Burk, *Angew. Chem., Int. Ed.*, 2012, **51**, 10938.
11. H. Fujihara, H. Mima and N. Furukawa, *J. Am. Chem. Soc.*, 1995, **117**, 10153.
12. (a) S. Murata and T. Suzuki, *Chem. Lett.*, 1987, **16**, 849; (b) S. Murata and T. Suzuki, *Tetrahedron Lett.*, 1987, **28**, 4297; (c) S. Murata and T. Suzuki, *Tetrahedron Lett.*, 1987, **28**, 4415.
13. M. Tiecco, L. Testaferri, M. Tingoli and D. Bartoli, *Tetrahedron Lett.*, 1989, **30**, 1417.
14. (a) *The Chemistry of Organic Selenium and Tellurium Compounds*, ed. S. Patai and Z. Rappoport, Wiley, New York, 1986, vol. 1, 1987, vol. 2; (b) C. Paulmier, *Selenium Reagents and Intermediates in Organic Synthesis*, Pergamon Press, Oxford, 1986.
15. H. Poleschner and K. Seppelt, *Chem. – Eur. J.*, 2004, **10**, 6565.
16. A. Pop, A. Silvestru, E. J. Juárez-Pérez, M. Arca, V. Lippolis and C. Silvestru, *Dalton Trans.*, 2014, **43**, 2221.
17. A. Beleaga, V. A. Bojan, A. Pöllnitz, C. I. Rat and C. Silvestru, *Dalton Trans.*, 2011, **40**, 8830.
18. (a) J. Jeske, W.-W. du Mont, F. Ruthe, P. G. Jones, L. M. Mercuri and P. Deplano, *Eur. J. Inorg. Chem.*, 2000, 1591; (b) J. Jeske, W.-W. du Mont and P. G. Jones, *Angew. Chem., Int. Ed.*, 1997, **36**, 2219.
19. L. C. Forfar, M. Green, M. F. Haddow, S. Hussein, J. M. Lynam, J. M. Slattery and C. A. Russell, *Dalton Trans.*, 2015, **44**, 110.
20. (a) R. A. Varga, M. Kulcsar and A. Silvestru, *Acta Crystallogr., Sect. E: Struct. Rep. Online*, 2010, **E66**, o771; (b) A. Pop, A. Silvestru, E. J. J. Pérez, M. Arca, V. Lippolis and C. Silvestru, *Dalton Trans.*, 2014, **43**, 2221.

21. V. P. Singh, H. B. Singh and R. J. Butcher, *Eur. J. Inorg. Chem.*, 2010, **4**, 637.
22. P. Rakesh, H. B. Singh and R. J. Butcher, *Organometallics*, 2013, **32**, 7275.
23. (a) J. L. Dutton, H. M. Tuononen and P. J. Ragogna, *Angew. Chem., Int. Ed.*, 2009, **48**, 4409; (b) C. D. Martin and P. J. Ragogna, *Inorg. Chem.*, 2012, **51**, 2947.
24. A. B. Bergholdt, K. Kobayashi, E. Horn, O. Takahashi, S. Sato, N. Furukawa, M. Yokoyama and K. Yamaguchi, *J. Am. Chem. Soc.*, 1998, **120**, 1230.
25. (a) J. L. Dutton, R. Tabeshi, M. C. Jennings, A. J. Lough and P. J. Ragogna, *Inorg. Chem.*, 2007, **46**, 8594; (b) J. L. Dutton, A. Sutrisno, R. W. Schurkob and P. J. Ragogna, *Dalton Trans*, 2008, 3470.
26. C. D. Martin, M. C. Jennings, M. J. Ferguson and P. J. Ragogna, *Angew. Chem., Int. Ed.*, 2009, **121**, 2244; (b) J. L. Dutton, H. M. Tuononen, M. C. Jennings and P. J. Ragogna, *J Am Chem Soc*, 2006, **128**, 12624.
27. C. D. Martin, C. M. Le and P. J. Ragogna, *J. Am. Chem. Soc.*, 2009, **131**, 15126.
28. A. Kozma, J. Petuškova, C. M. Lehmann and M. Alcarazo, *Chem. Commun.*, 2013, **49**, 4145.
29. D. H. Evans, N. E. Gruhn, J. Jin, B. Li, E. Lorance, N. Okumura, N. A. M. Ruvalcaba, U. I. Zakai, S.-Z. Zhang, E. Block and R. S. Glass, *J. Org. Chem.*, 2010, **75**, 1997.
30. S. Egalahewa, M. Albayer, A. Aprile and J. L. Dutton, *Inorg. Chem.*, 2017, **56**, 1282.
31. (a) T. Wirth, G. Fragale and M. Spichty, *J. Am. Chem. Soc.*, 1998, **120**, 3376; (b) L. Uehlin, G. Fragale and T. Wirth, *Chem. – Eur. J.*, 2002, **8**, 1125; (c) T. Wirth, *Angew. Chem., Int. Ed.*, 2000, **39**, 3740; (d) D. M. Browne and T. Wirth, *Curr. Org. Chem.*, 2006, **10**, 1893; (e) D. M. Freudendanl, S. A. Shahzad and T. Wirth, *Eur J. Org. Chem.*, 2009, 1649.
32. S. E. Denmark, W. R. Collins and M. D. Cullen, *J. Am. Chem. Soc.*, 2009, **131**, 3490.
33. H. Poleschner and K. Seppelt, *Angew. Chem., Int. Ed.*, 2008, **47**, 6461.
34. F. R. Knight, R. A. M. Randall, K. S. A. Arachchige, L. Wakefield, J. M. Griffin, S. E. Ashbrook, M. Bühl, A. M. Slawin and J. D. Woollins, *Inorg. Chem.*, 2012, **51**, 11087.
35. H.-Y. Wang, F. Yang, X.-L. Li, X.-M. Yan and Z.-Z. Huang, *Chem. – Eur. J.*, 2009, **15**, 3784.
36. W. H. Midura, J. Scianowski, A. Banach and A. Zaja, *Tetrahedron Asymmetry*, 2014, **25**, 1488.
37. H. Poleschner and K. Seppelt, *Angew. Chem., Int. Ed.*, 2013, **52**, 12838.
38. B. Mueller, T. T. Takaluoma, R. S. Laitinen and K. Seppelt, *Eur. J. Inorg. Chem.*, 2011, 4970.
39. S. Yadav, S. T. Manjare, H. B. Singh and R. J. Butcher, *Dalton Trans.*, 2016, **45**, 12015, also see references on diorgano diselenium dications cited in this reference.

40. C. Adamo, F. Demartin, P. Deplane, F. A. Devillanova, F. Isaia, F. Lelj, V. Lippolis, P. Lukesc and M. L. Mercuric, *Chem. Commun.*, 1996, 873.
41. J. Beck and S. J. Eck, *Z. Anorg. Allg. Chem.*, 2010, **636**, 1910.
42. J.-F. Poon, V. P. Singh, J. Yan and L. Engman, *Chem. – Eur. J.*, 2015, **21**, 2447.
43. F. V. Singh and T. Wirth, *Org. Lett.*, 2011, **13**, 6504.
44. (a) T. G. Back, B. P. Dyck and M. Parvez, *J. Chem. Soc., Chem. Commun.*, 1994, 515; (b) T. G. Back, B. P. Dyck and M. Parvez, *J. Org. Chem.*, 1995, **60**, 703; (c) T. G. Back and B. P. Dyck, *J. Chem. Soc., Chem. Commun.*, 1996, 2567; (d) T. G. Back and S. Nan, *J. Chem. Soc., Chem. Commun.*, 1998, 3123; (e) T. G. Back, B. P. Dyck and S. Nan, *Tetrahedron*, 1999, **55**, 3191.
45. M. Tiecco, L. Testaferri, C. Santi, C. Tomassini, F. Marini, L. Bagnoli and A. Temperini, *Angew. Chem., Int. Ed.*, 2003, **42**, 3131.
46. S. E. Denmark, D. Kalyani and W. R. Collins, *J. Am. Chem. Soc.*, 2010, **132**, 15752.
47. W. Niu and Y.-Y. Yeung, *Org. Lett.*, 2015, **17**, 1660.

CHAPTER 6

Organoselenide Fluorophores for Probing Reactive Oxygen Species and Heavy Metals

SANDIP V. MULAY,[a,b] YOUNGSAM KIM,[a,b] WOO HYUN LEE,[a] SHUBHANGI SINGH[a] AND DAVID G. CHURCHILL*[a,b]

[a] Korea Advanced Institute of Science and Technology (KAIST), Department of Chemistry, 373-1 Guseong-dong, Yuseong-gu, Daejeon 305-701, Republic of Korea; [b] Institute for Basic Science (IBS), Center for Catalytic Hydrocarbon Functionalizations, 373-1 Guseong-dong, Yuseong-gu, Daejeon 305-701, Republic of Korea

*Email: dchurchill@kaist.ac.kr

6.1 Introduction

Molecular fluorescence is a widely used technique in the biological sciences. It is a convenient, rapid, relatively safe, and simple method for imaging a wide variety of biological systems. Molecular fluorophores of various classes are readily synthesized and continue to be reported in the literature. Considering the fluorogenic core in question, as well as the chosen substituents and their locations, the changes that the π-delocalized system undergo are important (see later). Different designs of closely related systems help unravel the photophysical attributes and mechanisms at play; the photomechanism can give rise to a useful optical response based on the chemical input.[1] While there is no "perfect sensor", the parameters we often consider relate to sensitivity, selectivity, and reversibility. Selenide groups

Organoselenium Compounds in Biology and Medicine: Synthesis, Biological and Therapeutic Treatments
Edited by Vimal Kumar Jain and K. Indira Priyadarsini

Published by the Royal Society of Chemistry, www.rsc.org

that can work in a nucleophilic or electrophilic manner have become useful functional groups.[2–19] Organoselenides are reactive, and in particular, can covalently react with and bind oxygen-containing species, in particular reactive oxygen species (ROS); alternatively, a selenide may serve as a chelating atom—a lone pair donor (L-type ligand) or fasten to a metal *via* a single bond (X-type bond). Thus, the selenium atom and selenides are able to be readily oxidized and metalated. The mode of detection is fluorescent in origin, for example, simple change of –Se– to –Se(O)–, and thus it can change the fluorescence-quenching properties of the Se towards the chromophore to which it is attached. A closer analysis can be made when considering the highest occupied molecular orbital (HOMO)–lowest unoccupied molecular orbital (LUMO) energy levels and how they change as a function of oxidation. Accordingly, this versatility allows for capability as an active site in chemosensing frameworks. The utility of selenium in synthetic compounds is furthered nicely by direct isosteric comparisons to sulfur and tellurium analogues; where synthetic possibility has given way to the actual isolation and study of these derivatives, the comparisons are always fruitful.

The area of selenium-based chemosensors has been recently reviewed in the realm of ROS detection and the detection of related analytes by established chemosensing research groups such as those of Yoon, Han, Panda, Zade, and Churchill.[8–10] This chapter brings out certain reports that have helped shape the recent trajectory of the field and have helped turn the corner for the field. Selenium-containing compound research fits conveniently into the needs of both environmental and medical science. The common road with any chemosensing work is to first achieve interesting cuvette-based fluorescence responses followed by verification in assays with living cells, and further with tissue and living animals where possible and when prudent. Nowadays, animal models allow for excellent proofs of principle and sometimes involve disease models.

The current chapter considers two main classes of analytes, as follows. (i) ROS involve a host of discrete low molecular mass compounds; some of these involve very short lifetimes and for these reasons and others (species diversity, reactivity, and analyte size/shape) are not always easy to study exclusively without interference from other closely related ROS species. Note that in addition to a litany of ROS species, there are related compounds such as reactive nitrogen species (RNS) and reactive sulfur species (RSS). (ii) Research regarding selenium-containing probes for toxic heavy metal ions such as mercuric (Hg^{2+}) and lead(II) (Pb^{2+}) are also described herein.[10]

6.2 ROS Detection

ROS such as hydroxyl radical, hydrogen peroxide, superoxide, hypochlorous acid, nitric oxide, and peroxynitrite play significant roles in various biological processes. These species are not just detrimental to the health of living organisms; in many cases, they are required and part of the natural function of biological systems. However, *excess* production of ROS and RNS causes oxidative stress, which has been well established to be manifestly involved in various diseases and disorders such as cardiovascular disease, atherosclerosis, rheumatoid arthritis, neurodegenerative disease, cancer, and diabetes. Due to this aetiology and a host of other factors, it is important to monitor and trace these small and transient species through their often compartmental course through human living cells and tissue.

Despite the interest surrounding complexes such as ebselen, synthetic efforts have revolved around the selenide functional group and the incorporation of the [C-Se-C] motif into chemical systems that are π-delocalized. The Ar–Se–Ar species (Ar denotes aromatic ring) rose to prominence because of its manageability, stability crystallinity, and limited odour compared to those of its aliphatic counterparts.[13] The important isosterism between the chalcogen/chalcogenides, *e.g.* comparison with Ar–S–Ar and Ar–O–Ar, has been reported thoroughly in the literature (see below).

The presence of Se atom in organoselenium chemistry takes on two principal forms: the selenide and the selenoxide. Despite an abundance of historical organochalcogenide literature, it had not been exploited for chemosensing until recently. The $R_2E \rightarrow R_2E{=}O$ (E = heavier chalcogen) motif was only combined with fluorescence in recent years.[10] In early work, inorganic selenium was explored as an analyte; it was used as an additive to organic molecules to create a fluorogenic species—the irreversible covalent modification imparted by the selenium onto the organic moiety allowed for clear optical changes.

The first example of connecting fluorescence responses with the inherent redox signalling with selenium- or group 16 congener-containing systems was performed by Detty *et al.*[20] The chalcogenapyrylium dye (**1**) becomes chemically oxidized based on analysis of emission characteristics (near-infrared region) observed in microscopy; additionally, the imaging of the cell allowed for subcellular location of the dye and the presence of components of oxidative stress (ROS) in the mitochondria were thought to be the causative agents for the TURN-ON fluorescence response (Figure 6.1).

Considering the closely related sulfur-based literature for a moment, a benzothiophene chemosensing species (**2**) reported by Churchill and coworkers in 2008[21] involved a *meso*-aryl group to serve as the probe nucleus. In this report, *m*-CPBA was used to oxidize the sulfur atom. The sulfide was

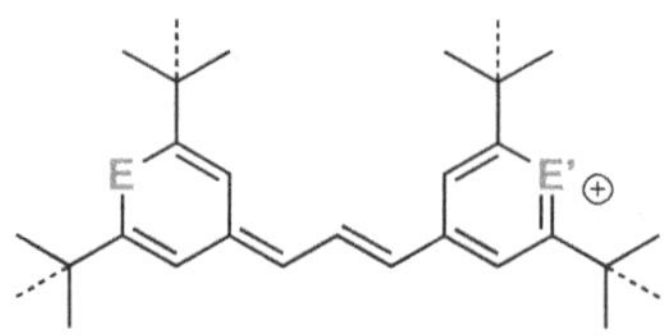

1

Figure 6.1 Chalcogenopyrylium-based dye design reported by Detty *et al.*[20]

duly converted to the sulfoxide (**3**) as a means of a proof of principle that oxidation imparts an optical change to the dyad dye system (both ultraviolet-visible and fluorescence characteristics were obtained) (Figure 6.2).

Another milestone in demonstrating practicality for fluorophore/chromophore chemosensing systems comes from a report by Jain and coworkers.[22] In this report, an Ar–Se–Ar system undergoes chemical oxidation; conversion of the diarylselenide (**4**) to the selenoxide (**5**) upon treatment with H_2O_2 was clearly monitored using ^1H nuclear magnetic resonance spectroscopy. Subsequent chemical reduction is enacted by the addition of a bulky thiophenol derivative (Figure 6.3). While it is not much of a fluorophore, the Ar–Se–Ar unit (**4**) is a high-energy chromophore.

At about the same time, an article by Kim *et al.* reported the detection of hypochlorite with a BODIPY system (**6**) in which sulfur is chemically oxidized to the sulfoxide (**7**) (Figure 6.4).[23] This system also utilized a stable BODIPY dyad (Phenyl:BODIPY) core. However, in this example a single chalcogenide centre is placed on the electron donor moiety of the dyad (**6**).

It is necessary to understand how the chemical oxidation unfolds in the selenium versions of the chalcogen species discussed so far. When oxidized, selenium has options; it could, in fact, undergo a change to become $Se(OH)_2$

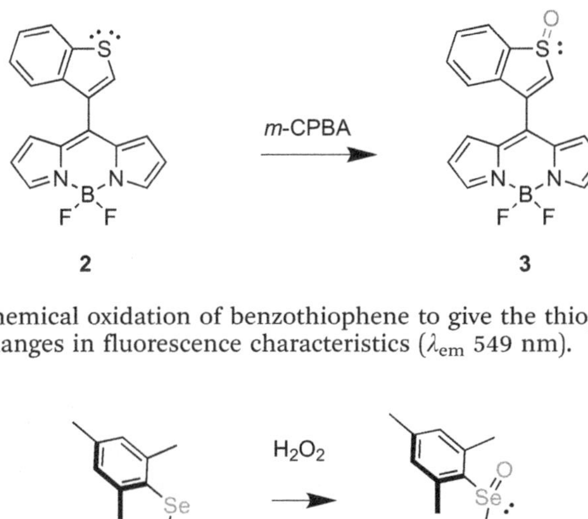

Figure 6.2 Chemical oxidation of benzothiophene to give the thiophene oxide and changes in fluorescence characteristics (λ_{em} 549 nm).

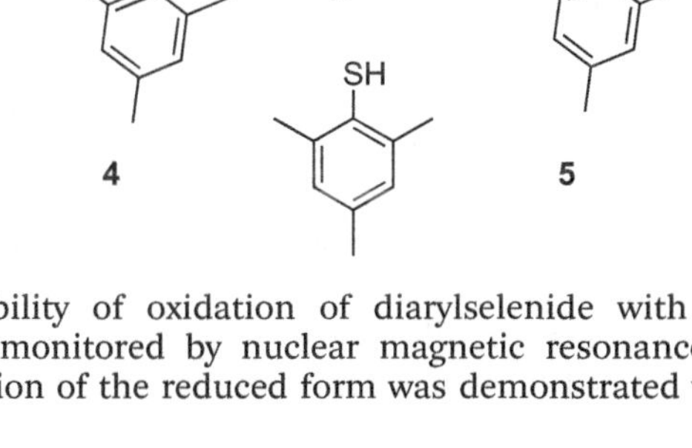

Figure 6.3 Reversibility of oxidation of diarylselenide with hydrogen peroxide (H_2O_2) monitored by nuclear magnetic resonance spectroscopy. The generation of the reduced form was demonstrated with arylsulfide.

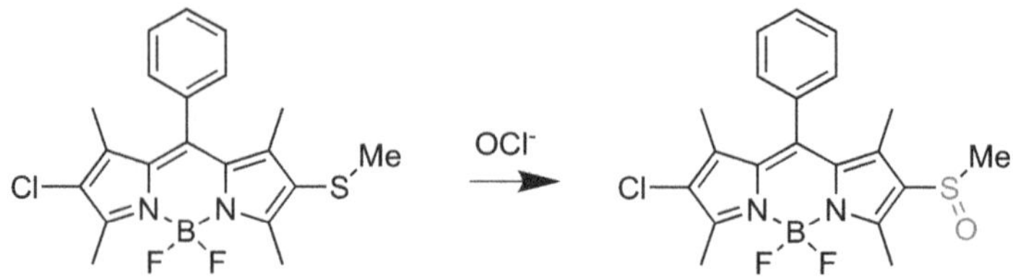

Figure 6.4 Sulfur mono-oxidation led to a fluorescence response in a BODIPY species.

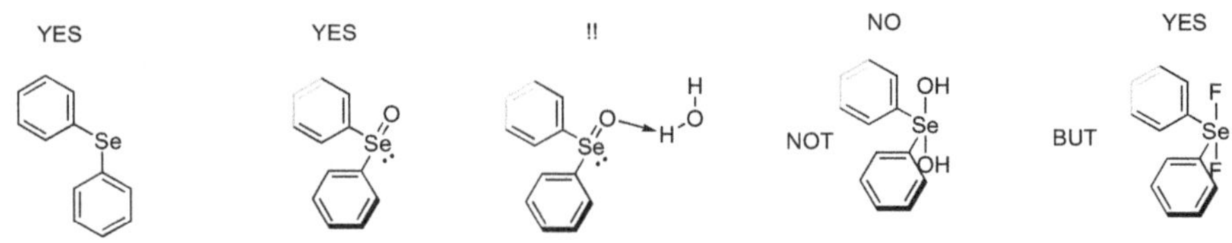

Figure 6.5 Related structures involving diphenylselenide in their tetravalent oxidized forms and the selenium IV difluoride form.

(**11**) instead of forming a terminal selenoxide (Se=O) (**10**). The selane (**11**) and other intermediates may be present in different distributions at different reaction times; the selane intermediate is a higher energy species (see later). The chemosensing becomes complicated because there exist many important intermediates: kinetic products, as well as thermodynamic products. Figure 6.5 provides structures of the related species used in the present discussion. Interestingly, there is an overall crystal formulation of [SeO_2H_2] (**10** or **11**), but not in the manner we might expect: instead of the existence of $Se(OH)_2$, an intact discrete water molecule is present and hydrogen-bonded to the [O=Se] group (X-ray structure) (**10**). Having mentioned the possibility of obtaining a stable version of the selane, a real challenge in which both ligands are hydroxo species (**11**), we should note that **12** does exist stably in some conditions, and is routinely isolated and characterized. In recent years, new attention has been brought to selane chemistry, and should be considered in the context of hydroxylation.

The occurrence of multiple oxidation events that are sequential at a single probe is an important aim in the pursuit of *multiple input detection*. Here, we present a sulfur-based chemosensor reported by Singh *et al.*;[24] the chalcogen-based substituents in the *meso*-aryl group (**13**) were demonstrated to be chemically oxidized; multiple oxygenation led to stepwise differences in fluorescence characteristics (Figure 6.6).

Reversibility, which has already been introduced, is an essential feature for future probing. In 2013, two separate reports, the first by the Han group

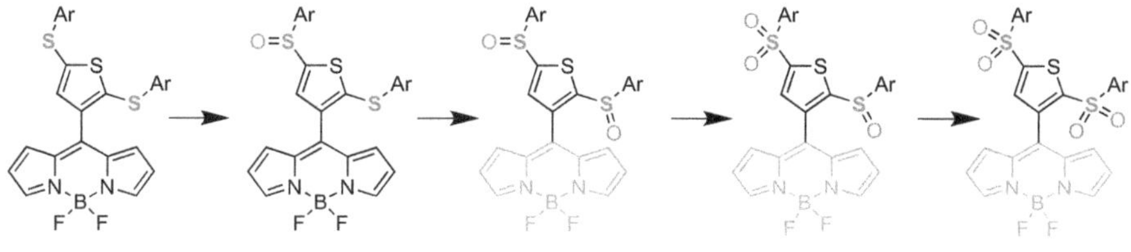

Figure 6.6 Report of multiple oxidation and stepwise addition of oxygen atoms. The chromophores in **15–17** are those that exhibit heighted fluorescence because of the PET being "turned-off".

(Dalian, China)[25] and the second from the Wu group (Taiwan),[26] described oxidation at a phenylselenide species that involved a *meso*-aryl-substituted BODIPY species. Like Jain's report,[22] here these systems showed reversibility.

6.2.1 ROS Detection Since 2012

Since 2012, numerous reports based on BODIPY and other fluorescent frameworks (**18a–p**) have emerged, in which the use of a selenium centre as a reactive centre allows for turn OFF–ON of the fluorescence capacity of the fluorophore based on the turn ON–OFF of the influence of the strong heavy atom effect. Chemical oxidation at the selenium centre (**19**) helps lower the electron richness of the system and forbids the electron richness to quench the fluorescence: the electrons of the lone pair readily quench the electronically excited state in the fluorophore system (Figure 6.7).[27–43] The location of the chalcogen atom alters the propensity for emission turn OFF–ON or turn ON–OFF events. In certain cases, the mechanism does differ from the simple oxidation scenario [$R_2Se \rightarrow R_2Se{=}O$] (**18** to **19**).[10] Again, corresponding tellurium analogues are often available for comparison. Many different chemosensor derivatives were prepared, reported, and are cited herein and feature prominently in various recent reviews (Figure 6.7).[8–10] Diselenides are also featured, and can become chemically oxidized (**18a** and **18j**) (Table 6.1). The fate of the Se–Se bond and its potential cleavage in some solvent media and under the influence of other analytes is an important aspect for future studies.

6.3 Ebselen Derivativization in ROS Chemosensing

As mentioned at the beginning of this chapter, the structure and function of ebselen is prominent in the mind of the chemist; various functional molecules have emerged (**21–23**). The opening/closing of the five-membered ebselen heterocyclic ring is often demonstrated (Figure 6.8),[44] and the heterocyclic chemistry for Se atom can accommodate both a lone pair donations and direct N–Se bonding. The chemistry of spirocyclic opening has

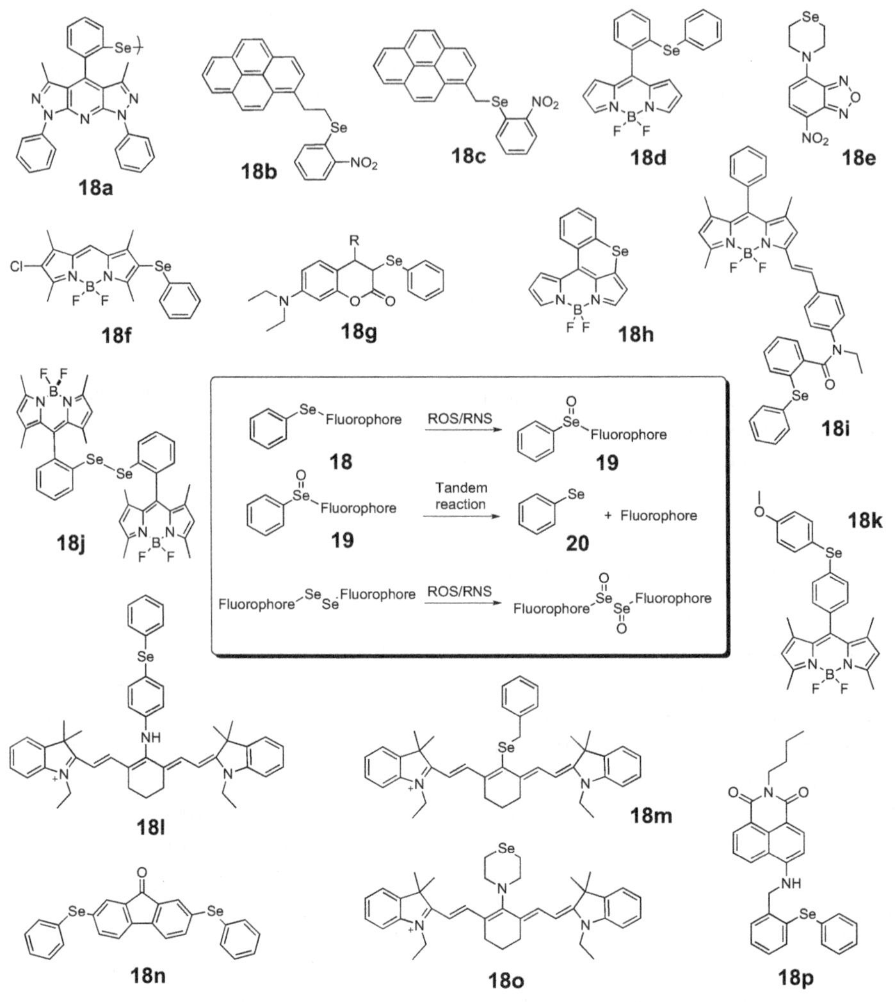

Figure 6.7 Organochalcogen-based fluorescent chemosensors reported in the literature, based on the selenium atom present undergoing oxidation. Compound **18g**, R = CH_3 or simply H.

also been exploited. Ebselen and the pursuit of related selenium chemistry for health benefits predates the quest for biological-based chemosensing with regard to medicinal properties. The formation of the Se–N bond was available for the detection of hydrogen peroxide.[32] Rodamine containing selenolactone when treated with nitric oxide as an analyte leads to the enhancement of the fluorescence intensity due to the opening of spirocyclic selenolactone instead of selenium oxidation. However, through all of these reports, the oxidized analyte species is scavenged instead of binding reversibly.[8–10] Earnest efforts for clean binding and reversibility of ROS species are still forthcoming.[15]

Table 6.1 Various selinides with their emission and excitation maxima.

Probe structure	λ_{ex}	λ_{em}	Analyte
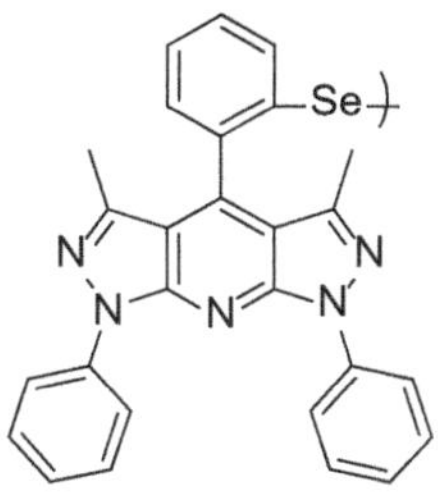 **18a**	311 nm	436 nm	Hypochlorite
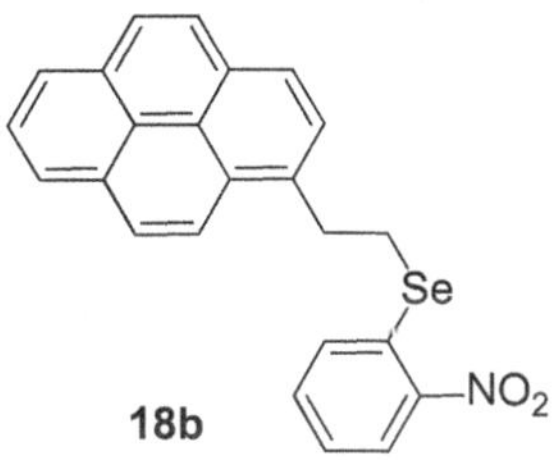 **18b**	365 nm	392 nm	Hypochlorite
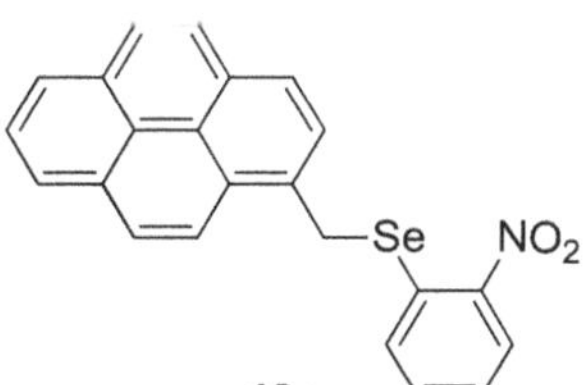 **18c**	337 nm	392 nm	Hypochlorite
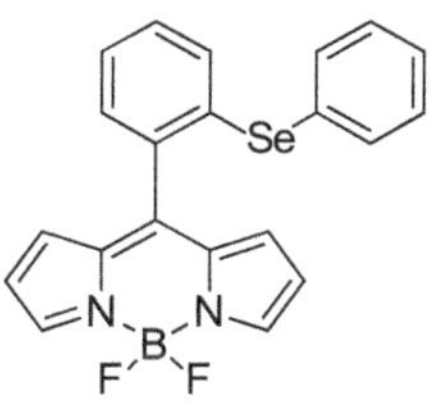 **18d**	510 nm	526 nm	Hypochlorite
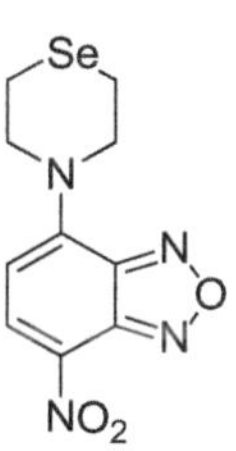 **18e**	468 nm	544 nm	Hypochlorite

Table 6.1 (*Continued*)

Probe structure	λ_{ex}	λ_{em}	Analyte
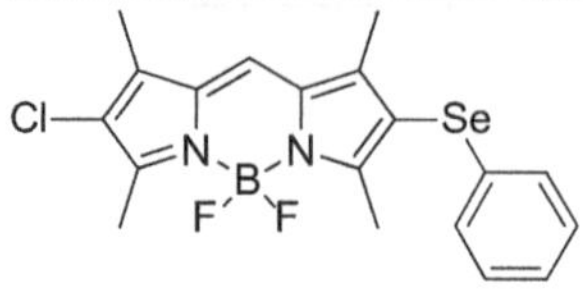 **18f**	511 nm	526 nm	Hypochlorite
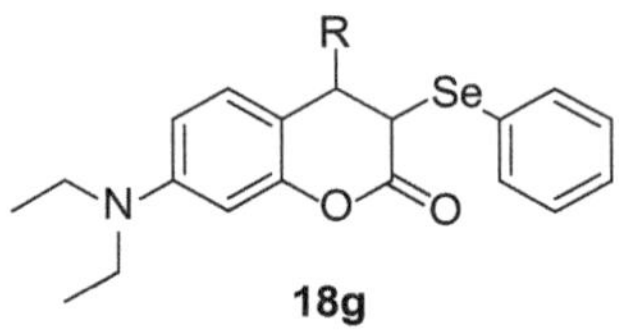 **18g**	405 nm (R = H, CH3)	480 nm (R = H) 468 nm (R = CH3)	Hypochlorite
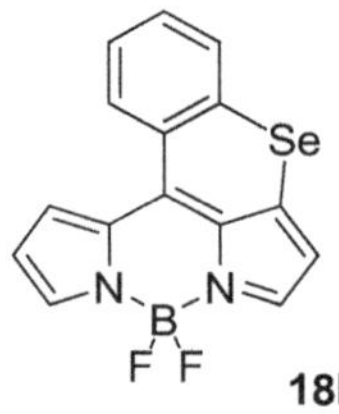 **18h**	572 nm	597 nm	Hypochlorite
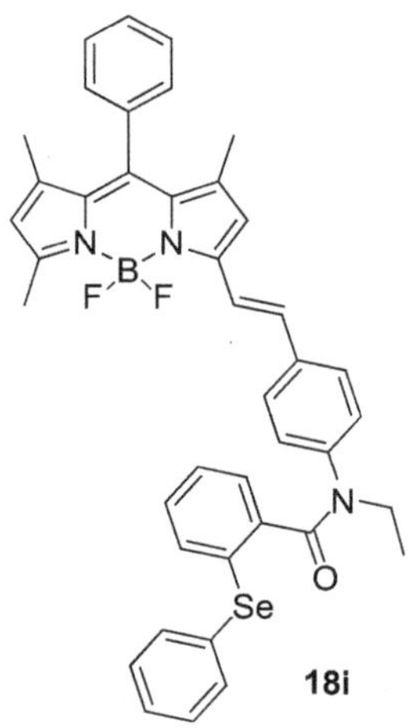 **18i**	559 nm	572 nm	Peroxynitrite
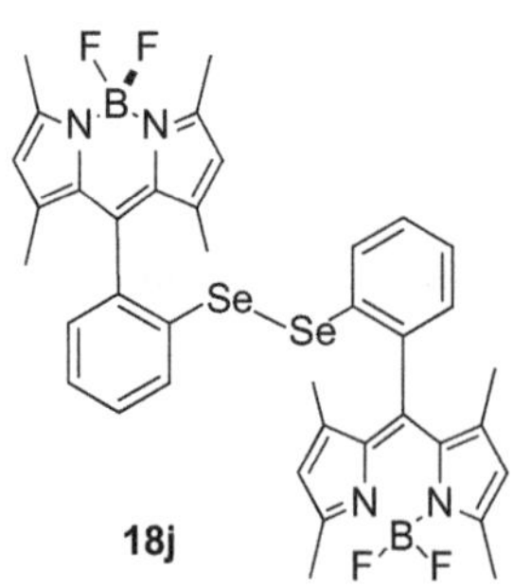 **18j**	504 nm	514 nm	Superoxide

Table 6.1 (*Continued*)

Probe structure	λ_{ex}	λ_{em}	Analyte
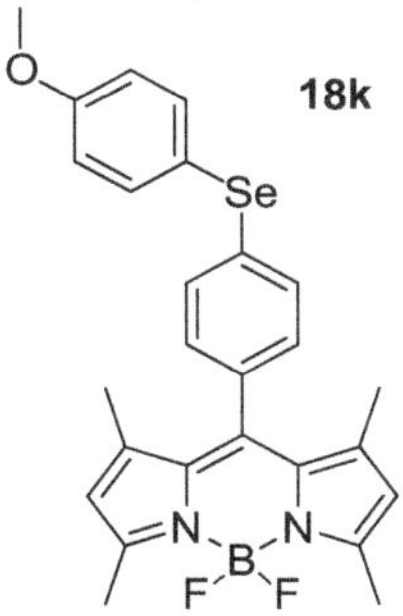 18k	460 nm	510 nm	Hypochlorite
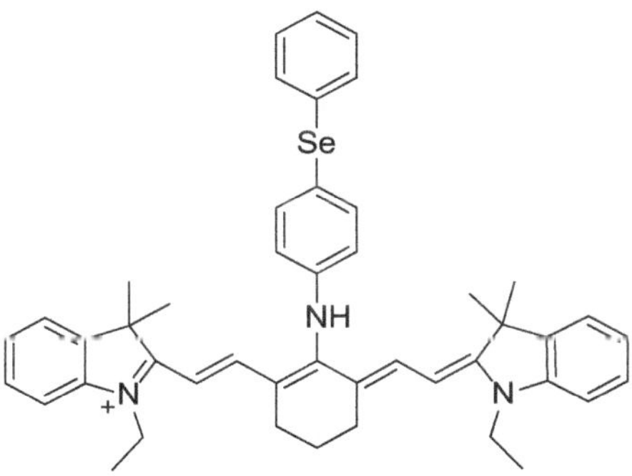 18l	758 nm	800 nm	Peroxynitrite
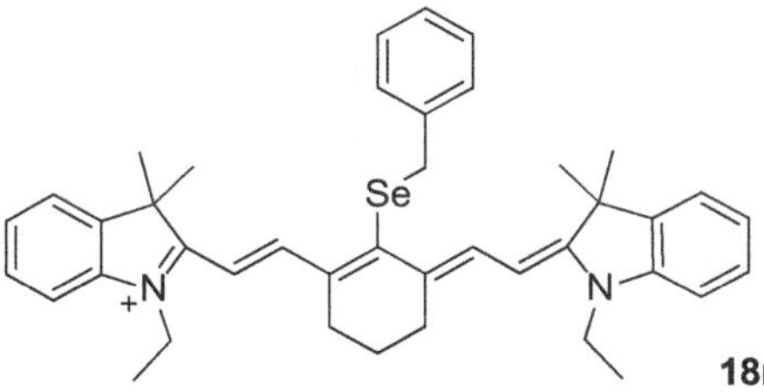 18m	770 nm	800 nm	Peroxynitrite
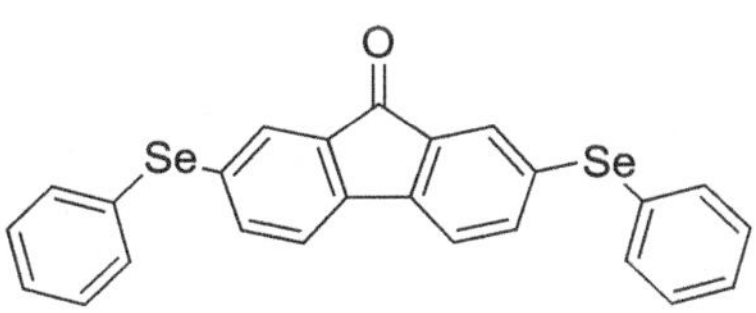 18n	800 nm	500 nm	Hypochlorite (two-photon fluorescence probe)
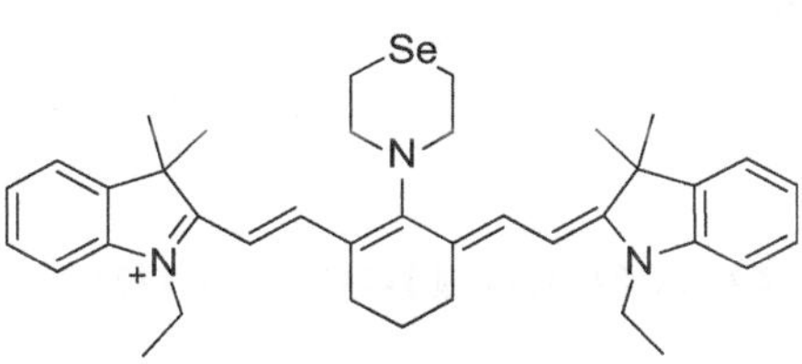 18o	690 nm	786 nm	Hypochlorite

Table 6.1 (*Continued*)

Probe structure	λ_{ex}	λ_{em}	Analyte
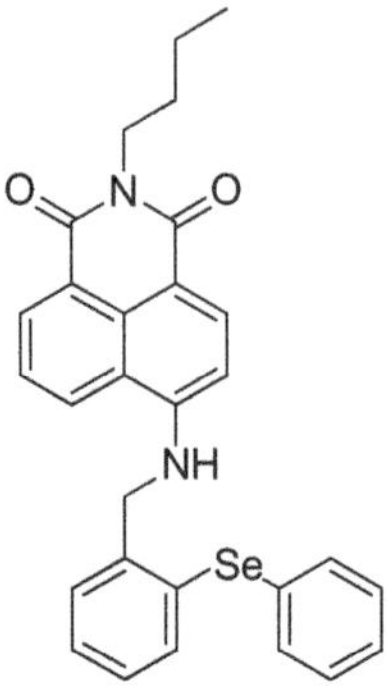 **18p**	450 nm	523 nm	Hypochlorite

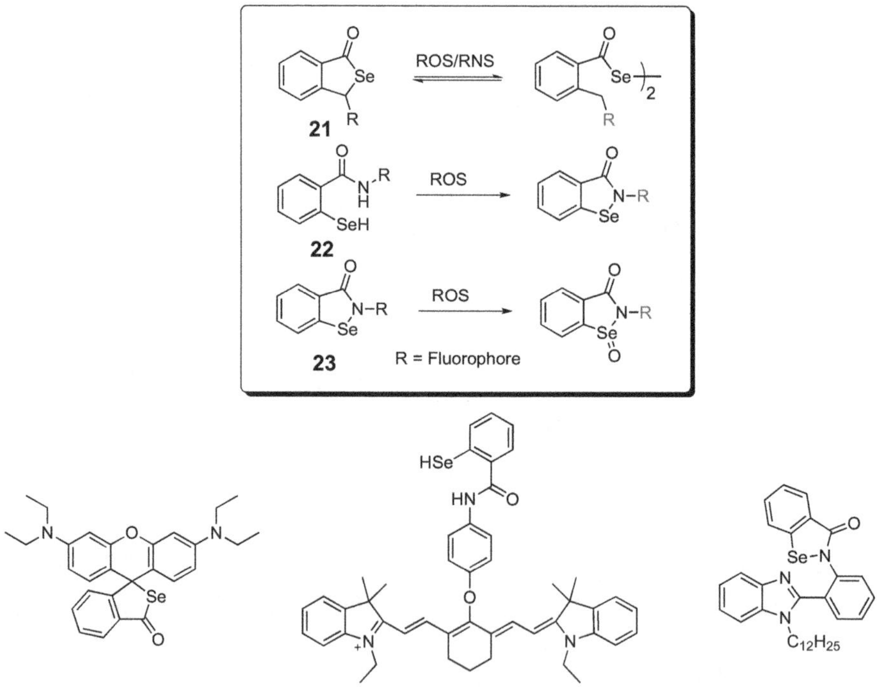

21a **22a** **23a**

Figure 6.8 Organoselenium chemosensors based on Ebselen activity and analogous moieties.[32,37,44]

6.4 Heavy Metals and their Detection Through the use of Selenium

In terms of heavy metal detection, selenium holds tremendous potential;[45–55] however, precautions are required, because of the "ore-forming" tendency of the group 16 atom, which acts like a thermodynamic sink.[55] As with sulfur, the heavier chalcogen atom naturally has a "soft–soft" binding propensity which can strongly steer chemistry as well as chemosensing-based approaches from a thermodynamic standpoint. There are numerous reports involving mercury and selenide interactions that involve the synthetic vantage point and that do not necessarily involve optical changes of interest. In terms of chemosensing, there have been a number of mercury-based sensors bearing a selenide unit as an intended design parameter. A notable and important early report by Yoon and coworkers involved the use of a selenium-containing spirolactone (**24**, Figure 6.9).[45] A separate lactam framework was reported by a Chinese research group.[54]

For the next examples of chemosensing, we step outside the formal [C-Se-C] motif scope to introduce a report by Tang *et al.* (Figure 6.10).[47] In this research, mercury was studied in water; the researchers utilized the fluorescein framework. The fluorescein-based probe (**25**) featured phosphinate-protecting groups [–OP(=Se)Ph_2], which allows for the substitution of these groups to give turn OFF–ON fluorescence enhancement generated by the reactivity of Hg^{2+} with selenium. With the probe in the presence of mercuric ion, the number of interfering metal cations were few.

Kumar and Singh reported a new molecule involving an adaptation of a TREN-like organometallic skeleton (**27**, Figure 6.11).[49] The molecule serves in the capacity of a tripodal ligand (tris(2-(phenylselanyl)ethyl)amine) and can be involved in metal chelation. This unassuming system was able to be exploited for 1:1 mercury binding and allowed for enhancement of fluorescence by chelation.

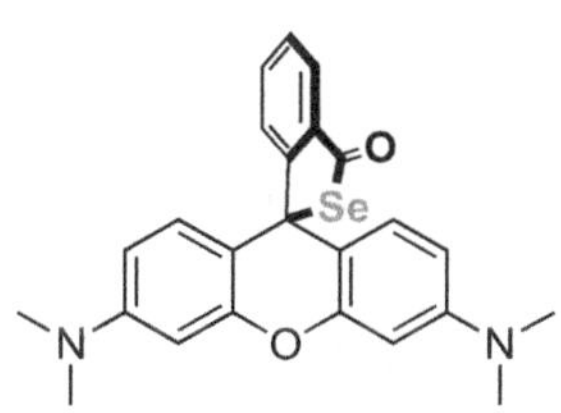

24

Figure 6.9 An early example of a selenium-based probe for mercuric sensing.

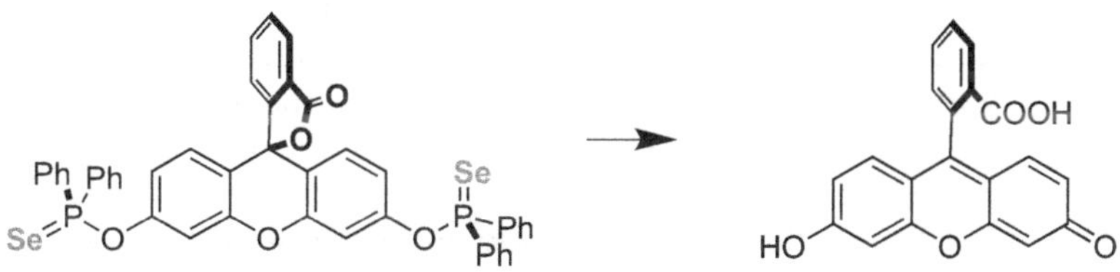

25 **26**

Figure 6.10 Disruption of double-protection groups of the spirolactone version of fluorescein for mercury sensing.[47]

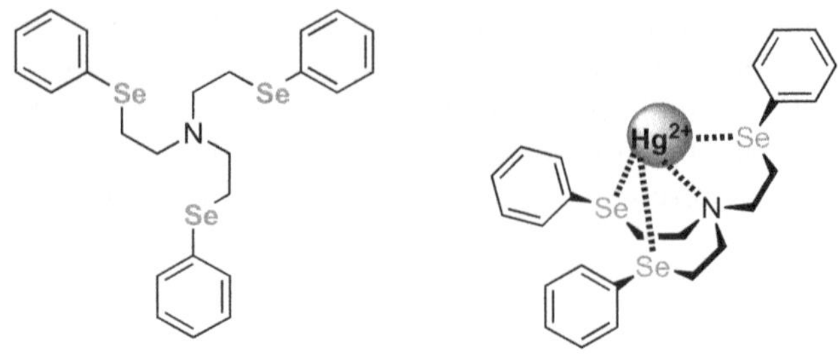

27

Figure 6.11 Tripodal design for mercuric ion sensing.[49]

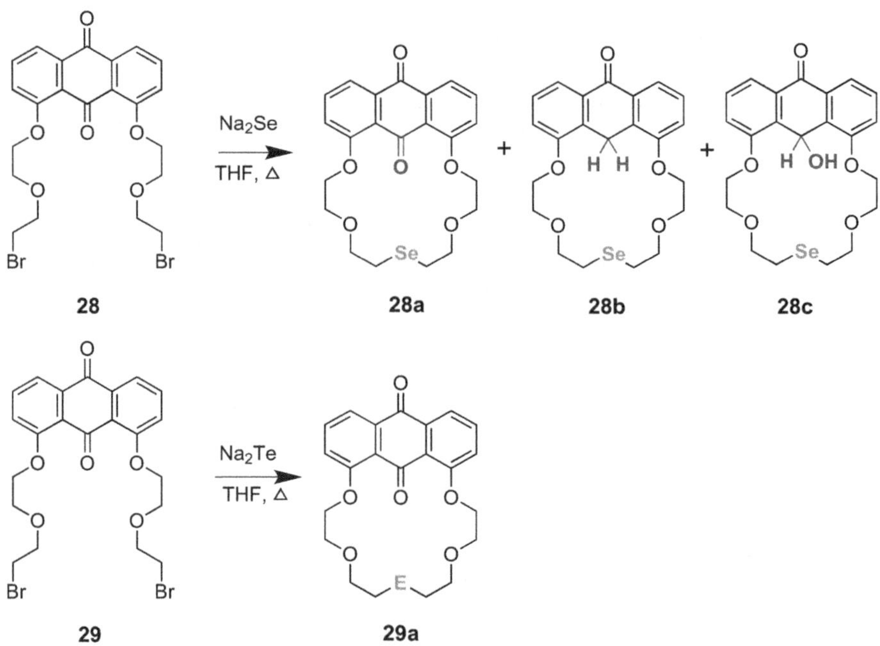

Figure 6.12 Isosteric binding cavities with single-atom differences based on an anthraquinone core.
Adapted from ref. 46 with permission from the Royal Society of Chemistry.

It is worth mentioning a scientific report by Haq *et al.* which discusses organoselenium-based sensing from a different vantage point.[50] In this report, mercury in the form of the mercury electrode, was used in an analytical manner. Selenium and non-selenium compounds were tested at this electrode. The differential pulse voltammetric behaviour of the different compounds was recorded; the organometallic compounds did not perturb the Hg(*l*) at the outer layer of the electrode. However, mercury selenide (HgSe) was generated when free selenide (inorganic source) was used in solution.

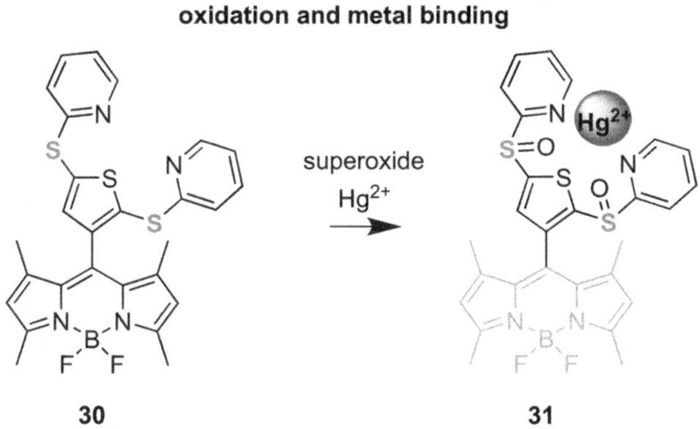

Figure 6.13 Addition of reactive oxygen species and a metal ion to give an increase in fluorescence.

A synthetic report by Mariappan *et al.* described details of a successfully prepared set of derivatives (**28a** and **29a**) and their structural characterization, as well as a discussion of interesting side products.[46] The report is a complete study ranging from ligand design, coordination chemistry, and a novel chemosensing perspective. The demonstration of the utility of Na_2E (E = Se, Te) as a versatile starting material was further established (Figure 6.12).

The relevant organochalcogen-based chemosensing literature of toxic metal ions does not extend much beyond the detection of the mercury ion. Cadmium, for example, would be predicted to strongly bind to selenium in a sample entirely absent of Hg^{2+}, and perhaps of Ag^{+}. A survey of the literature reveals no previous reports that include all these three key words: "organoselenium", "cadmium", and "probe". Other considerations for future probe design includes *f*-element chemistry (see later).

Tandem reports, in which both heavy elements and ROS were analysed together, are few. A report by Singh *et al.* describes organosulfur centers that bind to mercuric ion and the probe's reactivity towards superoxide ion (**30**, Figure 6.13).[52] In this system, there is a fluorescence TURN-ON detection in which both analytes are used together; the order of analyte addition does not matter in this case.

6.5 Conclusion and a Glance at the Future

Fluorescence continues to dominate the scientific landscape as an important tool for the excellent and state-of-the-art analysis of biological systems, especially in the simple and straight-forward analysis of living cells which are the building blocks of life. Molecular fluorescence detection is used broadly in research; novel functional synthetic molecules that might also be fluorescent have the potential to be biologically active as well, in ways that are desirable. Group 16 elements such as selenium, both in inorganic (elemental, SeO_2, *etc.*) and organoselenium (R-Se-R) forms, serve as sacrificial and chemically active centres: as reductants or scavenging centres. Selenium facilitates analyte detection in which the analyte is either ROS or metal-based species. Unlike sulfur, the chemistry and coordination geometry of selenium is extremely versatile, owing to its various valence states and geometric capacity.

We urge the student, researcher, and expert in the field to take time to understand the rudimentary chemistry of selenium and analogous chemistry of sulfur and tellurium. Then, with a certain organic skeleton in mind, consider the notion of whether the positioning of the group external to the ring or within an aromatic ring may be more effective for the purposes in mind.

In terms of different skeletons that could be further researched, we feel that more synthetic work needs to be undertaken with the selenophene motif as part of the chemosensing system. Separately, there is much literature in the environmental sciences involving cations of heavy metals and organoselenium-based organics. Therefore, heavy metals such as *f*-elements are absent from the literature of organoselenium-based chemosensing. We think that the future is bright for chalcogen-containing probes and that there are numerous future combined analyte studies including metal ion studies and other analytes that can emerge with selenium-containing organic species as a reactive interface in the sensor. The most interesting cases of chemosensing are ones in which the science of a technological system has not yet been worked out. To further consider the concept stemming from Figure 6.5, at longer times and at higher temperatures, ROS will act on such organic or other systems to give thermodynamic products; these products may be of the same or similar identity, and for this reason it is important to fully understand the geometry and sterics around the reactive chalcogen center for a given system and in a given solvent medium. However, in biological imaging research pursuits, the temperature and the experimental time frame are often fixed. The fixed temperature of 37 °C is an undeniable working temperature. Additionally, the relatively short reaction (induction) times of <2 days are considered; therefore, an important goal for future work will be to address production of existing kinetic products which contribute to the fluorescence. It is important to consider the rudimentary coordination chemistry of selenium itself further and what can be mustered with sterical groups within organic or organometallic systems.

There is much more work to be done with species that involve selective ROS detection and fluorescence. Additionally, targeting ions of elements other than mercury and lead, and with better selectivity than exists currently is a necessary consideration. In future research, the selectivity for other classes of analytes can be worked out. Future probe design will result from inspired connections between functional groups, binding groups, fluorophore designs, multimer design, and multiple differential binding sites. There exist opportunities to couple any of these modes to enhance the capacity of aspects of chemosensing, which is also manifested in supramolecular chemistry, molecular machine, and synthetic and natural polymer contexts. Thus, fundamental synthetic (main group) organic and inorganic chemistry, as well as analytical chemistry[1] are crucial. In addition, the future of the field of chemosensing with organselenium seems to be about the preparation of different chromophores. However, novel starting materials and access to test animals in complex and expensive studies are often invoked as part of a research study in order to try to complete the bigger picture and to bring a particular novel compound forward as a real candidate for future research and medicine.

Acknowledgements

David G. Churchill acknowledges research support from (i) the Mid-Career Researcher Program through the NRF (National Research Foundation) of Korea (NRF-2014R1A2A1A11052980) funded by MEST, (ii) the Center for Catalytic Hydrocarbon Functionalizations, Institute for Basic Science (IBS) for financial support (IBS-R010), and HRHR Program (N11160081) from KAIST. Sandip V. Mulay and Youngsam Kim acknowledge IBS research Fellowships and KAIST for providing research facilities. In closing, the authors should like to apologize for any omissions of scientific work that may have been made in preparing and in shaping this original contribution.

References

1. D. Harris, *Quantitative Analysis*, W. H. Freeman and Company, New York, 7th edn, 2007.
2. For reviews published in *Chemical Reviews* see references **2–7**: A. J. Mukherjee, S. S. Zade, H. B. Singh and R. B. Sunoj, *Chem. Rev.*, 2010, **110**, 4357.
3. G. Perin, E. J. Lenardão, R. G. Jacob and R. B. Panatieri, *Chem. Rev.*, 2009, **109**, 1277.
4. G. Zeni, D. S. Lüdtke, R. B. Panatieri and A. L. Braga, *Chem. Rev.*, 2006, **106**, 1032.
5. C. W. Nogueira, G. Zeni and J. B. T. Rocha, *Chem. Rev.*, 2004, **104**, 6255.
6. T. G. Chasteen and R. Bentley, *Chem. Rev.*, 2003, **103**, 1.
7. G. Mugesh, W.-W. du Mont and H. Sies, *Chem. Rev.*, 2001, **101**, 2125.
8. For recent reviews see the following reviews **8–10**: D. Wu, L. Chen, N. Kwon and J. Yoon, *Chem*, 2016, **1**, 674.
9. Z. Lou, P. Li and K. Han, *Acc. Chem. Res.*, 2015, **48**, 1358.
10. (a) S. Panda, A. Panda and S. S. Zade, *Coord. Chem. Rev.*, 2015, **300**, 86; (b) S. T. Manjare, Y. Kim and D. G. Churchill, *Acc. Chem. Res.*, 2014, **47**, 2985.
11. K. P. Bhabak and G. Mugesh, *Acc. Chem. Res.*, 2010, **43**, 1408.
12. G. Mugesh and H. B. Singh, *Acc. Chem. Res.*, 2002, **35**, 226.
13. More distant reviews are as follows **16–20**: T. W. Campbell, H. G. Walker and G. M. Coppinger, *Chem. Rev.*, 1952, **50**, 279.
14. L. Engman, *Acc. Chem. Res.*, 1985, **18**, 274.

15. F. Wudl, *Acc. Chem. Res.*, 1984, **17**, 227.
16. D. Liotta, *Acc. Chem. Res.*, 1984, **17**, 28.
17. H. J. Reich, *Acc. Chem. Res.*, 1979, **12**, 22.
18. Important reviews that cover chemosensing broadly are as follows **21–22**: R. McRae, P. Bagchi, S. Sumalekshmy and C. J. Fahrni, *Chem. Rev.*, 2009, **109**, 4780.
19. X. Chen, F. Wang, J. Y. Hyun, T. Wei, J. Qiang, X. Ren, I. Shin and J. Yoon, *Chem. Soc. Rev.*, 2016, **45**, 2976.
20. M. R. Detty, P. B. Merkel, R. Hilf, S. L. Gibson and S. K. Powers, *J. Med. Chem.*, 1990, **33**, 1108.
21. S. Choi, K. Kim, J. Jeon, B. Meka, D. Bucella, K. Pang, S. Khatua, J. Lee and D. G. Churchill, *Inorg. Chem.*, 2008, **47**, 11071.
22. N. Ghavale, P. P. Phadnis, A. Wadawale and V. K. Jain, *Indian J. Chem., Sect. A: Inorg., Bio-inorg., Phys., Theor. Anal. Chem.*, 2011, **50**, 22.
23. T.-I. Kim, S. Park, Y. Choi and Y. Kim, *Chem. – Asian J.*, 2011, **6**, 1358.
24. A. P. Singh, K. M. Lee, D. P. Murale, T. Jun, H. Liew, Y. H. Suh and D. G. Churchill, *Chem. Commun.*, 2012, **48**, 7298.
25. B. Wang, P. Li, F. Yu, P. Song, X. Sun, S. Yang, Z. Lou and K. Han, *Chem. Commun.*, 2013, **49**, 1014.
26. S.-R. Liu and S.-P. Wu, *Org. Lett.*, 2013, **15**, 878.
27. G. Li, D. Zhu, Q. Liu, L. Xue and H. Jiang, *Org. Lett.*, 2013, **15**, 2002.
28. W. Chen, W. P. Bay, M. W. Wong and D. Huang, *Tetrahedron Lett.*, 2012, **53**, 3843.
29. X. Chen, K. A. Lee, E. M. Ha, K. M. Lee, Y. Y. Seo, H. K. Choi, H. N. Kim, M. J. Kim, C. S. Cho, S. Y. Lee, W. J. Lee and J. Yoon, *Chem. Commun.*, 2011, **47**, 4373.
30. G. Cheng, J. Fan, W. Sun, J. Cao, C. Hu and X. Peng, *Chem. Commun.*, 2014, **50**, 1018.
31. Y. Kim, M. Choi, S. T. Manjare, S. Jon and D. G. Churchill, *RSC Adv.*, 2016, **6**, 32013.
32. Y. X. Liao, K. Li, M. Y. Wu, T. Wu and X. Q. Yu, *Org. Biomol. Chem.*, 2014, **12**, 3004.
33. Z. Lou, P. Li, Q. Pan and K. Han, *Chem. Commun.*, 2013, **49**, 2445.
34. S. T. Manjare, J. Kim, Y. Lee and D. G. Churchill, *Org. Lett*, 2014, **16**, 520.
35. S. T. Manjare, S. Kim, W. D. Heo and D. G. Churchill, *Org. Lett*, 2014, **16**, 410.
36. Y. Shen, X. Zhang, X. Huang, S. Wen, M. Liu, Y. Deng, Y. Zhang, C. Zhang, J. Jin, H. Li and S. Yao, *RSC Adv.*, 2015, **5**, 79519.
37. C. Sun, W. Shi, Y. Song, W. Chen and H. Ma, *Chem. Commun.*, 2011, **47**, 8638.
38. B. Wang, P. Li, F. Yu, P. Song, X. Sun, S. Yang, Z. Lou and K. Han, *Chem. Commun.*, 2013, **49**, 1014.
39. B. Wang, F. Yu, P. Li, X. Sun and K. Han, *Dyes Pigm.*, 2013, **96**, 383.
40. K. Xu, H. Chen, J. Tian, B. Ding, Y. Xie, M. Qiang and B. Tang, *Chem. Commun.*, 2011, **47**, 9468.

41. F. Yu, P. Li, G. Li, G. Zhao, T. Chu and K. Han, *J. Am. Chem. Soc.*, 2011, **133**, 11030.
42. W. Zhang, W. Liu, P. Li, J. kang, J. Wang, H. Wang and B. Tang, *Chem. Commun.*, 2015, **51**, 10150.
43. Z. Lou, P. Li and K. Han, *Methods Mol. Biol*, 2015, **1208**, 97.
44. K. Xu, M. Qiang, W. Gao, R. Su, N. Li, Y. Gao, Y. Xie, F. Kong and B. Tang, *Chem. Sci.*, 2013, **4**, 1079.
45. X. Chen, K.-H. Baek, Y. Kim, S.-J. Kim, I. Shin and J. Yoon, *Tetrahedron*, 2010, **66**, 4016.
46. K. Mariappan, M. Alaparthi, M. Hoffman, M. A. Rama, V. Balasubramanian, D. M. John and A. G. Sykes, *Dalton Trans.*, 2015, **44**, 11774.
47. B. Tang, K. Ding, Xu and L. Tong, *Chem. – Eur. J.*, 2009, **15**, 3147.
48. E. G. Sormo, T. M. Ciesielski, I. B. Overjordet, S. Lierhagen, G. S. Eggen, T. Berg and B. M. Jenssen, *Environ. Sci. Technol.*, 2011, **45**, 6561.
49. A. Kumar and J. D. Singh, *Inorg. Chem.*, 2012, **51**, 772.
50. I. Haq, M. R. Jan and W. F. Smyth, *J. Chem. Soc. Pak.*, 1987, **9**, 375.
51. E. S. Gould and J. D. McCullough, *J. Am. Chem. Soc.*, 1951, **73**, 3196.
52. A. P. Singh, D. P. Murale, Y. Ha, H. Liew, K. M. Lee, A. Segev, Y.-H. Suh and D. G. Churchill, *Dalton Trans.*, 2013, **42**, 3285.
53. W. Shi, S. Sun, X. Li and H. Ma, *Inorg. Chem.*, 2010, **49**, 1206.
54. W. Fischer, *J. Chem. Educ.*, 2001, **78**, 133.
55. K. Xu, M. Qiang, W. Gao, R. Su, N. Li, Y. Gao, Y. Xie, F. Kong and B. Tang, *Chem. Sci.*, 2013, **4**, 1079.

Bio-physics, Chemistry and Effect on Oxidative Stress of Organoselenium Compounds

CHAPTER 7

Non-bonded Selenium· · ·Heteroatom Interactions in Selenoenzyme Glutathione Peroxidase and Mimics

M. IWAOKA* AND K. ARAI

Tokai University, Department of Chemistry, Kitakaname, Hiratsuka-shi, Kanagawa 259-1292, Japan
*Email: miwaoka@tokai.ac.jp

7.1 Introduction

Glutathione peroxidase (GPx) is the representative and most extensively studied selenocysteine (Sec)-containing enzyme.[1–3] This unique anti-oxidative selenoenzyme catalyses the reduction of hydroperoxides, such as H_2O_2 and ROOH, by use of glutathione (GSH) as a reducing cofactor. The active site (Figure 7.1A) comprises three genetically conserved amino acid residues, *i.e.* Sec(U), Gln(Q), and Trp(W), among which Sec plays a key role as the reaction center.[4] The other two are located at hydrogen-bonding distances from Sec, constituting a catalytic triad. Although the importance of this catalytic triad on the enzymatic antioxidative function is well evidenced by phylogenetic analysis[5,6] as well as site-directed mutational studies,[7,8] the

Organoselenium Compounds in Biology and Medicine: Synthesis, Biological and Therapeutic Treatments
Edited by Vimal Kumar Jain and K. Indira Priyadarsini

Published by the Royal Society of Chemistry, www.rsc.org

modes of interaction between the triad amino acid residues as well as the mechanistic roles of the Gln and Trp are still not yet clear.

In the meantime, Tosatto *et al.*[9] suggested the presence of a catalytic tetrad at the active site of GPx. They proposed that another proximate amino acid residue, *i.e.* Asn (N) next to Trp of the catalytic site, participates in the enzymatic activity together with the triad amino acid residues. This was inferred based on the sequence homology as well as the mutational substitution of Asn136 by the other amino acids, such as alanine (Ala), for *Drosophila melanogaster* GPx (*Dm*GPx), the active site of which comprises Cys45, Gln80, and Trp135.[9] Indeed, the X-ray structure of a U46C mutant of human GPx4[10] shows that the four amino acid residues, *i.e.* Cys46, Gln81, Trp136, and Asn137, construct a tetrahedral cluster at the active site (Figure 7.1B). Nevertheless, there is no straightforward evidence for the importance of such a tetrad on the enzymatic function of Sec-dependent GPx. Thus, investigating the interactions among the plausible tetrad amino acid residues, *i.e.* Sec, Gln, Trp, and Asn, is also an intriguing subject of current research. This chapter provides an overview of progress in the design of GPx mimics and related research areas, to address the possible correlation

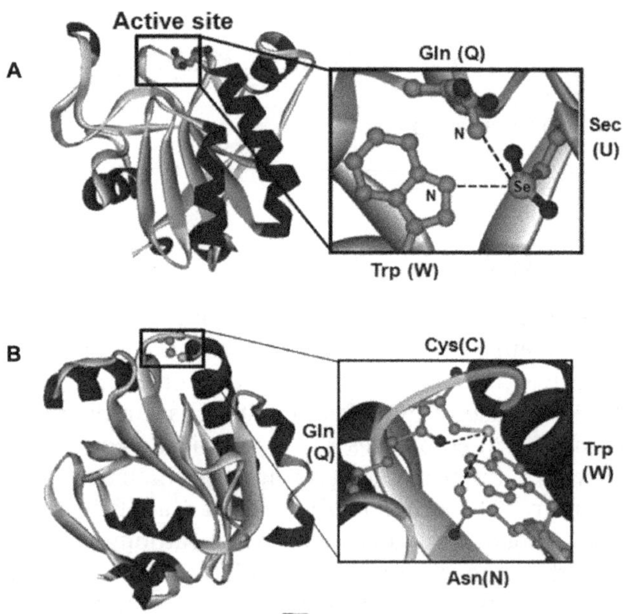

Figure 7.1 Active site structures of glutathione peroxidase (GPx). (A): a catalytic triad of bovine GPx (pdb code: 1gp1). Se(U)···N(Q) = 3.27 Å, Se(U)···N(W) = 3.56 Å, ∠N(Q)···Se(U)···N(W) = 70.2°. Reprinted with permission from ref. 59. Copyright 2014 American Chemical Society. (B): a proposed catalytic tetrad of Sec46Cys mutant of human GPx4 (pdb code: 2obi). S(C)···O(Q) = 3.33 Å, S(C)···N(W) = 3.92 Å, S(C)···N(N) = 3.63 Å.

between the weak interactions among the amino acid residues present at the GPx active site and antioxidative enzymatic function.

The selenium (Se) atom in organic molecules usually exists in a divalent state with two covalently bonded substituents and two lone pairs of valence electrons. However, because of its hypervalent nature, the Se atom frequently shows weak secondary bonding interactions with nearby heteroatoms.[11] At the GPx active site (Figure 7.1), two modes of such interactions would be possible between the Sec and the other amino acid residues. One is the NH· · ·Se hydrogen bond, in which the Se atom plays hydrogen-bond acceptor and is slightly donating the electrons to the NH moiety, and the other is the non-bonded O· · ·Se interaction, in which the Se atom accepts electrons from the interacting O atom. The latter interaction mode, which can be called a chalcogen bond[12] or σ-hole interaction[13] has already been studied elaborately[14,15] and successfully applied to asymmetric and/or catalytic reactions[16–20] and modelling of GPx mimics.[21–23] Features of these two types of weak interactions are explored in Section7.2.

Another characteristic nature of selenium is the variable oxidation state of a Se atom, which is more easily oxidized and reduced by redox agents than a sulfur (S) atom. This redox behavior of selenium is essential to the antioxidative function of GPx. In the GPx catalytic cycle, the oxidation number of the active-center Se atom turns around between −2 and 0 (or +2 when overoxidized).[24] Depending on the oxidation state, the modes of weak interactions at the GPx active site would change. This possibility is discussed in Section 7.3. Controlling the redox reactivity of the Se atom by surrounding functional groups has long been a central issue of the design of GPx mimics. Thus, various organoselenium compounds as well as selenocysteine derivatives have been synthesized and tested for their GPx-like capacity to detoxify harmful hydroperoxides.[25–31] Representative GPx mimics are classified into four groups in Section 7.4 in association with how exactly or briefly they imitate the local structure of the GPx active site. Finally, the current status of this research area and its perspectives are outlined in Section 7.5.

7.2 Modes of Weak Interactions Involving a Selenium Atom

Various modes of weak secondary bonding interactions are possible for a divalent Se atom because selenium is a fourth-row element and the valence electrons, which are weakly bound to the nucleus, are significantly polarizable. However, to understand the molecular mechanism of the GPx function, only the NH· · ·Se hydrogen bond and the non-bonded O· · ·Se interaction would be important in the light of the structure of the active site (Figure 7.1). In this section, chemical properties of these weak interactions are overviewed along with the related interactions, such as OH· · ·Se and CH···Se hydrogen bonds and non-bonded X···Se interactions, where X is N, F, Cl, *etc.*

7.2.1 NH···Se and Related Hydrogen Bonds

Chemical properties of the NH···Se hydrogen bond have seldom been studied in the literature. Instead, the analogous OH···Se hydrogen bond has repeatedly been analysed experimentally and theoretically. Byun and Kang analysed conformations of a Sec dipeptide (*i.e.* Ac-Sec-NHMe) by *ab initio* calculation in water,[32] which showed that the preferred conformation in water does not have a back-bone hydrogen bond, but has a five-membered NH···Se hydrogen bond (Section 7.4.3). Since the acid dissociation constant of the SeH group is much larger than water ($pK_a = 5.24$),[33] a Sec residue should be deprotonated under physiological conditions and the resulting negative charge on the Se atom would strengthen the NH···Se hydrogen bond. Similar NH···Se^- hydrogen bonds can be formed at the GPx active site in the selenol intermediate state (Section 7.3.1).

Harada *et al.*[34] reported that an intramolecular OH···Se hydrogen bond can control the molecular structure of 2-(methylseleno)ethanol ($HOCH_2CH_2SeCH_3$) in the gas phase. Madzhidov and Chmutova[35] estimated the OH···Se hydrogen bonding energy between dimethyl selenide (CH_3SeCH_3) and methanol (CH_3OH) to be ~4 kcal mol^{-1} by *ab initio* calculation. These experimental and theoretical studies suggest that the OH···Se hydrogen bond is intrinsically as strong as usual hydrogen bonds. The same can be applied to an analogous NH···Se hydrogen bond. Yet the energy may significantly decrease in aqueous medium due to the electrostatic character.

CH···Se hydrogen bonding is frequently found in crystals of organoselenium compounds.[36,37] This weak interaction has a less electrostatic character and is much weaker than NH/OH···Se hydrogen bonds (1~2 kcal mol^{-1}).[38] Therefore, CH···Se hydrogen bonds would be less important in controlling the GPx activity than NH/OH···Se hydrogen bonds. Nonetheless, it can be present at the active site as CH moieties are abundant in proteins.

7.2.2 Non-bonded O···Se and Related Interactions

Non-bonded interactions between a divalent Se atom and various heteroatoms, such as N, O, F, Cl, *etc.* have been well characterized.[39–44] For this type of interactions, characteristic structural propensities are commonly observed. When the interaction is stronger (*i.e.* the non-bonded X···Se distance is shorter), the backside Se—Y bond becomes longer and the three atoms of the X···Se—Y unit align more linearly. These propensities suggest the importance of the orbital interaction between an electron-donating X atom (*i.e.* n_X) and an electron-accepting Se atom (*i.e.* σ^*_{SeY}) for stabilization. This is consonant with the observation that the ^{77}Se nuclear magnetic resonance (NMR) signal apparently shifts downfield as the interaction becomes strong.[45] Interestingly, there is an almost linear correlation between a logarithm of the interaction energy ($E_{X\cdots Se}$), which was calculated as the $n_X \rightarrow \sigma^*_{SeY}$ orbital interaction energy by natural bond orbital analysis,[46] and the relative X···Se atomic distance ($d = r_{X\cdots Se}/(vdw_X + vdw_{Se})$), where *vdw* means van der Waals radius (Figure 7.2).[45] This correlation clearly confirms that non-bonded X···Se interactions have a common mechanistic feature despite the magnitude of the stabilization energy falling in a wide range. The obtained equation, $\log E_{X\cdots Se} = 5.855(1-d) - 0.302$ would be useful to estimate the energy of the non-bonded X···Se interaction on the basis of the X···Se atomic distance, which can be determined by X-ray analysis.

The strength of the interaction decreases in the order of $O(sp^2)\cdots Se$ (6 to 36 kcal mol^{-1})[14,15] >$N(sp^3)\cdots Se$ (3 to 30 kcal mol^{-1})[42,47] >$O(sp^3)\cdots Se$ (2 to 15 kcal mol^{-1})[14,15] >F···Se (2 to 7 kcal mol^{-1})[48,49] >Cl···Se ≈ Br···Se (0.2 to 2 kcal mol^{-1}).[50,51] The stabilization energy is monotonously controlled by the electron-accepting potential of Se and the electron-donating potential of X. However, in the case of weak Cl···Se and Br···Se interactions, the effect of electron correlation would be more significant than the $n_X \rightarrow \sigma^*_{SeY}$ orbital interaction.[50,51] Thus, there is some deviation from the correlation line (a broken line in Figure 7.2).

At the active site of GPx, the oxidation state of the Se atom turns around. When it is oxidized, the Se atom has a less negative (or more positive) charge,

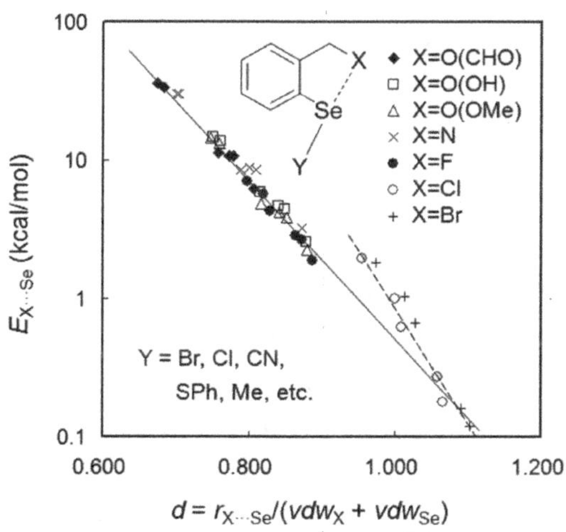

Figure 7.2 Correlation plots between a logarithm of the $n_X \rightarrow \sigma^*_{SeY}$ orbital interaction energy ($E_{X\cdots Se}$) and the relative X···Se atomic distances [$d = r_{X\cdots Se}/(vdw_X + vdw_{Se})$] obtained for a series of aromatic selenium compounds.
Adapted from Physical- and Bio-Organic Chemistry of Nonbonded Selenium-Oxygen Interactions, M Iwaoka and S Tomoda, Phosphorus, Sulfur, and Silicon, and the Related Elements, 23rd February 2005, Taylor & Francis, reprinted by permission of the publisher, Taylor and Francis Ltd.[45]

hence a non-bonded O···Se interaction rather than the NH···Se hydrogen bond would become more important. On the contrary, when the Se atom is reduced, it should have a more negative (or less positive) charge, hence the NH···Se hydrogen bond would become dominant. These predictions are examined in further detail in the next section.

7.3 Catalytic Cycle of GPx and the Intermediates

According to extensive kinetic studies,[52,53] it has been revealed that the catalytic cycle of GPx can be delineated by a ping–pong mechanism, in which three active intermediates, *i.e.* selenol (GPx-SeH), selenenic acid (GPx-SeOH), and selenenylsulfide (GPx-SeSG), are involved (Figure 7.3). TheGPx-SeH state, which would be predominantly deprotonated to GPx-Se$^-$ under physiological conditions, should be a major state and is first oxidized with hydroperoxide (ROOH) to afford GPx-SeOH and ROH. In the second step, the generated GPx-SeOH intermediate reacts with GSH to give the GPx-SeSG intermediate and H_2O. Finally, the GPx-SeSG intermediate reacts with another GSH to reproduce the GPx-SeH state and oxidized glutathione (GSSG). Thus, GPx catalyses the reduction of harmful hydroperoxides to water and/or alcohol using GSH, which is abundant in cells. However, when oxidative stress is severe or antioxidative potency is impaired, the GPx-SeOH intermediate would be over-oxidized to the stable seleninic acid (GPx-SeO_2H) state (Figure 7.3). Indeed, bovine GPx was thus isolated as a stable seleninic acid form, and the molecular structure was determined using X-ray crystallography (Figure 7.1A).[54] In this section, possible interaction modes between the Se atom and nearby functional groups at the GPx active site are discussed for each intermediate state.

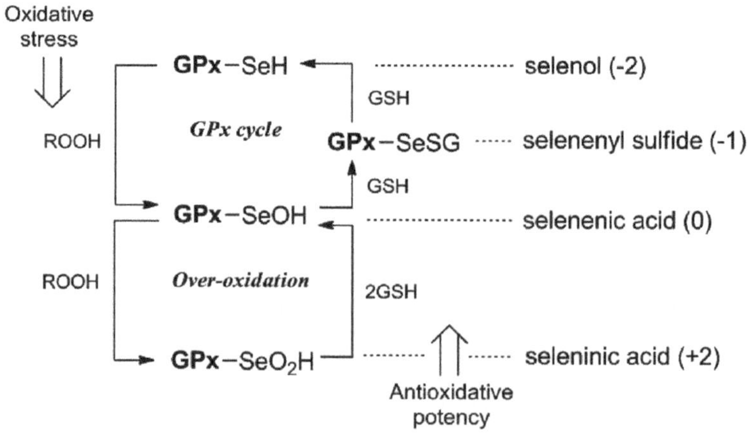

Figure 7.3 A catalytic cycle of glutathione peroxidase (GPx). Oxidation numbers of the Se atom are shown in parentheses.

7.3.1 GPx-SeH

According to the X-ray structure of GPx (Figure 7.1A),[54] the nitrogen atoms of Q80 and W158 side-chains exist close to the Se atom of U45 (3.27 and 3.56 Å, respectively), suggesting the presence of NH···Se hydrogen bonding interaction at the active site in the reduced GPx-SeH state. The interaction modes in this state were studied theoretically by Morokuma and Musaev and co-workers.[55,56] Based on the X-ray structure of human plasma GPx,[57] they made a truncated model and performed density functional theory (DFT) calculations.[55,56] The results showed that the two NH groups of Q and W side-chains stay close to the active center Se atom with one NH of Q forming a NH···Se hydrogen bond and the other NH of W forming NH···O hydrogen bond with the amide oxygen of the side chain of Q (Figure 7.4A). According to this truncated GPx active-center model, there is a hydrogen-bond network around the SeH group. The network includes the backbone NH group of the glycine residue next to U and a water molecule present near the active center. It was therefore proposed that the amide side-chain of Q would play a more important role in enzymatic function than the indole side-chain of W. A similar but slightly different hydrogen-bond network was suggested by ONIOM study by the same authors.[58]

Iwaoka and co-workers modelled the GPx active site using designed Sec-containing peptides.[59] They selected the amino acid sequence of GQAUAWG as a possible catalytic triad model and showed experimentally that the GPx-like antioxidative function is cooperatively enhanced by the presence of both Q and W near U. The structure of the selenolate state, which was obtained by *ab initio* calculation at HF/6-31G(d), showed the presence of two NH···Se

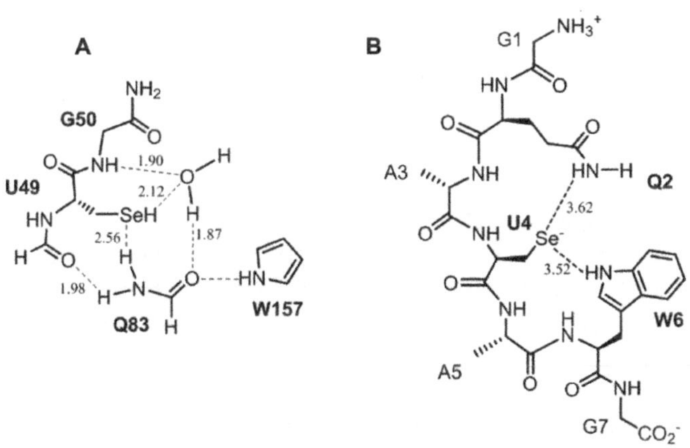

Figure 7.4 Networks of weak interactions around the Se atom at the glutathione peroxidase (GPx) active site proposed for the selenol intermediate. (A): Morokuma and Musaev's truncated model.[55] (B): Iwaoka and co-workers' selenopeptide model.[59] For model B, distances (Å) between Se and N atoms are shown.

hydrogen bonds (Figure 7.4B). The N···Se distances (3.62 and 3.52 Å) and the N···Se···N angle (74.5°) were similar to those of the X-ray structure (Figure 7.1A), supporting the adequacy of the selenopeptide to model the catalytic triad.

The two GPx active-site models mentioned indicate different interaction networks around the Se atom. However, there are common characteristic features, such as both Q and W keep stay near U, the NH of Q forms a NH···Se hydrogen bond, and fluctuation of the Se atom is constrained to some extent by a hydrogen-bond network. These features will be useful for the design of GPx mimics using small organoselenium compounds.

7.3.2 GPx-SeOH

Selenenic acid is an extremely labile species, so GPx-SeOH cannot be detected or observed experimentally. It should be promptly converted to GPx-SeSG by the reaction with GSH or over-oxidized to GPx-SeO_2H in the presence of excess amounts of oxidants. However, when appropriate models are chosen, selenenic acid species can be stabilized and characterized by various spectroscopic methods as well as by single-crystal X-ray crystallography.[60–62] By using a sterically bulky terphenyl unit with a bowl shape, Goto *et al.*[63] achieved direct oxidation of the selenol with H_2O_2 to the kinetically stabilized selenenic acid. The X-ray structure showed that the SeOH moiety is surrounded by benzene rings. No secondary interaction was observed for this Se atom. The SeOH species was then reduced stepwise back to the selenol *via* the selenenyl sulfide intermediate by the reaction with a thiol substrate. Thus, the GPx catalytic cycle (Figure 7.3) was fully evidenced.

Interaction modes around the Se atom at this oxidation state have been investigated theoretically. DFT studies by Morokuma and Musaev and co-workers[55,56] showed the presence of a similar hydrogen-bond network to that of the truncated GPx-SeH model shown in Figure 7.4A. In the hydrogen-bond network (Figure 7.5A), the NH of Q forms a NH···Se hydrogen bond despite the positive atomic charge that the Se atom should have now. Conversely, by performing *ab initio* calculations at the MP2 level for simple cluster models, such as MeSeCl···$MeCONH_2$, Bayse *et al.*[64] demonstrated possible rotation of the side-chain amide group of Q to form non-bonded O···Se interaction in the GPx-SeOH state.

Such rotation was also suggested by Iwaoka and co-workers[59] by *ab initio* calculation using a triad selenopeptide model (Figure 7.5B). Two structures were obtained for the selenenic acid intermediate. In both structures, the indole ring of W remains in the proximity of U, forming a NH···Se and/or CH···Se hydrogen bond. In contrast, the amide group of Q2 is rotated and pushed a little away from U4. In structure B1, a non-bonded O···Se interaction is found between the amide O atom of Q2 and the Se atom of U4. In structure B2, a non-bonded O···Se interaction is found between the main

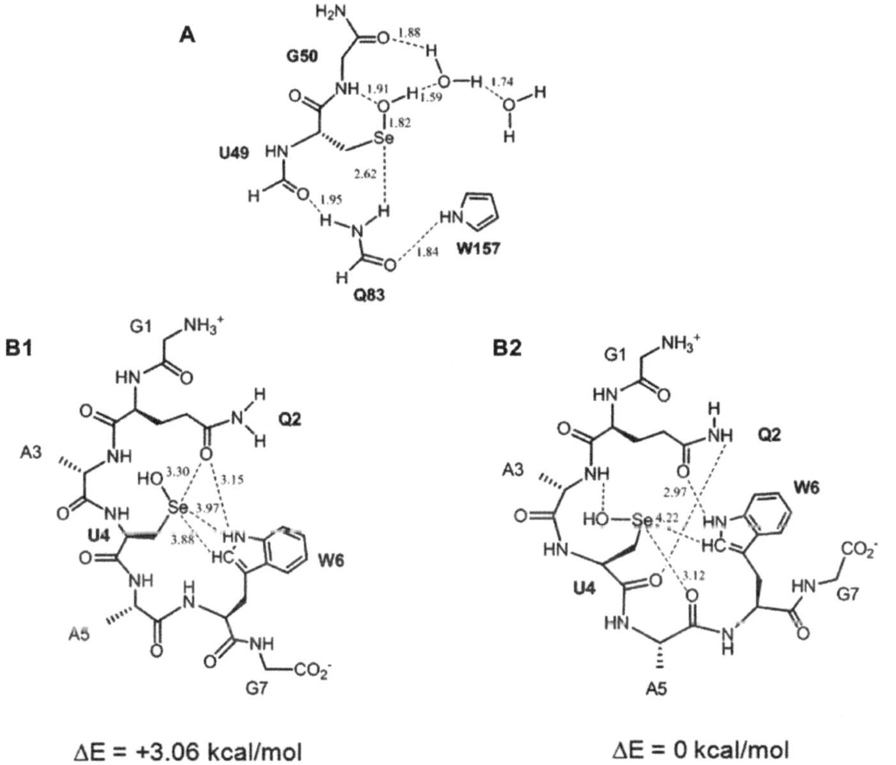

Figure 7.5 Networks of weak interactions around the Se atom at the glutathione peroxidase (GPx) active site proposed for the selenenic acid intermediate. (A): Morokuma and Musaev's truncated model.[55] (B): Iwaoka and co-workers' selenopeptide model.[59] For model B, distances (Å) between Se and O or C atoms and those between N and O atoms are shown.

chain O atom of A5 and the Se atom of U4. Thus, O···Se interaction is predominant in this selenopeptide model. These interaction networks around the Se atom would prevent the selenenic acid intermediate from undesirable decomposition or further oxidation by thermodynamically stabilizing the labile SeOH moiety.

7.3.3 GPx-SeSG

The selenenyl sulfide intermediate is thermodynamically the most stable species of the three intermediates involved in the GPx catalytic cycle. This means that the reaction step from GPx-SeSG to GPx-SeH is the rate-determining step. In the GPx active site, this step should be significantly accelerated by the substantial affinity toward the second GSH molecule, which selectively attacks the S atom of the Se–S bond, not the intrinsically more reactive Se atom, to produce GPx-SeH and GSSG. However, simplified

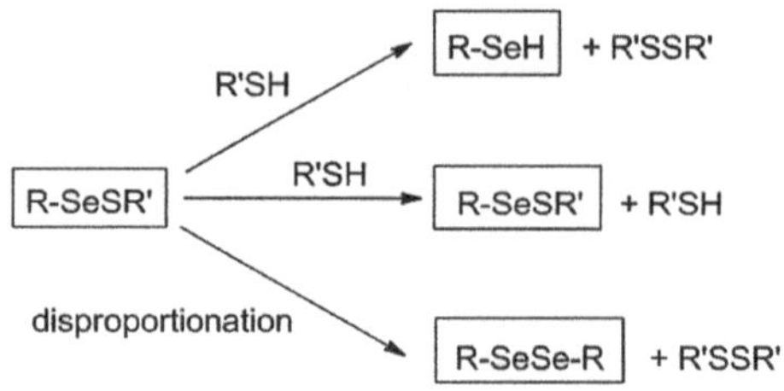

Figure 7.6 Three reaction paths for selenenyl sulfide intermediates of glutathione peroxidase (GPx) mimics.

GPx mimics do not have such affinity. Therefore, when they are employed in activity assays, selenenyl sulfide species should be accumulated in the reaction mixture. The interaction modes at this stage would be especially important for design of effective GPx mimics.

There are three possible reaction paths for the selenenyl sulfide of GPx mimics (Figure 7.6). The first path is the desirable reaction with a thiol substrate at the S atom to produce a selenol species, but this is energetically unfavourable because the lowest unoccupied molecular orbital is more localized on the Se atom than on the S atom. The second is the undesirable reaction with a thiol substrate at the Se atom to produce the same selenenyl sulfide species. Sarma and Mugesh[65] demonstrated that preference between these two reaction paths can be reversed by using a thiol substrate with a non-bonded O/N···S interaction. It was also suggested that the presence of a strong O···Se interaction in the GPx mimic is detrimental for the desirable first reaction path.[66] Thus, the strength of the non-bonded O···Se interaction would be an important factor to determine the GPx-like catalytic activity of synthetic GPx mimics. The third path is bimolecular disproportionation to produce a diselenide and a disulfide. Pertinent control of these three reaction paths is possible by adjusting the strength of O···Se interaction and by employing an appropriate thiol substrate.

The interaction networks suggested by Morokuma and Musaev's truncated GPx model and Iwaoka and co-workers' selenopeptide model are shown in Figure 7.7. They are very different from each other. In the former model, no direct interaction was observed between U and Q or W. Instead, a backbone NH of a leucine residue forms a NH···Se hydrogen bond. In contrast, in the latter model it is seen that the close contact between the Se atom and the indole group remains through the formation of a CH···Se hydrogen bond as observed for the selenenic acid intermediate. Two additional O···Se interactions are present, which are formed in the backside directions of the Se–S and Se–C covalent bonds as usually observed in such non-bonded interactions. It should be noted that the amide group of Q is again rotated to form a direct O···Se interaction.

As discussed above, a network of weak interactions in the GPx active site can be significantly changed, not only by the oxidation states of the Se atom, but also by the employed active-site models and a thiol substrate. Therefore,

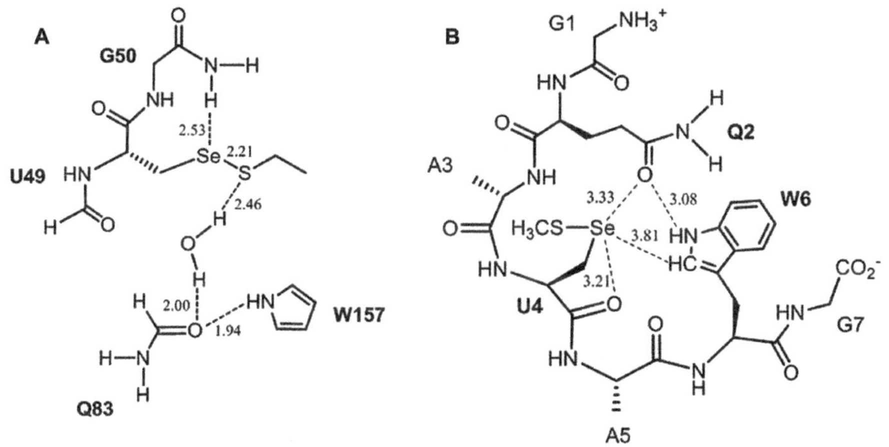

Figure 7.7 Networks of weak interactions around the Se atom at the glutathione peroxidase (GPx) active site proposed for the selenenyl sulfide intermediate. (A): Morokuma and Musaev's truncated model.[55] (B): Iwaoka and co-workers' selenopeptide model.[59] For model B, distances (Å) between Se and O or C atoms and those between N and O atoms are shown.

further studies are still necessary to clearly address the correlation between the interaction modes and the antioxidative function of GPx.

7.4 Design of GPx Mimics

A wide range of organoselenium compounds have been investigated to date as mimics of GPx, which have been frequently updated in the literature.[25–31,67–69] Herein, these GPx mimics are classified into four groups (Figure 7.8). An overview of the cumulative data for each GPx-mimic group is given from the viewpoint of the interaction modes between the Se active center and the nearby heteroatom-containing functional groups.

There are two main streams of the design of GPx mimics. In the first stream, a rigid benzene ring is utilized instead of the bulky protein structure, and a Se atom and various functional groups (X and/or Y) are substituted in juxtapositions on the benzene ring (group 2). Conversely, a flexible peptide fragment containing a Sec residue is extracted from the active site in the second stream (group 4). Such selenopeptide mimics can be further simplified into selenocysteine itself and the other water-soluble selenium compounds with a polar functional group (group 3). Ebselen (**1**) and its analogues (group 1) would be located in the confluence of the two streams, as they are not only aromatic selenium compounds, but also have an amide group like a selenopeptide model, although the amide group of ebselen is cyclized to make a five-membered isoselenazole ring.

Since selenol, selenenic acid, and selenenyl sulfide species, which are involved in the GPx catalytic cycle, are usually unstable, they cannot be

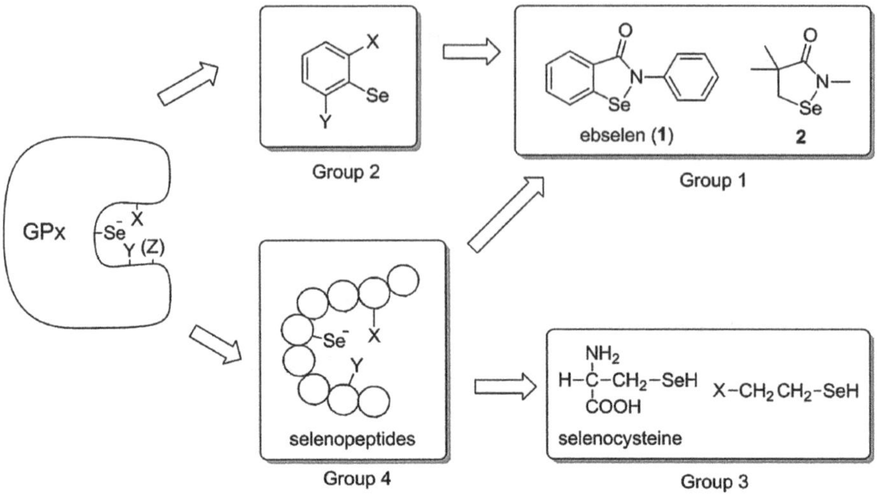

Figure 7.8 Two streams of glutathione peroxidase (GPx) mimics and their classification.

employed directly as GPx mimics. Therefore, diselenides or their equivalents, such as a cyclized selenenamide like ebselen (**1**), are employed as pre-catalysts of GPx mimics. These pre-catalysts are converted into the corresponding selenols by reduction with a thiol substrate or into labile selenenic acids by oxidation with H_2O_2. The active species thus generated get into the catalytic cycle.

7.4.1 Ebselen and Related GPx Models

Ebselen (**1**),[70–72] with a five-membered isoselenazole ring fused to a benzene skeleton, is the first organoselenium compound that was investigated as a possible GPx mimic. Various catalytic cycles, depending on the reaction conditions, have been proposed for this famous pre-catalyst. Details of the catalytic cycles were summarized in a recent review.[29] The initial step for activation of **1** is probably the reaction with a thiol to produce an open-chain selenenyl sulfide, which has an amide moiety close to the Se atom. The selenenyl sulfide produced would be subsequently converted into a selenol or a diselenide, depending on the thiol substrate employed (see Figure 7.6). These active species would then be oxidized with H_2O_2 to a selenenic acid or seleninic acid intermediate. Thus, ebselen (**1**) catalyzes the reduction of H_2O_2 with a thiol. Cyclic selenenamide **2** developed by Reich and Jasperse[73] can be considered as a simplified ebselen model, in which the two benzene rings of **1** are replaced by three methyl groups. However, the catalytic cycle suggested for **2** was different from those proposed for **1**.

A wide range of ebselen analogues has been developed and examined for GPx-like antioxidative functions *in vitro*. For example, analogues with various substituents on the amide nitrogen atom[66,74,75] and those modified

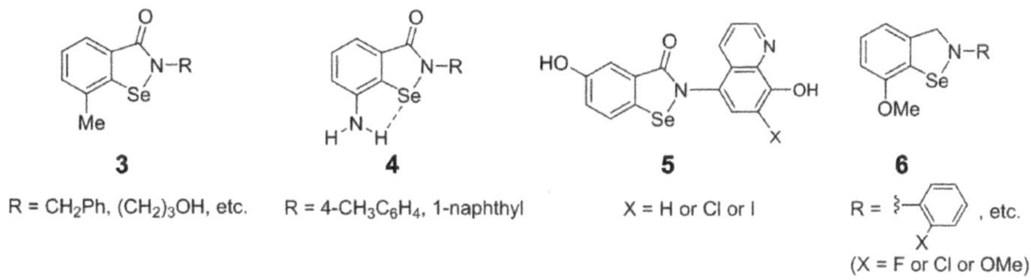

Figure 7.9 Analogues of ebselen (**1**) with substitution at the *ortho* or *para* position with respect to the Se atom.

with amino acid synthons[76,77] have been reported. Introduction of an additional substituent at the *ortho* or *para* position with respect to the Se atom significantly affected their GPx-like activity. Examples are shown in Figure 7.9. Substitution with a functional group of *o*-Me (**3**),[78] *o*-NH_2 (**4**),[79] or *p*-OH (**5**)[80] resulted in enhancement of the GPx-like activity compared to ebselen (**1**), as a consequence of weaker O···Se interactions in the open-chain selenenyl sulfide intermediate state due to steric and/or electronic substituent effects. Such non-bonded interactions, rather than NH···Se hydrogen bonds, have been observed frequently in the diselenide state (Section 7.4.2). Recently, novel isoselenazolidine derivatives (**6**) fused to a benzene ring were reported to have both GPx and peroxiredoxin activities in *in vitro* assays.[81]

7.4.2 Aromatic Diselenides with Functional Groups

From a historical point of view, the second generation of GPx mimics would be aromatic diselenides with intramolecular coordinating functional groups (X and Y) close to the Se atom (Figure 7.10). Diselenide **7** with a tertiary amino group at the *ortho* position was designed by Wilson *et al.*[82] Similar diselenides, such as **8–10**, were later synthesized by Iwaoka and Tomoda,[21] Mugesh *et al.*,[83,84] and Wirth and co-workers,[85,86] respectively. In these diselenides, the roles of intramolecular non-bonded N/O···Se interactions were remarkable. Such interactions would stabilize a labile selenenic acid intermediate through $n_{N/O} \rightarrow \sigma^*_{SeOH}$ orbital interaction.[21] In contrast, in the selenenyl sulfide intermediate state, two opposite effects of the interaction were pointed out. Firstly, the non-bonded O/N···Se interaction would sterically block the nucleophilic attack of a thiol on the Se atom leading the desired nucleophilic attack on the S atom.[21] Secondly, the interaction would enhance the electrophilicity of the Se atom, which misleads a thiol substrate to approach the Se atom. Bhabak and Mugesh[87] demonstrated that the introduction of an additional methoxy group at the *ortho* position has a significant effect of weakening the non-bonded N···Se interaction in the selenenyl sulfide state, thereby enhancing the GPx-like activity of **11**.

The early studies described suggested that regulation of the strength of non-bonded O/N···Se interaction is an effective strategy to optimize the

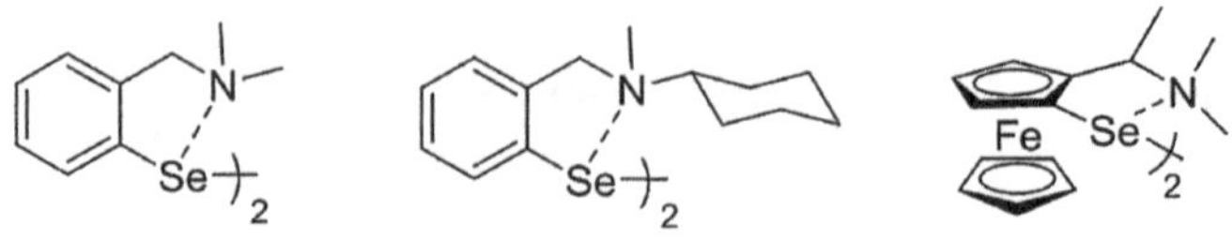

7 8 9

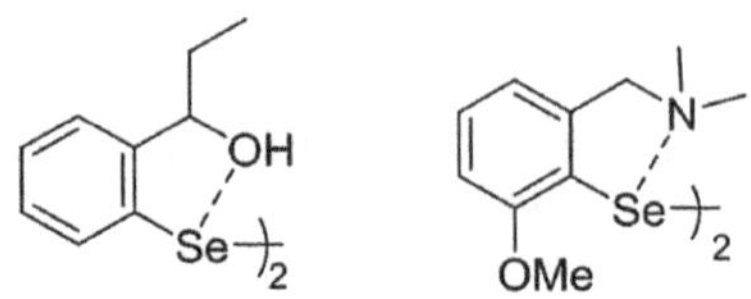

10 11

Figure 7.10 Aromatic diselenides with a coordinating functional group at the *ortho* position.

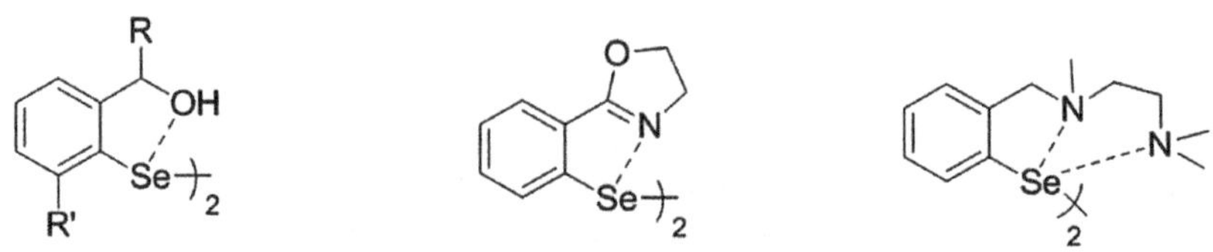

12 13 14

R = Et
R' = H or Me or CF_3 or CH(OH)Et

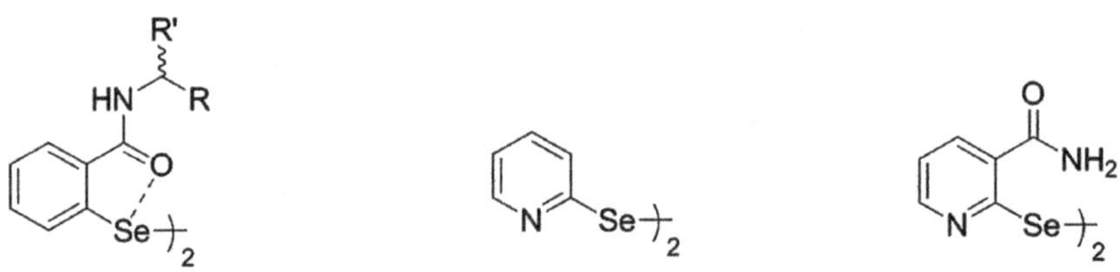

15 16 17

R' = Me, R = CH_2Ph
R' = H, R = Py

Figure 7.11 Aromatic diselenides modified with various functional groups to regulate the strength of non-bonded N/O···Se interactions.

GPx-like activity of the mimics of this group. Thus, numerous aromatic diselenides with various functional groups at the *ortho* position(s) have been developed to date. Representative recent examples are summarized in Figure 7.11. Diselenides with a hydroxy group (**12**),[85] an oxazoline ring (**13**),[88] two tertiary amino groups to mimic the catalytic triad (**14**),[89] and a secondary amide group (**15**)[90,91] have been investigated extensively. In addition, di(2-pyridyl) diselenide derivatives (**16** and **17**) have been reported to have significant antioxidant activities.[92,93]

7.4.3 Selenocysteine and Related GPx Models

Selenocysteine is utilized as a redox active center by GPx.[4] However, its GPx-like redox behaviors had not been extensively studied until recently. Singh *et al.*[94] investigated in detail the GPx-like catalytic cycle of selenocystine (Sec_2), which is an oxidized dimer of Sec having a diselenide (SeSe) bond, by analyzing the kinetics of the reaction between three kinds of hydroperoxides (H_2O_2, CumOOH, and *t*BuOOH) and two kinds of thiols (GSH and dithiothreitol) in the presence of a catalytic amount of Sec_2. It was revealed that the active intermediates generated from Sec_2 have more affinity toward peroxides than thiols. In the meantime, ^{77}Se NMR analysis on the reaction mechanism[94] suggested that the initial activation step of this pre-catalyst is the oxidation with H_2O_2, rather than the reduction with a thiol, to afford the seleninic acid (Sec-SeO_2H; δ_{Se} 1183 ppm), which would be formed by overoxidation of the highly reactive selenenic acid intermediate (Sec-SeOH). By treatment of Sec-SeO_2H with three equivalents of cysteine (Cys-SH) as a thiol substrate, the selenenyl sulfide (Sec-SeS-Cys; δ_{Se} 349 ppm) was almost quantitatively obtained. The catalytic cycle of Sec_2 was thus elucidated. By *ab initio* calculation, it was further suggested that the observed intermediates, as well as Sec_2, have intramolecular hydrogen bonds in water, as shown in Figure 7.12.

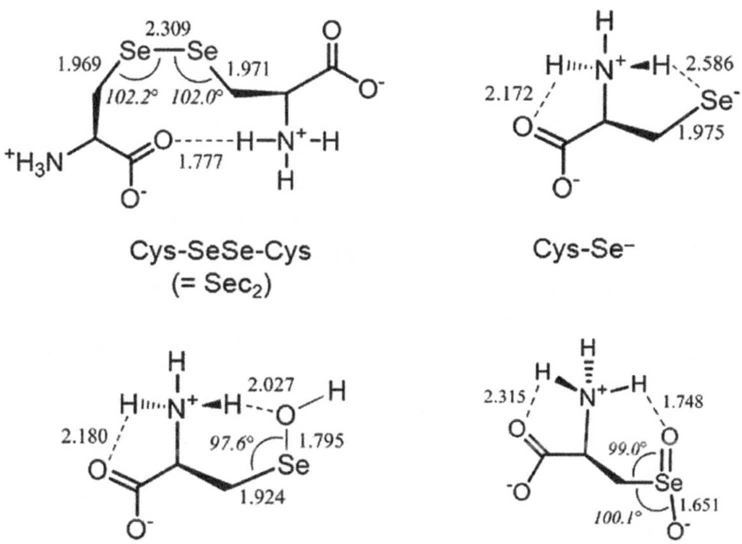

Figure 7.12 Stable conformations of Sec_2 and its derivatives by *ab initio* calculation.[94]

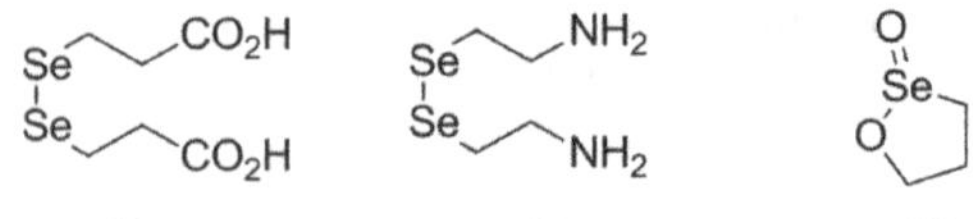

18 **19** **20**

Figure 7.13 Molecular structures of water-soluble diselenides and a cyclic seleninate.

By removing either of the amino or carboxy groups from Sec_2, 3,3′-diselenodipropionic acid (**18**) and selenocystamine (**19**) are obtained, respectively (Figure 7.13). These water-soluble diselenides were tested by Priyadarsini and co-workers as possible GPx mimics[95] and non-toxic antioxidants.[96,97] Interesting redox behaviors of similar seleninic acid derivatives were reported by Back and co-workers.[98–101] For example, the simple cyclic seleninate ester **20** exhibited high GPx-like activity through a unique catalytic cycle, which includes open-chain thioseleninate (HO–$CH_2CH_2CH_2$–Se(O)SBn) and selenenic acid (HO–$CH_2CH_2CH_2$–SeOH) intermediates.[102]

7.4.4 Designed Sec-containing Peptides

The GPx-like antioxidative function of some Sec-containing peptides have been investigated with the expectation that they would serve as useful probes or models for understanding the molecular mechanism of GPx enzymatic function. To synthesize designed selenopeptides with an arbitrary amino acid sequence, particular Sec derivatives are required, but they were not easily available previously. Now, several effective synthetic methods have been developed for the preparation of such Sec derivatives.[103–110] At the same time, the solid-phase peptide synthesis technique for selenopeptides has been improved.[111–116] Such progress has allowed us to utilize various short selenopeptides as mechanistic probes for the antixodative function of GPx.

Selenoglutathione (GSeH; γEUG),[117–120] a selenium analogue of GSH, is one of the interesting short selenopeptides, as it exhibits unusual redox properties.[114,121,122] Yoshida *et al.* investigated the GPx activity of selenoglutathione diselenide (GSeSeG) (**21**) and two other short selenopeptides (**22** and **23**), which have the same amino acid sequence as the GPx active site (*i.e.* LUG and SLUGT, respectively) (Figure 7.14). GSeSeG (**21**) exhibited more than two-fold higher GPx activity than Sec_2, whereas **22** and **23** showed much less activity.[122] The results suggested that the local amino acid sequence at the GPx active site is not important for the antioxidative function. The unusually high GPx activity observed for **21** was attributed to the presence of alternative reaction paths in addition to the normal GPx cycle, as shown in Figure 7.15. The alternative paths are possible because **21** can behave as a substrate of glutathione reductase,[114,121,122] which is added to the assay solution to reduce GSSG to GSH. However, a detailed examination of the GPx activity of **21** revealed that the presence of the alternative reaction paths is not enough to fully explain the high GPx activity.[122] The high activity may

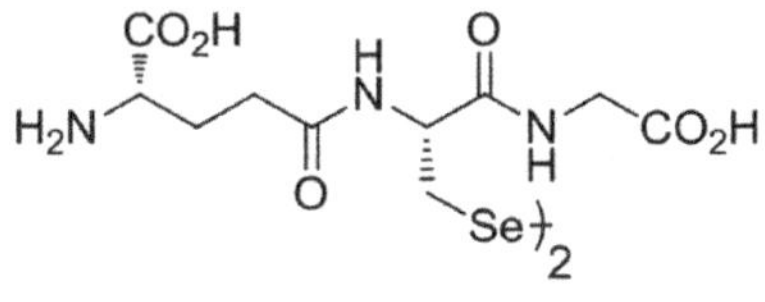

GSeSeG (**21**)

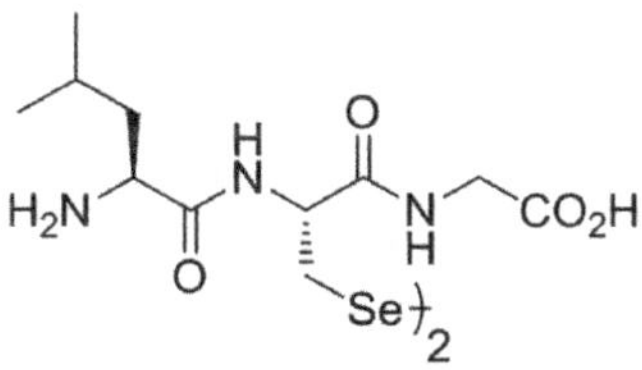

22

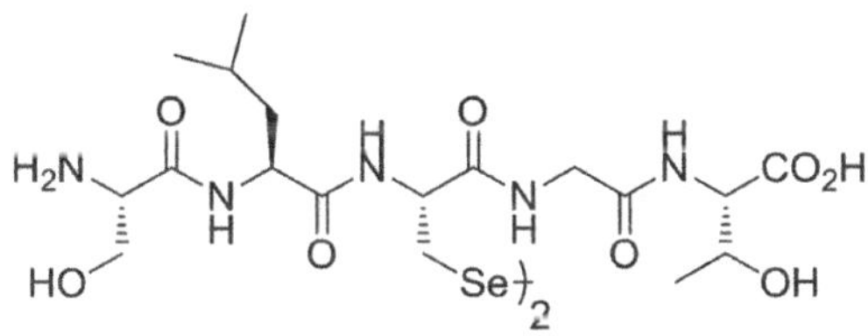

23

Figure 7.14 Short selenocysteine-containing peptides as glutathione peroxidase (GPx) active-site models.

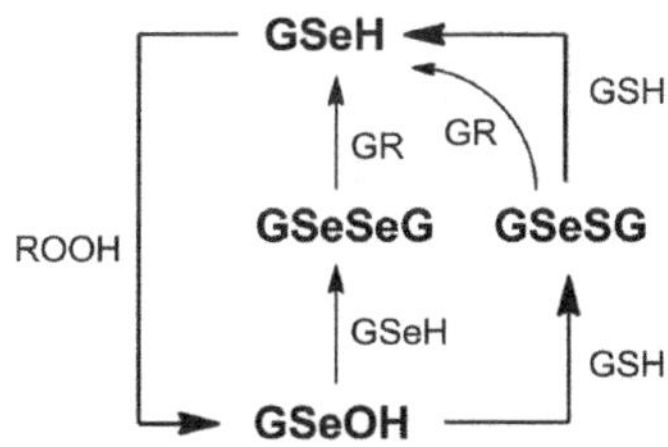

Figure 7.15 Catalytic cycles of selenoglutathione. In addition to the glutathione peroxidase (GPx) cycle (outer circle), two bypass routes (inner paths) are present. GR: glutathione reductase.

also be due to stereoelectronic effects of the γ-glutamic acid residue in **21**. Nevertheless, the interaction modes around the Se atom of GSeSeG (**21**) and its active intermediates are still unknown.

Since the primary amino acid sequence around the Sec residue of the GPx active site did not have any significant effect on antioxidant activity,[122] the next target of the GPx model was a selenopeptide emulating the catalytic triad structure.[123] One simple model, *i.e.* a short selenopeptide containing Sec (U), Gln (Q), and Trp (W) residues along the amino acid sequence, has been designed by Iwaoka and co-workers (Section 7.3.1).[59] Based on the results from molecular simulation, they synthesized a selenopeptide with an amino acid sequence of GQAUAWG, in which U, Q, and W are separated by a single Ala residue (Figures 7.4, 7.5, and 7.7). Compared with GAAUAAG, the GPx-like activity was enhanced by the introduction of Q, whereas it was decreased by the introduction of W, suggesting that Q is more important than W for GPx activity. Of interest, the activity was cooperatively increased by introduction of both Q and W. Thus, importance of the GPx catalytic triad was strongly supported.[59] Applying similar approaches not only for the putative catalytic tetrad, but also for the active sites of other selenoenzymes would be worthwhile. Other short selenopeptides with different amino acid sequences have also been investigated as possible GPx mimics.[124,125]

Chemical modification of proteins with a selenating reagent, such as H_2Se and NaHSe, has been applied for the synthesis of semi-synthetic selenoproteins. The first example was reported by Wu and Hilvert in 1989,[126] who demonstrated that chemical modification of Ser at the active site of subtilisin to Sec altered the enzymatic function from protease to peroxidase, like GPx.[127,128] Similar transformation of the enzymatic activity by chemical modification or site-directed mutagenesis was achieved for other proteins.[124,129,130] Luo *et al.* devised engineered selenoabzymes by introducing U, Q, and W into the antibody scFv2F3 at the pertinent positions, and showed that it exhibits significant GPx activity by collaboration among the triad amino acid residues.[131]

7.5 Conclusions and Perspectives

In this chapter, on the basis of cumulative achievements described in the literature regarding the structure and enzymatic functions of GPx and the structure–activity relationship of various GPx mimics, we describe possible interaction modes between the selenium atom of the GPx active center and the nearby amino acid residues. Although GPx is the representative and most extensively studied Sec-containing enzyme, the modes of interaction between the catalytic triad amino acid residues, *i.e.* Sec, Gln, and Trp, are still not yet well elucidated. Nonetheless, the presence of a catalytic tetrad has also been suggested, but there is no straightforward evidence for the importance of such a tetrad on the enzymatic function of Sec-dependent GPx. How to control the reactivity of a redox-active selenium atom by surrounding functional groups has been a central issue of the design of GPx mimics by applying small organoselenium compounds. Representative GPx mimics were classified here into four groups in association with how exactly or briefly they imitate the local structure of the GPx active site. These are ebselen (**1**) and its analogues (group 1), aromatic selenium compounds with functional groups (group 2), selenocysteine and the related water-soluble selenium compounds (group 3), and designed selenopeptide mimics (group 4).

Two modes of weak interactions would be possible at the GPx active site. One is the NH···Se hydrogen bond, in which the Se atom plays a hydrogen-bond acceptor and slightly donates electrons to the NH moiety, and the other is the non-bonded O···Se interaction, in which the Se atom accepts electrons from the interacting O atom. These two modes of interactions would pertinently blend at the GPx active site, depending on the oxidation state of the intermediates, *i.e.* GPx-SeH, GPx-SeOH, and GPx-SeSG. Possible models of such interaction networks around the selenium atom at the GPx active site have been proposed (Figures 7.4, 7.5, and 7.7). However, there is no general consensus regarding the exact interaction networks, which would be significantly complicated in the real enzyme.

Although the exact networks of the interaction modes and the mechanistic roles of the catalytic triad and putative tetrad are still beyond our understanding, new experimental techniques are available now to approach this difficult issue, such as the Sec-mediated native chemical ligation (Sec-NCL) for selenopeptide synthesis,[111–113,115] site-directed mutagenesis for selenoproteins,[132–137] Cys auxotrophic expression of selenoproteins in the presence of Sec,[130,138–140] and so on. With these methodologies, more convincing interaction networks and more detailed mechanistic roles of the catalytic triad (and/or tetrad) will be elucidated in near future. Furthermore, it is now clear that to develop GPx mimics with high antioxidant activity, the control of the interaction modes around the selenium atom is essential. In particular pertinent regulation of the strength of the non-bonded O···Se interaction in the selenenyl sulfide intermediate state is important, as discussed in Sections 7.3.3 and 7.4.2. Accordingly, the design of new GPx-like antioxidant drugs should be pursued. However, another aspect of GPx mimics, *i.e.* the pro-oxidant nature,[31] should be borne in mind.

References

1. L. Flohe, W. A. Günzler and H. H. Schock, *FEBS Lett.*, 1973, **32**, 132.
2. S. Toppo, L. Flohé, F. Ursini, S. Vanin and M. Maiorino, *Biochim. Biophys. Acta, Gen. Subj.*, 2009, **1790**, 1486.
3. R. Brigelius-Flohé and M. Maiorino, *Biochim. Biophys. Acta, Gen. Subj.*, 2013, **1830**, 3289.
4. J. W. Forstrom, J. J. Zakowski and A. L. Tappel, *Biochemistry*, 1978, **17**, 2639.
5. F. Ursini, M. Maiorino, R. Brigelius-Flohé, K. D. Aumann, A. Roveri, D. Schomburg and L. Flohé, *Methods Enzymol.*, 1995, **252**, 38.
6. D. Dimastrogiovanni, M. Anselmi, A. E. Miele, G. Boumis, L. Petersson, F. Angelucci, A. Di Nola, M. Brunori and A. Bellelli, *Proteins: Struct., Funct., Bioinf.*, 2010, **78**, 259.
7. C. Rocher, J.-L. Lalanne and J. Chaudiere, *Eur. J. Biochem.*, 1992, **205**, 955.
8. M. Maiorino, K.-D. Aumann, R. Brigelius-Flohé, D. Doria, J. van den Heuvel, J. McCarthy, A. Roveri, F. Ursini and L. Flohé, *Biol. Chem. Hoppe-Seyler*, 1995, **376**, 651.

9. S. C. E. Tosatto, V. Bosello, F. Fogolari, P. Mauri, A. Roveri, S. Toppo, L. Flohé, F. Ursini and M. Maiorino, *Antioxid. Redox Signaling*, 2008, **10**, 15156.
10. P. Scheerer, A. Borchert, N. Krauss, H. Wessner, C. Gerth, W. Höhne and H. Kuhn, *Biochemistry*, 2007, **46**, 9041.
11. A. J. Mukherjee, S. S. Zade, H. B. Singh and R. B. Sunoj, *Chem. Rev.*, 2010, **110**, 4357.
12. M. Iwaoka, *Noncovalent Forces*, Springer International Publishing, 2015, pp. 265–289.
13. J. S. Murray, P. Lane and P. Politzer, *Int. J. Quantum Chem.*, 2008, **108**, 2770.
14. H. Komatsu, M. Iwaoka and S. Tomoda, *Chem. Commun.*, 1999, 205.
15. M. Iwaoka, H. Komatsu, T. Katsuda and S. Tomoda, *J. Am. Chem. Soc.*, 2004, **126**, 5309.
16. M. Iwaoka and S. Tomoda, *J. Chem. Soc., Chem. Commun.*, 1992, 1165.
17. K. Fujita, K. Murata, M. Iwaoka and S. Tomoda, *Tetrahedron*, 1997, **53**, 2029.
18. G. Fragale, T. Wirth, G. Fragale and M. Neuburger, *Chem. Commun.*, 1998, 1867.
19. G. Fragale and T. Wirth, *Eur. J. Org. Chem.*, 1998, **1998**, 1361.
20. M. Tiecco, L. Testaferri, C. Santi, C. Tomassini, S. Santoro, F. Marini, L. Bagnoli, A. Temperini and F. Costantino, *Eur. J. Org. Chem.*, 2006, **2006**, 4867.
21. M. Iwaoka and S. Tomoda, *J. Am. Chem. Soc.*, 1994, **116**, 2557.
22. G. Mugesh, A. Panda, H. B. Singh, N. S. Punekar and R. J. Butcher, *J. Am. Chem. Soc.*, 2001, **123**, 839.
23. S. S. Zade, H. B. Singh and R. J. Butcher, *Angew. Chem., Int. Ed.*, 2004, **43**, 4513.
24. H. J. Reich and R. J. Hondal, *ACS Chem. Biol.*, 2016, **11**, 821.
25. K. P. Bhabak and G. Mugesh, *Acc. Chem. Res.*, 2010, **43**, 1408.
26. E. E. Alberto, V. do Nascimento and A. L. Braga, *J. Braz. Chem. Soc.*, 2010, **21**, 2032.
27. X. Huang, X. Liu, Q. Luo, J. Liu and J. Shen, *Chem. Soc. Rev.*, 2011, **40**, 1171.
28. C. Santi, C. Tidei, C. Scalera, M. Piroddi and F. Galli, *Curr. Chem. Biol.*, 2013, 7, 25.
29. G. Mugesh, *Curr. Chem. Biol.*, 2013, **7**, 47.
30. A. J. Pacuła, F. Mangiavacchi, L. Sancineto, E. J. Lenardão, J. Ścianowski and C. Santi, *Curr. Chem. Biol.*, 2016, **9**, 97.
31. M. Iwaoka, in *Organoselenium Chemistry Between Synthesis and Biochemistry*, ed. C. Santi, Bentham Science Publishers, 2014, pp. 361–378.
32. B. J. Byun and Y. K. Kang, *Biopolymers*, 2011, **95**, 345.
33. R. E. Huber and R. S. Criddle, *Arch. Biochem. Biophys.*, 1967, **122**, 164.
34. T. Harada, H. Yoshida, K. Ohno, H. Matsuura, J. Zhang, M. Iwaoka and S. Tomoda, *J. Phys. Chem. A*, 2001, **105**, 4517.
35. T. I. Madzhidov and G. A. Chmutova, *J. Mol. Struct.: THEOCHEM*, 2010, **959**, 1.

36. M. Iwaoka and S. Tomoda, *J. Am. Chem. Soc.*, 1994, **116**, 4463.
37. S. J. Narayanan, B. Sridevi, T. K. Chandrashekar, A. Vij and R. Roy, *Angew. Chem., Int. Ed.*, 1998, **37**, 3394.
38. M. Iwaoka, H. Komatsu and S. Tomoda, *Bull. Chem. Soc. Jpn.*, 1996, **69**, 1825.
39. B. M. Goldstein, S. D. Kennedy and W. J. Hennen, *J. Am. Chem. Soc.*, 1990, **112**, 8265.
40. F. T. Burling and B. M. Goldstein, *J. Am. Chem. Soc.*, 1992, **114**, 2313.
41. D. H. R. Barton, M. B. Hall, Z. Lin, S. I. Parekh and J. Reibenspies, *J. Am. Chem. Soc.*, 1993, **115**, 5056.
42. M. Iwaoka and S. Tomoda, *J. Am. Chem. Soc.*, 1996, **118**, 8077.
43. W. Nakanishi, S. Hayashi, A. Sakaue, G. Ono and Y. Kawada, *J. Am. Chem. Soc.*, 1998, **120**, 3635.
44. W. Nakanishi, S. Hayashi and S. Toyota, *J. Org. Chem.*, 1998, **63**, 8790.
45. M. Iwaoka and S. Tomoda, *Phosphorus, Sulfur Silicon Relat. Elem.*, 2005, **180**, 755.
46. F. Weinhold and C. R. Landis, *Chem. Educ. Res. Pract.*, 2001, **2**, 91.
47. M. Iwaoka and S. Tomoda, *Phosphorus, Sulfur Silicon Relat. Elem.*, 1992, **67**, 125.
48. M. Iwaoka, H. Komatsu and S. Tomoda, *Chem. Lett.*, 1998, **27**, 969.
49. M. Iwaoka, H. Komatsu, T. Katsuda and S. Tomoda, *J. Am. Chem. Soc.*, 2002, **124**, 1902.
50. M. Iwaoka, T. Katsuda, S. Tomoda, J. Harada and K. Ogawa, *Chem. Lett.*, 2002, **31**, 518.
51. M. Iwaoka, T. Katsuda, H. Komatsu and S. Tomoda, *J. Org. Chem.*, 2005, **70**, 321.
52. D. T. Y. Chiu, F. H. Stults and A. L. Tappel, *Biochim. Biophys. Acta, Enzymol.*, 1976, **445**, 558.
53. F. Ursini, M. Maiorino and C. Gregolin, *Biochim. Biophys. Acta, Gen. Subj.*, 1985, **839**, 62.
54. O. Epp, R. Ladenstein and A. Wendel, *Eur. J. Biochem.*, 1983, **133**, 51.
55. R. Prabhakar, T. Vreven, K. Morokuma and D. G. Musaev, *Biochemistry*, 2005, **44**, 11864.
56. R. Prabhakar, K. Morokuma and D. G. Musaev, *Biochemistry*, 2006, **45**, 6967.
57. B. Ren, W. Huang, B. Åkesson and R. Ladenstein, *J. Mol. Biol.*, 1997, **268**, 869.
58. R. Prabhakar, T. Vreven, M. J. Frisch, K. Morokuma and D. G. Musaev, *J. Phys. Chem. B*, 2006, **110**, 13608.
59. T. Takei, Y. Urabe, Y. Asahina, H. Hojo, T. Nomura, K. Dedachi, K. Arai and M. Iwaoka, *J. Phys. Chem. B*, 2014, **118**, 492.
60. T. Saiki, K. Goto and R. Okazaki, *Angew. Chem. Int. Ed. Engl.*, 1997, **36**, 2223.
61. A. Ishii, S. Matsubayashi, T. Takahashi and J. Nakayama, *J. Org. Chem.*, 1999, **64**, 1084.
62. S. Sase, R. Kakimoto and K. Goto, *Angew. Chem., Int. Ed.*, 2015, **54**, 901.

63. K. Goto, M. Nagahama, T. Mizushima, K. Shimada, T. Kawashima and R. Okazaki, *Org. Lett.*, 2001, **3**, 3569.
64. C. A. Bayse, R. A. Baker and K. N. Ortwine, *Inorg. Chim. Acta*, 2005, **358**, 3849.
65. B. K. Sarma and G. Mugesh, *J. Am. Chem. Soc.*, 2005, **127**, 11477.
66. K. P. Bhabak and G. Mugesh, *Chem. – Eur. J.*, 2007, **13**, 4594.
67. G. Mugesh and H. B. Singh, *Chem. Soc. Rev.*, 2000, **29**, 347.
68. G. Mugesh, W.-W. du Mont and H. Sies, *Chem. Rev.*, 2001, **101**, 2125.
69. D. Bhowmick and G. Mugesh, *Org. Biomol. Chem.*, 2015, **13**, 10262.
70. A. Müller, E. Cadenas, P. Graf and H. Sies, *Biochem. Pharmacol.*, 1984, **33**, 3235.
71. A. Wendel, M. Fausel, H. Safayhi, G. Tiegs and R. Otter, *Biochem. Pharmacol.*, 1984, **33**, 3241–3245.
72. H. Sies, *Free Radic. Biol. Med.*, 1993, **14**, 313.
73. H. J. Reich and C. P. Jasperse, *J. Am. Chem. Soc.*, 1987, **109**, 5549.
74. A. J. Pacuła, J. Ścianowski and K. B. Aleksandrzak, *RSC Adv.*, 2014, **4**, 48959.
75. M. Elsherbini, W. S. Hamama, H. H. Zoorob, D. Bhowmick, G. Mugesh and T. Wirth, *Heteroat. Chem.*, 2014, **25**, 320.
76. K. Satheeshkumar and G. Mugesh, *Chem. – Eur. J.*, 2011, **17**, 4849.
77. K. Selvakumar, P. Shah, H. B. Singh and R. J. Butcher, *Chem. – Eur. J.*, 2011, **17**, 12741.
78. S. J. Balkrishna, S. Kumar, G. K. Azad, B. S. Bhakuni, P. Panini, N. Ahalawat, R. S. Tomar, M. R. Detty and S. Kumar, *Org. Biomol. Chem.*, 2014, **12**, 1215.
79. V. P. Singh, J. Poon, J. Yan, X. Lu, M. K. Ott, R. J. Butcher, P. J. Gates and L. Engman, *J. Org. Chem.*, 2017, **82**, 313.
80. Z. Wang, Y. Wang, W. Li, Z. Liu, Z. Luo, Y. Sun, R. Wu, L. Huang and X. Li, *Dalton Trans.*, 2015, **44**, 20913.
81. D. Bhowmick, S. Srivastava, P. D'Silva and G. Mugesh, *Angew. Chem., Int. Ed.*, 2015, **54**, 8449.
82. S. R. Wilson, P. A. Zucker, R. R. C. Huang and A. Spector, *J. Am. Chem. Soc.*, 1989, **111**, 5936.
83. G. Mugesh, A. Panda, H. B. Singh, N. S. Punekar and R. J. Butcher, *Chem. Commun.*, 1998, **0**, 2227.
84. G. Mugesh, A. Panda, H. B. Singh, N. S. Punekar and R. J. Butcher, *J. Am. Chem. Soc.*, 2001, **123**, 839.
85. T. Wirth, *Molecules*, 1998, **3**, 164.
86. M. L. Jauslin, T. Wirth, T. Meier and F. Schoumacher, *Hum. Mol. Genet.*, 2002, **11**, 3055.
87. K. P. Bhabak and G. Mugesh, *Chem. – Eur. J.*, 2008, **14**, 8640.
88. G. Mugesh, A. Panda, H. B. Singh and R. J. Butcher, *Chem. – Eur. J.*, 1999, **5**, 1411.
89. D. Bhowmick and G. Mugesh, *Org. Biomol. Chem.*, 2015, **13**, 9072.
90. V. Nascimento, N. L. Ferreira, R. F. S. Canto, K. L. Schott, E. P. Waczuk, L. Sancineto, C. Santi, J. B. T. Rocha and A. L. Braga, *Eur. J. Med. Chem.*, 2014, **87**, 131.

91. J. Rafique, S. Saba, R. F. S. Canto, T. E. A. Frizon, W. Hassan, E. P. Waczuk, M. Jan, D. F. Back, J. B. T. Da Rocha and A. L. Braga, *Molecules*, 2015, **20**, 10095.
92. C. A. Collins, F. H. Fry, A. L. Holme, A. Yiakouvaki, A. Al-Qenaei, C. Pourzand and C. Jacob, *Org. Biomol. Chem.*, 2005, **3**, 1541.
93. C. Parashiva Prabhu, P. P. Phadnis, A. P. Wadawale, K. Indira Priyadarsini and V. K. Jain, *J. Organomet. Chem.*, 2012, **713**, 42.
94. B. G. Singh, P. P. Bag, F. Kumakura, M. Iwaoka and K. I. Priyadarsini, *Bull. Chem. Soc. Jpn.*, 2010, **83**, 703.
95. A. Kunwar, B. Mishra, A. Barik, L. B. Kumbhare, R. Pandey, V. K. Jain and K. I. Priyadarsini, *Chem. Res. Toxicol.*, 2007, **20**, 1482.
96. B. Mishra, A. Barik, A. Kunwar, L. B. Kumbhare, K. I. Priyadarsini and V. K. Jain, *Phosphorus, Sulfur Silicon Relat. Elem.*, 2008, **183**, 1018.
97. A. Kunwar, P. Bansal, S. J. Kumar, P. P. Bag, P. Paul, N. D. Reddy, L. B. Kumbhare, V. K. Jain, R. C. Chaubey, M. K. Unnikrishnan and K. I. Priyadarsini, *Free Radical Biol. Med.*, 2010, **48**, 399.
98. T. G. Back, D. Kuzma and M. Parvez, *J. Org. Chem.*, 2005, **70**, 9230.
99. D. J. Press, E. A. Mercier, D. Kuzma and T. G. Back, *J. Org. Chem.*, 2008, **73**, 4252.
100. N. M. R. McNeil, M. C. Matz and T. G. Back, *J. Org. Chem.*, 2013, **78**, 10369.
101. N. M. R. McNeil, D. J. Press, D. M. Mayder, P. Garnica, L. M. Doyle and T. G. Back, *J. Org. Chem.*, 2016, **81**, 7884.
102. T. G. Back and Z. Moussa, *J. Am. Chem. Soc.*, 2002, **124**, 12104.
103. N. M. Okeley, Y. Zhu and W. A. van der Donk, *Org. Lett.*, 2000, **2**, 3603.
104. O. Flögel, G. Casi, D. Hilvert and D. Seebach, *Helv. Chim. Acta*, 2007, **90**, 1651.
105. M. Iwaoka, R. Ooka, T. Nakazato, S. Yoshida and S. Oishi, *Chem. Biodiversity*, 2008, **5**, 359.
106. L. A. Wessjohann and A. Schneider, *Chem. Biodivers.*, 2008, **5**, 375.
107. H. Liu and A. Dömling, *Chem. Biol. Drug Des.*, 2009, **74**, 302.
108. A. L. Braga, L. A. Wessjohann, P. S. Taube, F. Z. Galetto and F. M. de Andrade, *Synthesis*, 2010, **2010**, 3131.
109. S. Flemer, *Protein Pept. Lett.*, 2014, **21**, 1257.
110. S. Shimodaira and M. Iwaoka, *ARKIVOC*, 2016, **2017**, 260.
111. M. D. Gieselman, L. Xie and W. A. van der Donk, *Org. Lett.*, 2001, **3**, 1331.
112. R. J. Hondal, B. L. Nilsson and R. T. Raines, *J. Am. Chem. Soc.*, 2001, **123**, 5140.
113. R. Quaderer, A. Sewing and D. Hilvert, *Helv. Chim. Acta*, 2001, **84**, 1197.
114. J. Beld, K. J. Woycechowsky and D. Hilvert, *Biochemistry*, 2007, **46**, 5382.
115. N. A. McGrath and R. T. Raines, *Acc. Chem. Res.*, 2011, **44**, 752.
116. S. Flemer Jr., *Molecules*, 2011, **16**, 3232.
117. S. H. Van Dorst and P. J. Peterson, *J. Sci. Food Agric.*, 1984, **35**, 601.
118. T. Tamura, T. Oikawa, A. Ohtaka, N. Fujii, N. Esaki and K. Soda, *Anal. Biochem.*, 1993, **208**, 151.
119. Y.-C. Park, J.-B. Kim, Y. Heo, D.-C. Park, I.-S. Lee, H.-W. Chung, J.-H. Han, W.-G. Chung, S. C. Vendeland and P. D. Whanger, *Biol. Trace Elem. Res.*, 2004, **98**, 143.

120. Y. Rao, M. McCooeye, A. Windust, E. Bramanti, A. D'Ulivo and Z. Mester, *Anal. Chem.*, 2010, **82**, 8121.
121. J. Beld, K. J. Woycechowsky and D. Hilvert, *Biochemistry*, 2009, **48**, 4662.
122. S. Yoshida, F. Kumakura, I. Komatsu, K. Arai, Y. Onuma, H. Hojo, B. G. Singh, K. I. Priyadarsini and M. Iwaoka, *Angew. Chem., Int. Ed.*, 2011, **50**, 2125.
123. M. Iwoaka, *Biochalcogen Chemistry: The Biological Chemistry of Sulfur, Selenium, and Tellurium*, American Chemical Society, 2013, vol. 1152, pp. 163–177.
124. Y. Sun, T. Li, H. Chen, K. Zhang, K. Zheng, Y. Mu, G. Yan, W. Li, J. Shen and G. Luo, *J. Biol. Chem.*, 2004, **279**, 37235.
125. G. Casi and D. Hilvert, *J. Biol. Chem.*, 2007, **282**, 30518.
126. Z. P. Wu and D. Hilvert, *J. Am. Chem. Soc.*, 1989, **111**, 45134.
127. Z. P. Wu and D. Hilvert, *J. Am. Chem. Soc.*, 1990, **112**, 5647.
128. I. M. Bell, M. L. Fisher, Z. P. Wu and D. Hilvert, *Biochemistry*, 1993, **32**, 3754.
129. X. Ren, P. Jemth, P. G. Board, G. Luo, B. Mannervik, J. Liu, K. Zhang and J. Shen, *Chem. Biol.*, 2002, **9**, 789.
130. S. Boschi-Muller, S. Muller, A. Van Dorsselaer, A. Böck and G. Branlant, *FEBS Lett.*, 1998, **439**, 241.
131. Q. Luo, C. Zhang, L. Miao, D. Zhang, Y. Bai, C. Hou, J. Liu, F. Yan, Y. Mu and G. Luo, *Amino Acids*, 2013, **44**, 1009.
132. F. Zinoni, A. Birkmann, T. C. Stadtman and A. Böck, *Proc. Natl. Acad. Sci.*, 1986, **83**, 4650–4654.
133. K. E. Sandman, J. S. Benner and C. J. Noren, *J. Am. Chem. Soc.*, 2000, **122**, 960.
134. Z. Jiang, E. S. J. Arnér, Y. Mu, L. Johansson, J. Shi, S. Zhao, S. Liu, R. Wang, T. Zhang, G. Yan, J. Liu, J. Shen and G. Luo, *Biochem. Biophys. Res. Commun.*, 2004, **321**, 94.
135. O. Rengby, L. Johansson, L. A. Carlson, E. Serini, A. Vlamis-Gardikas, P. Kårsnäs and E. S. J. Arnér, *Appl. Environ. Microbiol.*, 2004, **70**, 5159.
136. V. A. Shchedrina, S. V. Novoselov, M. Y. Malinouski and V. N. Gladyshev, *Proc. Natl. Acad. Sci.*, 2007, **104**, 13919.
137. T. Pan, Y. Liu, C. Si, Y. Bai, S. Qiao, L. Zhao, J. Xu, Z. Dong, Q. Luo and J. Liu, *ACS Catal.*, 2017, **7**, 1875.
138. M.-P. Strub, F. Hoh, J.-F. Sanchez, J. M. Strub, A. Böck, A. Aumelas and C. Dumas, *Structure*, 2003, **11**, 1359.
139. H. Yu, J. Liu, A. Böck, J. Li, G. Luo and J. Shen, *J. Biol. Chem.*, 2005, **280**, 11930.
140. Y. Ge, Z. Qi, Y. Wang, X. Liu, J. Li, J. Xu, J. Liu and J. Shen, *Int. J. Biochem. Cell Biol.*, 2009, **41**, 900.

CHAPTER 8

Carbohydrate-derived Organoselenium Compounds: Synthesis and Application in the Structural Analysis of Biomolecules

CHANDRABABUNAIDU KONA,[a,c] JUNPEI SHIMABUKURO,[a,c] MAKOTO KISO[a,c] AND HIROMUNE ANDO*[a,b,c]

[a] Gifu University, Department of Applied Bioorganic Chemistry, 1-1 Yanagido, Gifu-shi, Gifu 501-1193, Japan; [b] Gifu University, Center for Highly Advanced Integration of Nano and Life Sciences (G-CHAIN), 1-1 Yanagido, Gifu-shi, Gifu 501-1193, Japan; [c] Kyoto University, Institute for Integrated Cell-Material Sciences (WPI-iCeMS), Yoshida Ushinomiya-cho, Sakyo-ku, Kyoto 606-8501, Japan
*Email: hando@gifu-u.ac.jp

8.1 Introduction

Selenocarbohydrates are useful mimics of biologically significant carbohydrates because of the inherent chemical and physical properties of selenium. However, their use in biological studies has been limited. Chemical modifications and structural studies have helped to reveal many insights into their biological functions and mechanisms. Selenocarbohydrates serve as protein ligand mimics in carbohydrate–protein complexes. Taking

Organoselenium Compounds in Biology and Medicine: Synthesis, Biological and Therapeutic Treatments
Edited by Vimal Kumar Jain and K. Indira Priyadarsini

Published by the Royal Society of Chemistry, www.rsc.org

advantage of the anomalous dispersion of selenium in response to X-ray irradiation, protein structures can be more easily determined with multi-wavelength anomalous dispersion (MAD) phasing.[1,2] Three-dimensional (3-D) structures of the carbohydrate–protein complexes could be determined by the single-wavelength anomalous dispersion (SAD)/MAD[2–11] phasing method without a SeMet-labeled protein.[12]

Thus, determining high-resolution 3-D structures of carbohydrate–protein complexes is invaluable for understanding the biological processes that are mediated through carbohydrate–protein interactions on the cell membrane, including cell adhesion, cell proliferation, cell differentiation, and pathogen–host cell interactions. Selenium has also been used to derivatize nucleic acids for crystal structure determination.[13] Thus, in the recent years, the introduction of selenium within the carbohydrate framework has received substantial attention due to its potential pharmaceutical applications and structural applications.[13] The incorporation of selenium into molecules gives a fingerprint for identifying the compounds in mass spectrometry and nuclear magnetic resonance (NMR) spectroscopy, because of its isotopic distribution and the presence of the ^{77}Se isotope with nuclear spin of $\frac{1}{2}$.

8.2 Selenium in Carbohydrate Chemistry

During the past two decades, selenium incorporation into carbohydrates has received substantial attention from the synthetic community, due to its interesting biological and structural properties. Depending on the position of selenium in the carbohydrate, selenocarbohydrates are named as selenoether pseudo-carbohydrates, selenosugars and selenoglycosides (Scheme 8.1).[2b]

8.2.1 Selenoether Pseudo-carbohydrates

Cumpstey's group reported the synthesis of the pyranose- and furanose-based selenoether pseudo-disaccharides *via* a non-glycosidic selenium linkage of the two monosaccharides by treating nucleophilic selenium carbohydrates **1** with a carbohydrate sulfonate **5**. The authors have investigated the synthesis of the primary (C-6) **2**, **3** and secondary (C-3) **4** selenides.[14] At the same time, Lüdtke and co-workers reported the synthesis

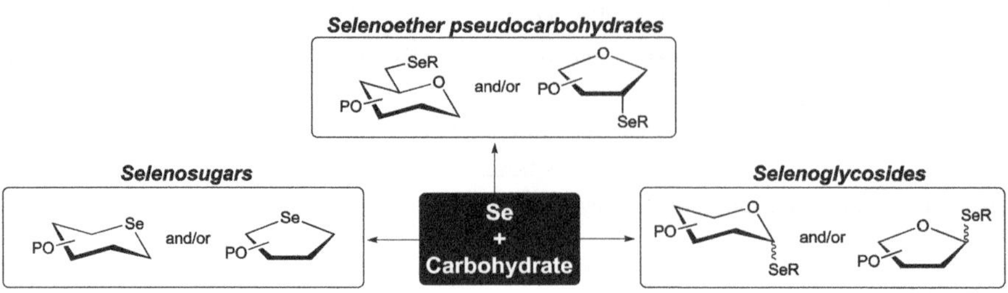

Scheme 8.1 Selenium in carbohydrates.

of C-6-linked selenoether pseudo-disaccharides **7–10** and selenoether pseudo-monosaccharides **11–13**, such as C6-seleno alkyl, aryl, acyl and propargyl analogues, in a similar approach to that taken by the Cumpstey group using the diselinide **14**. In contrast, the selenium-linked neoglycoconjugates **16** were synthesized through a nucleophilic ring opening of chiral *N*-bocaziridines **17** and/or by the treatment of the diselinide **15** with the tosylate **6** in the presence of $NaBH_4$ (Scheme 8.2).[15]

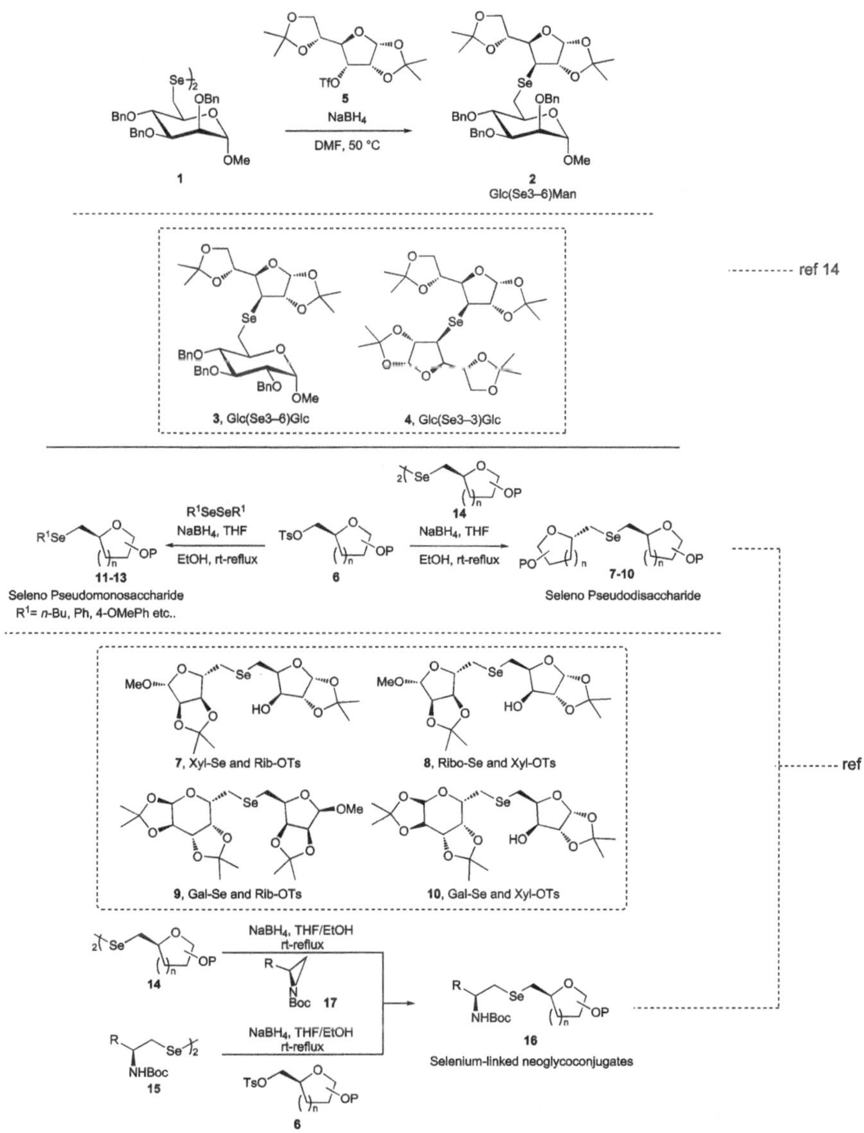

Scheme 8.2 Selenoether pseudodisaccharides and neoglycoconjugates from Lüdtke and co-workers.

8.2.2 Selenosugars

Displacement of the sugar ring oxygen by selenium leads to the formation of the selenosugars. The research groups of Jeong,[16] Pinto,[17] and Matsuda[18] independently disclosed the first synthesis of selenosugar and seleno-nucleoside derivatives with a unified strategy using Pummerer rearrangement as the key transformation. They observed that the steric effect of bulky selenium atom in the 4′-selenoribonuelcosides caused the different sugar puckering (2′-endo/3′-exo, S-type) when compared with the 4′-oxo- and 4′-thionucleosides (2′-exo/3′-endo, N-type) (Scheme 8.3). These conformational differences provided an access point to 4′-selenonucleosides as biological tools or as drugs.[2b]

The synthesis of 4′-selenonucleoside from different purine and pyrimidine bases is depicted in Scheme 8.4. Matsuda and Jeong's synthesis involved the displacement of dibromide or dimesylate derived from the corresponding diol derivative of D-ribitol **18** with sodium selenide to deliver the selenosugar **19**, which is further transformed to the cyclic silyl protected selenosugar **20** by treatment with the TIPDSCl and pyridine. The selenosugar **20** is oxidized to the corresponding selenoxide **21** under ozonolysis. Now the stage was set for Pummerer-type rearrangement with the selenoxide derivative and with the nucleobases to give the 4′-selenonucleoside **22**, which is further transformed to 4′Se uridine **23** and 4′Se cytidine derivative **24**. Likewise, the selenoxide **25** was converted to the nucleosides **26** and **27**.[16,18] A similar strategy was used by Pinto and co-workers for the synthesis of 4′-selenonucleoside **24**, **29–31** from the selenoxide **28**.[17]

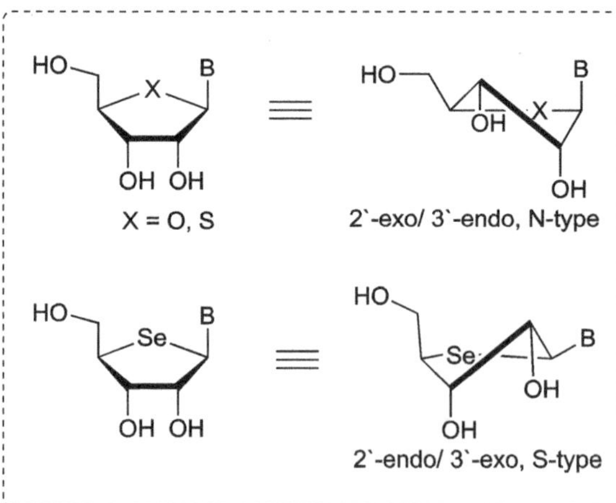

Scheme 8.3 Conformations of 4′-oxo- and 4′-thionucleosides (N-type) and 4′-selenonucleosides (S-type).

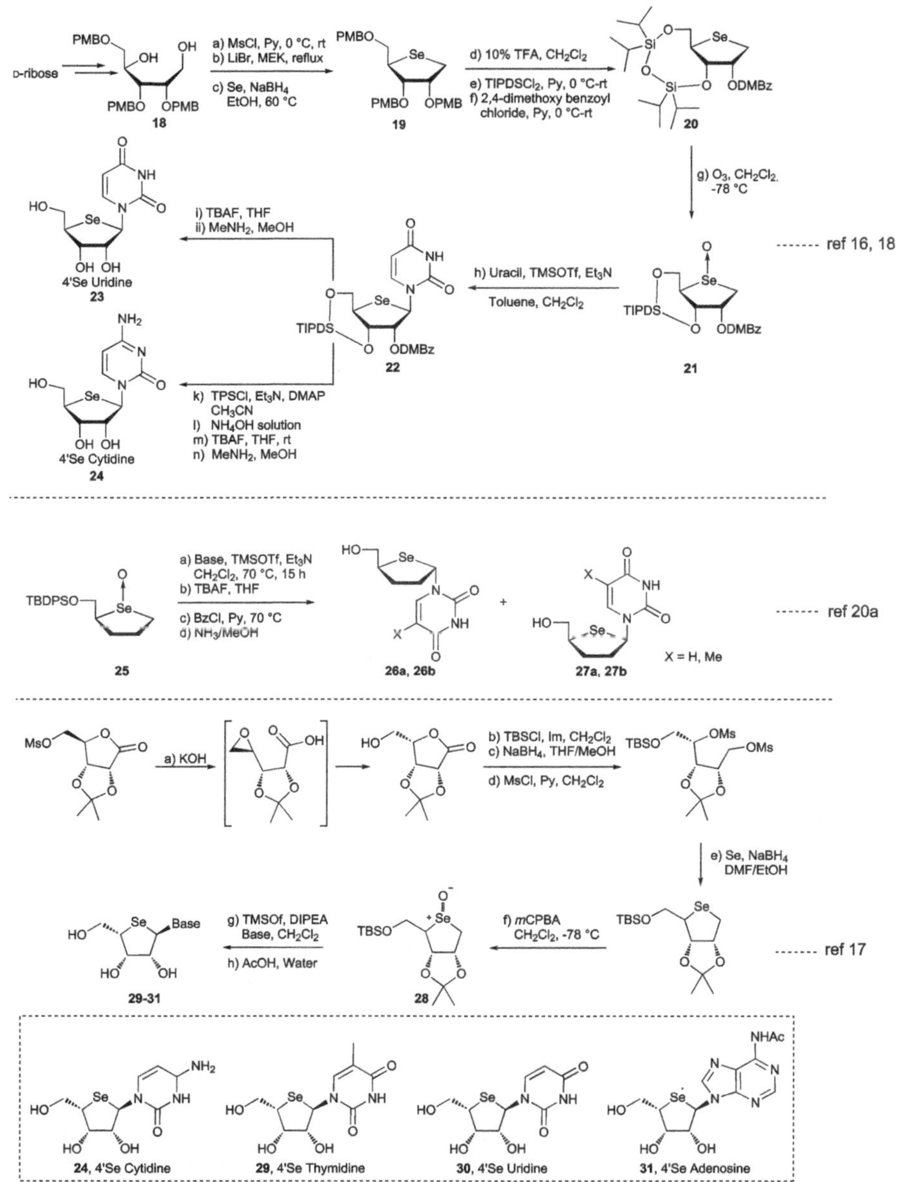

Scheme 8.4 4′-Selenoribonucleosides by the Matsuda, Jeong, and Pinto groups.

8.2.2.1 Biological Importance

Unlike 4′-oxo- and 4′-thionucleosides, most of the 4′-selenonucleosides did not show any promising antiviral or anticancer activity.[19,20] Taking into account the S-type conformation, it is evident that the lack of activity is due to

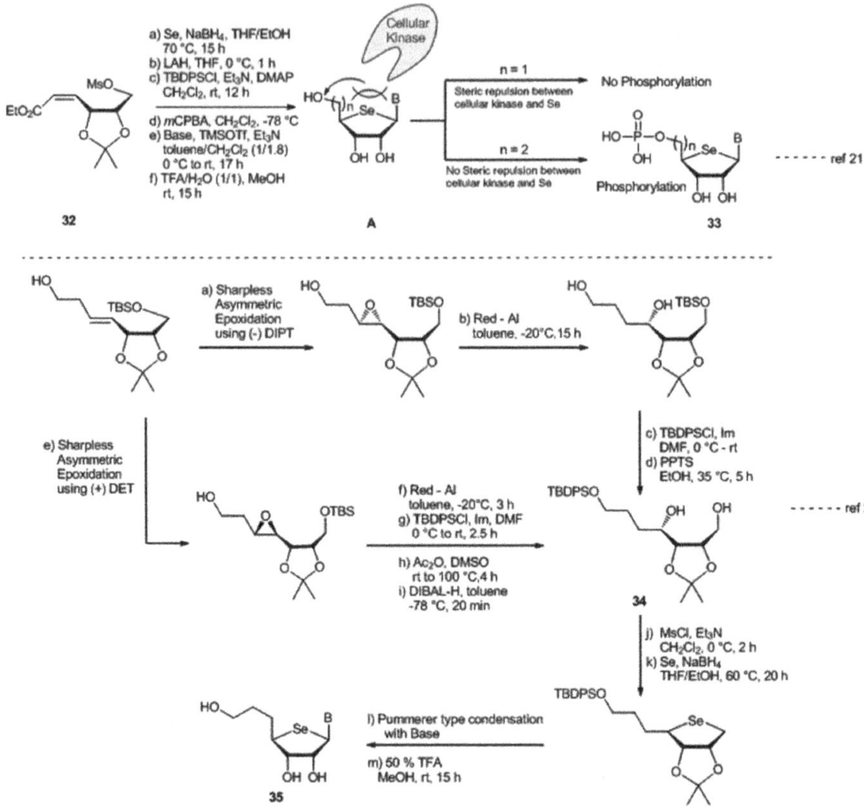

Scheme 8.5 Stereoselective synthesis of 4′-selenonucleosides and 5′-homo-4′-selenoribose *via* seleno-Michael reaction.

the bulky selenium atom impeding the phosphorylation of the 5′-hydroxyl group by cellular kinases as shown in **A.** In efforts to overcome the aforementioned limitations, Jeong and co-workers[20] synthesized the 5′-homo-4′-selenonucleosides **35** *via* a novel seleno-Michael reaction and provided a one-carbon and two-carbon homologation at the 5′ position. As depicted in Scheme 8.5, the mesylate **32** was subjected to seleno-Michael reaction with sodium selenide in THF and EtOH at 70 °C. The C6 ester was further transformed into the 5′-homo-4′-selenonucleoside **A** by means of Pummerer rearrangement and global deprotection. This one-carbon homologation resulted in potent antiviral activity by allowing the 5′-homo-4′-selenonucleosides to be phosphorylated by cellular kinases to give **33** with no steric repulsion.[21] The authors have further extended their studies by synthesizing the two-carbon homologated 5′-homo-4′-selenonucleoside **35** from an intermediate **34** by using the seleno-Michael reaction as the key transformation.[22]

Selenosugars can also act as antioxidants. Oxidative stress in biological systems causes an imbalance between the production of reactive oxygen

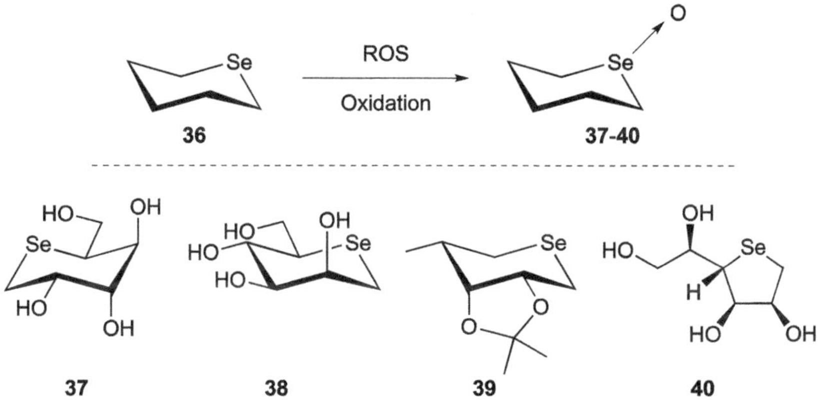

Scheme 8.6 Representative examples of selenosugar antioxidants.

species (ROS) and antioxidant systems, resulting in tissue damage and a number of inflammatory diseases and neurodegenerative disorders. Glutathione peroxidase one compound which can reduce ROS generation by acting as a line of cellular defense against oxidative damage.[2b]

To mimic these properties and to treat the diseases associated with oxidative damage, simple 5-selenopyranose and 4-selenofuranose monosaccharides have been prepared as water-soluble scavengers for hypohalous acids and peroxides. Storkey *et al.* synthesized sulfur and selenium derivatives of L-1-deoxynojirimycin and D-1-deoxymannojirimycin and evaluated their properties as scavengers for hypochlorous acid.[23] The selenosugars **37** and **38** were more potent scavengers than the corresponding thiosugars. Merino-Montiel *et al.* synthesized L-isofucoselenofagomine derivatives **39** to mimic the function of glutathione peroxidase by scavenging H_2O_2 (Scheme 8.6).[24] Davies and co-workers reported that the 4-selenofuranose derivative **40** could be the potent modulator of peroxidase-mediated damage at sites of acute and chronic inflammation and in multiple human pathologies.[25]

8.2.3 Selenoglycosides

The incorporation of selenium in place of the anomeric oxygen in sugars yields selenoglycosides. Selenoglycosides have several applications in carbohydrate chemistry and are useful intermediates for the preparation of *C*-glycosides,[26] glycoconjugates,[27] functionalized glycals,[28] glycosylhalides,[29] hemiacetals,[30] orthoesters, and deoxyglycosides.[31] Aryl selenoglycosidesare the powerful glycosyl donors in the multistep oligosaccharide synthesis, because they could be chemoselectively activated in the presence of other nucleosides such as thioglycosides.[32] Selenoglycosides are stable to the hydrolytic action of glycosidase enzymes, suggesting that they could have a potential application in the development of new carbohydrate-based drugs.[33]

Due to its wide applicability, Pinto and co-workers[34] reported TESOTf-mediated glycosylation of 4-selenol **42** with a trichloroacetimidate donor **41** to give selenodisaccharide **43** as a 4.5 : 1 α : β mixture. Subsequently, Czernecki and Randriamandimby reported the synthesis of 1→6 seleno-disaccharides **47** by reducing symmetric diglycosyldiselenides **44** and reacting the corresponding selenolates with either 6-*O*-tosyl **45** or 6-deoxy-6-iodo **46** monosaccharides (Scheme 8.7).[35]

The alkyl and aryl selenoglycosides **50** have also been prepared by the reaction of alkyl/aryl selenolates (generated upon treating the corresponding diselenides **49** with a hydride reducing agent) with glycosyl halides **48**.[33] Tiwari and Misra have developed a novel protocol for the synthesis of selenoglycosides **51** or **52** by using the indium(I) iodide-assisted cleavage of aryl diselenides as a key reaction.[36] This method has been used for the preparation of a number of acetylated saccharides in high yields with retention of configuration at the anomeric position.[37] Chandrasekaran and co-workers[38] provided the rongalite ($HOCH_2SO_2Na$)-mediated cleavage of aryl diselenides for the glycosylation of acetylated α-glycosyl bromides **48** to give aryl β-selenoglycosides **51** (Scheme 8.8).

Ishihara and co-workers have reported the synthesis of glucose β-selenoglycosides **54** from a β-glycosyl *p*-methylbenzoselenoate **53**.[39] The selenolate anion precursor **53** is selectively activated with a secondary amine

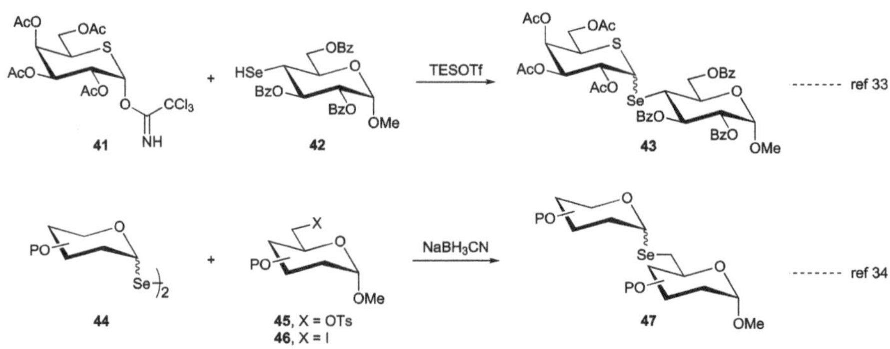

Scheme 8.7 Pinto's and Czernecki's synthesis of selenodisaccharides.

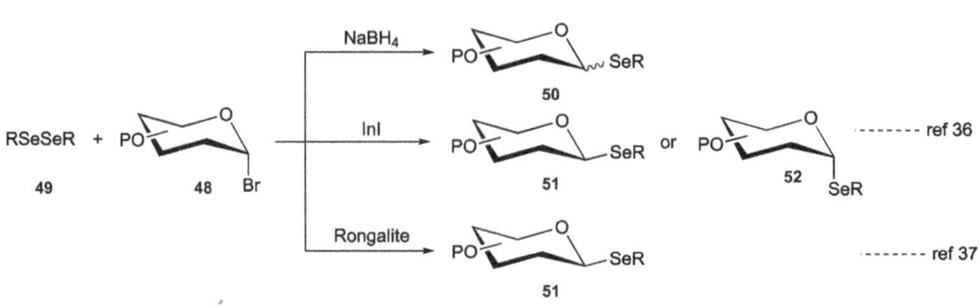

Scheme 8.8 Synthesis of selenoglycosides from diselenides and glycosyl bromides.

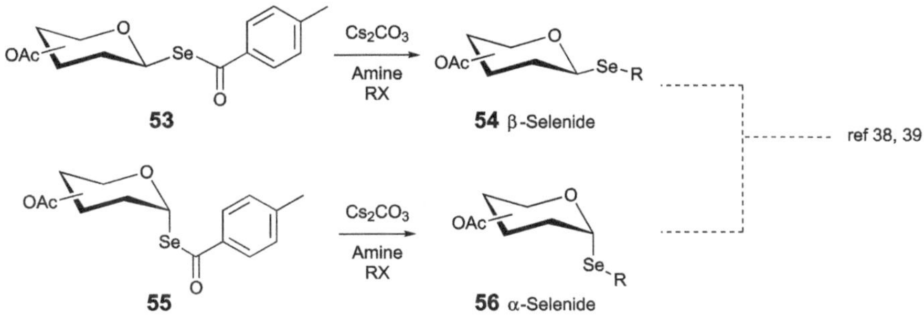

Scheme 8.9 Ishihara and co-workers' synthesis of β- and α-selenoglycosides.

to produce a β-selenolate anion, which is then stabilized by a cesium counterion. Addition of an electrophile followed by a subsequent nucleophilic displacement gives the β-selenoglycoside **54**. The generality of this protocol was explored through several selenoglycosides such as alkyl and aryl selenoglycosides, selenodisaccharides and selenoglycosyl amino acids. Encouraged by these observations, the authors further extended their studies for the synthesis of α-selenoglycosides **56** *via* the formation of an α-selenolate anion intermediate.[39] The α-glycosyl *p*-methylbenzoselenoate **55** was synthesized by reacting a β-glycosyl chloride with potassium *p*-methylselenobenzoate. The reaction produced a large variety of α-selenoglycosides in excellent yields (Scheme 8.9).[2b]

Subsequently, Szilágyi and co-workers explored an iterative approach for the synthesis of the β-selenoglycosides **59** by using glycosyl-isoselenuronium salts **57** and **58** as glycosyl donors.[40] These isoselenuronium salts were prepared by reacting α-glycosyl bromides with selenourea in refluxing acetone and subsequent treatment with an electrophile in the presence of acetonitrile and triethylamine to afford β-selenide **59**. Kiso and co-workers developed a novel TMSOTf-mediated transacetalization approach between a glycosyltrichloroacetimidate **60** and a selenoacetal for the synthesis of selenoglycosides **61** (Scheme 8.10).[41] Gratifyingly, under the same reaction conditions the authors evidenced the exclusive formation of alkyl and aryl β-selenoglycosides **61** from the acetylated saccharides. More recently, McDonagh and coworkers have studied the anomeric effect of the selenoglycosides **62** and reported the synthesis of the selenium-incorporated sphingolipid **63**.[2b]

8.3 Seleninates and Selenonates

Selenium-based biomimetic compounds play a crucial role in biology,[42] medicine,[43] and the crystallographic study of proteins and oligonucleotides.[44] Their continual development contributes towards understanding and manipulating the normal action of enzymes and other proteins, as well as in the development of new drugs. Synthetic chemists are tasked with

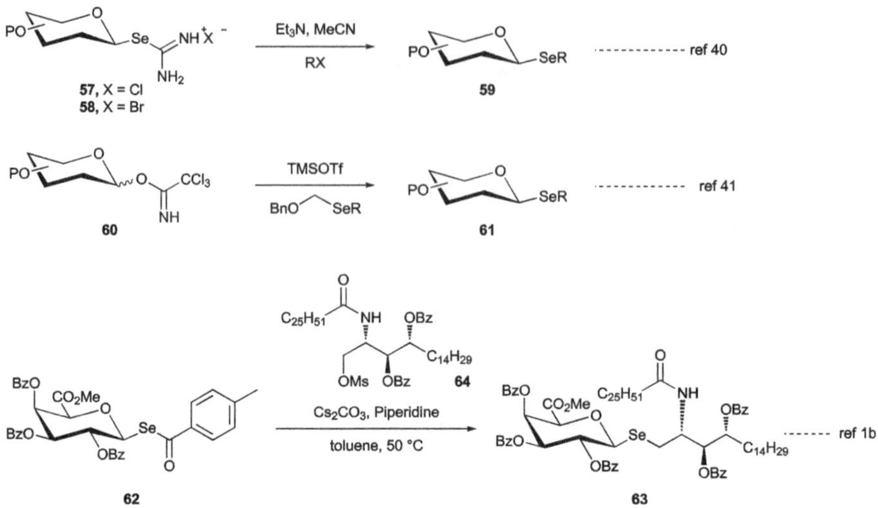

Scheme 8.10 Synthesis of β-selenoglycosides by using isoselenuronium salts and transacetalization.

developing cleaner and more efficient reactions for achieving their transformations. In this context, here we discuss a novel synthesis of seleninates, selenonates, and various other organoselenium compounds that can mimic the *O*-phosphate functionality of the lysophosphatidic acid (LPA) **67**.[45] LPA is a potent signaling agent and is involved in neurogenesis, wound healing, and cancer progression[46] as *N*-acetylneuraminic acid **65**, D-glucono-1,5-lactone 6-*O*-phosphate **66**, lipoic acid **70**, heparin **68**, and 2′-deoxythymidine 5-*O*-phosphate **69**.[47] It is known that the seleninic and selenonic acids **71**, **72** and their salts can act as isosteres of the biologically important *O*-phosphate, *O*-sulfate, and carboxylate functionalities (Scheme 8.11).[48]

Knapp and co-workers reported a mild and efficient protocol for the synthesis of pyranose-, nucleoside-, polyhydric-, and amino acid-based aliphatic seleninates **74a**, **74b** and selenonates **75a–75i** by the oxidation of their corresponding selenoester precursors **73**. The selenoesters **73** were synthesized by the treatment of primary halides with the *in situ*-generated selenocarboxylate anion **76** [from the corresponding carboxylic acid **77** with Woollins' reagent **78** (Scheme 8.12; Figure 8.1)] under the Mitsunobu conditions (Scheme 8.13).[49]

8.3.1 Reactions with Active Site Functionality

Seleninic acids are highly reactive towards thiols over a wide range of pH values. The redox coupling of glucopyranoside-based seleninate **74a** with the *N*-Boc-cysteine methyl ester **79** gave a selenosulfide **80** within a minute at 23 °C (Scheme 8.14).[50] A number of enzyme active sites contain a cysteine

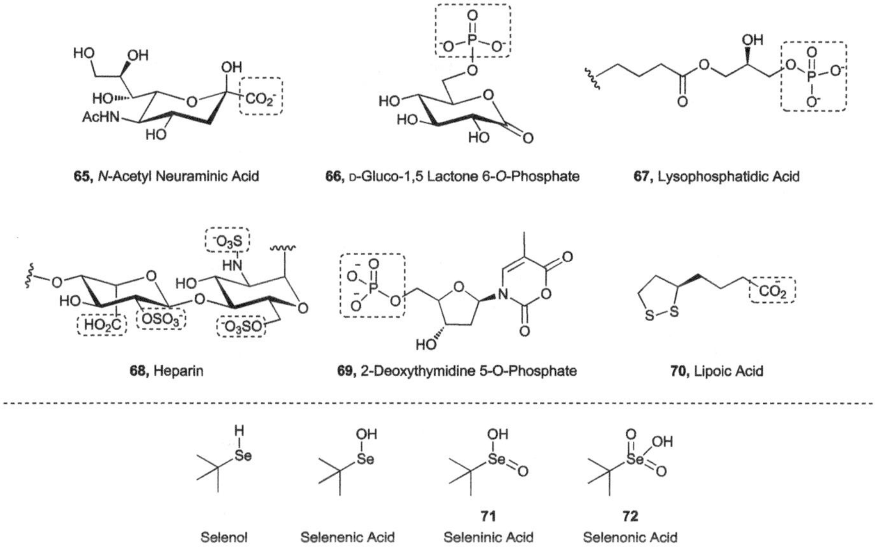

Scheme 8.11 Natural products containing *O*-phosphate, *O*-sulfate and carboxylate and their possible selenium isosteres.

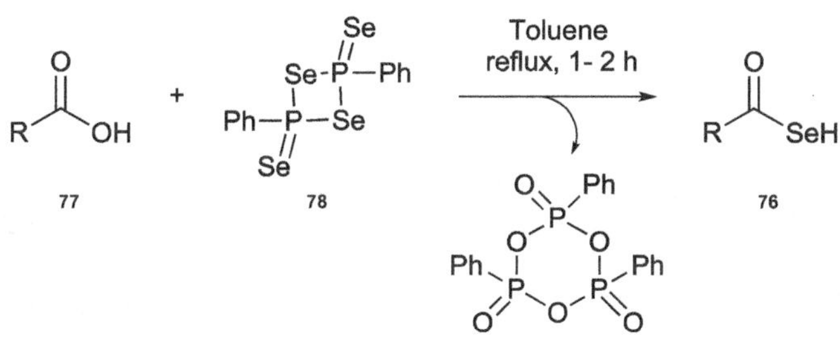

Scheme 8.12 Synthesis of selenocarboxylic acids.

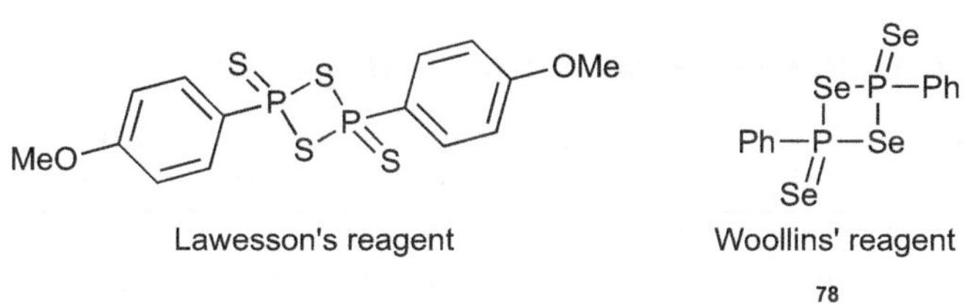

Figure 8.1 Lawesson's and Woollins' reagents.

sulfhydryl, so given the appropriate seleninate-containing substrate mimic, this reaction is a potential avenue for irreversible inhibition by covalent attachment.[48f,51]

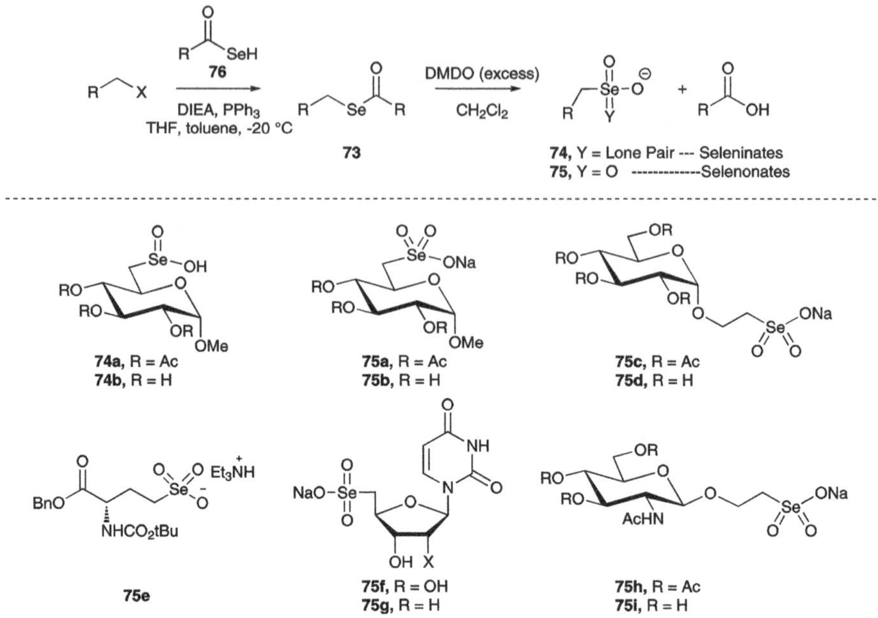

Scheme 8.13 Synthesis of aliphatic seleninates and selenonates and representative examples.

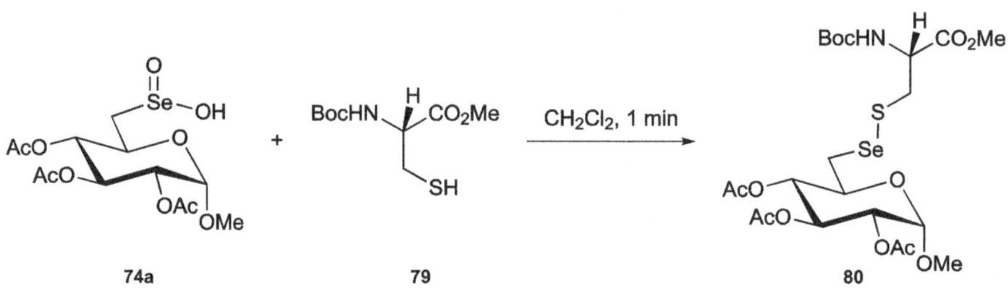

Scheme 8.14 Redox coupling of *N*-boc-cysteine methyl ester with seleninate.

8.3.2 Aromatic Organoselenium Compounds

Selenocarboxylates are good reagents for the introduction of a selenium atom into various aliphatic organic structures. In the case of the aromatic compounds, especially the electron-rich aromatic compounds, these reagents are not suitable for selenium incorporation. A highly electrophilic selenium reagent such as seleninic acids is required. When the aromatic or hetero-aromatic compounds are subjected to seleninic acid, alkyl aryl selenides **81** are formed instantly, which are further transformed to their oxidized adducts as selenoxide **82** or selenone **85**. The selenoxide **82** undergoes *syn* elimination under heating conditions to yield the respective selenenic acid **83**, which in turn reacts *in situ* with other nucleophiles and yields aromatic organoselenium compounds **84a–84f**. The selenones **85** afford the corresponding seleninates **86a–86e** in good yields upon treatment with the nucleophiles *via* the Se-dealkylation (Scheme 8.15).[48f,49]

N-Benzoyltyrosine ethyl ester, *N*-acetyltryptophan ethyl ester, and *N*-boc-histidine benzyl ester are the three other amino acids which are selenylated on the aromatic ring and imidazole ring, as shown in Figure 8.2.[49b,49d,50]

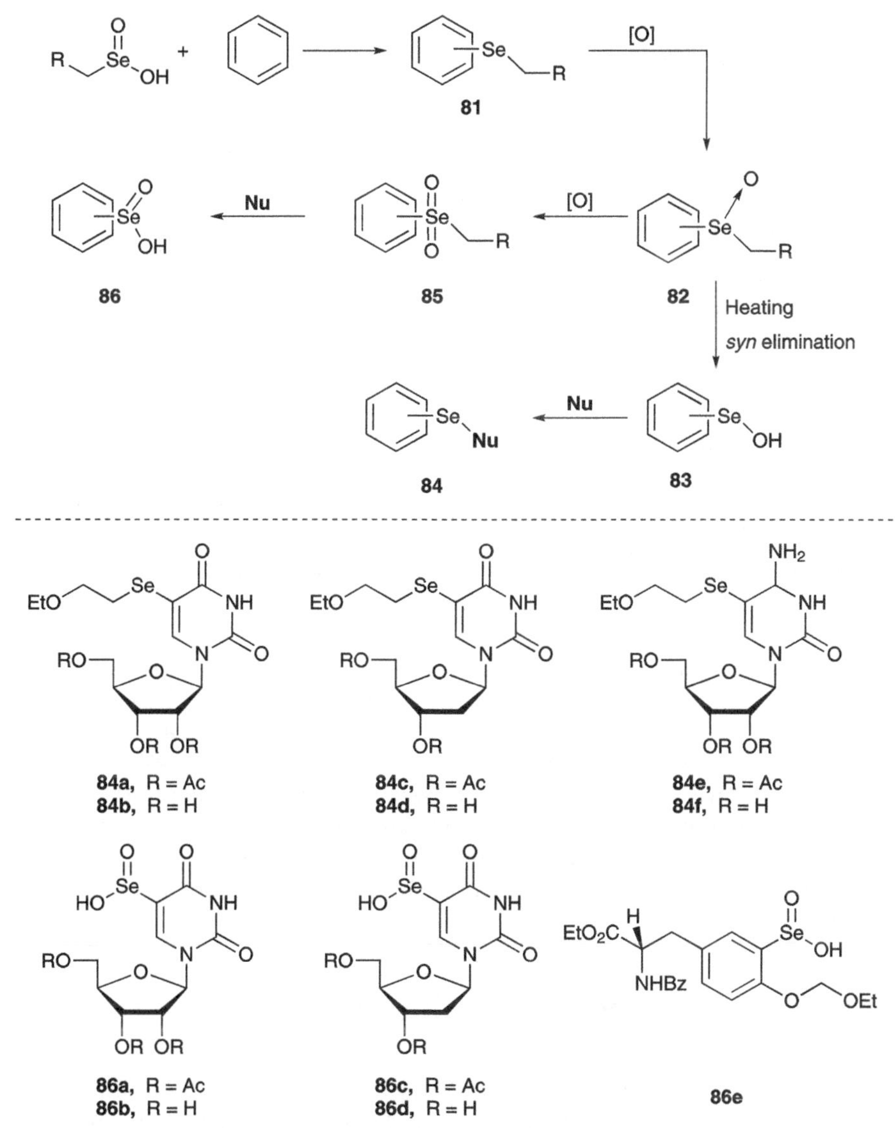

Scheme 8.15 Synthesis of aromatic organoselenium compounds.

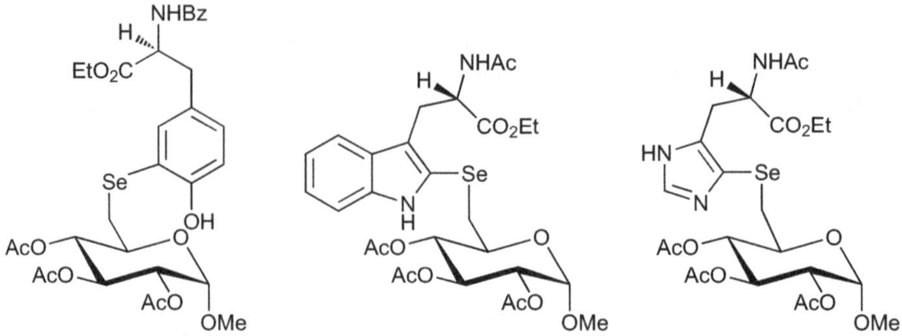

87 **88** **89**

Figure 8.2 Coupling of seleninate with protein side chain functionality.

8.4 Selenium in DNA/RNA

Selenium derivatization of nucleic acids generates the modified biopolymers with novel biophysical properties. The possible selenium functionalization at different positions in the sugar, nucleobase, and phosphate group of the nucleotide is depicted in Figure 8.3.[52]

There has been an explosion in the number of reports involving the concept of selenium incorporation in DNA. Huang and co-workers have pioneered the synthesis and analyzed the structure–function activities of selenium-derivatized (or -modified) nucleic acids. The authors reported the synthesis of 5′-Se nucleosides **93a–93e** and **94** by activating the 5′-OH groups of **90** with good leaving groups such as Br, Ms, and Ts, **91**, **92**, followed by a S_N2 substitution with selenide nucleophiles in the presence of a phase-transfer catalyst (Scheme 8.16).[53]

Following this success, the authors focused on Se incorporation at the 2′-α-position of nucleosides because the 2′-derivatization may allow retention of the RNA, as well as A-form DNA sugar puckers. In this regard, they have synthesized the 2′-Me-Se-U **95**, 2′-MeSe-C **96** and 2′-MeSe-T **97** phosphoramidites and have synthesized many 2′-Se-derivatized DNAs and RNAs **98** as depicted in Scheme 8.17.[2b,44,54,55] 2′-MeSe-A **99** and 2′-MeSe-G **100** phosphoramidites were synthesized by Micura and co-workers (Scheme 8.18).[56] Synthesis of the 4′Se modifications has already been discussed in Section 2.2.

8.4.1 Nucleobase-modified Analogues

The substitution of oxygen by selenium in the nucleobases opens a new era for nucleic acid base-pairing and stacking interactions at the atomic level. Huang and co-workers have documented the first report in these Se-modified nucleobases by introducing Se at the 4-position of thymidine **101** by a sequence of transformations as depicted in Scheme 8.19, affording the

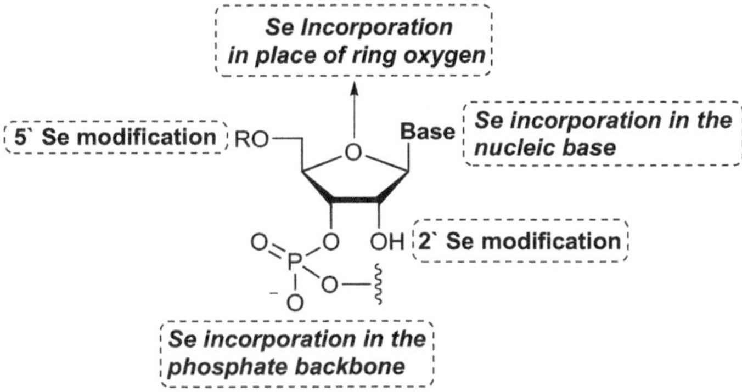

Figure 8.3 Selenium modifications sites in DNA and RNA.

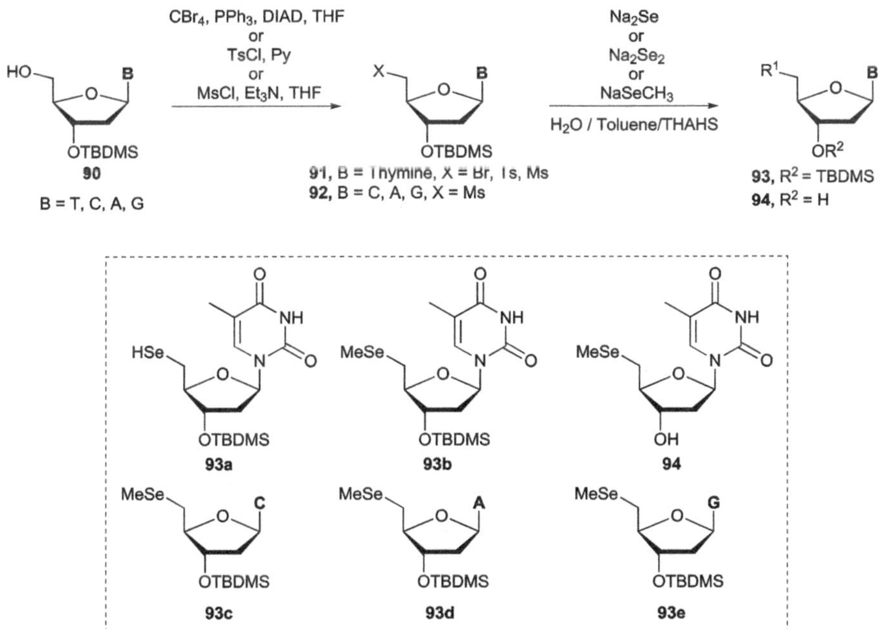

Scheme 8.16 Synthesis of 5′-Se nucleosides.

Se4T phosphoramidite **102.**[57] The synthesized Se4T phosphoramidite was incorporated into the RNA oligonucleotide **103,** through standard solid-phase synthesis. In a similar manner, the authors extended the protocol for the synthesis of diverse building units 2-Se-thymidine (Se2T)[57] **106,** 5-Se-thymidine (Se5T)[58] **107,** 6-Se-guanosine (Se6G)[59] **108,** and 2-Se-uridine (Se2U)[60] **104** phosphoramidites and incorporated them into DNA and RNA by solid-phase synthesis, affording the oligonucleotides **105, 109** and **110** (Scheme 8.19).

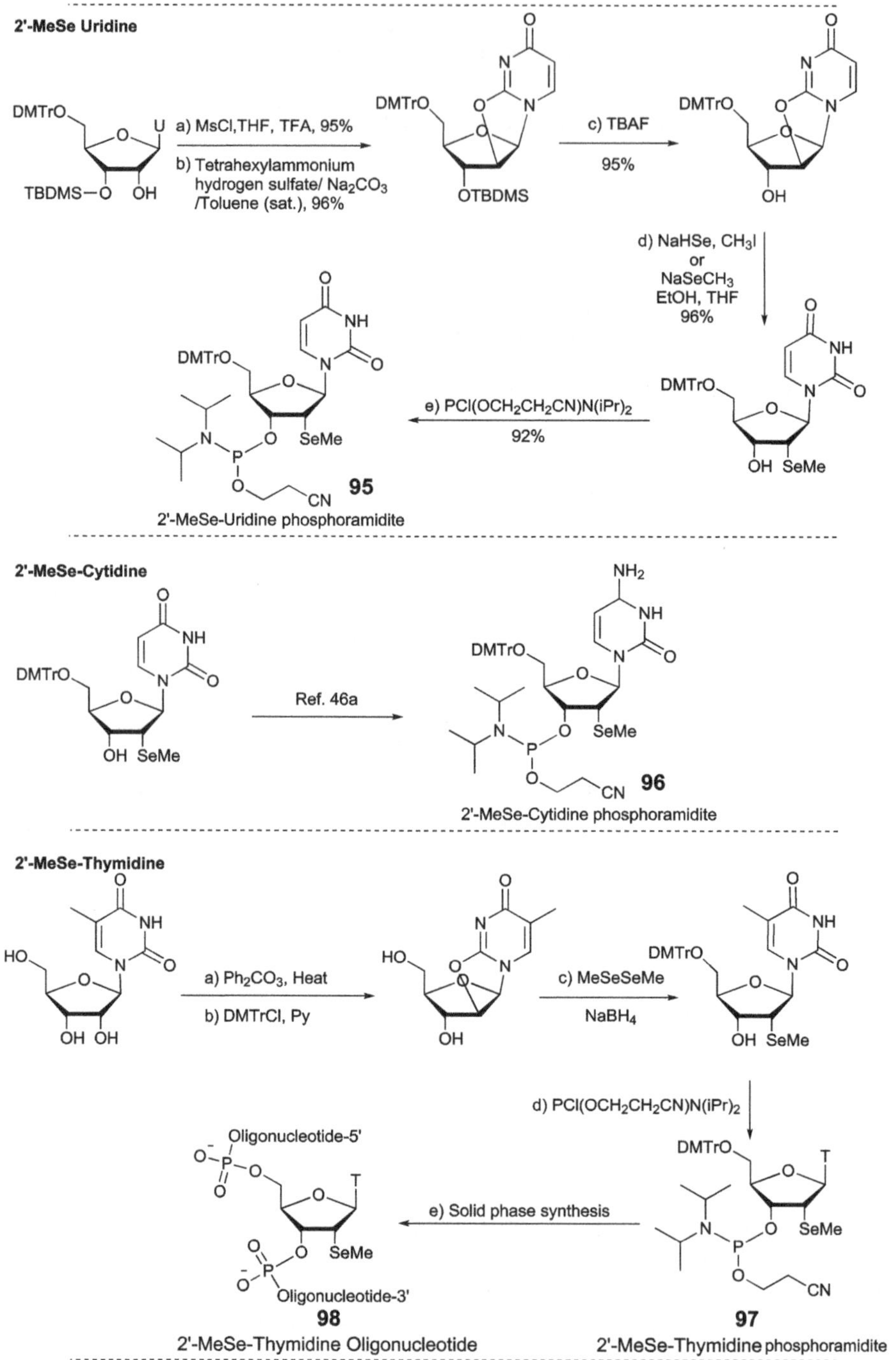

Scheme 8.17 2′-Me-Se-U/C/T phosphoramidites and 2′-Se-derivatized DNAs and RNAs.

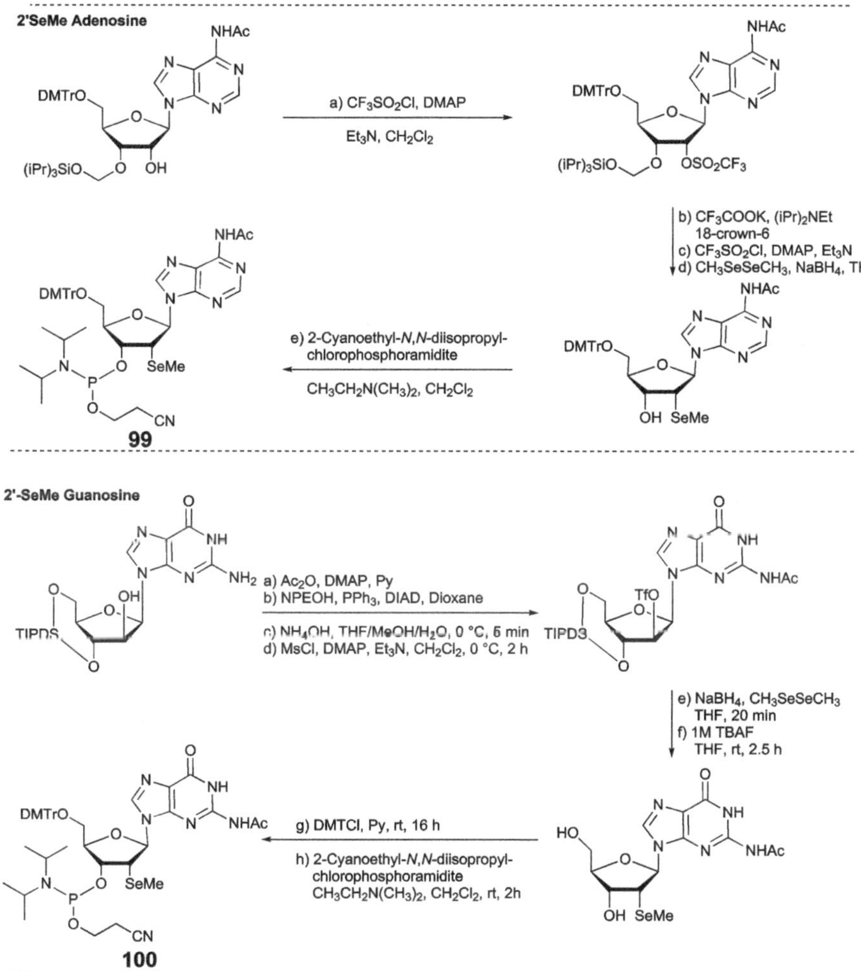

Scheme 8.18 2′-Se-A/G, synthesized by Micura and co-workers.

8.4.2 Phosphate Backbone-modified Analogues

Phosphoroselenoates (PSe) are oligonucleotides in which one of the oxygen atoms of the linking phosphate is replaced with selenium.[61] Egli and co-workers[62,63] reported the synthesis of a mixture of P-diasteromers **112** by incorporating the selenium in the phosphate backbone of the DNA **111** by using potassium selenocyanate as the selenating reagent (Scheme 8.20). The PSe-modified DNA is stable to oxidation and is a suitable candidate for structure determination by MAD analysis on a crystallographic time scale.[62,63]

Diastereomerically-pure phosphoroselenoate synthesis is the most challenging task for chemists. In this context, Stec and co-workers[64] reported a novel method for the stereo-controlled chemical synthesis of PSe

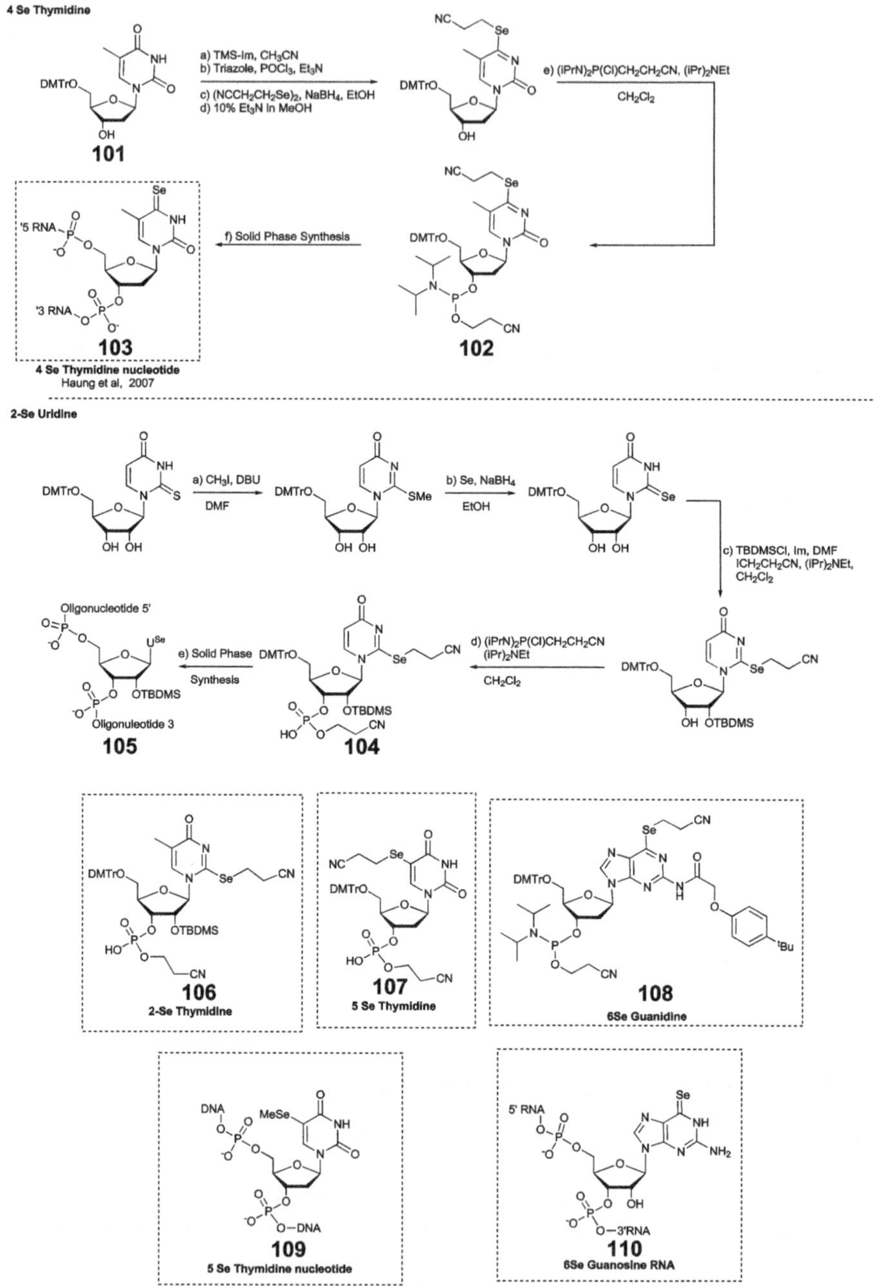

Scheme 8.19 Se-nucleobase-modified analogues.

oligonucleotides based on the oxathiophospholane approach. By using chromatographic techniques, phosphoroselenoate diasteromers can be separated in cases of oligonucleotides containing a single phosphoroselenoate.[65]

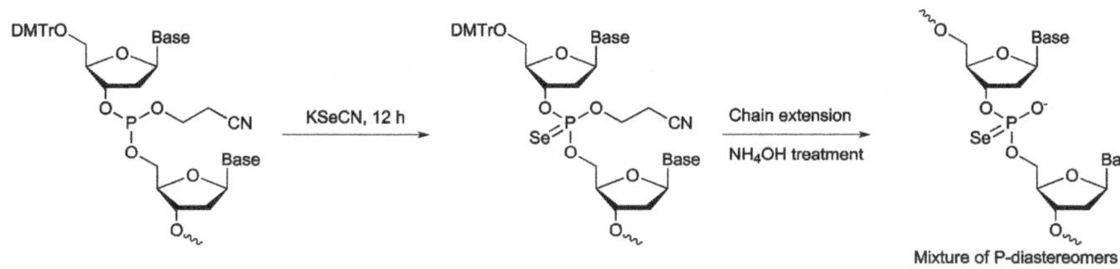

111 **112**

Scheme 8.20 Se incorporation in the phosphate backbone of the nucleotide.

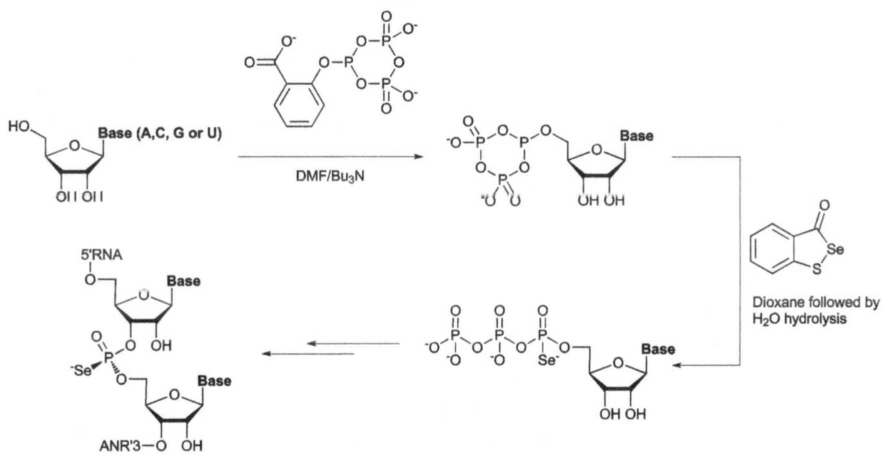

114 **113**

Scheme 8.21 Synthesis of 5′-(α-P-seleno)triphosphates (NTPαSe, Rp or Sp) and phosphoroselenoate RNA (PSe-RNA, Rp).

However, for oligonucleotides containing more than one phosphoroselenoate group, separation of diastereomers is challenging and impractical.

In an effort to overcome these limitations, Huang's research group provided a solution by performing the very first enzymatic synthesis for the stereo-defined phosphoroselenoates based on the recognition of the SP and/or RP diastereomers of α-Se-dNTPs or α-Se-NTPs by polymerases.[66] They developed a novel and efficient one-pot synthesis for α-Se-NTPs **113** without protecting the nucleosides and used it for the large-scale synthesis of α-Se-NTPs for PSe-RNA **114** transcription (Scheme 8.21).[67]

8.5 Application in Structural Analysis

Selenium modification of molecules offers a valuable advantage by facilitating crystallization and phasing in X-ray crystallography for 3-D structural determination. Selenium offers favorable scattering properties with a K absorption edge of 0.9795 Å, which is readily accessible with synchrotron radiation.[68] Selenium is an ideal anomalous scattering center for MAD and SAD analysis and has facilitated the crystallographic analysis of biomacromolecules.[69]

These Se-modifications in the nucleoside and nucleotide are stable and do not cause significant structural and functional perturbations, because they possess the same periodic properties as oxygen.[70] Thus, the systematic incorporation of selenium in place of oxygen atoms generates modified nucleic acids for X-ray crystallography using the MAD and SAD techniques.[71] These selenium nucleosides and nucleotides are better and fine models for the native DNA and RNA structures, and provide useful 3-D structural information for comprehensive understanding of nucleic acids and protein–nucleic acid complexes.[72]

The 2′-MeSe modification was a more successful and valuable modification for the 3-D structural determination of several RNAs and DNAs. This modification provided the first X-ray crystal structure of a Se-modified DNA *via* Se MAD phasing technique used by Huang and co-workers in collaboration with the Egli research group.[73] The crystal structure of the Se-modified DNA decamer [(5′-GCGTAdU2′-SeACGC-3′)2] was determined at 1.30 Å resolution *via* Se MAD phasing, comparing with the corresponding native one at 1.8 Å resolution. From data analysis, it was clear that the 2′SeMe modification in the DNA does not cause any structural changes in the DNA or RNA.[74] Thus, the 2′SeMe modification crystallizes the nucleotides for the MAD phasing; the structures were analyzed using the X-ray crystallography.

The first protein–nucleic acid complex structure was determined by Se-derivatized nucleic acids as the co-crystals of Se6G-DNA with RNA and RNase H using MAD phasing.[75] Subsequently, Micura and co-workers reported the crystal contact model of the Se-RNA 16mer and its structure was compared with the native one, which provided evidence that the RNA helices were packed in head-to-tail fashion with a few side-to-side contacts.[76]

The X-ray crystal structure of the 4′-Se-uridine was reported by Jeong *et al.* who synthesised it by seleno-Pummerer rearrangement of 4′Se uridine nucleosides. They also undertook a NMR conformational study by coupling-constant and nuclear Overhauser effect analysis for the 4′Se uridine derivative, which is suggestive for a C3′-endo to a C2′-endo pucker conformational change after selenium incorporation in the solution state.[77] Such conformational changes in 4′-Se-pyrimidine and 4′-Se-purine nucleosides with high β-selectivity and their corresponding DNAs and RNAs have been investigated by Pinto and co-workers.[78]

Interestingly, Egli and co-workers reported selenium derivatization in the phosphate backbone of DNA and RNA for structural determination of oligonucleotide. The crystal structure of the Z-form of the DNA was determined with one of the pure PSe diastereomers. From crystal data analysis, it was clear that the Se incorporation in the phosphate backbone does not alter the native structure. Thus, it is a powerful modification to determine the complete crystal structure of the homo-DNA [dd(CGAATTCG)2].[79]

8.5.1 Se–Carbohydrate–Protein Complexation

The chemical modification of carbohydrates is the most useful approach for solving crystal structures of carbohydrate-binding proteins. Bouckaert and co-workers solved the crystal structure of the 17 kDa F17-G lectin domain with the help of a selenoglycoside, Me-SeGlcNAc, **115** using the three-wavelength MAD technique. The selenium incorporation at the anomeric position gave the potential ligand Me-SeGlcNAc for the protein, because the anomeric oxygen participates most in the protein interactions (Scheme 8.22).[80]

Imberty and co-workers solved the crystal structure of the *Ralstonia solanacearum* lectin with the help of selenofucoside ligand and using the MAD technique. Solving the its structure revealed that the 90 amino acids are arranged in two consecutive and very similar four-stranded antiparallel sheets, lying side by side in parallel and connected by one long loop in a six-bladed β-propeller crystal structure.[81] The authors later solved the β propeller crystal structure of *Psathyrella velutina* lectin by using the monomeric structure of the β-Me-SeGlcNAc complex as a molecular replacement probe using the program MOLREP.[82]

Methylselenofucoside was used as a ligand to solve the crystal structure of the N-terminal BC2L-C domain.[83] The selenomethionine N-terminal BC2L-C domain (SeMet-BC2L-C-nt) contains only two methionine residues; anomalous phasing power was increased by crystallizing in the presence of methylseleno-α-L-fucopyranoside (αMeSeFuc). The crystal structure was solved using the multiple anomalous dispersion method and refined at 1.42 Å resolution. The authors further extended their studies by solving the crystal structure of fucose-specific lectin from *Aspergillus fumigatus conidia* by co-crystallizing with the methyl-α-L-selenofucoside (MeSeFuc) using the SAD technique.[84]

The co- and post-translational glycosylation of proteins plays a crucial role in protein folding and stability. In this context, Davis and co-workers[85] developed a new strategy for post-translational modification by selenenylsulfide (**117** or **118**)-mediated protein glycoconjugation technique with mono-and oligosaccharides of up to seven saccharide units in size at single and multiple sites in a variety of proteins (Scheme 8.23). The authors investigated two parallel strategies for this in which the protein cysteine **116** residue plays both electrophilic and nucleophilic roles.[85]

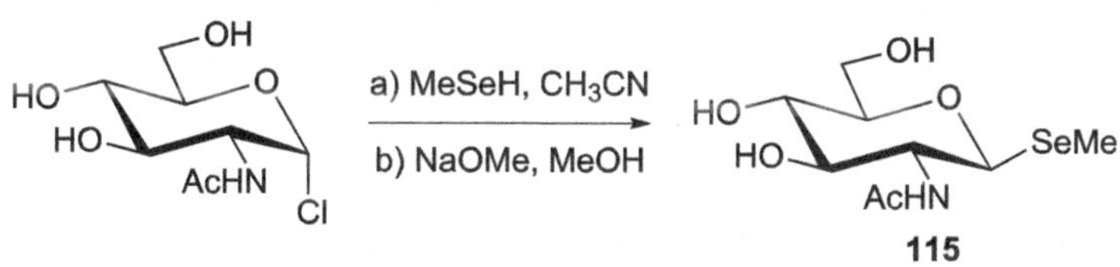

Scheme 8.22 Synthesis of the selenium derivative of GlcNAc.

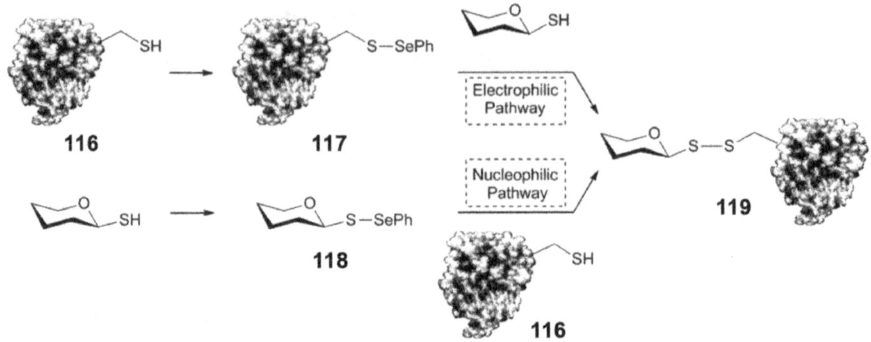

Scheme 8.23 Pathways for the glycoconjugation of protein cysteine.

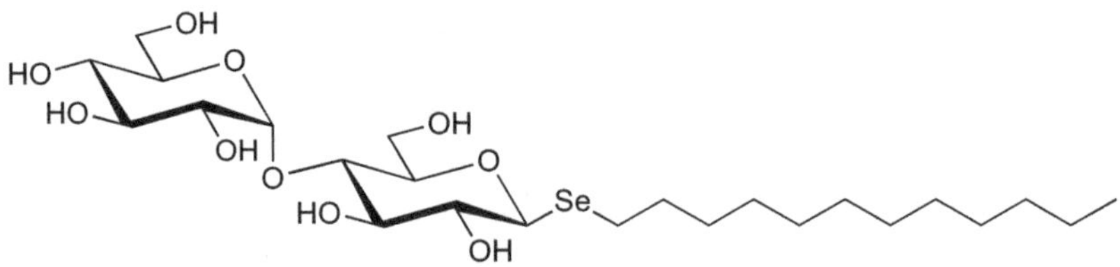

120

Figure 8.4 Dodecyl β-D-selenomaltoside (SeDDM).

Seleno-detergents are the specific targets for heavy-atom agents in experimental phasing of membrane proteins in protein crystallography. In this context, Miyano and co-workers studied the crystal structure of human leukotriene C4 synthase by using the dodecyl-D-selenomaltoside (SeDDM) **120** as a seleno-detergent for MAD phasing.[86] In the data analysis, the anomalous phasing of Se more accurately defined the position of SeDDM. The long alkyl chain environment causes hydrophobic interactions between the SeDDM and LTC4S (Figure 8.4).

Subsequently, Kiso and co-workers envisioned the 6-SeMe lactose **121** derivative to act as a molecular tool in the X-ray structural determination of human galectin-9 NCRD (Figure 8.5). The carbohydrate–protein complex with 6-SeMe lactose **121** and galectin-9 NCRD was co-crystallized and the structure could be determined with SAD/MAD phasing.[87] Furthermore, they also reported the synthesis of five regioisomers of seleno-fucoses **122–126** with respect to the substitution position of MeSe and 3-D structural determination of *Aspergillus oryzae* lectin (AOL) (Figure 8.5). They also validated the effect of substitution position on the interaction between AOL and seleno-fucose **123**.[88,89]

8.5.2 Selenium in NMR Structural Analysis

Of the six naturally occurring isotopes, only ^{77}Se is NMR-active, with a spin quantum number $I = 1/2$ and sensitivity three times greater than ^{13}C. The

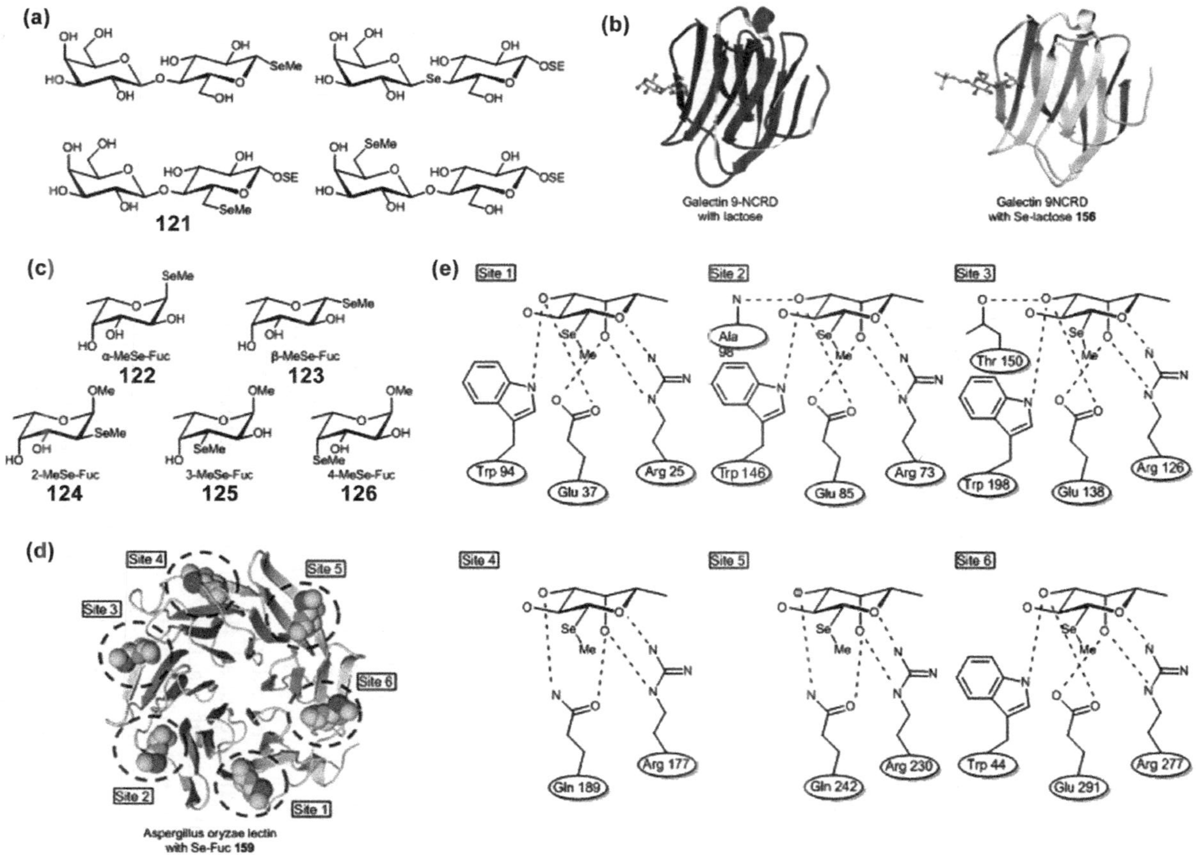

Figure 8.5 (a) Structures of seleno-lactoses. SE: 2-(trimethylsilyl)ethyl. (b) Ribbon diagrams of human galectin-9 NCRD with α-lactose (left) (PDB_ID: 2EAK) and with 6-MeSe-lactose **156** (right) (PDB ID: 3WLU). (c) Structures of seleno-fucoses. (d) Overall structure of *Aspergillus oryzae* lectin (AOL) with β-Se-Fuc **159**, represented as a ribbon model. (e) Schematic representation of the substrate recognition pattern of AOL in each binding site.

^{77}Se nucleus has a very large chemical shift range covering approximately 3000 ppm. ^{77}Se NMR is a powerful tool to examine the nature and interactions of selenium in macromolecules.[90]

Gettins and co-workers have examined the relaxation properties of selenium in a constant selenium sulfide (Se–S) linkage. For this purpose, they incorporated Se–S linkage with the ^{77}Se in glutathione, reduced ribonuclease, and bovine hemoglobin by using the 6,6′-diseleno bis (3-nitrobenzoic acid) (SeNB) as the reagent.[90]

The ^{77}Se NMR spectrum of SeNB-labeled bovine hemoglobin at 9.4 T shows a major resonance peak for the cysteine-attached SeNB at 574 ppm, while a minor peak at 474 ppm is for the residual reagent (DSeNB). The extremely broad signal at 574 ppm representing SeNB corresponds to the hemoglobin tetramer by attaching the approximately four cysteine units and with a half-height line width of ~1000 Hz, whereas the same sample gave the four-fold reduction in half-height line width of 220 Hz at 574 ppm.[90] This result indicated that the low-field instruments detected the poor signal-to-noise ratio. In a similar manner, ^{77}Se/SeNB-labeled glutathione and ^{77}Se-labeled ribonuclease A were synthesized and their relaxation properties were examined using ^{77}Se NMR spectroscopy. The other class of macromolecules for which selenium NMR might be useful is selenonucleotide-containing tRNAs, such as those found in *Escherichia coli*, *Clostridium sticklandii*, and *Methanococcus vannielii*.[91]

It is well known that sulfur is a key contributor to biological reactivity, but it is difficult to investigate the isotopic labelling using NMR spectroscopy. By using selenium as a surrogate, Rozovsky and co-workers developed a generally applicable method for ^{77}Se isotopic enrichment for heterologous proteins expressed in *Escherichia coli* and demonstrated the ^{77}Se NMR spectroscopy of multiple selenocysteine and selenomethionine residues in the sulfhydryl oxidase augmenter of liver regeneration (ALR). From the NMR comparison profile, it was clear that the peaks appearing at 412 and 426 ppm disappeared and a new peak was identified at 651 ppm by adding a reducing agent to the flavin. Interestingly, after removal of the reducing agent, a reversible phenomenon was observed.[92] Thus, with the help of selenium labelling the active sites were located by NMR spectra of the ALR bound to oxidized and reduced flavin adenine dinucleotide. Finally, the electronic environments of reactive sulfur sites were located by using the ^{77}Se NMR spectroscopy during catalysis.

Widmalm and co-workers[55] reported a method for finding the interactions between proteins and selenium-labeled carbohydrate ligands **127–129**, by using a new technique called SEAL by NMR. It was demonstrated with seleno-glycosides binding to lectins and the selenium nucleus serves as an NMR-active handle in the ^{77}Se NMR spectroscopy to study the binding interactions. This approach is very sensitive to binding as well as tolerant to the additives. In addition, the authors claimed that in future this technique would enable the potential improvement of the use of selenium in biology, medicine, structural elucidation, and the

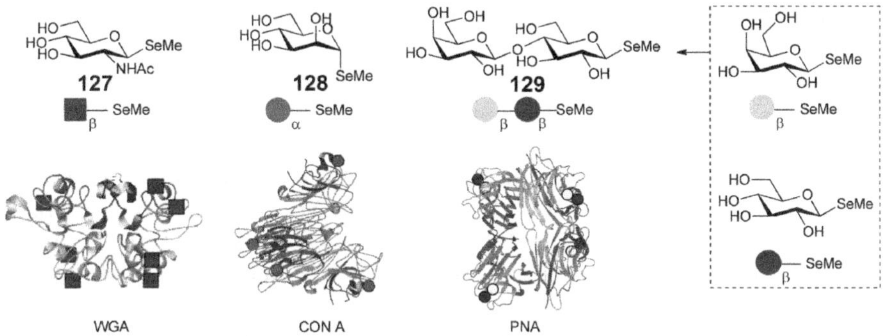

Figure 8.6 Schematic and CFG representation of compounds **163–167** studied using SEAL by NMR. The lectin models based on X-ray crystal structural analysis were originated as follows: WGA:[93] PDB ID 2UVO; ConA:[94] PDB ID 5CNA; PNA:[95] PDB ID 2PEL).

development of novel selenium-based glycomimetic drug candidates and bioactive molecules (Figure 8.6).

References

1. A. M. Deacon and S. E. Ealick, *Structure*, 1999, **7**, R161(b) Q. Cheng, T. Sandalova, Y. Lindqvist and E. S. J. Arnér, *J. Biol. Chem.*, 2009, **284**, 3998.
2. (a) L. Ronconi and P. J. Sadler, *Coord. Chem. Rev.*, 2008, **252**, 2239; (b) A. W. McDonagh, Phd Thesis, Synthesis of Novel Glycolipids and Investigation into the Anomerisation of Selenium Glycosides, National University of Ireland, 2016.
3. (a) Z. J. Witczak and S. Czernecki, *Adv. Carbohydr. Chem. Biochem.*, 1998, **53**, 143; (b) H. J. Reich and R. J. Honda, *ACS Chem. Biol.*, 2016, **11**, 821.
4. (a) C. Santi and S. Santoro, in *Organoselenium Chemistry: Synthesis and Reactions*, ed. T. Wirth, Wiley-VCH Verlag GmbH & Co. KGaA, 1st edn, 2012, ch. 1, pp. 1–51; (b) F. W. Bürgler, G. Fragale and T. Wirth, *ARKIVOC*, 2007, 21.
5. (a) P. A. Grieco, S. Gilman and M. Nishizawa, *J. Org. Chem.*, 1976, **41**, 1485; (b) K. B. Sharpless and M. W. Young, *J. Org. Chem.*, 1975, **40**, 947; (c) Z. Meng and S. J. Danishefsky, *Angew. Chem., Int. Ed.*, 2005, **44**, 1511; T. Wirth, S. Häuptli and M. Leuenberger, *Tetrahedron: Asymmetry*, 1998, **9**, 547; (d) Y. Nishibayashi and S. Uemura, in *Organoselenium Chemistry: Synthesis and Reactions*, T. Wirth, Wiley-VCH Verlag GmbH & Co. KGaA, 1st edn, 2012, ch. 7, pp. 287–320.
6. (a) L. M. Stephenson and D. R. Speth, *J. Org. Chem.*, 1979, **44**, 4683; (b) G.-J. ten Brink, J. Martijn Vis, I. W. C. E. Arends and R. A. Sheldon, *Tetrahedron*, 2002, **58**, 3977; (c) B. M. Trost, C. Pissot-Soldermann, I. Chen and G. M. Schroeder, *J. Am. Chem. Soc.*, 2004, **126**, 4480; (d) J. Młochowski and H. Wójtowicz-Młochowska, *Molecules*, 2015,

20, 10205; (e) M. A. Kipp, E. E. Stüeken, A. Bekker and R. Buick, *Proc. Natl. Acad. Sci.*, 2017, **114**, 875.
7. (a) J. L. Kice and T. W. S. Lee, *J. Am. Chem. Soc.*, 1978, **100**, 5094; (b) M. Gruttadauria, P. L. Meo and R. Noto, *Tetrahedron*, 1999, **55**, 14097.
8. K. Ajiki, M. Hirano and K. Tanaka, *Org. Lett.*, 2005, **7**, 4193.
9. (a) R. Gilmour, T. J. Prior, J. W. Burton and A. B. Holmes, *Chem. Commun.*, 2007, **43**, 3954; (b) N. M. R. McNeil, M. C. Matz and T. G. Back, *J. Org. Chem.*, 2013, **78**, 10369; (c) N. M. R. McNeil, D. J. Press, D. M. Mayder, P. Garnica, L. M. Doyle and T. G. Back, *J. Org. Chem.*, 2016, **81**, 7884.
10. (a) W. P. Jackson, S. V. Ley and A. J. Whittle, *J. Chem. Soc., Chem. Commun.*, 1980, 1173; (b) M. Besev and L. Engman, *Org. Lett.*, 2000, **2**, 1589; (c) F. V. Singh and T. Wirth, *Org. Lett.*, 2011, **13**, 6504; (d) Y. Kawamata, T. Hashimoto and K. Maruoka, *J. Am. Chem. Soc.*, 2016, **138**, 5206, and the reference are there in.
11. H. Duddeck, in *Chemical Shifts and Coupling Constants for Selenium-77*, ed. R. R. Gupta and M. D. Lechner, Springer, Berlin, Heidelberg, 2004, ch. 1, vol. 35G, pp. 1–11.
12. (a) L. Buts, R. Loris, E. De Genst, S. Oscarson, M. Lahmann, J. Messens, E. Brosens, L. Wyns, H. De Greve and Bouckaert, *J. Acta Crystallogr.*, 2003, **D59**, 1012; (b) N. Kostlánová, E. P. Mitchell, H. Lortat-Jacob, S. Oscarson, M. Lahmann, N. Gilboa-Garber, G. Chambat, M. Wimmerová and A. Imberty, *J. Biol. Chem.*, 2005, **280**, 27839.
13. T. Suzuki, H. Makyio, H. Ando, N. Komura, M. Menjo, Y. Yamada, A. Imamura, H. Ishida, S. Wakatsuki, R. Kato and M. Kiso, *Bioorg. Med. Chem.*, 2014, **22**, 2090.
14. V. Fournière and I. Cumpstey, *Tetrahedron Lett.*, 2010, **51**, 2127.
15. (a) H. C. Braga, A. D. Wouters, F. B. Zerillo and D. S. Lüdtke, *Carbohydr. Res.*, 2010, **345**, 2328; (b) R. F. Affeldt, H. C. Braga, L. L. Baldassari and D. S. Lüdtke, *Tetrahedron*, 2012, **68**, 10470.
16. L. S. Jeong, D. K. Tosh, H. O. Kim, T. Wang, X. Hou, H. S. Yun, Y. Kwon, S. K. Lee, J. Choi and L. X. Zhao, *Org. Lett.*, 2008, **10**, 209.
17. K. Jayakanthan, B. D. Johnston and B. M. Pinto, *Carbohydr. Res.*, 2008, **343**, 1790.
18. Y. Inagaki, N. Minakawa and A. Matsuda, *Nucleic Acids Symp. Ser.*, 2007, **51**, 139.
19. J. K. Watts, B. D. Johnston, K. Jayakanthan, A. S. Wahba, B. M. Pinto and M. J. Damha, *J. Am. Chem. Soc.*, 2008, **130**, 8578.
20. (a) L. S. Jeong, Y. N. Choi, D. K. Tosh, W. J. Choi, H. O. Kim and J. Choi, *Bioorg. Med. Chem.*, 2008, **16**, 9891; (b) L. S. Jeong, D. K. Tosh, W. J. Choi, S. K. Lee, Y.-J. Kang, S. Choi, J. H. Lee, H. Lee, H. W. Lee and H. O. Kim, *J. Med. Chem.*, 2009, **52**, 5303.
21. P. K. Sahu, G. Kim, J. Yu, J. Y. Ahn, J. Song, Y. Choi, X. Jin, J.-H. Kim, S. K. Lee, S. Park and L. S. Jeong, *Org. Lett.*, 2014, **16**, 5796.
22. G. Kim, Y. Choi, P. K. Sahu, J. Yu, S. Qu, D. Lee and L. S. Jeong, *Org. Lett.*, 2015, **17**, 4636.

23. C. Storkey, M. J. Davies, J. M. White and C. H. Schiesser, *Chem. Commun.*, 2011, **47**, 9693.
24. P. Merino-Montiel, Ó. López and J. G. Fernández-Bolaños, *Tetrahedron*, 2012, **68**, 3591.
25. C. Storkey, D. I. Pattison, J. M. White, C. H. Schiesser and M. J. Davies, *Chem. Res. Toxicol.*, 2012, **25**, 2589.
26. (a) A. Mallet, J.-M. Mallet and P. Sinay, *Tetrahedron. Asymmetry*, 1994, **5**, 2593; (b) L. Grant, Y. Liu, K. E. Walsh, D. S. Walter and T. Gallagher, *Org. Lett.*, 2002, **4**, 4623.
27. O. Boutureira, G. J. L. Bernardes, M. Fernández-González, D. C. Anthony and B. G. Davis, *Angew. Chem., Int. Ed.*, 2012, **51**, 1432.
28. (a) D. J. Chambers, G. R. Evans and A. J. Fairbanks, *Tetrahedron*, 2004, **60**, 8411; (b) F. Bravo, M. Kassou, Y. Dıa, Z and S. Castillón, *Carbohydr. Res.*, 2001, **336**, 83; (c) D. J. Chambers, G. R. Evans and A. J. Fairbanks, *Tetrahedron Lett.*, 2003, **44**, 5221.
29. (a) R. M. van Well, T. S. Kärkkäinen, K. P. R. Kartha and R. A. Field, *Carbohydr. Res.*, 2006, **341**, 1391; (b) S. Yamago, T. Yamada, O. Hara, H. Ito, Y. Mino and J.-I. Yoshida, *Org. Lett.*, 2001, **3**, 3867.
30. S. Mehta and B. M. Pinto, *J. Org. Chem.*, 1993, **58**, 3269.
31. K. C. Nicolaou, H. J. Mitchell, K. C. Fylaktakidou, H. Suzuki and R. M. Rodríguez, *Angew. Chem., Int. Ed.*, 2000, **39**, 1089.
32. (a) K. D. Randell, B. D. Johnston, P. N. Brown and B. M. Pinto, *Carbohydr. Res.*, 2000, **325**, 253; (b) P. Grice, S. V. Ley, J. Pietruszka and H. W. M. Priepke, *Angew. Chem., Int. Ed. Engl.*, 1996, **35**, 197; (c) S. Yamago, T. Yamada, H. Ito, O. Hara, Y. Mino and J.-I. Yoshida, *Chem. - Eur. J.*, 2005, **11**, 6159.
33. (a) D. Crich, D.-H. Suk and S. Sun, *Tetrahedron: Asymmetry*, 2003, **14**, 2861; (b) S. Valerio, A. Iadonisi, M. Adinolfi and A. Ravidà, *J. Org. Chem.*, 2007, **72**, 6097; (c) R. Benhaddou, S. Czernecki and D. Randriamandimby, *Synlett*, 1992, 967.
34. S. Mehta, J. S. Andrews, B. Svensson and B. M. Pinto, *J. Am. Chem. Soc.*, 1995, **117**, 9783.
35. S. Czernecki and D. Randriamandimby, *J. Carbohydr. Chem.*, 1996, **15**, 183.
36. (a) P. Tiwari and A. K. Misra, *Tetrahedron Lett.*, 2006, **47**, 2345; (b) W. R. Bowman in *Organoselenium Chemistry: Synthesis and Reactions*, ed. T. Wirth, Wiley-VCH Verlag GmbH & Co. KGaA, 1st edn, 2012, ch. 3, pp. 111–146.
37. Y. Kawai, H. Ando, H. Ozeki, M. Koketsu and H. Ishihara, *Org. Lett.*, 2005, **7**, 4653.
38. C. Venkateswarlu, V. Gautam and S. Chandrasekaran, *Carbohydr. Res.*, 2014, **396**, 48.
39. M. Nanami, H. Ando, Y. Kawai, M. Koketsu and H. Ishihara, *Tetrahedron Lett.*, 2007, **48**, 1113.
40. A. A. Kumar, T. Z. Illyés, K. E. Kövér and L. Szilágyi, *Carbohydr. Res.*, 2012, **360**, 8.

41. T. Suzuki, N. Komura, A. Imamura, H. Ando, H. Ishida and M. Kiso, *Tetrahedron Lett.*, 2014, **55**, 1920.
42. (a) Y. Kobayashi, Y. Ogra, K. Ishiwata, H. Takayama, N. Aimi and K. T. Suzuki, *Proc. Natl. Acad. Sci. U. S. A.*, 2002, **99**, 15932; (b) K. El-Bayoumy and R. Sinha, *Mutat. Res.*, 2004, **551**, 181; (c) P. D. Whanger, *Br. J. Nutr.*, 2004, **91**, 11.
43. (a) G. Mugesh, W.-W. du Mont and H. Sies, *Chem. Rev.*, 2001, **101**, 2125; (b) M. Soriano-Garcia, *Curr. Med. Chem.*, 2004, **11**, 1657.
44. (a) C. Hobartner and R. Micura, *J. Am. Chem. Soc.*, 2004, **126**, 1141; (b) Y. Buzin, N. Carrasco and Z. Huang, *Org. Lett.*, 2004, **6**, 1099; (c) D. M. Hatch, J. O. Boles, Z. Z. Li and L. A. Silks, *Curr. Org. Chem.*, 2004, **8**, 47.
45. W. H. Moolenar, *J. Biol. Chem.*, 1995, **270**, 12949.
46. G. B. Mills and W. H. Moolenar, *Nat. Rev. Cancer*, 2003, **3**, 582.
47. (a) C. W. Carreras and D. V. Santi, *Annu. Rev. Biochem.*, 1995, **64**, 721; (b) R. Schauer, *Glycoconjugate. J.*, 2000, **17**, 485; (c) R. N. Perham, *Annu. Rev. Biochem.*, 2000, **69**, 961; (d) N. J. Kruger and A. V. Schaewen, *Curr. Opin. Plant Biol.*, 2003, **6**, 236; (e) D. R. Coombe and W. C. Kett, *Cell. Mol. Life Sci.*, 2005, **62**, 410.
48. (a) G. A. Patani and E. J. LaVoie, *Chem. Rev.*, 1996, **96**, 3147; (b) D. Krehan, B. Frolund, P. Krogsgaard-Larsen, J. Kehler, G. A. R. Johnson and M. Chebib, *Neurochem. Int.*, 2003, **42**, 561; (c) C. S. Rye and J. B. Baello, *Curr. Med. Chem.*, 2005, **12**, 3127; (d) H. Yuan and R. B. Silverman, *Bioorg. Med. Chem.*, 2006, **14**, 1331; (e) V. Ahmed, Y. Liu, C. Silvestro and S. D. Taylor, *Bioorg. Med. Chem.*, 2006, **14**, 8564; (f) M. Abdo, Biomimetic Enzyme Inhibitors, The State University of New Jersey, 2010.
49. (a) S. Knapp and E. Darout, *Org. Lett.*, 2005, **7**, 203; (b) M. Abdo and S. Knapp, *J. Am. Chem. Soc.*, 2008, **130**, 9234; (c) M. Abdo, S. Liu, B. Zhou, D. C. Walls, L. Wu, S. Knapp and Z. Zhang, *J. Am. Chem. Soc.*, 2008, **130**, 13196; (d) M. Abdo, Y. Zhang, V. L. Schrammand and S. Knapp, *Org. Lett.*, 2010, **12**, 2982; (e) M. Abdo and S. Knapp, *J. Org. Chem.*, 2012, **77**, 3433.
50. (a) Selenylation of phenols: J. Oddershede, L. Henriksen and S. Larsen, *Org. Biomol. Chem.*, 2003, **1**, 1053; (b) Selenylation of indoles: D. Crich and J. W. Davies, *Tetrahedron Lett.*, 1989, **30**, 4307, and references cited therein; (c) Selenylation of imidazoles: T. G. Back and R. G. Kerr, *Can. J. Chem.*, 1986, **64**, 308.
51. R. B. Silverman, *The Organic Chemistry of Drug Design and Drug Action*, Elsevier, Amsterdam, 2nd edn, 2004, (Enzyme Inhibition and Inactivation), ch. 5, pp. 274–321.
52. A. M. Kaur, A. Rob, J. C. Williams and Z. Huang, in *Biochalcogen Chemistry: The Biological Chemistry of Sulfur, Selenium, and Tellurium*, ed. C. A. Bayse and J. L. Brumaghim, ACS Symp. Ser., American Chemical Society, Washington, DC, 2013, ch. 5, vol. 1152, pp. 89–126.

53. N. Carrasco, D. Ginsburg, Q. Du and Z. Huang, *Nucleosides, Nucleotides Nucleic Acids*, 2001, **20**, 1723.
54. (a) Q. Du, N. Carrasco, M. Teplova, C. J. Wilds, M. Egli and Z. Huang, *J. Am. Chem. Soc.*, 2002, **124**, 24; (b) M. Teplova, C. J. Wilds, Z. Wawrzak, V. Tereshko, Q. Du, N. Carrasco, Z. Huang and M. Egli, *Biochimie*, 2002, **84**, 849; (c) N. Carrasco, Y. Buzin, E. Tyson, E. Halpert and Z. Huang, *Nucleic Acids Res.*, 2004, **32**, 1638; (d) J. Sheng and Z. Huang, *Int. J. Mol. Sci.*, 2008, **9**, 258; (e) Y. Buzin, N. Carrasco and Z. Huang, *Org. Lett.*, 2004, **6**, 1099; (f) J. Salon, G. Chen, Y. Portilla, M. W. Germann and Z. Huang, *Org. Lett.*, 2005, **7**, 5645; (g) J. Sheng, J. Jiang, J. Salon and Z. Huang, *Org. Lett.*, 2007, **9**, 749; (h) T. Murai, in *Organoselenium Chemistry: Synthesis and Reactions*, ed. T. Wirth, Wiley-VCH Verlag GmbH & Co. KGaA, 1st edn, 2012, ch. 6, pp. 257–285.
55. C. Hamark, J. Landstrçm and G. Widmalm, *Chem. - Eur. J.*, 2014, **20**, 13905.
56. (a) C. Hobartner, R. Rieder, C. Kreutz, B. Puffer, K. Lang, A. Polonskaia, A. Serganov and R. Micura, *J. Am. Chem. Soc.*, 2005, **127**, 12035; (b) H. Moroder, C. Kreutz, K. Lang, A. Serganov and R. Micura, *J. Am. Chem. Soc.*, 2006, **128**, 9909; (c) J. Salon, J. Sheng, J. Gan and Z. Huang, *J. Org. Chem.*, 2010, **75**, 637.
57. J. Salon, J. Sheng, J. Gan and Z. Huang, *J. Org. Chem.*, 2010, **75**, 637.
58. A. E. Hassan, J. Sheng, W. Zhang and Z. Huang, *J. Am. Chem. Soc.*, 2010, **132**, 2120.
59. (a) J. Salon, J. Jiang, J. Sheng, O. O. Gerlits and Z. Huang, *Nucleic Acids Res.*, 2008, **36**, 7009; (b) J. Salon, J. Gan, R. Abdur, H. Liu and Z. Huang, *Org. Lett.*, 2013, **15**, 3934.
60. H. Sun, J. Sheng, A. E. Hassan, S. Jiang, J. Gan and Z. Huang, *Nucleic Acids Res.*, 2012, **40**, 5171.
61. K. Mori, C. Boiziau, C. Cazenave, M. Matsukura, C. Subasinghe, J. Cohen, S. Broder, J. Toulme and C. Stein, *Nucleic Acids Res.*, 1989, **17**, 8207.
62. (a) C. J. Wilds, R. Pattanayek, C. Pan, Z. Wawrzak and M. Egli, *J. Am. Chem., Soc.*, 2002, **124**, 14910; (b) M. Egli, P. S. Pallan, R. Pattanayek, C. J. Wilds, P. Lubini, G. Minasov, M. Dobler, C. J. Leumann and A. Eschenmoser, *J. Am. Chem. Soc.*, 2006, **128**, 10847.
63. P. S. Pallan and M. Egli, *Nat. Protoc.*, 2007, **2**, 640.
64. P. Guga, A. Maciaszek and W. J. Stec, *Org. Lett.*, 2005, **7**, 3901.
65. P. Guga, *Curr. Top. Med. Chem.*, 2007, **7**, 695.
66. (a) N. Carrasco, J. Caton-Williams, G. Brandt, S. Wang and Z. Huang, *Angew. Chem., Int. Ed.*, 2006, **45**, 94; (b) G. Brandt, N. Carrasco and Z. Huang, *Biochemistry*, 2006, **45**, 8972.
67. L. Lin, J. Caton-Williams, M. Kaur, A. M. Patino, J. Sheng, J. Punetha and Z. Huang, *RNA*, 2011, **17**, 1932.
68. W. A. Hendrickson, J. R. Horton and D. M. LeMaster, *EMBO J.*, 1990, **9**, 1665.
69. A. M. Deacon and S. E. Ealick, *Structure*, 1999, **7**, R161.

70. (a) N. Carrasco and Z. Huang, *J. Am. Chem. Soc.*, 2004, **126**, 448; (b) J. Salon, J. Sheng, J. Jiang, G. Chen, J. Caton-Williams and Z. Huang, *J. Am. Chem. Soc.*, 2007, **129**, 4862; (c) J. Salon, J. Jiang, J. Sheng, O. O. Gerlits and Z. Huang, *Nucleic Acids Res.*, 2008, **36**, 7009; (d) A. E. Hassan, J. Sheng, W. Zhang and Z. Huang, *J. Am. Chem. Soc.*, 2010, **132**, 2120; (e) L. Lin, J. Sheng and Z. Huang, *Chem. Soc. Rev.*, 2011, **40**, 4591; (f) H. Sun, J. Sheng, A. E. Hassan, S. Jiang, J. Gan and Z. Huang, *Nucleic Acids Res.*, 2012, **40**, 5171; (g) J. Sheng, J. Gan and Z. Huang, *Med. Res. Rev.*, 2013, **33**, 1119.
71. (a) J. Jiang, J. Sheng, N. Carrasco and Z. Huang, *Nucleic Acids Res.*, 2007, **35**, 477; (b) J. Sheng and Z. Huang, *Int. J. Mol. Sci.*, 2008, **9**, 258; (c) L. Lin, J. Sheng and Z. Huang, *Chem. Soc. Rev.*, 2011, **40**, 4591; (d) J. Sheng, J. Gan and Z. Huang, *Med. Res. Rev.*, 2013, **33**, 1119; (e) J. Sheng, A. E. Hassan, W. Zhang, J. Zhou, B. Xu, A. S. Soares and Z. Huang, *Nucleic Acids Res.*, 2011, **39**, 3962; (f) L. Lin and Z. Huang, Se-derivatized RNAs for X-ray crystallography, *Methods Mol. Biol.*, 2012, **941**, 213; (g) J. Sheng, W. Zhang, A. E. Hassan, J. Gan, A. S. Soares, S. Geng, Y. Ren and Z. Huang, *Nucleic Acids Res.*, 2012, **40**, 8111; (h) H. Sun, J. Sheng, A. E. Hassan, S. Jiang, J. Gan and Z. Huang, *Nucleic Acids Res.*, 2012, **40**, 5171; (i) J. Salon, J. Gan, R. Abdur, H. Liu and Z. Huang, *Org. Lett.*, 2013, **15**, 3934; (j) H. Sun, S. Jiang, J. Caton-Williams, H. Liu and Z. Huang, *RNA*, 2013, **19**, 1309; (k) W. Zhang, E. A. Hassan and Z. Huang, *Sci. China: Chem.* 2013, **56**, 273.
72. (a) V. Olieric, U. Rieder, K. Lang, A. Serganov, C. Schulze-Briese, R. Micura, P. Dumas and E. Ennifar, *RNA*, 2009, **15**, 707; (b) J. Sheng and Z. Huang, *Chem. Biodiversity*, 2010, **7**, 753; (c) L. Lin, J. Sheng and Z. Huang, *Chem. Soc. Rev.*, 2011, **40**, 4591.
73. Q. Du, N. Carrasco, M. Teplova, C. J. Wilds, M. Egli and Z. Huang, *J. Am. Chem. Soc.*, 2002, **124**, 24.
74. M. Teplova, C. J. Wilds, Z. Wawrzak, V. Tereshko, Q. Du, N. Carrasco, Z. Huang and M. Egli, *Biochimie*, 2002, **84**, 849.
75. (a) A. E. Hassan, J. Sheng, W. Zhang and Z. Huang, *J. Am. Chem. Soc.*, 2010, **132**, 2120; (b) H. Sun, J. Sheng, A. E. Hassan, S. Jiang, J. Gan and Z. Huang, *Nucleic Acids Res.*, 2012, **40**, 5171.
76. B. Puffer, H. Moroder, M. Aigner and R. Micura, *Nucleic Acids Res.*, 2008, **36**, 970.
77. L. S. Jeong, D. K. Tosh, H. O. Kim, T. Wang, X. Hou, H. S. Yun, Y. Kwon, S. K. Lee, J. Choi and L. X. Zhao, *Org. Lett.*, 2008, **10**, 209.
78. (a) K. Jayakanthan, B. D. Johnston and B. M. Pinto, *Carbohydr. Res.*, 2008, **343**, 1790; (b) J. K. Watts, B. D. Johnston, K. Jayakanthan, A. S. Wahba, B. M. Pinto and M. J. Damha, *J.* Am. Chem. Soc., 2008, **130**, 8578.
79. (a) M. Egli, P. S. Pallan, R. Pattanayek, C. J. Wilds, P. Lubini, G. Minasov, M. Dobler, C. J. Leumann and A. Eschenmoser, *J. Am. Chem. Soc.*, 2006, **128**, 10847; (b) P. S. Pallan and M. Egli, *Nat. Protoc.*, 2007, **2**, 640.

80. L. Buts, R. Loris, E. De Genst, S. Oscarson, M. Lahmann, J. Messens, E. Brosens, L. Wyns, H. De Greve and J. Bouckaert, *ActaCryst*, 2003, **D59**, 1012.
81. N. Kostlánová, E. P. Mitchell, H. Lortat-Jacob, S. Oscarson, M. Lahmann, N. Gilboa-Garber, G. Chambat, M. Wimmerova and A. Imberty, *J. Biol. Chem.*, 2005, **280**, 27839.
82. G. Cioci, E. P. Mitchell, V. Chazalet, H. Debray, S. Oscarson, M. Lahmann, C. Gautier, C. Breton, S. Perez and A. Imberty, *J. Mol. Biol.*, 2006, **357**, 1575.
83. O. Šulák, G. Cioci, M. Delia, M. Lahmann, A. Varrot, A. Imberty and M. Wimmerova, *Structure*, 2010, **18**, 59.
84. J. Houser, J. Komarek, N. Kostlanova, G. Cioci, A. Varrot, S. C. Kerr, M. Lahmann, V. Balloy, J. V. Fahy, M. Chignard, A. Imberty and M. Wimmerova, *PLoS ONE*, 2013, **8**, e83077.
85. O. Boutureira, G. J. L. Bernardes, M. Fernández-González, D. C. Anthony and B. G. Davis, *Angew. Chem., Int. Ed.*, 2012, **51**, 1432.
86. H. Saino, H. Ago, Y. Ukitaa and M. Miyano, *ActaCryst*, 2011, **F67**, 1666.
87. T. Suzuki, H. Makyio, H. Ando, N. Komura, M. Menjo, Y. Yamada, A. Imamura, H. Ishida, S. Wakatsuki, R. Kato and M. Kiso, *Bioorg. Med. Chem.*, 2014, **22**, 2090.
88. H. Makyio, J. Shimabukuro, T. Suzuki, A. Imamura, H. Ishida, M. Kiso, H. Ando and R. Kato, *Biochem.Biophys. Res.Commun.*, 2016, **477**, 477.
89. J. Shimabukuro, H. Makyio, T. Suzuki, Y. Nishikawa, M. Kawasaki, A. Imamura, H. Ishida, H. Ando, R. Kato and M. Kiso, *Bioorg. Med.Chem.*, 2017, **25**, 1132.
90. P. Gettins and S. Wardlaw, *J. Biol. Chem.*, 1991, **266**, 3442.
91. A. J. Wittwer, L. Tsai, W. Ching and T. C. Stadtman, *Biochemistry*, 1984, **23**, 4650.
92. (a) S. A. Schaefer, M. Dong, R. P. Rubenstein, W. A. Wilkie, B. J. Bahnson, C. Thorpe and S. Rozovsky, *J. Mol. Biol.*, 2013, **425**, 222; (b) S. Rozovsky, in Biochalcogen Chemistry: The Biological Chemistry of Sulfur, Selenium, and *Tellurium*, ed. C. A. Bayse and J. L. Brumaghim, ACS Symp. Ser., American Chemical Society, Washington, DC, 2013, ch. 6, vol. 1152, pp. 127–142.
93. D. Schwefel, C. Maierhofer, J. G. Beck, S. Seeberger, K. Diederichs, H. M. Möller, W. Welte and V. Wittmann, *J. Am. Chem. Soc.*, 2010, **132**, 8704.
94. J. H. Naismith, C. Emmerich, J. Habash, S. J. Harrop, J. R. Helliwell, W. N. Hunter, J. Raftery, A. J. Kalb and J. Yariv, *Acta Crystallogr. D*, 1994, **50**, 847.
95. R. Banerjee, K. Das, R. Ravishankar, K. Suguna, A. Surolia and M. Vijayan, *J. Mol. Biol.*, 1996, **259**, 281.

CHAPTER 9

Reaction of Selenium Compounds with Reactive Oxygen Species and the Control of Oxidative Stress

LUKE D. CARROLL AND MICHAEL J. DAVIES*

Dept. of Biomedical Sciences, Panum Institute, University of Copenhagen, Blegdamsvej 3, Copenhagen 2200, Denmark
*Email: davies@sund.ku.dk

9.1 Introduction: Oxidant Formation During Inflammation

Oxidants are continually generated in biological systems by normal physiological processes, during ageing, and as a response to external stimuli.[1] A number of oxidants, including radicals (species with unpaired electrons, *e.g.* nitric oxide), and two-electron oxidants (*e.g.* hydrogen peroxide, other peroxides, peroxynitrous acid, and hypochlorous acid), are intentionally generated by enzymes to perform biological functions. In other cases, oxidants are formed as byproducts of electron transfer (*e.g.* by electron leakage from mitochondria, endoplasmic reticulum, and plasma membrane electron transport chains) or enzymatic reactions (*e.g.* oxidase and peroxidase enzymes).

Activated white blood cells employ intracellular NADPH and NADPH oxidase complexes (NOxs) present on the plasma membrane to generate

Organoselenium Compounds in Biology and Medicine: Synthesis, Biological and Therapeutic Treatments
Edited by Vimal Kumar Jain and K. Indira Priyadarsini

Published by the Royal Society of Chemistry, www.rsc.org

superoxide radicals ($O_2^{\bullet -}$) and hydrogen peroxide (H_2O_2) in a directed and coordinated response to invading pathogens (*e.g.* bacteria, yeast, viruses, and fungi[2,3]). $O_2^{\bullet -}$ is known to inactivate several enzymes containing Fe–S clusters or react with other metal ions. However, these processes are limited in scope, and are slow compared to reaction with another $O_2^{\bullet -}$ (*i.e.* dismutation) or reaction with other radicals, such as $NO^{\bullet}$ which results in the formation of peroxynitrite ($ONOO^-$) and its protonated form, peroxynitrous acid (ONOOH; for further information see later), and some other long-lived radicals (*e.g.* Tyr phenoxyl radicals).

In contrast, H_2O_2 and other peroxides (*e.g.* those formed on lipids, amino acids, peptides, proteins, and DNA) can induce significant biological damage *via* multiple mechanisms. Peroxides can undergo direct two-electron oxidation reactions (*e.g.* with thiols or thioethers, although such reactions are generally slow), or react with redox active metal ions (primarily iron and copper, in most biological systems) *via* one-electron reactions to give radicals.[1] The reaction of H_2O_2 with Fe^{2+} (the Fenton reaction[4]), gives rise to hydroxyl radicals ($HO^{\bullet}$), with analogous reactions occurring with Cu^+. In the case of other peroxides (ROOH), one-electron reduction of the O–O bond gives alkoxyl radicals ($RO^{\bullet}$) and HO^-. Alkoxyl radicals typically undergo either rapid hydrogen atom abstraction reactions (intra- or inter-molecularly) to give alcohols (ROH) or fragmentation reactions that typically result in the generation of carbon-centred radicals ($R^{\bullet}$).[5] The reaction of many radicals with biological targets, *via* hydrogen atom abstraction, addition, or electron transfer, results in carbon-centred radicals ($R^{\bullet}$), with subsequent formation of peroxyl radicals ($ROO^{\bullet}$) *via* reaction with O_2 at close to diffusion-controlled rates.

Peroxidase enzymes, such as myeloperoxidase (MPO), use H_2O_2 to generate a family of powerful two-electron oxidants. These include hypochlorous acid (HOCl, from chloride ions), hypobromous acid (HOBr, from bromide ions) and hypothiocyanous acid (HOSCN, from thiocyanate ions). These oxidants react with a variety of biological targets (particularly sulfur- and selenium-containing species), but reaction of HOCl and HOBr with amines (RNH_2) results in formation of chloramines (RNHCl) and bromamines (RNHBr) that are themselves oxidants, albeit of lower reactivity than HOCl and HOBr.[6–8] Other members of the heme peroxidase superfamily, including eosinophil, salivary, gastric, thyroid, lacto, and vascular peroxidases or peroxidasins are also able to produce some of these species, although the nature of the species formed varies between different members of the superfamily.[6,9] Most of these peroxidase enzymes also generate radicals *via* classical peroxidase reactions using readily oxidized substrates such as phenols, indoles, urate, and ascorbic acid.[6,10,11] A significant number of other enzymes also generate radicals, in a controlled manner, to effect metabolic transformations (*e.g.* prostaglandin synthases, cytochrome P450s, and ribonucleotide reductases; reviewed in ref. 12).

Peroxynitrous acid (ONOOH) is a powerful oxidant produced by reaction of $O_2^{\bullet -}$ with nitric oxide ($NO^{\bullet}$). $O_2^{\bullet -}$ is produced not only by NADPH oxidase

(NOX) enzymes (of which there are several isoforms[2,3]), but also *via* multiple other cellular enzymes and processes (*e.g.* electron leakage from mitochondrial and other electron transport chains, xanthine oxidase, and uncoupled nitric oxide synthases). $NO^{\bullet}$ is generated by nitric oxide synthase enzymes (NOS) that are present in endothelial cells (eNOS), macrophages (iNOS), and neuronal cells (nNOS), as well as other cell types. ONOOH can react with dissolved CO_2 to form the peroxynitrosocarbonate anion ($ONOOCO_2^-$) and both of these species can induce direct two-electron oxidation reactions, as well as initiate radical processes *via* decomposition to hydroxyl ($HO^{\bullet}$) and $NO_2^{\bullet}$ radicals (from ONOOH) or $NO_2^{\bullet}$ and $CO_3^{-\bullet}$ (from $ONOOCO_2^-$).[13]

A number of peroxidase-catalysed reactions, as well as photochemical processes (*i.e.* from light absorption) and chemical reactions (*e.g.* reaction of HOCl with H_2O_2, or termination reactions of peroxyl radicals) can give rise to species in high-energy (excited) states. These species, which include singlet oxygen ($^1O_2$) and triplet ketones (RR′C=O*) can induce both one- and two-electron oxidation reactions.[5,14]

This short summary of oxidant formation during inflammation highlights that (a) a diverse mixture of both one- and two-electron oxidants are generated at sites of inflammation; and (b) formation of a single initial species can result in subsequent secondary reactions that generate multiple additional oxidants. Thus, the oxidant milieu at sites of inflammation is both highly complex and dynamic.

9.2 Antioxidant and Protective Systems

In physiological systems, the formation and reactions of oxidants are tightly regulated by a battery of defensive and repair systems[1] including (a) low molecular mass radical/oxidant scavengers (*e.g.* ascorbic acid, tocopherols, urate, or thiols); (b) enzymes that remove oxidants or their precursors (*e.g.* superoxide dismutases (SODs) that convert $O_2^{\bullet -}$ to O_2 and H_2O_2, catalase which converts H_2O_2 to H_2O, peroxiredoxins (Prx), and glutathione peroxidases (GPx) that detoxify a range of peroxides); (c) repair enzymes (*e.g.* methionine sulfoxide reductases, disulfide reductases, and sulfiredoxins); and (d) enzymes and organelles that remove non-repairable damage (*e.g.* proteasomes, lysosomes, DNA repair enzymes, and phospholipases).

However, despite this diversity of protective systems, there is considerable evidence that oxidative damage is widespread in most organisms exposed to O_2 (*i.e.* nearly all mammals, plants, and microorganisms).[1] This may arise from an increased rate or extent of oxidant generation, a failure or decrease in activity of defense systems, or both. This altered balance between formation and removal/repair, in favour of enhanced oxidation, is often termed "oxidative stress". Inappropriate control or activation of the oxidant-generating systems of the immune response is of particular importance in human disease. Thus inappropriate, mistimed, or excessive stimulation of white cells, and consequent generation of oxidants at elevated levels has been associated with a large number of pathologies linked with acute or chronic inflammation; examples include atherosclerosis, arthritis, asthma, some cancers, systemic lupus erythematosus, and some neurodegenerative conditions among many others.[6,15]

Many of the protective systems described above require sulfur- or selenium-containing residues to carry out their functions. Both sulfur and selenium centres are readily oxidized, but as selenium is located directly below sulfur in the periodic table it is typically more reactive due to its more favourable redox and nucleophilic properties.[16] Thus, at neutral pH the thiol (sulfhydryl, RSH) group of Cys residues is usually present in its uncharged neutral form as it has a $pK_a \sim 8.7$ (for the free amino acid; this value varies considerably in proteins and with thiol structure). In contrast, the selenol (RSeH) group of selenocysteine (Sec) has a much lower pK_a (typically ~ 5.2), and is therefore usually ionized (RSe^-) at physiological pH.[17] The greater electron density and increased nucleophilicity of the ionized selenium species results in much greater reactivity with most oxidants, and considerable differences in the reactivity of the oxidized species formed.[16]

These data support the hypothesis that sulfur- and selenium-containing species modulate the activity of both radical and molecular oxidants and provide protection against inflammation-induced damage. These effects appear to arise from multiple modes of action including direct oxidant scavenging, as cofactors for cellular enzymes, as key residues in protective enzymes and repair enzymes, and *via* their important role in redox-based cell signaling processes.[18–21] This review focuses primarily on the chemistry of low molecular mass selenium compounds (Scheme 9.1), together with selected comparisons with sulfur analogues as appropriate, their potential role in the defense against inflammatory oxidants, and a brief summary of the recent developments in the use of selenium compounds to treat or alleviate disease.

9.3 Oxidative Chemistry of Sulfur and Selenium Compounds

The enhanced ability of selenium compounds to scavenge oxidants over their sulfur-containing counterparts is readily exemplified by their absolute rate constants with biological oxidants. For low molecular mass materials, the second-order rate constants (k) for the reactions of a variety of oxidants with selenium species are typically 10- to 100-fold greater than for their sulfur analogues (Tables 9.1 and 9.2). On the basis of these data it might be expected that selenocysteine (Sec) and selenomethionine (SeMet) would be capable of substituting for some of the functions of sulfur-containing amino acids, while also being more reactive.[17] Studies that have tested this hypothesis in mutated proteins[22–26] are discussed in more detail later. However, while this enhanced reactivity is typical for low molecular mass compounds, other factors in Sec-containing proteins, such as tertiary structure, local environment, and specific amino acid interactions and reactions can also play key roles in modulating their reactivity.[27,28]

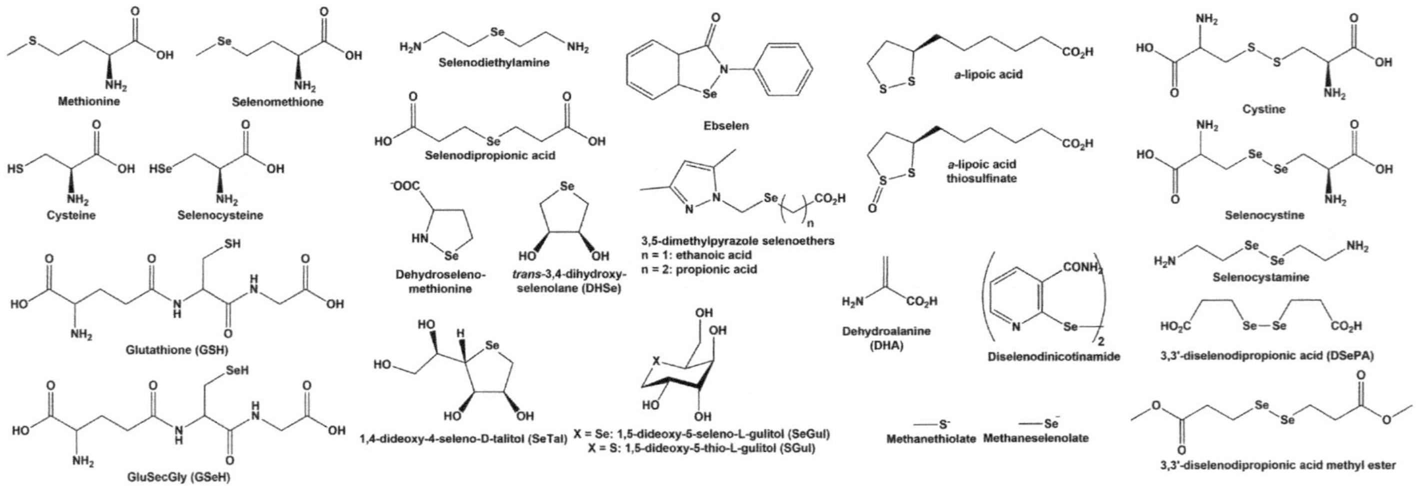

Scheme 9.1 Chemical structures of novel selenium- and sulfur-containing compounds and related species discussed in this chapter.

Table 9.1 Summary of the second-order rate constants (*k*) for the reactions of a range of inflammatory oxidants with selected selenium- and sulfur-containing low molecular mass compounds. All values are reported at pH 7.2–7.8 and at *ca.* 22 °C unless otherwise stated. Errors have not been quoted but are available in the original articles referenced.

Compound	$k(HOCl)$/ $M^{-1} s^{-1}$	$k(HOBr)$/ $M^{-1} s^{-1}$	$k(HOSCN)$/ $M^{-1} s^{-1}$	$k(Gly\text{-}NCl)$/ $M^{-1} s^{-1}$	$k(ONOOH)$/ $M^{-1} s^{-1}$	$k(H_2O_2)$/ $M^{-1} s^{-1}$	$k(^1O_2)$/ $M^{-1} s^{-1}$
Selenomethionine (SeMet)	3.2×10^{8} [a]	1.4×10^{7} [a]	2.8×10^{3} [a]	2.3×10^{3} [b]	2.5×10^{3} [c]		
Methionine (Met)	3.4×10^{7} [d]	3.6×10^{6} [e,f]	Slow	2.0×10^{2} [g]	3.6×10^{2} [h]		
					1.8×10^{2} [j]		
Selenocysteine (Sec)			1.2×10^{6} [j]		1.6×10^{5} [j]		
Cysteine (Cys)	3.6×10^{8} [d]	1.2×10^{7} [e,f]	7.8×10^{4} [p]	3.5×10^{2} [h]	3.8×10^{3} [h]		
					2.4×10^{3} [i]		
Selenoglutathione (GSeH)			1.7×10^{6} [j]				
Glutathione (GSH)			2.5×10^{4} [k]	2.3×10^{2} [g]	1.36×10^{3} [r]		
					7.9×10^{2} [i]		
α-Lipoic acid					1.4×10^{3} [l]		1.38×10^{8} [m]
Ebselen			30 [j]		2×10^{6} [n]	4.8 [o]	2.5×10^{6} [p]
SeTal	1.0×10^{8} [a]	1.5×10^{7} [a]	$\sim1\times10^{2}$ [a]	4.3×10^{2} [b]	2.3×10^{3} [i]		
Dihydroxyselenolane						~0.6 [r]	
Methanethiolate						10.2 [q]	
Methaneselenolate						34.4 [r]	
3,3′-diselenodipropionic acid					1.3×10^{3} [i]	1.4×10^{-2} [s]	
Selenocystamine					7.2×10^{2} [i]	8.9×10^{-4} [s]	

[a] From ref. 119.
[b] From ref. 159.
[c] From ref. 113.
[d] From ref. 65.
[e] From ref. 66.
[f] For *N*-α-acetyl-Met-OMe.
[g] From ref. 170.
[h] From ref. 171 at 37 °C.
[i] From ref. 70.
[j] From ref. 62.
[k] From ref. 63.
[l] From ref. 69 at 37 °C.
[m] From ref. 128.
[n] From ref. 121 at pH > 8.
[o] From ref. 122.
[p] From ref. 172.
[q] From ref. 91.
[r] From ref. 61.
[s] From ref. 150.

Much of the kinetic data available for low molecular mass selenium compounds is for reactions of selenoethers (RSeR′; *e.g.* SeMet) and diselenide (RSeSeR′) species, rather than the selenols (RSeH; *e.g.* Sec), due to the very rapid autoxidation of selenols in the presence of O_2, which complicates studies of these species. Thus, although selenols are likely to be highly reactive and efficacious oxidant scavengers, information for these species is incomplete. A summary of the available kinetic and product data is given below.

9.3.1 Selenols

9.3.1.1 *Radical Reactions*

One-electron oxidation of thiols typically yields thiyl radicals ($RS^•$).[29,30] Thus, glutathione (GSH) reacts rapidly with $HO^•$, $NO_2^•$, $CO_3^{•-}$, and radicals formed on proteins, to give the thiyl radical $GS^•$. $GS^•$ can react with O_2 to form a peroxyl radical in a reversible manner, or with GS^- to form $GSSG^{•-}$.[29–32] This radical-anion undergoes rapid electron transfer with O_2 to form superoxide ($O_2^{•-}$),[33] which can be rapidly detoxified by SOD. As GSH is typically present at millimolar concentrations in cells (typically 2–10 mM), formation of $GS^•$ and the subsequent radical chain reactions to form GSSG and $O_2^{•-}$ may be a viable mechanism for the detoxification of protein- and lipid-bound radicals *in vivo*.[34,35] An alternative pathway for formation of $RS^•$ is the reaction of HOCl with a thiol to form a sulfenyl chloride (RS–Cl) intermediate, which typically undergoes rapid hydrolysis to give a sulfenic acid (RSOH), but can also undergo one-electron reduction with metal ions such as Fe^{2+} to give $RS^•$.[36]

The reactions of free thiols with $HO^•$ occur at the diffusion limit (Table 9.2),[29,30,37,38] but to our knowledge no rate constants for $HO^•$ with free selenols have been determined. A recent report compared the reactivity of Cys and Sec with Tyr-derived free radicals.[39] At pH 7.4, the Tyr phenoxyl radical ($TyrO^•$) generated on *N*-acetyl-Tyr-amide oxidizes Sec residues with a k value that is approximately 1000 times higher than for Cys residues (Table 9.2);[39] the selenium-substituted glutathione (GSeH) also reacted rapidly (Table 9.2). Oxidation of Sec by insulin-bound $TyrO^•$ was decreased relative to the *N*-acetyl-Tyr-amide derived radical, but was still very fast (k, $1.6\times10^8\ M^{-1}\ s^{-1}$); the increased bulk of GSeH resulted in a slower reaction (k, $4\times10^6\ M^{-1}\ s^{-1}$) with the insulin-bound radical, although this was still approximately 400 times faster than the corresponding GSH reaction.[39] These kinetic data suggest that free selenols may be potent scavengers of protein-bound free radicals, with the reactions having similar rate constants to ascorbate and urate.[39,40]

Thiyl radicals can also abstract hydrogen atoms from susceptible sites to form carbon-centred radicals,[41] thereby propagating oxidative damage.[29,30] Thiyl radicals formed on GSH can mediate intramolecular abstraction of hydrogen atoms from multiple sites of the GSH molecule, forming a variety

Table 9.2 Summary of the second-order rate constants (k) for the reactions of a range of radical species with selected selenium- and sulfur-containing low molecular mass compounds. All values are reported at pH 7.2–7.8 and at *ca.* 22 °C unless otherwise stated. Errors have not been quoted but are available in the original articles referenced.

Compound	$k(HO^•)$/ $M^{-1} s^{-1}$	$k(TyrO^•)$/ $M^{-1} s^{-1}$	$k(CO_3^{•-})$/ $M^{-1} s^{-1}$	$k(CCl_3OO^•)$/ $M^{-1} s^{-1}$
Selenomethionine (SeMet)	1.4×10^{10} [a]			
Methionine (Met)	2.3×10^{10} [b]			
Selenocysteine (Sec)		8×10^{8} [c]		
Cysteine (Cys)	4×10^{10} [d]	6×10^{5} [c]		
Selenoglutathione (GSeH)		5×10^{8} [c]		
Glutathione (GSH)	1.4×10^{10} [e]			
α-Lipoic acid	4.7×10^{10} [f]			1.8×10^{8} [f]
Ebselen				2.9×10^{8} [g]
Dihydroxyselenolane	9×10^{9} [h]			
Cystine	2.1×10^{9} [i]			
GSSG	9.3×10^{9} [j]			
Selenocystine	8.1×10^{9} [a]		2.1×10^{9} [a]	
3,3′-Diselenodipropionic acid			3.4×10^{9} [k]	2.7×10^{8} [l]
Selenocystamine			8.4×10^{8} [k]	8.3×10^{6} [k]
Diselenopropionic acid methyl ester			9.6×10^{8} [k]	1.1×10^{8} [k]
Selenodiethylamine				1×10^{8} [m]
Selenodipropionic acid				4.2×10^{8} [m]
3,5-Dimethylpyrazole selenopropionic acid	1.4×10^{10} [n]			7.8×10^{8} [n]

[a]From ref. 93 at pH 7.
[b]From ref. 173.
[c]For $TyrO^•$ generated on *N*-acetyl-Tyr-amide, from ref. 39.
[d]From ref. 37.
[e]From ref. 38.
[f]From ref. 128.
[g]From ref. 95.
[h]From ref. 110.
[i]From ref. 123 at pH 6.5.
[j]From ref. 124 at pH 7.
[k]From ref. 127.
[l]From ref. 147 at pH 7.
[m]From ref. 99.
[n]From ref. 109.

of products.[41] Alternatively, thiyl radicals may abstract hydrogen atoms from proteins, which can lead to protein damage and potentially perturb protein function.[42–44] As an example, thiyl radicals can cause protein cross-linking and aggregation when they dimerize to give disulfides.[45,46] Thiyl radicals can also undergo reversible addition reactions with unsaturated fatty acid side-chains of lipids and triglycerides (as well as hydrogen abstraction reactions), which can lead to *cis–trans* isomerization. As natural fatty acids are found exclusively in the *cis* form, the resulting increase in the *trans* : *cis* ratio has a number of important implications, including perturbation of enzyme function, cellular metabolism, membrane structure and signaling.[47–50]

One-electron oxidation of selenols yields selenyl radicals (RSe•). These have a lower reduction potential than thiyl radicals and are therefore less capable of removing a hydrogen atom from other substrates.[51] It has therefore been suggested that the formation of selenyl rather than thiyl radicals may be less damaging, as selenyl radicals do not propagate radical chain reactions[51,52] and decay primarily *via* radical recombination reactions.

Thiyl radicals can also undergo a desulfonization pathway to form dehydroalanine (DHA), providing an alternative mechanism for DHA formation to the two-electron oxidation of selenols (see later).[41,42] DHA is itself reactive and can undergo Michael addition with other thiols to form covalent thioether adducts.[53] GSH can undergo both desulfonization and addition to other DHA residues, leading to covalent thioether adduction that is often referred to as irreversible glutathionylation. Irreversible glutathionylation of eye lens proteins has been detected, and the levels correlate with the incidence and severity of age-related cataracts.[54,55] DHA residues generated by either one- or two-electron oxidation of Cys and Sec residues have been detected on albumin,[56] GAPDH,[57] GPx,[58] and other selenoproteins,[59] but irreversible glutathionylation of these proteins *via* Michael addition of GSH has not been detected in tissue or *in vivo*. It has been suggested that irreversibly glutathionylated proteins are rapidly turned over by most cells, but due to an absence, or very low rate, of catabolism of lens proteins this tissue accumulates this modification, allowing its detection.[60]

9.3.1.2 Non-radical Reactions

As indicated earlier, obtaining absolute kinetic data for free selenols (RSeH) is complicated due to their rapid auto-oxidation to diselenides (RSe-SeR), but some experimental data and theoretical calculations are available. Computational studies on the reactivity of sulfur- *versus* seleno- molecules (methanethiolate *versus* methaneselenolate) with H_2O_2 suggested only moderate differences in the rate constants (approximately three-fold; 10.2 *vs.* 35.4 $M^{-1}\,s^{-1}$, respectively),[61] and small changes in reaction energy barriers in aqueous solution (~1.8 kJ mol^{-1}).[28] In contrast, experimental rate constants for the reaction of HOSCN with free Sec and Sec-containing peptides (Gly-Sec-Gly, selenoglutathione) are in the range of 10^4–10^6 $M^{-1}\,s^{-1}$ (pH 7.4 and 22 °C)[62] and are 16–80-fold higher than the corresponding thiols (Cys, GSH; Table 9.1).[63] Rate constants for HOCl and HOBr with selenols have not yet been reported in the literature, but initial studies[64] indicate that these reactions are very fast (see Table 9.1). The corresponding rate constants for thiols are $>10^8$ and *ca.* 10^7 $M^{-1}\,s^{-1}$ for HOCl and HOBr, respectively (Table 9.1).[65–67] ONOOH also exhibits enhanced reactivity with selenols compared to thiols (Table 9.1), with rate constants for thiols (*e.g.* GSH) of the order of 10^3 $M^{-1}\,s^{-1}$,[68,69] and for the corresponding selenols, k is in the range 10^5–10^6 $M^{-1}\,s^{-1}$ and up to 830-fold faster than for the corresponding thiols.[70]

Rate constants for the reactions of H_2O_2, ONOOH, and HOSCN with the active site Sec residue in GPx have been determined to be 10^5–10^7 $M^{-1}\,s^{-1}$.[62,71–73] These rate constants, particularly for H_2O_2 ($k \sim 10^7$ M^{-1} s^{-1}),[71] are markedly higher than the predicted reactions with free selenols, and this rate enhancement compared to free Sec or Cys has been suggested to be due to hydrophobic effects and hydrogen bonding interactions.[28]

Two-electron oxidation of thiols by ONOOH and HOCl results in the initial formation of sulfenic acid (RSOH) species;[74] oxidation of selenols is believed to form the corresponding selenenic acids (RSeOH). In the case of reaction with HOCl, chloramines, and HOSCN, intermediates containing a sulfur/selenium-(pseudo)halogen (Cl/SCN) bond (*e.g.* RS–Cl and RSeCl) are believed to be formed and undergo rapid hydrolysis to the sulfenic acid, but there is little direct evidence for these intermediates on low molecular mass materials under physiologically relevant conditions (*i.e.* in aqueous solution at physiological pH).[7,75–80] These intermediates undergo further reactions to yield disulfides/diselenides (RSe-SeR), mixed seleno-sulfur species (RSe-SR), and higher sulfur or selenium oxyacids (*i.e.* sulfinic (RSO_2H) and sulfonic (RSO_3H) acids and their corresponding selenium derivatives).[81] In the case of the selenenic acids, these reactions are rapid[82] and no direct evidence for their formation has been obtained under aqueous conditions for low molecular mass species. However, the chemical properties of a stable low molecular mass selenenic acid have been reported in organic solvents.[83] In addition, a selenenic acid intermediate in the active site of bovine GPx1 has recently been trapped with dimedone and characterized by liquid chromatography mass spectrometry following tryptic protein digestion.[84] This protein-bound selenenic acid displayed unexpected stability in the absence of oxidizing or reducing agents, and could still be trapped >10 min after formation, suggesting the protein structure effectively shields it from further reactions.[84]

Oxidation of Sec residues on proteins has provided evidence for the formation of a further product, dehydroalanine (DHA; detected *via* a blotting method[85]), *via* formation of seleninic ($RSeO_2H$) and selenonic ($RSeO_3H$) acids and subsequent β-elimination (Scheme 9.1).[81] Se-phenylselenocysteine (Scheme 9.1) and related low molecular mass selenium-containing species can also form DHA residues by reacting with H_2O_2 and undergoing selenoxide elimination.[86–89] The factors that control the product distribution arising from oxidation of Sec have yet to be fully established, but are likely to include steric and electronic factors (see ref. 51).

9.3.2 Selenoethers and Thioethers

9.3.2.1 Radical Reactions

Selenoethers (selenides) with the general structure R-Se-R′ are some of the most common selenium species encountered, with selenomethionine

(SeMet) being the most widely distributed in nature. SeMet can be non-selectively incorporated into proteins in place of Met, and is thought to play a protective role *in vivo*. The major product of selenoether oxidation appears to be the selenoxide in most cases. The latter are readily reduced (*e.g.* by thiols, see later), which allows for catalytic recycling of the selenoether, and efficient scavenging of oxidants. As such, there has been interest in developing selenides as potential therapeutics in inflammatory conditions, with cyclic selenides demonstrating rapid oxidant consumption, low toxicity, and efficacy in ameliorating inflammatory conditions.[90–92]

SeMet reacts rapidly with one-electron oxidants including $HO^•$, $ROO^•$, and inorganic radicals.[93–95] The reaction with the $HO^•$ occurs at diffusion-controlled rates.[93,94] One-electron reactions result in the formation of the SeMet cation radical, which can be stabilized by a three-electron interaction between the Se centre and the amino acid nitrogen, or *via* interaction with another SeMet to give a dimeric three-electron bonded species.[96,97] Subsequent reactions of these species with O_2 gives selenomethionine selenoxide (SeMetO) as a major product, although this is not a quantitative process.[93,95] Related studies have been undertaken with model compounds.[98] Selenoethers with the general formula $Se(CH_2CH_2X)_2$, where X is NH_3, OH, or CO_2^- react rapidly with model peroxyl radicals such as $CCl_3OO^•$ with rate constants of 1×10^8 M^{-1} s^{-1} where $X = NH_3^+$ and 4.2×10^8 for $X = CO_2^-$.[99] The reasons for increased rate of reaction for $X = CO_2^-$ have been investigated; it has been shown that a three-electron bond is formed between the carboxyl group and the selenium atom, resulting in a stabilization of the intermediate in a similar manner to SeMet. Dimeric radical-cations are formed which then undergo hydrolysis yielding the selenoxide as a major product. These species can also undergo decarboxylation reactions, resulting in an alkyl radical.[99]

Ebselen, a benzisoselenazone, has been investigated extensively as a possible therapeutic compound against stroke and other inflammatory disease (reviewed in ref. 19, 100, 101). This species reacts with $CCl_3OO^•$ with $k = 2.9 \times 10^8$ M^{-1} s^{-1};[95] however it was found to be ineffective in preventing the oxidation of lipids induced by free radical initiators, suggesting a much slower rate of reaction with other peroxyl radicals.[102,103] Ebselen derivatives have been shown to scavenge peroxyl radicals with an efficacy greater than α-tocopherol, with a selenoxide product proposed.[104] However, ebselen undergoes metabolism, ring opening, and degradation, which can give rise to toxicity, and as a consequence multiple analogues have been developed to increase its efficacy and reduce toxicity. This area has been extensively reviewed.[19,105–108] Selenoether derivatives of 3,5-dimethylpyrazole have been developed, which show enhanced oxidant reactivity, while minimizing undesired side reactions such as decarboxylation and deamination.[109] A propionic acid derivative has been shown to react with $HO^•$ with $k = 1.4 \times 10^{10}$ M^{-1} s^{-1} and with $CCl_3OO^•$ with $k = 7.8 \times 10^8$ M^{-1} s^{-1}.[109] *Trans*-3,4-dihydroxyselenolane (DHS) is a five-membered cyclic selenoether (selenide) which reacts with $HO^•$ with $k = 9 \times 10^9$ M^{-1} s^{-1},[110] and has significant GPx activity.[90,111] This compound also gives rise to accelerated healing of indomethacin-derived stomach ulcers, although it is as yet unclear whether this is due to an oxidant scavenging/removal activity or other pathways.[90] A related five-membered selenosugar, SeTal, has recently been reported to react with $O_2^{\bullet -}$ and H_2O_2 though the kinetics of these reactions were not determined; these reactions are likely to be slow.[92]

9.3.2.2 *Non-radical Reactions*

SeMet reacts rapidly with a wide variety of two-electron oxidants, including the MPO-derived hypohalous acids HOCl, HOBr, and HOSCN, with rate constants of 3.2×10^8, 1.4×10^7, and 2.8×10^3 $M^{-1}\,s^{-1}$, respectively.[65] SeMet also reacts at significant rates with chloramines (RNHCl species) that are formed on reaction of HOCl and amine groups, with rate constants k in the range 10^2–10^3 $M^{-1}\,s^{-1}$.[112] SeMet reacts with ONOOH with a rate constant of 2.5×10^3 $M^{-1}\,s^{-1}$,[113] as well as H_2O_2 and other organic peroxides, although the rate constants for these reactions have not been determined.[114] The major product arising from two-electron oxidation of SeMet is the selenoxide, selenomethionine selenoxide (SeMetO), with this being formed in near-stoichiometric yields.[112–116] However, oxidation of SeMet by HOCl can also result in the formation of a cyclic dehydromethionine species,[112] analogous to the species detected for the parent amino acid Met.[117]

A series of stable, highly water soluble selenium sugars have been developed, and the reactivity of these species with two-electron oxidants has been explored.[118,119] The five-membered ring species SeTal reacts rapidly with HOCl and HOBr with rate constants of 1.0×10^8 and 1.5×10^7 $M^{-1}\,s^{-1}$, respectively, but much more slowly with HOSCN, with a rate constant of 2.8×10^3 $M^{-1}\,s^{-1}$.[62,65,119] SeTal also reacts with chloramines with rate constants of a similar magnitude to SeMet (*i.e.* 10^2–10^3 $M^{-1}\,s^{-1}$).[112] Some of the six-membered ring selenocarbohydrate species, such as SeGul, are somewhat less reactive with the MPO-derived oxidants and chloramines, with this thought to be due to a lower level of steric strain around the selenium centre.[65,112,119] These cyclic selenosugars also react at appreciable rates with ONOOH, with the rate constants for these reactions being around 2.5×10^3 $M^{-1}\,s^{-1}$, and approximately 10-fold higher than for the corresponding sulfur species.[70] The rate constants for reaction of these selenosugars with H_2O_2 have not been determined, but the related species DHS reacts with $k\sim0.6$ $M^{-1}\,s^{-1}$.[91] The major oxidation products of these reactions of SeTal and DHS are the corresponding selenoxides.[111,112,120] Ebselen has also been shown to be an effective scavenger of two-electron oxidants, with this species reacting with ONOOH with k 2×10^6 $M^{-1}\,s^{-1}$,[121] although its reaction with H_2O_2 is relatively slow ($k=4.8$ $M^{-1}\,s^{-1}$; compare with the value reported above for DHS), with this resulting in the formation of the selenoxide.[122] In addition, ebselen has been reported to be capable of reacting with other peroxides, with this resulting in an inhibition of lipid oxidation, although the kinetics of these reactions have not been characterized.[102]

9.3.3 Diselenides and Disulfides

Less is known about the reactivity of diselenides, compared to the selenoethers/selenides; however, there is growing interest in this class of compounds as potential therapeutic agents, due to their greater stability and relative efficacy as antioxidants.

9.3.3.1 *Radical Reactions*

One-electron oxidation of the disulfides cystine and GSSG by $HO^•$ occurs with $k = 2.1 \times 10^9$ M^{-1} s^{-1} (at pH 6.5[123]) and 9.3×10^9 M^{-1} s^{-1} (at pH 7[124]), respectively, which is slightly slower than for the corresponding thiols. Rate constants for reaction of a number of diselenides have also been reported, with k for selenocystine and $HO^•$ being 8.1×10^9 M^{-1} s^{-1} at pH 7,[125] which is approximately four-fold faster than for the sulfur analogue. The rate constant for reaction of 3,3′-diselenodipropionic acid (DSePA) with the model peroxyl radical ($CCl_3OO^•$) has been reported as 2.7×10^8 M^{-1} s^{-1} at pH 7.[126] The latter value indicates that DSePA reacts with similar rate constants to those for chain-breaking antioxidants such as α-tocopherol and ascorbate, suggesting that diselenides may be effective radical scavengers in some circumstances. These reactions generate transient species that have absorption maxima in the range 450–550 nm.[125,126] This absorption band and the kinetic behaviour of the intermediates that give rise to it are consistent with the formation of a $RSeSeR^{•+}$ radical-cation, with this being stabilized by interactions with the carboxylate oxygen atoms.[125] These species are stabilized relative to the corresponding disulfide radical-cations, with the first-order rate constant for the decay of the selenocystine radical-cation being 2.4×10^3 s^{-1}.[125] Selenocystine and the related diselenides selenocystamine, diselenopropionic, and diselenopropionic acid methyl ester, also react rapidly with $CO_3^{-•}$ with $k = 2.1$, 3.4, 0.84, and 0.96×10^9 M^{-1} s^{-1}, respectively.[127] These diselenides also react with $CCl_3OO^•$, although with a greater variation in rate constant, with the values for selenocystamine, diselenopropionic, and diselenopropionic acid methyl ester being 8.3×10^6, 1.7 and 1.1×10^8 M^{-1} s^{-1}, respectively.[127]

Reaction of α-lipoic acid, a cyclic disulfide compound, with $HO^•$, model peroxyl radicals ($CCl_3OO^•$; $k = 1.8 \times 10^8$ M^{-1} s^{-1}),[128] and singlet oxygen are all fast processes (Tables 9.1 and 9.2), but the reactions with H_2O_2 and $O_2^{•-}$ are orders of magnitude slower and have not been determined accurately.[128]

9.3.3.2 *Non-radical Reactions*

Disulfide and diselenide oxidation typically occurs at slower rates than for thiols/selenols and thio-/seleno-ethers due to their partially oxidized state. Oxidation of disulfides such as oxidized glutathione (GSSG) has been investigated due to the potential implications of loss of GSSG on the maintenance of cellular thiol/disulfide pools and redox balance. Cyclic disulfides such as α-lipoic acid and α-lipoamide have received considerable attention due to their purported antioxidant properties, and their use in several clinical trials.[69,128]

Many two-electron oxidants react relatively slowly with disulfides such as GSSG and cystine, with the rates of endogenous decay of ONOOH and HOSCN not perturbed by the presence of most disulfides.[63,69] In contrast, HOCl and HOBr react readily with disulfides such as 3,3′-dithiodipropionic acid and *N,N*′-acetyl-cystine, with $k \sim 2\times10^5$ and 10^6 $M^{-1}s^{-1}$, respectively.[65,66,129] Furthermore, recent data indicate that there is considerable variation (by up to four orders of magnitude) in the reactivity of disulfides as a result of their conformation, and stabilization by suitably positioned lone pairs of electrons. The reasons for these large changes in reactivity have been elucidated.[130] ONOOH also reacts at a moderate rate with α-lipoic acid to generate a disulfide-*S*-oxide (thiosulfinate; Scheme 9.1) that can be readily reduced by thiols.[69]

Oxidation of unstrained disulfides by H_2O_2,[131] singlet oxygen,[132,133] and ONOOH[134] have also been reported to give disulfide mono- and di-oxides as initial products. The disulfide-*S*-oxides (RS(O)SR) are weak oxidants that can oxidize thiols in dehydrogenases (like the lipoic acid species, see earlier),[135] and disrupt zinc–sulfur clusters, resulting in enzyme inactivation.[136] Disulfide-*S*-oxides are also capable of thionylating proteins,[137] such as protein kinase C, resulting in inhibition of kinase function,[138] and potential modulation of redox signaling pathways.[137,139]

Diselenides have been gaining increasing interest, as these provide a stable alternative to selenols and selenoethers for the development of organic selenium compounds as therapeutic antioxidants and GPx mimetics (*e.g.* ref. 18, 96, 140–148). Studies investigating the effects of substitution of functional groups on aromatic diselenides demonstrate that the reactivity of the selenium centre is dependent on the surrounding electronic environment.[149] Selenocystamine and related diselenides react with ONOOH with second-order rate constants in the range of 500–1100 $M^{-1}s^{-1}$.[70] The reaction of diselenide species selenocystamine and diselenodipropionic acid with H_2O_2 is slow, with second-order rate constants of 8.9×10^{-4} and 1.4×10^{-2} $M^{-1}s^{-1}$, respectively.[150] Oxidation of GSeSeG and Fmoc-selenocystine by HOCl cleaves the diselenide bond and results in the formation of seleninic acid products.[150]

Although the absolute rate constants for direct reaction of diselenides with H_2O_2 do not appear to have been reported, the ability of GSeSeG (the selenium analogue of GSSG) and other diselenide-containing peptides to consume H_2O_2 has been shown *via* an enzyme-coupled assay.[151] A study has shown that diselenodinicotinamide exhibits potent GPx mimetic activity and can prevent oxidant-induced damage *in vitro*.[152] Direct scavenging of H_2O_2 by this diselenide is slow, and it has been reported that it is the formation of a stabilized selone species (RC=Se) on reduction of diselenodinicotinamide by GSH that drives the GPx activity.[153] It has also been shown that non-bonding interactions between Se centres and nucleophilic groups such as amines, amides or alcohols can modulate the reactivity of many selenium groups,[154] as shown between H_2O_2 and an aromatic diselenide.[155]

9.4 Recycling and Repair of Selenium and Sulfur Oxidation Products

The primary oxidation products of selenides by both one- and two-electron oxidants are selenoxides. Selenoxides can rapidly be reduced by thiol species *via* a two-step reduction process.[156–158] The proposed mechanism involves an initial thiol that reacts with the selenium, forming a selenosulfide species. The second thiol then reacts with the selenosulfide bond, reforming the parent selenide, and producing a disulfide and water. The rate constant for the initial attack of GSH onto SeMetO and related selenoxides ranges between 10^2 and 10^3 $M^{-1}s^{-1}$.[159] Other reducing agents, such as GSH, L-cysteine, *N*-acetyl-L-cysteine, ascorbic acid, and methimazole can also reduce SeMetO.[158]

Ebselen is able to act as a GPx mimetic and undergoes a catalytic oxidation cycle in the presence of thiols, with two potential mechanisms. GSH can react with oxidized ebselen to form a selenosulfide species, which can subsequently be reduced by a second thiol, reforming GSSG and reduced ebselen.[121] Alternatively, GSH can react with reduced ebselen to form a selenosulfide, which can react with a second GSH to form GSSG and a selenol. The ebselen selenol species can then react with oxidants form a selenenic species. The selenenic acid can either reform ebselen eliminating water or react with GSH to form the selenosulfide species.[160–162] Substitution on the phenyl group has been investigated in attempt to enhance the GPx-mimetic capability, by varying the strength of the Se–N bond to stabilize the selenol while maintaining easy oxidation of the selenium centre.[107,108,163]

SeMetO can also be reduced by thioredoxin reductase at the expense of NADPH.[114] In the presence of thioredoxin, the rate of NADPH consumption increases, suggesting an enhanced removal of selenoxide.[114] However, the cyclic selenoxide SeTalO formed on oxidation of the selenosugar SeTal is not reduced at a significant rate by these enzymes.[159] As GSSG is a product of the reduction of selenoxides, coupling the glutathione reductase system allows for the rapid reduction of GSSG to GSH, facilitating a catalytic reduction cycle. Addition of SeMet to the glutathione reductase system increases the rate of consumption of NADPH upon addition of taurine chloramine, suggesting that SeMet enhances the capacity of glutathione reductase to remove oxidants.[159]

Diselenides are also capable of interaction with antioxidant enzymes. Selenocystine can be reduced by Trx to form selenocysteine, which can then rapidly react with oxidants to reform the diselenide.[164–166] GSH is also able to initiate this cycle for selenocystine as well as other diselenides, including GSeSeG and a nicotinamide-derived diselenide.[148,152] Ebselen also reacts with Trx, although the bond that is reduced in this case is the Se–N bond, with this giving a selenol.[161,164]

9.5 Conclusions

There is now a vast body of evidence supporting the hypothesis that sulfur- and selenium-containing species are crucial to maintaining redox balance in living organisms. The kinetic and chemical data reviewed in this chapter indicate that low molecular mass selenium-containing compounds offer potential as oxidant removal systems due to their favourable kinetic, nucleophilic, and reduction/recycling properties, provided that potential issues of toxicity and non-specificity can be overcome (reviewed in ref. 18, 19, 21, 167). A number of studies have been reported on the design and synthesis of low molecular mass selenium-containing compounds that might act as effective oxidant scavengers, and a number of these have provided strong supporting data on such actions. Many of these developments have arisen through careful consideration of the non-bonding interactions between the selenium atom and nucleophilic nitrogen or oxygen centres, and how this may affect reactivity, stability, and metabolism.[105,154,168] These strategies, and others (*e.g.* ref. 111, 118, 142, 148, and 169), may limit unwanted metabolism of selenium compounds, and thereby limit downstream toxicity. However, further work is clearly needed in this area.

Abbreviations

DHA	dehydroalanine
DSePA	3,3′-diselenodipropionic acid
GAPDH	glyceraldehyde-3-phosphate dehydrogenase
GPx	glutathione peroxidase
GSH	L-glutathione
GSSG	glutathione disulfide
MetSO	methionine sulfoxide
Prx	peroxiredoxin
Sec	selenocysteine
SeGul	1,5-dideoxy-5-seleno-L-gulitol
SGul	1,5-dideoxy-5-thio-L-gulitol
SeMet	seleno-L-methionine
SeMetO	seleno-L-methionine selenoxide
SeTal	1,4-dideoxy-4-seleno-D-talitol
Trx	thioredoxin
TrxR	thioredoxin reductase

Acknowledgements

The authors are grateful to the Novo Nordisk Foundation (grants NNF13OC0004294 and NNF15OC0018300 to MJD) for financial support. The authors do not have any conflicts of interest to declare.

References

1. B. Halliwell and J. M. C. Gutteridge, *Free Radicals in Biology & Medicine*, Oxford University Press, Oxford, 5th edn, 2015.
2. G. R. Drummond and C. G. Sobey, *Trends Endocrinol. Metabolism*, 2014, **25**, 452–463.
3. E. Cifuentes-Pagano, D. N. Meijles and P. J. Pagano, *Antioxid. Redox Signaling*, 2014, **20**, 2741–2754.
4. H. J. H. Fenton, *J. Chem. Soc.*, 1894, **65**, 899–910.
5. M. J. Davies, *Photochem. Photobiol. Sci.*, 2004, **3**, 17–25.
6. M. J. Davies, C. L. Hawkins, D. I. Pattison and M. D. Rees, *Antioxid. Redox Signaling*, 2008, **10**, 1199–1234.
7. C. L. Hawkins, D. I. Pattison and M. J. Davies, *Amino Acids*, 2003, **25**, 259–274.
8. D. I. Pattison and M. J. Davies, *Curr. Med. Chem.*, 2006, **13**, 3271–3290.
9. T. J. Barrett and C. L. Hawkins, *Chem. Res. Toxicol.*, 2012, **25**, 263–273.
10. C. C. Winterbourn, *Nat. Chem. Biol.*, 2008, **4**, 278–286.
11. P. G. Furtmuller, M. Zederbauer, W. Jantschko, J. Helm, M. Bogner, C. Jakopitsch and C. Obinger, *Arch. Biochem. Biophys.*, 2006, **445**, 199–213.
12. J.-A. Stubbe, *Chem. Commun.*, 2003, **20**, 2511–2513.
13. S. Carballal, S. Bartesaghi and R. Radi, *Biochim. Biophys. Acta*, 2014, **1840**, 768–780.
14. D. I. Pattison, A. S. Rahmanto and M. J. Davies, *Photochem. Photobiol. Sci.*, 2012, **11**, 38–53.
15. B. S. van der Veen, M. P. de Winther and P. Heeringa, *Antioxid. Redox Signaling*, 2009, **11**, 2899–2937.
16. R. J. Hondal, S. M. Marino and V. N. Gladyshev, *Antioxid. Redox Signaling*, 2013, **18**, 1675–1689.
17. C. Jacob, G. I. Giles, N. M. Giles and H. Sies, *Angew. Chem., Int. Ed.*, 2003, **42**, 4742–4758.
18. L. Orian and S. Toppo, *Free Radical Biol. Med.*, 2014, **66**, 65–74.
19. G. K. Azad and R. S. Tomar, *Mol. Biol. Rep.*, 2014, **41**, 4865–4879.
20. C. M. Weekley and H. H. Harris, *Chem. Soc. Rev.*, 2013, **42**, 8870–8894.
21. E. E. Battin and J. L. Brumaghim, *Cell Biochem. Biophys.*, 2009, **55**, 1–23.
22. G. W. Snider, E. Ruggles, N. Khan and R. J. Hondal, *Biochemistry*, 2013, **52**, 5472–5481.
23. J. D. Chandler, D. P. Nichols, J. A. Nick, R. J. Hondal and B. J. Day, *J. Biol. Chem.*, 2013, **288**, 18421–18428.
24. L. Zhong and A. Holmgren, *J. Biol. Chem.*, 2000, **275**, 18121–18128.
25. J. R. Gasdaska, J. W. Harney, P. Y. Gasdaska, G. Powis and M. J. Berry, *J. Biol. Chem.*, 1999, **274**, 25379–25385.
26. J. Martinez, S. Lisa, R. Sanchez, W. Kowalczyk, E. Zurita, M. Teixido, E. Giralt, D. Andreu, J. Avila and M. Gasset, *PLos One*, 2011, **6**.
27. P. Nagy, A. Karton, A. Betz, A. V. Peskin, P. Pace, R. J. O'Reilly, M. B. Hampton, L. Radom and C. C. Winterbourn, *J. Biol. Chem.*, 2011, **286**, 18048–18055.

28. B. Cardey and M. Enescu, *J. Phys. Chem. A*, 2007, **111**, 673–678.
29. C. Schoneich, *Free Radical Res.*, 2016, **50**, 143–149.
30. M. Trujillo, B. Alvarez and R. Radi, *Free Radical Res.*, 2016, **50**, 150–171.
31. X. J. Zhang, N. Zhang, H. P. Schuchmann and C. Vonsonntag, *J. Phys. Chem.*, 1994, **98**, 6541–6547.
32. R. Zhao, J. Lind, G. Merenyi and T. E. Eriksen, *J. Am. Chem. Soc.*, 1994, **116**, 12010–12015.
33. R. L. Willson, *J. Chem. Soc. D*, 1970, 1425–1426.
34. C. C. Winterbourn, *Free Radical Biol. Med.*, 1993, **14**, 85–90.
35. W. H. Koppenol, *Free Radical Biol. Med.*, 1993, **14**, 91–94.
36. M. J. Davies and C. L. Hawkins, *Free Radical Res.*, 2000, **33**, 719–729.
37. G. Nucifora, B. Smaller, R. Remko and E. C. Avery, *Radiat. Res.*, 1972, **49**, 96–111.
38. T. E. Eriksen and G. Fransson, *J. Chem. Soc. Perkin Trans. 2*, 1988, 1117–1122.
39. D. Steinmann, T. Nauser, J. Beld, M. Tanner, D. Gunther, P. L. Bounds and W. H. Koppenol, *Biochemistry*, 2008, **47**, 9602–9607.
40. J. M. Gebicki, T. Nauser, A. Domazou, D. Steinmann, P. L. Bounds and W. H. Koppenol, *Amino Acids*, 2010, **39**, 1131–1137.
41. T. Nauser, W. H. Koppenol and C. Schoneich, *J. Phys. Chem. B*, 2012, **116**, 5329–5341.
42. O. Mozziconacci, B. A. Kerwin and C. Schoneich, *J. Phys. Chem., B*, 2011, **115**, 12287–12305.
43. O. Mozziconacci, B. A. Kerwin and C. Schoneich, *J. Phys. Chem. B*, 2010, **114**, 6751–6762.
44. O. Mozziconacci, T. D. Williams, B. A. Kerwin and C. Schoneich, *J. Phys. Chem. B*, 2008, **112**, 15921–15932.
45. D. A. Armstrong, *Applications of pulse-radiolysis for the study of short-lived sulfur species*, 1990.
46. P. Wardman and C. Vonsonntag, *Biothiols, Part A*, 1995, **251**, 31–45.
47. C. Ferreri, S. Kratzsch, L. Landi and O. Brede, *Cell. Mol. Life Sci.*, 2005, **62**, 834–847.
48. S. Kratzsch, K. Drossler, H. Sprinz and O. Brede, *Arch. Biochem. Biophys.*, 2003, **416**, 238–248.
49. C. Chatgilialoglu, C. Ferreri, M. Melchiorre, A. Sansone and A. Torreggiani, *Chem. Rev.*, 2014, **114**, 255–284.
50. I. N. Lykakis, C. Ferreri and C. Chatgilialoglu, *Angew. Chem. Int. Ed.*, 2007, **46**, 1914–1916.
51. T. Nauser, D. Steinmann and W. H. Koppenol, *Amino Acids*, 2012, **42**, 39–44.
52. T. Nauser, W. H. Koppenol and C. Schoneich, *Free Radical Biol. Med.*, 2015, **80C**, 158–163.
53. B. N. Naidu, M. E. Sorenson, T. P. Connolly and Y. Ueda, *J. Org. Chem.*, 2003, **68**, 10098–10102.
54. M. Linetsky and R. D. LeGrand, *Mol. Cell. Biochem.*, 2005, **272**, 133–144.
55. M. Linetsky, J. M. W. Hill, R. D. LeGrand and F. Hu, *Exp. Eye Res.*, 2004, **79**, 499–512.

56. R. Bar-Or, L. T. Rael and D. Bar-Or, *Rapid Commun. Mass Spectrom.*, 2008, **22**, 711–716.
57. J. Jeong, Y. Jung, S. Na, J. Jeong, E. Lee, M.-S. Kim, S. Choi, D.-H. Shin, E. Paek, H.-Y. Lee and K.-J. Lee, *Mol. Cell. Proteomics*, 2011, **10**.
58. C.-S. Cho, S. Lee, G. T. Lee, H. A. Woo, E.-J. Choi and S. G. Rhee, *Antioxid. Redox Signaling*, 2010, **12**, 1235–1246.
59. S. Ma, R. M. Caprioli, K. E. Hill and R. F. Burk, *J. Am. Soc. Mass Spectrom.*, 2003, **14**, 593–600.
60. D. M. Townsend, V. I. Lushchak and A. J. L. Cooper, in *Redox and Cancer Pt A*, ed. D. M. Townsend and K. D. Tew, 2014, vol. 122, pp. 177–198.
61. B. Cardey and M. Enescu, *ChemPhysChem*, 2005, **6**, 1175–1180.
62. O. Skaff, D. I. Pattison, P. E. Morgan, R. Bachana, V. K. Jain, K. I. Priyadarsini and M. J. Davies, *Biochem. J.*, 2012, **441**, 305–316.
63. O. Skaff, D. I. Pattison and M. J. Davies, *Biochem. J.*, 2009, **422**, 111–117.
64. M. T. Ignasiak and M. J. Davies, unpublished data.
65. C. Storkey, M. J. Davies and D. I. Pattison, *Free Radical Biol. Med.*, 2014, **73**, 60–66.
66. D. I. Pattison and M. J. Davies, *Biochemistry*, 2004, **43**, 4799–4809.
67. D. I. Pattison and M. J. Davies, *Curr. Med. Chem.*, 2006, **13**, 3271–3290.
68. W. H. Koppenol, J. J. Moreno, W. A. Pryor, H. Ischiropoulos and J. S. Beckman, *Chem. Res. Toxicol.*, 1992, **5**, 834–842.
69. M. Trujillo and R. Radi, *Arch. Biochem. Biophys.*, 2002, **397**, 91–98.
70. C. Storkey, D. I. Pattison, M. T. Ignasiak, C. H. Schiesser and M. J. Davies, *Free Radical Biol. Med.*, 2015, **89**, 1049–1056.
71. G. Takebe, J. Yarimizu, Y. Saito, T. Hayashi, H. Nakamura, J. Yodoi, S. Nagasawa and K. Takahashi, *J. Biol. Chem.*, 2002, **277**, 41254–41258.
72. K. Briviba, R. Kissner, W. H. Koppenol and H. Sies, *Chem. Res. Toxicol.*, 1998, **11**, 1398–1401.
73. S. Padmaja, G. L. Squadrito and W. A. Pryor, *Arch. Biochem. Biophys.*, 1998, **349**, 1–6.
74. L. Turell, H. Botti, S. Carballal, G. Ferrer-Sueta, J. M. Souza, R. Duran, B. A. Freeman, R. Radi and B. Alvarez, *Biochemistry*, 2008, **47**, 358–367.
75. T. J. Barrett, D. I. Pattison, S. E. Leonard, K. S. Carroll, M. J. Davies and C. L. Hawkins, *Free Radical Biol. Med.*, 2012, **52**, 1075–1085.
76. S. L. Algunindigue Nimmo, K. Lemma and M. T. Ashby, *Heteroat. Chem.*, 2007, **18**, 467–471.
77. M. T. Ashby and H. Aneetha, *J. Am. Chem. Soc.*, 2004, **126**, 10216–10217.
78. K. Lemma and M. T. Ashby, *J. Org. Chem.*, 2008, **73**, 3017–3023.
79. P. Nagy, K. Lemma and M. T. Ashby, *J. Org. Chem.*, 2007, **72**, 8838–8846.
80. P. Nagy, X. Wang, K. Lemma and M. T. Ashby, *J. Am. Chem. Soc.*, 2007, **129**, 15756–15757.
81. C. S. Cho, S. Lee, G. T. Lee, H. A. Woo, E. J. Choi and S. G. Rhee, *Antioxid. Redox Signaling*, 2010, **12**, 1235–1246.
82. R. J. Hondal and E. L. Ruggles, *Amino Acids*, 2011, **41**, 73–89.
83. Z. Zielinski, N. Presseau, R. Amorati, L. Valgimigli and D. A. Pratt, *J. Am. Chem. Soc.*, 2014, **136**, 1570–1578.

84. F. Li, J. Liu and S. Rozovsky, *Free Radical Biol. Med.*, 2014, **76**, 127–135.
85. N. M. Okeley, Y. Zhu and W. A. van Der Donk, *Org. Lett.*, 2000, **2**, 3603–3606.
86. J. Wang, S. M. Schiller and P. G. Schultz, *Angew. Chem. Inter. Ed.*, 2007, **46**, 6849–6851.
87. J. Guo, J. Wang, J. S. Lee and P. G. Schultz, *Angew. Chem., Int. Ed.*, 2008, **47**, 6399–6401.
88. K. Hashimoto, M. Sakai, T. Okuno and H. Shirahama, *Chem. Commun.*, 1996, 1139–1140.
89. M. Sakai, K. Hashimoto and H. Shirahama, *Heterocycles*, 1997, **44**, 319–324.
90. S. Chakraborty, S. K. Yadav, M. Subramanian, K. I. Priyadarsini, M. Iwaoka and S. Chattopadhyay, *Free Radical Res*, 2012, **46**, 1378–1386.
91. F. Kumakura, B. Mishra, K. I. Priyadarsini and M. Iwaoka, *Eur. J. Org. Chem.*, 2010, 440–445.
92. H. H. Ng, C. H. Leo, K. O'Sullivan, S. A. Alexander, M. J. Davies, C. H. Schiesser and L. J. Parry, *Biochem. Pharmacol.*, 2017, **128**, 34–45.
93. B. Mishra, K. I. Priyadarsini and H. Mohan, *J. Phys. Chem. A*, 2006, **110**, 1894–1900.
94. B. Mishra, A. Sharma, S. Naumov and K. I. Priyadarsini, *J. Phys. Chem. B*, 2009, **113**, 7709–7715.
95. C. Schoneich, V. Narayanaswami, K. D. Asmus and H. Sies, *Arch. Biochem. Biophys.*, 1990, **282**, 18–25.
96. K. I. Priyadarsini and B. Mishra, *Radiat. Phys. Chem.*, 2008, **77**, 1294–1299.
97. K. I. Priyadarsini and B. Mishra, *J. Indian Chem. Soc.*, 2010, **87**, 147–157.
98. T. Tobien, M. Bonifacic, S. Naumov and K. D. Asmus, *Phys. Chem. Chem. Phys.*, 2010, **12**, 6750–6758.
99. P. Prabhu, P. P. Bag, B. G. Singh, A. Hodage, V. K. Jain, M. Iwaoka and K. I. Priyadarsini, *Free Radical Res.*, 2011, **45**, 461–468.
100. A. Saeed, F. A. Larik and P. A. Channar, *Mini-Rev. Org. Chem.*, 2016, **13**, 312–324.
101. K. P. Bhabak and G. Mugesh, *Acc. Chem. Res.*, 2010, **43**, 1408–1419.
102. N. Noguchi, Y. Yoshida, H. Kaneda, Y. Yamamoto and E. Niki, *Biochem. Pharmacol.*, 1992, **44**, 39–44.
103. M. Morita, Y. Naito, T. Yoshikawa and E. Niki, *Bioorg. Med. Chem. Lett.*, 2016, **26**, 5411–5417.
104. S. Kumar, J. Yan, J.-F. Poon, V. P. Singh, X. Lu, M. Karlsson Ott, L. Engman and S. Kumar, *Angew. Chem., Int. Ed.*, 2016, **55**, 3729–3733.
105. K. Selvakumar, P. Shah, H. B. Singh and R. J. Butcher, *Chem. - Eur. J.*, 2011, **17**, 12741–12755.
106. S. J. Balkrishna, S. Kumar, G. K. Azad, B. S. Bhakuni, P. Panini, N. Ahalawat, R. S. Tomar, M. R. Detty and S. Kumar, *Org. Biomol. Chem.*, 2014, **12**, 1215–1219.
107. K. Satheeshkumar and G. Mugesh, *Chem. - Eur. J.*, 2011, **17**, 4849–4857.
108. K. P. Bhabak and G. Mugesh, *Chemistry*, 2007, **13**, 4594–4601.

109. A. Barik, B. G. Singh, A. Sharma, V. K. Jain and K. I. Priyadarsini, *J. Phys. Chem. A*, 2014, **118**, 10179–10187.
110. B. G. Singh, E. Thomas, F. Kumakura, K. Dedachi, M. Iwaoka and K. I. Priyadarsini, *J. Phys. Chem. A*, 2010, **114**, 8271–8277.
111. F. Kumakura, B. Mishra, K. I. Priyadarsini and M. Iwaoka, *Eur. J. Org. Chem.*, 2010, 440–445.
112. L. Carroll, D. I. Pattison, S. Fu, C. H. Schiesser, M. J. Davies and C. L. Hawkins, *Free Radical Biol. Med.*, 2015, **84**, 279–288.
113. S. Padmaja, G. L. Squadrito, J. N. Lemercier, R. Cueto and W. A. Pryor, *Free Radical Biol. Med.*, 1996, **21**, 317–322.
114. A. Suryo Rahmanto and M. J. Davies, *Free Radical Biol. Med.*, 2011, **51**, 2288–2299.
115. B. Gammelgaard, C. Cornett, J. Olsen, L. Bendahl and S. H. Hansen, *Talanta*, 2003, **59**, 1165–1171.
116. R. J. Krause, S. C. Glocke, A. R. Sicuri, S. L. Ripp and A. A. Elfarra, *Chem. Res. Toxicol.*, 2006, **19**, 1643–1649.
117. A. V. Peskin, R. Turner, G. J. Maghzal, C. C. Winterbourn and A. J. Kettle, *Biochemistry*, 2009, **48**, 10175–10182.
118. C. Storkey, M. J. Davies, J. M. White and C. H. Schiesser, *Chem. Commun.*, 2011, **47**, 9693–9695.
119. C. Storkey, D. I. Pattison, J. M. White, C. H. Schiesser and M. J. Davies, *Chem. Res. Toxicol.*, 2012, **25**, 2589–2599.
120. M. Iwaoka and S. Tomoda, *Chem. Lett.*, 2000, 1400–1401.
121. H. Masumoto, R. Kissner, W. H. Koppenol and H. Sies, *FEBS Lett.*, 1996, **398**, 179–182.
122. R. Morgenstern, I. A. Cotgreave and L. Engman, *Chem.-Biol. Interact.*, 1992, **84**, 77–84.
123. T. Masuda, S. Nakano and M. Kondo, *J. Radiat. Res.*, 1973, **14**, 339–345.
124. M. Quintiliani, R. Badiello, M. Tamba, A. Esfandi and G. Gorin, *Int. J. Radiat. Biol.*, 1977, **32**, 195–202.
125. B. Mishra, K. I. Priyadarsini and H. Mohan, *J. Phys. Chem. A*, 2006, **110**, 1894–1900.
126. A. Kunwar, B. Mishra, A. Barik, L. B. Kumbhare, R. Pandey, V. K. Jain and K. I. Priyadarsini, *Chem. Res. Toxicol.*, 2007, **20**, 1482–1487.
127. B. Mishra, A. Barik, A. Kunwar, L. B. Kumbhare, K. I. Priyadarsini and V. K. Jain, *Phosphorus, Sulfur Silicon Relat. Elem.*, 2008, **183**, 1018–1025.
128. L. Packer, E. H. Witt and H. J. Tritschler, *Free Radical Biol. Med.*, 1995, **19**, 227–250.
129. D. I. Pattison and M. J. Davies, *Chem. Res. Toxicol.*, 2001, **14**, 1453–1464.
130. M. Karimi, M. T. Ignasiak, B. Chan, A. K. Croft, L. Radom, C. H. Schiesser, D. I. Pattison and M. J. Davies, *Sci. Rep.*, 2016, **6**, 38572.
131. J. W. Finley, E. L. Wheeler and S. C. Witt, *J. Agric. Food Chem.*, 1981, **29**, 404–407.
132. T. P. A. Devasagayam, A. R. Sundquist, P. Di Mascio, S. Kaiser and H. Sies, *J. Photochem. Photobiol., B*, 1991, **9**, 105–116.

133. E. L. Clennan, D. Y. Wang, C. Clifton and M. F. Chen, *J. Am. Chem. Soc.*, 1997, **119**, 9081–9082.
134. T. Okamoto, T. Akaike, T. Sawa, Y. Miyamoto, A. van der Vliet and H. Maeda, *J. Biol. Chem.*, 2001, **276**, 29596–29602.
135. G. I. Giles, K. M. Tasker, C. Collins, N. M. Giles, E. O'Rourke and C. Jacob, *Biochem. J.*, 2002, **364**, 579–585.
136. G. I. Giles, K. M. Tasker and C. Jacob, *Gen. Physiol. Biophys.*, 2002, **21**, 65–72.
137. K.-P. Huang and F. L. Huang, *Biochem. Pharmacol.*, 2002, **64**, 1049–1056.
138. K.-P. Huang, F. L. Huang, P. K. Shetty and A. L. Yergey, *Biochemistry*, 2007, **46**, 1961–1971.
139. Y. Hu, T. W. Wang, X. Y. Liao, G. C. Du, J. A. Chen and J. G. Xu, *Protein Peptide Lett.*, 2010, **17**, 1234–1244.
140. K. P. Bhabak and G. Mugesh, *Acc. Chem. Res.*, 2010, **43**, 1408–1419.
141. A. F. de Bem, B. Fiuza, P. Calcerrada, P. M. Brito, G. Peluffo, T. C. Dinis, M. Trujillo, J. B. Rocha, R. Radi and L. M. Almeida, *Nitric Oxide*, 2013, **31**, 20–30.
142. B. S. Kumar, A. Kunwar, A. Ahmad, L. B. Kumbhare, V. K. Jain and K. I. Priyadarsini, *Radiat. Environ. Biophys.*, 2009, **48**, 379–384.
143. A. Kunwar, P. Bansal, S. J. Kumar, P. P. Bag, P. Paul, N. D. Reddy, L. B. Kumbhare, V. K. Jain, R. C. Chaubey, M. K. Unnikrishnan and K. I. Priyadarsini, *Free Radical Biol. Med.*, 2010, **48**, 399–410.
144. B. Mishra, L. B. Kumbhare, V. K. Jain and K. I. Priyadarsini, *J. Phys. Chem. B*, 2008, **112**, 4441–4446.
145. G. Mugesh, A. Panda, H. B. Singh, N. S. Punekar and R. J. Butcher, *J. Am. Chem. Soc.*, 2001, **123**, 839–850.
146. S. T. Stefanello, A. S. Prestes, T. Ogunmoyole, S. M. Salman, R. S. Schwab, C. R. Brender, L. Dornelles, J. B. Rocha and F. A. Soares, *Toxicol. In Vitro*, 2013, **27**, 1433–1439.
147. A. Kunwar, B. Mishra, A. Barik, L. B. Kumbhare, R. Pandey, V. K. Jain and K. I. Priyadarsini, *Chem. Res. Toxicol.*, 2007, **20**, 1482–1487.
148. S. Yoshida, F. Kumakura, I. Komatsu, K. Arai, Y. Onuma, H. Hojo, B. G. Singh, K. I. Priyadarsini and M. Iwaoka, *Angew. Chem., Int. Ed.*, 2011, **50**, 2125–2128.
149. M. Iwaoka and S. Tomoda, *J. Am. Chem. Soc.*, 1994, **116**, 2557–2561.
150. K. Gardiner, M. J. Davies and D. I. Pattison, unpublished data.
151. S. Yoshida, F. Kumakura, I. Komatsu, K. Arai, Y. Onuma, H. Hojo, B. G. Singh, K. I. Priyadarsini and M. Iwaoka, *Angew. Chem.*, 2011, **50**, 2125–2128.
152. P. C. Prabhu, B. Adhikari, P. P. Phadnis, S. Chakraborty, V. K. Jain and K. I. Priyadarsini, *Indian J. Chem. Sect. A*, 2014, **53**, 781–786.
153. P. Prabhu, B. G. Singh, M. Noguchi, P. P. Phadnis, V. K. Jain, M. Iwaoka and K. I. Priyadarsini, *Org. Biomol. Chem.*, 2014, **12**, 2404–2412.
154. B. J. Bhuyan, D. S. Lamani, G. Mugesh and T. Wirth, in *Handbook of Chalcogen Chemistry: New Perspectives in Sulfur, Selenium and Tellurium*,

ed. F. A. Devillanova and W. W. Du Mont, Royal Society of Chemistry, Cambridge, UK, 2nd edn, 2013, ch. 10.2, vol. 2.
155. M. Iwaoka and S. Tomoda, *J. Am. Chem. Soc.*, 1994, **116**, 2557–2561.
156. A. Assmann, M. Bonifacic, K. Briviba, H. Sies and K. D. Asmus, *Free Radical Res.*, 2000, **32**, 371–376.
157. H. Sies, L. O. Klotz, V. S. Sharov, A. Assmann and K. Briviba, *Zeitschrift fur Naturforschung. C*, 1998, **53**, 228–232.
158. R. J. Krause and A. A. Elfarra, *Biochem. Pharmacol.*, 2009, **77**, 134–140.
159. L. Carroll, D. I. Pattison, M. J. Davies and C. L. Hawkins, unpublished data.
160. M. Maiorino, A. Roveri, M. Coassin and F. Ursini, *Biochem. Pharmacol.*, 1988, **37**, 2267–2271.
161. A. Müller, E. Cadenas, P. Graf and H. Sies, *Biochem. Pharmacol.*, 1984, **33**, 3235–3239.
162. H. Sies, *Free Radical Biol. Med.*, 1993, **14**, 313–323.
163. S. S. Zade, S. Panda, S. K. Tripathi, H. B. Singh and G. Wolmershauser, *Eur. J. Org. Chem.*, 2004, 3857–3864.
164. G. E. Arteel, K. Briviba and H. Sies, *Chem. Res. Toxicol.*, 1999, **12**, 264–269.
165. M. Bjornstedt, M. Hamberg, S. Kumar, J. Xue and A. Holmgren, *J. Biol. Chem.*, 1995, **270**, 11761–11764.
166. M. Bjornstedt, S. Kumar and A. Holmgren, *J. Biol. Chem.*, 1992, **267**, 8030–8034.
167. C. M. Weekley and H. H. Harris, *Chem. Soc. Rev.*, 2013, **42**, 8870–8894.
168. K. Satheeshkumar and G. Mugesh, *Chem. – Eur. J.*, 2011, **17**, 4849–4857.
169. P. Prabhu, B. G. Singh, M. Noguchi, P. P. Phadnis, V. K. Jain, M. Iwaoka and K. I. Priyadarsini, *Org. Biomol. Chem.*, 2014, **12**, 2404–2412.
170. A. V. Peskin and C. C. Winterbourn, *Free Radical Biol. Med.*, 2001, **30**, 572–579.
171. B. Alvarez, G. Ferrer-Sueta, B. A. Freeman and R. Radi, *J. Biol. Chem.*, 1999, **274**, 842–848.
172. R. Scurlock, M. Rougee, R. V. Bensasson, M. Evers and N. Dereu, *Photochem. Photobiol.*, 1991, **54**, 733–736.
173. K. O. Hiller, B. Masloch, M. Gobl and K. D. Asmus, *J. Am. Chem. Soc.*, 1981, **103**, 2734–2743.

CHAPTER 10

Reactive Selenium Species: Redox Modulation, Antioxidant, Antimicrobial and Anticancer Activities

MUHAMMAD JAWAD NASIM,[a] WESAM ALI,[a]
ENRIQUE DOMÍNGUEZ-ÁLVAREZ,[b]
EUFRANIO N. DA SILVA JÚNIOR,[c]
RAHMAN SHAH ZAIB SALEEM[d] AND CLAUS JACOB*[a]

[a] Division of Bioorganic Chemistry, School of Pharmacy, Saarland University, Campus B 2.1, D-66123 Saarbruecken, Germany, Email: jawad.nasim@uni-saarland.de; s8wealii@stud.uni-saarland.de; [b] Instituto de Química Orgánica General, Consejo Superior de Investigaciones Científicas (IQOG-CSIC), Juan de la Cierva, 3, 28006, Madrid, Spain, Email: e.dominguez-alvarez@iqog.csic.es; [c] Institute of Exact Sciences, Department of Chemistry, Federal University of Minas Gerais, Belo Horizonte, MG 31270-901, Brazil, Email: eufranio@ufmg.br; [d] Department of Chemistry, School of Sciences and Engineering, Lahore University of Management Sciences, Opposite Sector-U, DHA, Lahore 54792, Pakistan, Email: rahman.saleem@lums.edu.pk
*Email: c.jacob@mx.uni-saarland.de

Organoselenium Compounds in Biology and Medicine: Synthesis, Biological and Therapeutic Treatments
Edited by Vimal Kumar Jain and K. Indira Priyadarsini

Published by the Royal Society of Chemistry, www.rsc.org

10.1 Selenium: 200 Years Before the Mast

Since its discovery 200 years ago, in 1817, by Jöns Jacob Berzelius (1779–1848), the element selenium has had a rather chequered relationship with biology and medicine, an often rather controversial liaison, mirrored by its position in the Periodic Table. Flanked by metalloids such as arsenic, antimony, tellurium, bromine and iodine, selenium occupies a rather "dark" corner shared by other elements with important, but often controversial and not always beneficial biological actions. Nonetheless, selenium is an important trace element essential for human and animal wellbeing and is found in many proteins and enzymes, often in the form of the amino acid selenocysteine (SeCys), a rather unusual amino acid which nowadays is considered the 21st essential amino acid in humans (see Figure 10.1 for chemical structures).[1] Apart from its presence in SeCys (**5**), selenium also occurs in proteins in the form of selenomethionine (SeMet, **8**), an amazing amino acid which seemingly freely substitutes for its sulfur analogue methionine (Met, **7**) and therefore provides an—often used—handle for "selenium enrichment". At the same time, those two selenium amino acids (and there are more natural selenium amino acid derivatives, such as selenoneine, **11**) and the kind of antioxidant redox behaviour often associated with them have stimulated a plethora of research into selenium and selenium compounds as potential antioxidants, chemopreventive and anticancer agents. Exemplified by the molecule ebselen (**9**), this interest in selenium triggered a first wave of biological selenium chemistry in the 1980s and 1990s, and an authoritative review by Mugesh *et al.*, published in 2001, bears witness to these developments.[2]

At that time, hopes were high that several of those compounds, among them ebselen (**9**), would soon sail through the various phases of clinical trials and enter the pharmaceutical market as powerful chemopreventive, antioxidant, anti-ischemic, anti-inflammatory, antidiabetic, anti-neurodegenerative or even anticancer drugs. Today, a glance at the field of "Organoselenium Compounds in Biology and Medicine", to quote the title of this book, is rather sobering. While some selenium preparations, such as sodium selenite (Na_2SeO_3), selenomethionine (**8**) and selenium-enriched yeast preparations are used widely in the form of selenium supplements, and ebselen (**9**) is still being considered as possible drug lead, most of the others have been "lost in translation", *i.e.* have failed to reach or pass successfully through the various stages of clinical trials. Indeed, there are rumours that selenium in cahoots with certain quinones and other "redox rogues" may even have ended up on the "most wanted" list of so-called PAINs (pan-assay interference compounds), considered by journals such as *Nature* and the *Journal of Medicinal Chemistry* as "promiscuous, assay-duping molecules polluting the literature and wasting resources".[3] Other selenium compounds may simply have been "forgotten" by the more biologically oriented community, or may have "faded away" over time, to quote one of the songs of the Irish metal band *Selene*.[4]

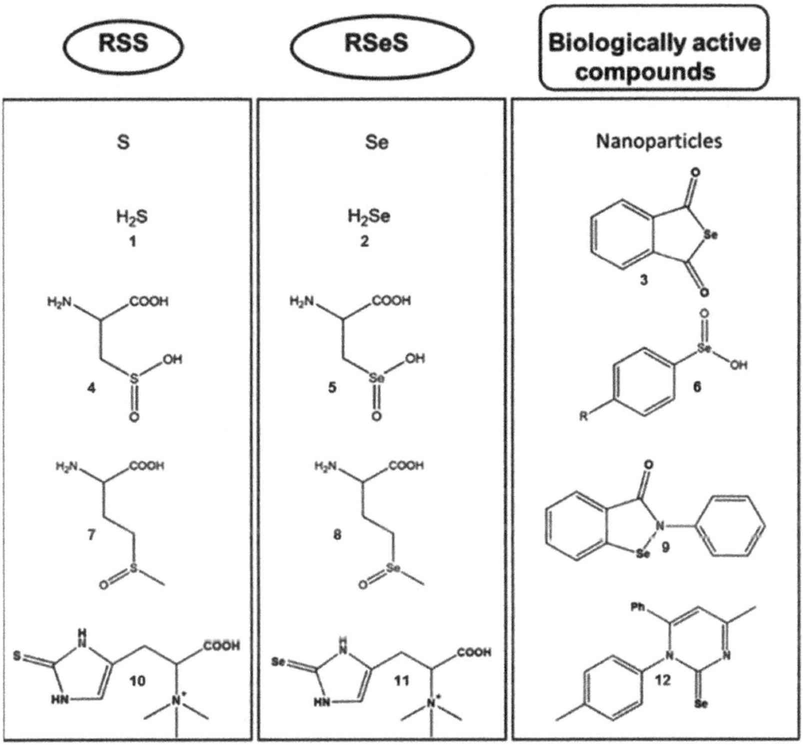

Figure 10.1 Biology provides a certain analogy between sulfur and selenium which may be used in the hunt for natural reactive selenium species (RSeS) and in the design of biologically active selenium compounds. This analogy begins with simple sulfur and selenium nanoparticles and H_2S and H_2Se release systems and has been useful, for instance, in the recent discovery of selenoneine. Yet there are also distinct differences, and selenium analogues are often considerably more biologically active than sulfur.

10.2 The Docs Bark but the Caravan Moves on

As part of this chapter, we depart from some of the better trodden parts of selenium research and turn our attention to more exotic compounds, actions and targets. In doing so, we move on from more traditional fields of selenium chemistry and biochemistry and enter a darker, more unchartered neck of the woods inhabited by lesser known selenium compounds with more curious activities. These compounds often defy the popular paradigm that selenium compounds are simply antioxidants, and this issue, in our opinion, requires a few friendly words of clarification.

From a basic redox perspective, we focus on reactive selenium species (RSeS). We have chosen this term intentionally, as it highlights the redox-modulating, often oxidizing nature of many of these selenium compounds, reflecting many well-known reactive sulfur species (RSS). Figure 10.1

provides a basic overview of the organoselenium compounds focused upon on in this chapter. This figure illustrates a certain analogy between sulfur and selenium in biology which may be used as a beginner's guide to tracking down or devising certain RSeS. Indeed, the concept of RSeS in biological redox control is just emerging, and, for the time being, it is wise to include simple, naturally occurring reactive selenium compounds, such as hydrogen selenide (H_2Se, **2**), as well as more complex organic selenium compounds of natural as well as synthetic origin under this umbrella. On a more biochemical level, these RSeS interact readily with cysteine residues of the cellular thiolstat, hence triggering signalling pathways that may eventually result in an activation of antioxidant responses, but may also lead to apoptosis, as has been shown for various RSS in the past.[5–8] Not surprisingly, these kinds of selenium compounds, which are oxidizing and often catalytic, in many ways represent the darker side of the element and its chemistry, and are designed to kill, effectively and selectively. Eventually, there may be a certain dichotomy between these RSeS and antioxidant glutathione peroxidase (GPx) mimics, yet literature reports describing the oxidizing action of certain selenium compounds, such as selenocystamine, selenobetaine and methyl seleninic acid date back to the 1980s, and here the research by Ip and Ganther and colleagues clearly stands out.[9,10] Still, we should emphasize that many of the molecules discussed here have only recently been synthesized or considered in a more biological context, and so far, the results associated with them are often of a more preliminary nature.

10.3 Natural Selenium Nanoparticles from Common Bacteria

Whilst red selenium itself is almost insoluble in water, it can be converted to or generated as selenium nanoparticles, which form stable nanosuspensions (Figure 10.2). Such red selenium particles can be produced surprisingly easily and in impressive yields and quantities with the help of bacteria, but also mechanically by milling and homogenization, and even chemically *via* the careful reduction of selenite (SeO_3^{2-}), whereby the red elemental form almost always dominates in the end compared to the grey form.[11,12] Besides "the importance of being nano", such materials are amazing because they (a) seemingly render elemental selenium bioavailable, (b) can be produced naturally by a range of—often common—bacteria such as *Klebsiella pneumoniae*, *Bacillus oryziterrae*, *Lactobacillus brevis* and *Staphylococcus carnosus*, (c) often carry an interesting protein-coat on their surface and (d) seem to act against relevant targets in the form of a pronounced antimicrobial activity.[12–15]

While the phenomenon of bacteria generating red selenium particles has been known for decades, the properties and potential uses of such particles have attracted due attention only recently. In essence, there are two main biological events associated with such elemental selenium micro- or

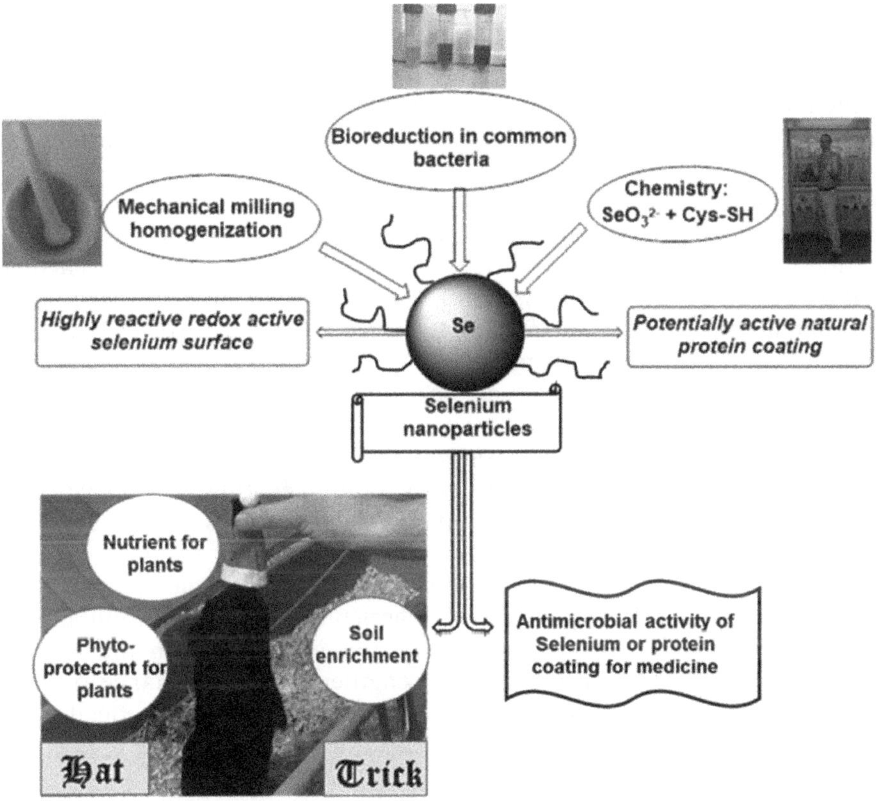

Figure 10.2 In the case of (red) selenium, biology endows us with some exquisite nanotechnology in form of particles generated by common bacteria. Unlike selenium micro- and nanoparticles generated mechanically or chemically, such natural particles also feature a coat of bacterial proteins. Initial studies on selenium particles of different origin have confirmed activity against microbes and nematodes. These particles may be useful in the context of medicine or agriculture, where a "hat trick" of simultaneous phyto-protection, plant nutrition and soil enrichment is feasible.

nanoparticles that are of major biological relevance. On one side, it seems that certain bacteria, such as *Staphylococcus aureus*, upon consumption of excess SeO_3^{2-} embark on a kind of "bioreductive Kamikaze mission", whereby the reduction of selenite to elemental selenium no longer serves as a simple means of detoxification, but eventually results in the formation of solid particles that cause severe stresses and mechanical damage to the cells affected. Unable to either dissolve or excrete such particles, those cells soon become sensitive to antibiotics or at some stage die from presence of the particles themselves. Hence the simple feeding of SeO_3^{2-} and subsequent formation of selenium nanoparticles can be employed to resensitize methicillin-resistant strains of *S. aureus*.[16] In contrast, mammalian cells usually bypass the "zero oxidation state" of selenium and focus on the formation of H_2Se (**2**) and a range of follow-on products which end up in amino acids or are excreted.

Such findings obviously raise the spectre that the particles themselves may be toxic, a notion supported by the fact that such naturally formed selenium particles not only consist of selenium, but are also covered by a—usually surprisingly well defined—protein coat, which is characteristic of the bacteria that the particles originate from.[17,18] Indeed, once isolated, those naturally generated selenium particles show an interesting biological activity against bacteria (*e.g. Escherichia coli*), yeast (*Saccharomyces cerevisiae*) and even nematodes (*Steinernema feltiae*). In contrast, similar particles have been generated by lactobacteria in milk enriched with SeO_3^{2-}, and the resulting dairy product has been considered a health-promoting cocktail fortified with elemental selenium.[15,19]

Therefore, it remains to be seen whether such particles act as pro- or antioxidants, if the selenium, or rather the protein coating, is responsible for certain biological activities and if and how such particles might be used in practice. Here, medical applications are feasible, for instance as anti-infective lotions or creams for skin. Alternatively, selenium nanoparticles may also provide a "hat trick" of functions in agriculture, *i.e.* as a combination of (a) a direct "green" phyto-protectant effective against certain moulds and pests, (b) an essential trace element strengthening the resistance of crops and, eventually, (c) selenium enrichment of crops which is beneficial to the nutritional chain, and the human end consumer (Figure 10.2).

10.4 Selenoneine or "pimp my RSS"

Moving from simple selenium nanoparticles to more complex organic molecules, the next item on our map is the compound selenoneine (**11**), an analogue of the sulfur-containing histidine derivative ergothioneine (**10**) (Figure 10.3). Ergothioneine (**10**), which is found naturally in mushrooms and can be synthesized fairly easily, has been studied extensively for decades, and its antioxidant actions are well established.[20,21] The biological activity of selenoneine (**11**), which was identified and subsequently isolated from the blood of tuna fish by Yamashita and co-workers in 2010, is less well established, but may also include certain antioxidant activity.[22,23]

In any case, selenoneine (**11**) is a fine, naturally occurring representative of organic selenourea derivatives, which have been investigated for their various interesting biological properties, including antioxidant and anticancer activities, for several years now. Interestingly, selenourea (**13**) itself has been described as a radioprotective agent and has been used as a phasing vehicle in the study of X-ray crystal structures of macromolecules.[24,25] Since the molecule is highly sensitive towards oxidation, organic derivatives have been designed to provide a more stable form of this unusual molecule with a C=Se double bond. Similar to thiones (*e.g.* ergothioneine), which often exist in equilibrium with thiols, the chemistry of such selenoureas is dominated

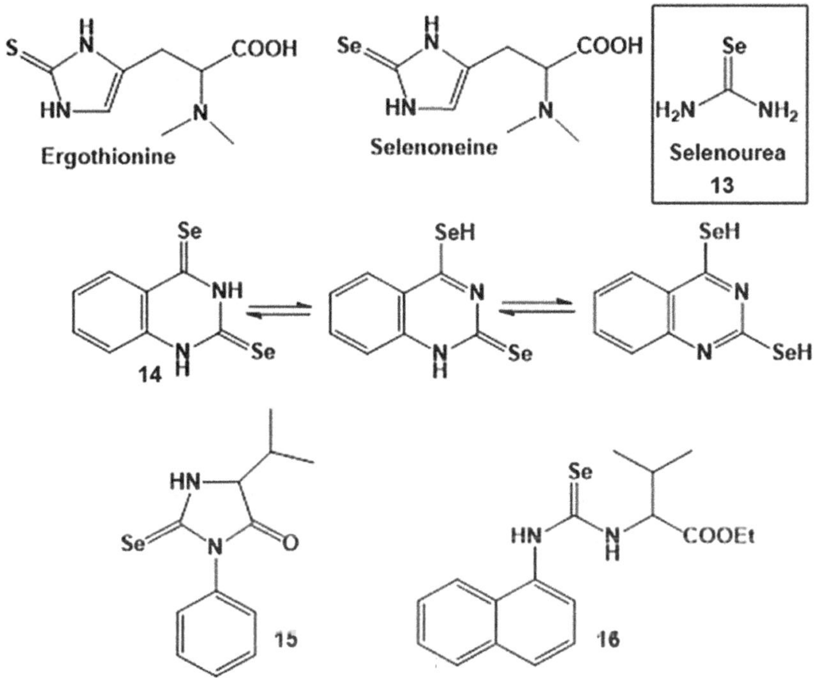

Figure 10.3 Ergothioneine is a well-known sulfur derivative of the amino acid histidine which occurs in ergot, a fungus frequently affecting rye and related crops. More recently, its selenium analogue, selenoneine, has been isolated from the blood of tuna fish. Chemically, these compounds represent a range of thio- and selenourea derivatives, respectively. As for thiones, selenones can enter a selenone/selenol equilibrium, which is crucial in the design of certain selenourea derivatives. Other *N*,*N*′-substituted derivatives do not undergo this tautomerism, but are also active biologically, for instance as antioxidants.

by the selenole–selenone equilibrium. Interestingly, this equilibrium of two tautomeric forms requires the presence of a lone pair at the adjacent nitrogen atom(s), and can be "frozen" by an adequate substitution pattern, for instance in selenohydantoins (**15**) (Figure 10.3).

Among the earliest reports of biological selenourea derivatives are studies by Ha and colleagues, who have synthesized four *N*,*N*-unsubstituted selenourea derivatives and investigated their ability to inhibit mushroom tyrosinase activity.[26] Merino-Montiel *et al.* have moved ahead with—structurally related—*O*-alkyl selenocarbamates, selenoureas and selenohydantoins, which *in vitro* show antioxidant activity against free radicals, hydrogen peroxide, alkyl peroxides and nitric oxide.[27]

Publications from Hussain and co-workers in 2014 and 2015 have dealt with selenourea derivatives with a ferrocene ring.[28,29] Besides a more traditional antioxidant activity, these ferrocene-derived selenourea derivatives also exhibit DNA binding, antibacterial and antifungal actions, while others have been used in the study of structure and function of α-peptides.[29] Considering this darker, more cytotoxic side of this specific chemistry, selenourea derivatives containing a steroidal diosgenin core combine both an antioxidant as well as an antiproliferative profile—when applied in the micromolar range—against various cancer cell lines.[30]

The underlying chemistry responsible for this spectrum of activities is still not fully understood and in any case, may vary, especially if the selenourea moiety is combined with other, equally biologically active motifs. It should also be noted that *N*,*N*′-unsubstituted derivatives may, for instance, react in their selenol form as strong reducing agents and/or nucleophiles. In contrast, *N*,*N*′-substituted derivatives are unable to form the selenol form, yet still tend to show some activity. Neganova *et al.* have recently studied such *N*,*N*′-substituted selenoureas for antioxidant activity and suggest that the activity could arise from mechanisms including a pronounced reduction capacity (*i.e.* electron transfer) and antiradical activity (*i.e.* proton transfer).[31] This kind of chemistry seems to be specific for selenium as no or only a slight activity has been observed in these studies for the relevant oxygen and sulfur analogues.

In fact, when considering different chalcogen forms of urea, it seems that the respective selenium derivatives are almost always more potent than the corresponding sulfur or oxygen analogues. Alcolea *et al.*, for instance, have demonstrated the anticancer potential of various selenourea derivatives against six human cancer cell lines, *i.e.* melanoma (1205Lu), lung carcinoma (A549), prostatic carcinoma (DU-145), colorectal carcinoma (HCT 116), pancreatic epithelioid carcinoma (PANC-1) and pancreatic adenocarcinoma (BxPC-3), while their thiourea counterparts were considerably less active.[32] More detailed mechanistic investigations indicate that these compounds may act *via* a combination of caspase activation and inhibition of anti-apoptotic proteins, a mode of action discussed later.

10.5 Selenomethionine, Selenoxides and Selenones

Following our selenium map down from the lesser known selenoneine (**11**) to selenomethionine (**8**), we arrive at the particularly stimulating chemistry of selenides (**17**), selenoxides (**8**) and selenones (**18**), which provides the basis for an intricate redox behaviour and biological activity associated with it. As mentioned in the introduction, SeMet (**8**) is a natural amino acid which can be incorporated into human proteins and enzymes, and hence has attracted attention as a possible nutritional selenium supplement. Besides supplying selenium in cases of deficiency, it is also thought to increase the overall selenium content in proteins and hence "cover" them with an antioxidant "coating". Still, SeMet (**8**) is not just an antioxidant able to sequester certain oxidizing species. Once oxidized, the resulting selenoxide or selenone may itself act as an oxidizing species, hence affecting proteins of the cellular thiolstat and possibly entering a specific catalytic cycle. This kind of

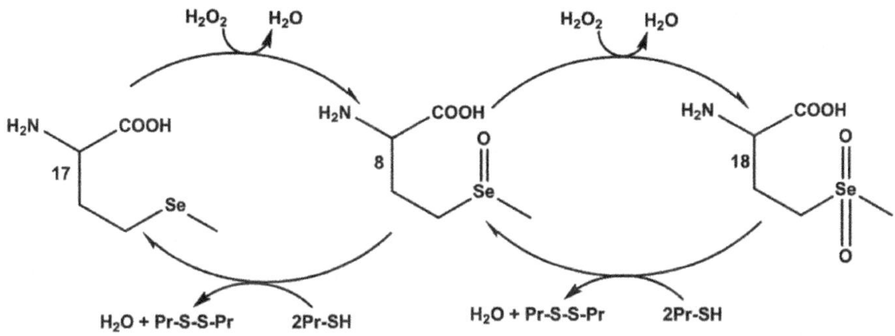

Figure 10.4 Selenomethionine is known to bestow proteins with considerable antioxidant activity. It can be oxidized to a selenoxide as part of a reversible, even catalytic cycle. Unlike sulfoxides, selenoxides do not seem to become "overoxidized" under physiological conditions, and reports of biological investigations of such selenones are surprisingly rare.

pro-oxidant and potentially catalytic behaviour is illustrated in Figure 10.4 and resonates throughout the realm of alkyl- or aryl-selenides. As for their related selenols (**23**), which exhibit a similar yet not identical catalytic behaviour, the eventual biological impact of such processes depends decisively on the presence of oxidants and thiols, and may result in pro- as well as antioxidant actions (an issue discussed more extensively later).

In biology, selenoxides remain an area of exploration. They are considered to participate in the selenium redox cycle that defends against oxidants, such as peroxynitrite and H_2O_2.[33] Selenides are closely related to selenols (**23**), chemically and in biology. The same notion also holds true for selenones (**18**) and seleninic acids (**19**), both of which share the formal selenium oxidation state of +2. Unlike their sulfur relatives, the sulfones, selenones ($RSe(O)_2R'$), which, curiously enough, in terminology but not in chemistry are equivocal to selenones ($RC(=Se)R'$), are fairly unstable molecules which do not seem to occur extensively in nature, even if one considers a certain analogy with the common "over-oxidation" of sulfides and sulfoxides to sulfones. There are some reports of a possible antibacterial activity, for instance of the compound dimethylselenone against *Pseudomonas fluorescens*, yet these reports are fairly sporadic.[34] In contrast, seleninic acids are fairly stable and very polar molecules, with a strong acid function centred around the selenium atom which, at the same time, is also highly reactive towards thiols. This unique combination of acidity and reactivity turns seleninic acids into promising agents that are not only highly—and selectively—reactive, but, at the same time, readily soluble in many solvents and media and able to penetrate biological membranes. Indeed, an "issue" often associated with organoselenium compounds is solubility and, in a more pharmaceutical context, bioavailability. Seleninic acids may address this problem. They are highly reactive, selective for thiols of the cellular thiolstat, catalytic rather than one-shot, stable and soluble. In contrast to sulfinic acids, seleninic acids react spontaneously, readily, fast and in several steps with thiols to form disulfides, selenylsulfides (**22**) and, in some cases, the reduced selenol (**23**). This chemistry is depicted in Figure 10.5.

Interestingly, in theory, one seleninic acid can oxidize and hence modify a total of four thiol groups, yet this considerable oxidizing power is fairly limited to thiols. At the same time, the selenol formed and some of the intermediates can enter catalytic cycles in the presence of a suitable oxidizing agent, such as hydrogen peroxide (H_2O_2). After an initial onslaught on thiols, this GPx-like catalysis can develop its own pro-oxidant dynamic by consuming additional thiols during the ensuing catalytic process(es). Eventually, a simple seleninic acid is therefore able to modify up to four

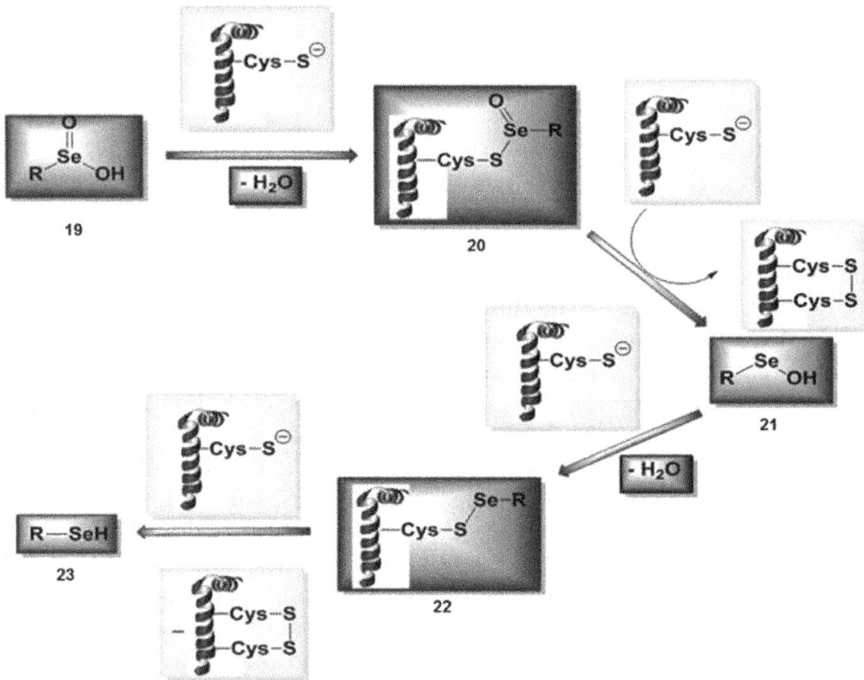

Figure 10.5 Due to their stability, selective chemical reactivity, solubility and good biological availability, seleninic acids form a particular and particularly interesting class of organoselenium compounds with an often pronounced biological activity. Initially, they may act as thiol-specific oxidants, and indeed are able to oxidatively modify a total of four cysteine residues (although their action often comes to a halt at the selenylsulfide state). Upon reduction to selenylsulfide, selenol or diselenide, such compounds turn into potent redox catalysts which, in the presence of reactive oxygen species (ROS) such as H_2O_2, may facilitate the widespread oxidation of cysteine (and selenocysteine) residues in peptides (such as glutathione), proteins and enzymes. They may therefore simultaneously act as antioxidants by removing H_2O_2 and pro-oxidants by modifying redox sensitive amino acid residues.

protein thiols, turn into an active pro-oxidant catalyst and in doing so, and attack the cellular thiolstat to a degree that eventually depends on the presence and concentration of oxidants, such as H_2O_2 in the cell. Together, the chemical reactivity of seleninic acids turns them into effective, yet selective, thiol-specific agents which, after an initial oxidative onslaught on cellular thiols, may metamorphize to a catalyst able to "sense" existing levels of reactive oxygen species (ROS) in cells and to facilitate the oxidation of further cellular thiols accordingly (Figure 10.5). Early reports on methylseleninic acid, for instance, reflect the specific biological activity and underlying biochemical modes of action that emanate from this oxidizing chemistry.[35–48] These reports bear witness to the complex biochemical signalling and control processes that are evoked or controlled by such simple yet aggressive selenium compounds with pronounced "sensor/effector" properties.

Since the seleninic acid group can serve as a reactive, polar group, various amphiphilic structures have been studied.[7,49] These molecules show particular promise as they are fairly straightforward to synthesize, chemically stable and, above all, readily soluble in aqueous media. As expected, these compounds rapidly and seemingly indiscriminately consume thiols and subsequently trigger a range of redox-related cellular pathways which, depending on the predisposition of the cell, may lead either to an antioxidant defence or to apoptosis. Such molecules are not only redox-active but also interact with membranes, an effect that in itself may evoke a biological activity. Predictably, these amphiphilic seleninic acids have been employed successfully against certain cancer cell lines, while "normal" cells seem to be less affected.[7] They are also active against various microorganisms.

From a synthetic chemical perspective, seleninic acids are of interest as potential ligands for metal ions, inhibitors of metalloproteins and because their synthesis involves selenocyanate precursors, which themselves are of considerable biological interest.[50–53]

10.6 Beyond the Traditional Se–N Bond

Besides SeMet and its pronounced dialkylselenide chemistry, selenocysteine (SeCys) is the main biological player of selenium. While its biological redox chemistry is often mentioned in concert with that of SeMet (**8**), there are several pronounced differences. Unlike SeMet (**8**), SeCys (**5**) is incorporated into proteins not at random, but as a result of a rather extensive insertion network composed of DNA, specific SECIS insertion sequences and tRNA.[54–56] Furthermore, SeCys (**5**) features a selenol/selenolate group which gives rise to a redox chemistry that resembles but also significantly differs from that of SeMet (**8**) and related organic disubstituted alkyl or aryl selenides. The evident hallmark of SeCys (**5**) is its ability to catalyse the conversion of peroxides to water and alcohols while consuming thiols in the process (Figure 10.6).

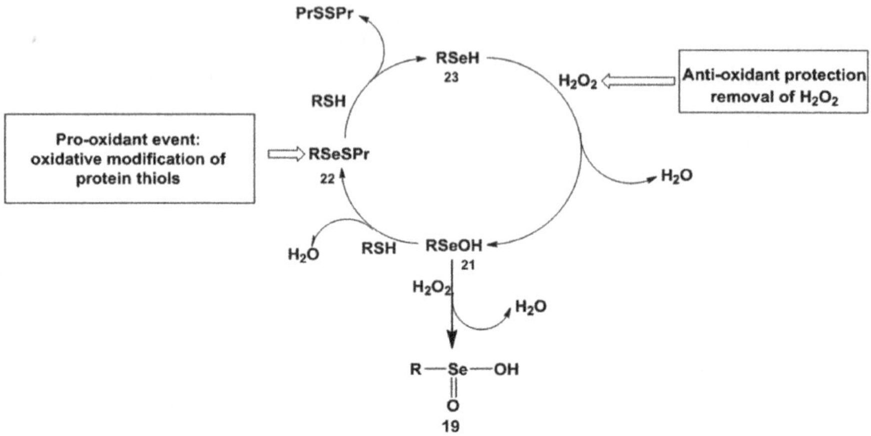

Figure 10.6 The basic chemical and biochemical principles behind catalytic sensor/effector agents. Inside cells, certain organoselenium compounds can act as redox catalysts that convert reactive oxygen species such as H_2O_2 to H_2O in the presence of suitable thiols. While this process may be directly "antioxidative" in the case of sacrificial and recyclable thiols, it can also represent a pro-oxidative signal if the thiols in question (PrSH) belong to redox-sensitive signalling proteins of the cellular thiolstat. Eventually, such signalling may result in the (indirect) activation of an antioxidant response and/or pro-apoptotic pathways.

In biology, this kind of catalysis has many facets, and is associated with several twist and turns. If one focuses solely on the removal of peroxide, this may be considered as the ultimate antioxidant, a notion supported by the fact that selenoenzymes such as GPx employ this very catalytic cycle for the effective removal of peroxides, such as H_2O_2. Unsurprisingly, therefore, compounds such as ebselen (**9**) have been developed in order to mimic this activity. In contrast, if one focuses on the consumption of thiols, such catalytic processes often trigger a pro-oxidant cascade which may, for instance, also result in apoptosis.

Seleno–nitrogen compounds are therefore of unique interest as they may interact in a variety of ways, both chemically as well as biochemically (Figure 10.7). Besides ebselen (**9**) and its various selenazole derivatives, various classes of compounds with different arrangements of selenium and nitrogen have been considered over the years, including selenodiazoles (**27**), selenocyanates, isoselenocyanates and charged selenazolinium salts (**24–26**).

Interestingly, these compounds often exhibit very specific shifts in ^{77}Se nuclear magnetic resonance (NMR) spectra, which may range from a few hundred to well over a thousand ppm. Since these chemical shifts reflect the electron density around the selenium atom, and hence may also be seen as a crude indicator of electrophilicity, they are sometimes used to rationalize and even predict reactivity and eventually biological activity. Compared to other relevant parameters, such as electrochemical oxidation and reduction

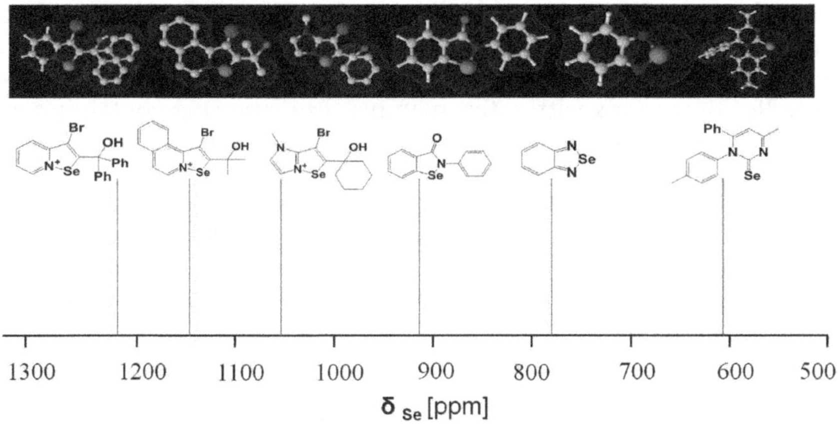

Figure 10.7 Ebselen, a potent mimic of the selenoenzyme glutathione peroxidase, has long been considered as the blueprint of direct and catalytic antioxidant activity. The selenium–nitrogen bond found in this selenazole reflects similar interactions between selenium and nitrogen in the enzyme structure and seems to bolster activity when compared, for instance, to other selenols or selenides. Recent investigations have aimed at an enhancement of this activity, for instance by increasing the electrophilicity of the reactive centre. A range of compounds have been synthesized and evaluated in biological test systems, including ebselen, selenadiazoles, selenazolinium salts and selenourea derivatives. Their distinct shifts in the ^{77}Se nuclear magnetic resonance spectrum have often been discussed as a means of ranking these compounds and eventually predicting chemical reactivity and biological activity.

potentials, such ^{77}Se NMR spectra are comparably straightforward to obtain and may even be predicted by the appropriate software. Nonetheless, the biological situation is considerably more complex, and other factors besides such a simple physico-chemical "ranking" of compounds also need to be considered. The following examples illustrate the benefits and limitations associated with lesser known RSeS encountered along this ^{77}Se NMR scale.

Selenazolinium compounds (**24–26**), for instance, somewhat resemble ebselen (**9**), yet contain an additional positive charge which significantly increases electrophilicity, and hence reactivity towards nucleophiles, such as thiols. Recent studies by Arsenyan *et al.* have demonstrated that these compounds are fairly easy to synthesize, are stable and also biologically rather active against cancer cells.[57] Nonetheless, their considerable reactivity may render them unsuitable for more targeted, *i.e.* selective, medical applications.

From a design perspective, starting with ebselen (**9**), selenadiazoles achieve "activation" by "adding" a second nitrogen atom, rather than a positive charge. As before, this reduces electron density and may promote the kind of electrophilic reactivity required for biological activity. Indeed, several selenazoles have been reported to increase intracellular ROS levels, leading to the kind of oxidative stress that ultimately induces apoptosis in cancer cells.[58]

It should be mentioned that there is a range of other selenium–nitrogen compounds that have been investigated, yet which contain aromatic ring systems or Se and N atoms separated by carbon atoms. Selenazoles are the heterocylic compounds with a five-member aromatic ring containing a nitrogen and a selenium atom, while selenazines are six-membered rings with selenium and nitrogen. These compounds have been the focus of extensive biological investigations, from the early studies published in 1983 by Lucas *et al.* on the inhibition of inosinate monophosphate dehydrogenase, a key enzyme of guanylate biosynthesis, to more recent reports on the cytotoxicity and DNA fragmentation caused by selenazoles and selenazines in human HT-1080 fibrosarcoma cells.[59,60]

10.7 Selenoesters and Anhydrides: Unstable Pipe Dreams or Elegant Pro-drugs

Interestingly, most of the compounds discussed so far contain a more or less stable, often aromatic selenium motif, and the Se atom is unlikely to escape due to the stability of the selenium–carbon bond and/or the aromatic ring system. Yet there are selenium compounds that entirely differ from this and contain "labile" selenium ready to escape and to develop its own biological action. Here, instability, either chemical or metabolic, may serve as part of a pro-drug concept, or as a release mechanism for biologically active selenium species, such as organic fragments and inorganic elemental selenium, for instance, elemental selenium, SeO_3^{2-} or indeed H_2Se (**2**).

One such labile compound is the phthalic selenoanhydride (**3**), the selenium isostere of phthalic anhydride (Figure 10.8). This compound, at least on paper, may hydrolyse to phthalic acid, liberating highly aggressive H_2Se (or HSe^-) in the process. Indeed, phthalic selenoanhydride exhibits an interesting redox profile, probably related to its high chemical reactivity, and has the capacity to induce apoptosis and to inhibit multidrug resistance (MDR) efflux pumps.[61,62]

A similar, also hydrolysable group is provided by the selenoester CO–Se, in which the carbonyl group is generally bound to an aryl/heteroaryl group with different substituents. On the other side of the structure, the Se atom is bound to an alkyl fragment which may contain different carbonyl groups. These selenium compounds (**28–31**) are labile, as the bond between the carbonyl and the selenium atom can break easily. This is important, as the supposed active species responsible of the biological activity are the selenium ions released during the breakdown of the selenoester, in analogy to the hydrolysis of the selenoanhydride (**3**) mentioned earlier. After the release, the selenium atoms, possibly initially as H_2Se or HSe^-, are free to exert their biological action, instead of being trapped in the molecular structure of the pro-drug. This pro-drug approach has a number of advantages. For instance, while small selenium molecules such as H_2Se (**2**) are notoriously sensitive to oxidation and may face difficulties in crossing

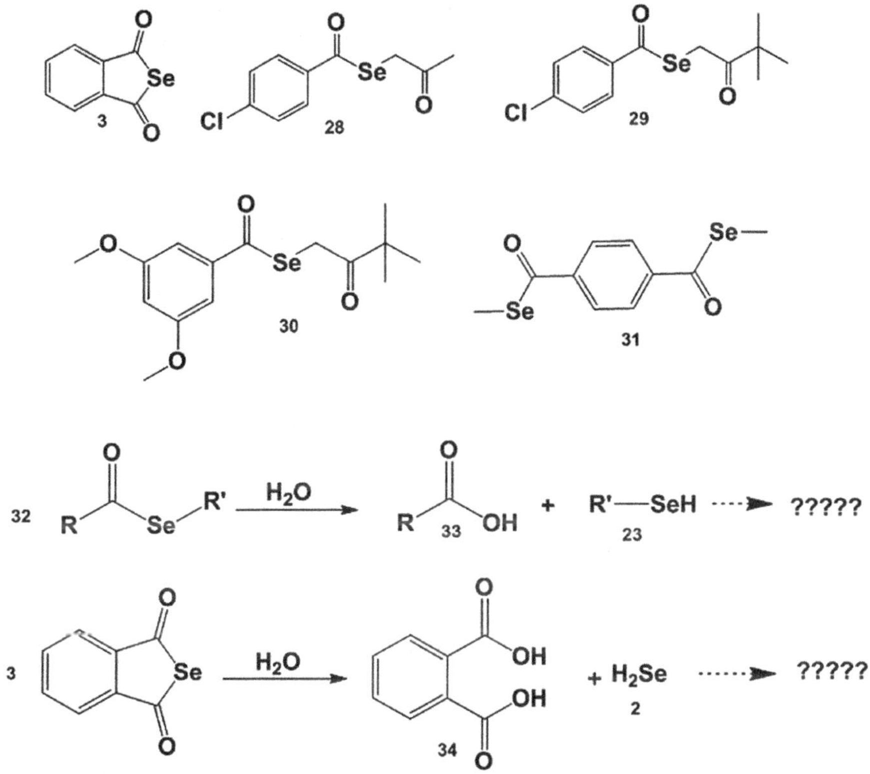

Figure 10.8 Selenoanhydrides and selenoesters are usually not particularly stable in aqueous media and hence have only occasionally been considered in a more biological or medical context. Still, hydrolysis of these compounds almost inevitably leads to the liberation of reactive selenium species, possibly including H_2Se and certain selenols. Considering the current interest in medically useful H_2S donors, such "selenium pro-drugs" or "selenium donors" may not be entirely uninteresting either.

cell membranes, pro-drugs are more stable and thus more likely to permeate such membranes. In line with this argument, the seleno-ester compounds not only showed a significant redox activity according to the data provided by electrochemical measurements using cyclic voltammetry, but also a spectrum of very pronounced and interesting biological activities, such as cytotoxic or cytostatic activity, chemopreventive activity, the ability to inhibit the MDR efflux pumps and the capacity to induce apoptosis.[63]

Antiproliferative activity in the (lower) micromolar range has been noted in the case of different cancer cell lines, such as PC-3, HT-29, MCF-7 and A549. Here, the cytostatic action of certain selenoesters was more potent than that exerted by etoposide and cisplatin, the two chemotherapy drugs considered as references.[63] Some of the most active derivatives even exerted a selective activity, such as the selenoesters **28–30**, which were selective for human Colo 320 MDR cancer cells with respect to MRC-5 human embryonic fibroblasts.[64] Rather than acting as antioxidants or chemopreventive agents, a possible mechanism which may explain the antiproliferative and/or cytotoxic activity of these compounds is the induction of apoptosis, as compounds **3** and **28–30** showed a strong capacity to induce apoptotic events, including early apoptosis and/or late apoptosis/necrosis, in different cancer cell lines.[61,62]

Eventually, it should be emphasized that the more traditional actions associated with selenium in the field of anticancer research, such as chemoprevention and selective cytotoxicity, are increasingly joined by the ability to overcome drug resistance, caused, for instance, by certain efflux pumps. Within this context, it is notable that the ketone-containing selenoesters **28–30** and the selenoanhydride **3** inhibit the ABCB1 efflux pump in two MDR cell lines that overexpress this MDR efflux pump, *i.e.* the mouse L5178Y MDR-derived T-lymphoma cell line and the human Colo 320/MDR LRP resistant colorectal adenocarcinoma cell line.[61,62] The strongest efflux pump inhibition was found for **28**, with a four-fold higher inhibition at just 2 μM concentration compared to that of the reference inhibitor verapamil at 20 μM concentration.[62] This exceptional inhibition suggests that these compounds, besides their more cytotoxic anticancer action, may also find potential applications as novel reversals of multidrug resistance. In other words, the combined administration of these active selenium compounds with chemotherapy drugs may add to and, at the same time, maintain or restore anticancer action against resistant tumours.

10.8 Selenium-based Redox Catalysts: Effective and Selective Rather than Fast and Furious

By considering the redox chemistry underlying the various biological actions of seleninic acids (**19**), the issue of redox catalysis has already come into focus (Figure 10.5). Upon successive reduction, seleninic acids (**19**) may eventually enter a catalytic cycle, which in its most basic form involves selenols (RSeH, **23**), selenylsulfides (RSeSR′, **22**) and selenenic acids (RSeOH, **21**). This cycle is commonly found in enzymes such as GPx. Some selenium compounds, such as ebselen (**9**), conduct a similar redox chemistry which may involve additional selenium species of a higher oxidation state. Even selenides, such as SeMet (**8**), are in principle amenable to this kind of catalytic redox chemistry. Yet as for any kind of catalysis, the catalyst itself is only part of the story. It requires substrates and generates products, and the nature, concentration and—in the context of biology—impact of all of those substrates and products on cellular processes must be considered. We will therefore, once more, deliberate on this central feature of the redox behaviour of many organoselenium compounds.

Inside living cells, such catalysis combines several exquisite characteristics which are highly desirable for drug design: efficiency at low concentrations, lack of inherent toxicity, specificity for the cellular thiolstat and selectivity for cells under pre-existing oxidative stress. In fact, certain organoselenium catalysts are active in sub-micromolar concentrations, yet only if their ROS substrate is present naturally in the target cell, which is frequently the case in cancer cells. In that particular environment, these catalysts facilitate the oxidative modification and hence activation of proteins and enzymes that eventually trigger an apoptotic cell death of such predisposed target cells (*e.g.* cancer cells), without causing excessive harm to normal cells which are naturally low in concentrations of ROS.[65–67]

Various studies support this idea of using catalysts as potential drugs which sense a particular biochemical cellular "signature" in the form of substrate and hence unleash their antioxidant or cytotoxic effects accordingly. Examples are not limited to selenium, but also include tellurium, and, curiously, the "no-no-quinones" and a couple of other "redox rogues".[68] Nonetheless, selenium—and tellurium—occupy a prominent position among the various redox catalysts considered in this context. While many "redox rogues" attack and oxidize more or less non-specifically various cellular components, such as membranes, metalloproteins and diverse organic redox systems, chalcogen-based catalysts, such as the ones shown as first-generation in Figure 10.9 (**35–37**), exhibit considerable selectivity for thiol (and selenol) groups, and hence only trigger very specific cellular signalling pathways, namely the ones of the cellular thiolstat.[69]

As for seleninic acids (**19**), there are not that many constraints as far as structural features are concerned, as catalytic activity, and not a specific binding or even "docking" is at the heart of the mode of action. In fact, selectivity for just one or two cellular cysteine proteins is not even desired, as pro-apoptotic signals are usually transduced *via* different pathways, and triggering several of them at the same time may be advantageous as it ensures that the signal eventually results in the desired effect, *i.e.* the killing of the target cell.

For this reason, a plethora of selenium-based redox catalysts is possible and multifunctional agents have been proposed, which combine ROS-generating redox sites, such as quinones, with the selenium redox site, which "uses" ROS such as H_2O_2 (**38, 39**).[70–73] These compounds are regarded as effective, as they add their own ROS to the already stressed target cell. Several sets of such multifunctional selenium catalysts have been synthesized successfully and have been evaluated extensively during the past 16 years. In some instances, such compounds have been deployed rather successfully against cancer cells, such as K562 (chronic myelogenous leukemia cells), CLL B-cells, HT29 (colon carcinoma cells), A549 (lung carcinoma cells), MCF7 (breast carcinoma cells) and various microorganisms including *Saccharomyces cerevisiae* and various strains of *Staphylococcus aureus*.[16,73]

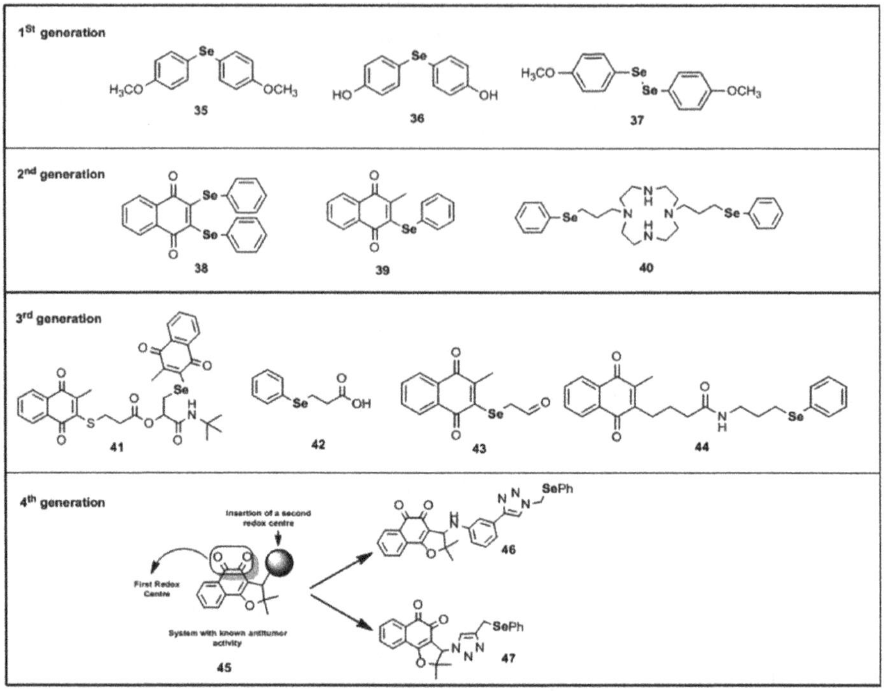

Figure 10.9 During the past two decades, the design, synthesis and subsequent biological evaluation of catalytic selenium-based "sensor/effector" agents has witnessed considerable progress. While simple mono- and diselenides have been used for initial "proof-of-concept" studies, chemical synthesis has soon moved on to multifunctional agents, which were initially generated by simple nucleophilic replacement reactions. More sophisticated methods include the linear and often sequential coupling of two or three components, multicomponent Passerini and Ugi reactions and, more recently, click reactions to combine suitable precursors to form highly biologically active agents.

10.9 Stick it and Click it like Beckham

Chemically speaking, multifunctional agents which combine different redox centres, reactivities and biological activities are complicated structures that often need to combine different chemical building blocks. It is therefore hardly surprising that their synthesis lends itself to synthetic reactions designed to assemble such individual blocks, either in the form of simple coupling reactions, multicomponent reactions (MCRs), such as the Passerini and Ugi reactions, or click reactions.[72,74–76]

Figure 10.9 provides an overview of the chemistry employed during the past decade and a half to assemble such multifunctional agents. Early strategies to generate a new generation of compounds have considered the combination of a selenium redox site with a quinone (see earlier) or metal binding site, for instance triazacyclononanes (TACN, **40**).[72] The chemistry underlying this is usually straightforward and involves simple substitution or coupling reactions. The agents obtained are not very complex, yet already show some promise in early *in vitro* and cell culture studies.[71,77] Notably, a related type of chemistry has also been employed to "piggy-back" selenium on proteins *via* covalent attachment to certain amino acid residues.[78]

More advanced, and more intricate strategies include the Passerini and Ugi MCR method, which requires the initial synthesis of a range of suitable building blocks. Once available, however, the "mix and match" combination of three or even four of these blocks swiftly leads to an entire library of interesting compounds (**41, 44**).

Alternatively, click chemistry may be used, which combines just two components (**42, 43**). It also enables a "mix and match" and leads to a similar small library of compounds, which in this case is more limited in numbers, but contains molecules that are less complex and perhaps more applicable in biology.

Indeed, click chemistry has a long tradition and is comparably straightforward. Since the groups of Meldal in Denmark[79] and Fokin and Sharpless in the United States[80] independently reported their pioneering works on the copper-catalysed azide-alkyne cycloaddition (CuAAC) reaction, the scope and importance of this reaction has experienced an exponential growth with an infinitude of creative and new applications.[81–84] Not surprisingly, the applicability of this reaction type for the synthesis of multifunctional agents has recently been explored.[85] By using lapachones (*i.e.* a quinoidal system) as prototype, the integration of selenium has been accomplished *via* click chemistry (**46, 47**).

By employing click chemistry and copper catalysis, *ortho*- and *para*-quinones have been prepared and evaluated against six types of cancer cell lines and three normal cell lines.[85] In general terms, such *ortho*-derivatives are highly active against a majority of the cell lines evaluated, with IC_{50} values usually below 2 μM. These results may be expected since β-lapachone analogues, just like the quinones discussed before, are able to generate ROS—which in turn may be used by the selenium redox centre—and hence tilt the cellular redox balance towards the process of programmed cell death.

In compounds with two redox centres, that is, a quinoidal moiety able to generate ROS and a selenium centre for peroxidation catalysis, naphthoquinones are able to generate ROS and the catalytically active organochalcogens are usually capable of using this ROS; such lethal cocktails of reactive species can push cancer cell lines over a critical redox threshold and finally kill them through apoptosis, as we and our colleagues recently proposed.[86–88]

10.10 Caveats Before Caviar

Since the rather amazing and probably slightly fictive use of selenium by professor Harry Block – the selenium rogue and remote spiritual relative of the authority on sulfur, professor Eric Block – against aliens in the movie *Evolution*, selenium has featured prominently in fact and fiction as a remedy against all kinds of diseases, (extra-terrestrial) pathogens included.[89] Yet with the champagne and caviar on ice, we should not conceal a few caveats associated with the idea of employing selenium compounds in biology and medicine in general, and as catalytic and therefore selective cytotoxic agents in particular.

The first warning, as usual, is concerned with the pharmacokinetics of such molecules, *i.e.* their application, uptake, distribution and eventual excretion. In most instances, such ADME (absorption–distribution–metabolism–elimination) studies are not performed initially, yet are obviously a crucial prerequisite in the context of any decent medical application. Here, issues related to metabolic stability, release of selenium and selenium toxicity are paramount. Indeed, the toxicity of such compounds against the organism in question, and possible side-effects, tend to be overlooked, at least during the initial, pre-clinical studies. While certain more widely available screens for "whole-organism toxicity" exist today, based, for instance, on fish embryos or nematodes, they are not used regularly. At the same time, such simple organisms are hard to "diagnose". A basic check of "normal" growth, appearance and even a quick blood test are not sufficient. Slow and hidden side-effects, such as those affecting behaviour, may be hard to detect. Computational approaches may provide some guidance, yet they should not replace empirical evidence.

The second issue with organoselenium compounds is related to activity. As indicated before, selenium is often not the best choice when it comes to activity. Tellurium analogues usually are more active, catalytically as well as in biological test systems. Still, selenium is a good compromise when it comes to the three chalcogen analogues: while it is more dangerous in such compounds compared to sulfur, it is also more active, and while it is less active than tellurium, it is also less dangerous.

10.11 Conclusions

As part of this chapter, we have briefly introduced the emerging concept of reactive selenium species (RSeS). The ideas behind this concept, and its various implications, are worth considering, not only in biology but also in medicine. We therefore conclude this part of the book with a brief appraisal of this concept and an outlook on developments we anticipate will manifest themselves during the years to come.

As far as the hypothesis of RSeS is concerned, we can distinguish two strands of investigation which are not necessarily mutually exclusive. Both deal with biologically active organoselenium compounds. On the one side, we find a growing interest in naturally occurring species such as SeCys and SeMet, but also Selenoneine, and it seems that many additional selenium metabolites are feasible, yet they have not been detected so far because of their low cellular concentrations, crude detection methods or simply because they are hidden in places where no one has looked for them. On the other side we witness a renewed focus on synthetic organoselenium agents which are considered as potential chemopreventive agents or drugs. This interest is fuelled by new data revealing considerable and facet-rich biological activities, often against targets that are otherwise hard to hit, such as resistant bacteria or cancer cells.

In both instances, analogies with sulfur in general, and with RSS in particular, may be instructive. As the example of the selenoneine (**11**)/ergothioneine (**10**) pair illustrates, such analogies may guide us in the search for and identification of organoselenium species stowed away in certain organisms. Similarly, they may inspire us in the design of new drug prototypes, in line with traditional bioisosteric replacement.

Nonetheless, there are also differences, and, sooner or later, we must allow the concept of RSeS to emancipate itself. Selenium is considerably less abundant in biology when compared to sulfur, both in concentration and (omni-)presence. Yet while there are more organic and inorganic RSS than RSeS, the selenium analogues are usually more reactive. SeMet (**8**), for instance, is no match for Met (7) as far as redox activity and (protective, antioxidant) activity are concerned, and we will probably observe a similar relationship between selenoneine (**11**) and ergothioneine (**10**). The same applies to synthetic agents. Hardly anyone of us has heard of ebsulfur, and the reason is very simple: it cannot undergo redox catalysis and is mostly inactive in the relevant biological activity assays.

We would therefore like to endorse the concept of RSeS in order to celebrate the bicentenary of the discovery of this amazing element. Vague as it may be, it may guide the next generation of selenium researchers and research in the field of naturally occurring, organic and inorganic selenium species with biological relevance. Eventually, the more natural RSeS present in living cells may form their own niche of redox biology, while the "intervening", often synthetic RSeS may provide leads for innovative new preventive agents and drugs. Together, these RSeS may provide a new impetus for those among us who have not given up hope that the first selenium-based pharmaceuticals will soon enter mainstream medicine. 2016 saw Selena Gomez's *Revival Tour*, and in 2017 and the years that follow, no doubt, there will be a revival tour of selenium as well. Happy birthday selenium!

Abbreviations

Cys	cysteine
GPx	glutathione peroxidase
GSH	glutathione (reduced form)
Met	methionine
ROS	reactive oxygen species
RSeS	reactive selenium species
RSS	reactive sulfur species
SeCys	selenocysteine
SeMet	selenomethionine

Acknowledgements

The authors acknowledge the financial support from the University of Saarland, Germany; the Landesforschungsförderungsprogramm of the Saarland (Grant No. WT/2 – LFFP 16/01); de Química Orgánica General, Consejo Superior de Investigaciones Científicas (IQOG-CSIC), Spain; Federal University of Minas Gerais, Brazil; and Lahore University of Management Sciences, Pakistan. The authors would like to express their sincerest gratitude to the many colleagues around the world contributing to the diverse field of selenium. Special thanks go to Ashfiq Ali, Ken Rory, Manfred M. Fuger, Wulgar Proll and many other colleagues from the Academiacs International network for their helpful discussions and advice.

References

1. R. L. Schmidt and M. Simonovic, *Croat. Med. J.*, 2012, **53**, 535.
2. G. Mugesh, W. W. du Mont and H. Sies, *Chem. Rev.*, 2001, **101**, 2125.
3. J. Baell and M. A. Walters, *Nature*, 2014, **513**, 481.
4. *Selene*, http://www.selene-theband.co.uk/.
5. N. E. Saidu, R. Touma, I. A. Asali, C. Jacob and M. Montenarh, *Biochim. Biophys. Acta*, 2013, **1830**, 2214.
6. N. E. Saidu, I. Abu Asali, B. Czepukojc, B. Seitz, C. Jacob and M. Montenarh, *Biochim. Biophys. Acta*, 2013, **1830**, 5267.
7. P. Du, N. E. Saidu, J. Intemann, C. Jacob and M. Montenarh, *Biochim. Biophys. Acta*, 2014, **1840**, 1808.
8. D. R. Allah, L. Schwind, I. A. Asali, J. Nasim, C. Jacob, C. Gotz and M. Montenarh, *Int. J. Oncol.*, 2015, **47**, 991.
9. C. Ip and H. E. Ganther, *Carcinogenesis*, 1992, **13**, 1167.
10. C. Ip, H. J. Thompson, Z. Zhu and H. E. Ganther, *Cancer Res.*, 2000, **60**, 2882.
11. T. Schneider, A. Baldauf, L. A. Ba, V. Jamier, K. Khairan, M. B. Sarakbi, N. Reum, M. Schneider, A. Roseler, K. Becker, T. Burkholz, P. G. Winyard, M. Kelkel, M. Diederich and C. Jacob, *J. Biomed. Nanotechnol.*, 2011, **7**, 395.
12. E. C. Estevam, S. Griffin, M. J. Nasim, P. Denezhkin, R. Schneider, R. Lilischkis, E. Dominguez-Alvarez, K. Witek, G. Latacz, C. Keck, K. H. Schafer, K. Kiec-Kononowicz, J. Handzlik and C. Jacob, *J. Hazard. Mater*, 2017, **324**, 22.
13. P. J. Fesharaki, P. Nazari, M. Shakibaie, S. Rezaie, M. Banoee, M. Abdollahi and A. R. Shahverdi, *Braz. J. Microbiol.*, 2010, **41**, 461.
14. P. Bao, K. Q. Xiao, H. J. Wang, H. Xu, P. P. Xu, Y. Jia, M. M. Haggblom and Y. G. Zhu, *Sci. Rep.*, 2016, **6**, 34054.
15. M. H. Yazdi, M. Mahdavi, N. Setayesh, M. Esfandyar and A. R. Shahverdi, *Daru*, 2013, **21**, 33.
16. E. C. Estevam, K. Witek, L. Faulstich, M. J. Nasim, G. Latacz, E. Dominguez-Alvarez, K. Kiec-Kononowicz, M. Demasi, J. Handzlik and C. Jacob, *Molecules*, 2015, **20**, 13894.

17. T. Wang, L. Yang, B. Zhang and J. Liu, *Colloids Surf., B*, 2010, **80**, 94.
18. M. Lenz, B. Kolvenbach, B. Gygax, S. Moes and P. F. Corvini, *Appl. Environ. Microbiol.*, 2011, **77**, 4676.
19. Y. Deng, C. Man, Y. Fan, Z. Wang, L. Li, H. Ren, W. Cheng and Y. Jiang, *Int. Dairy J.*, 2015, **44**, 31.
20. R. W. Li, C. Yang, A. S. Sit, Y. W. Kwan, S. M. Lee, M. P. Hoi, S. W. Chan, M. Hausman, P. M. Vanhoutte and G. P. Leung, *J. Pharmacol. Exp Ther.*, 2014, **350**, 691.
21. C. Jacob, *Nat. Prod. Rep.*, 2006, **23**, 851.
22. Y. Yamashita and M. Yamashita, *J. Biol. Chem.*, 2010, **285**, 18134.
23. Y. Yamashita, T. Yabu and M. Yamashita, *World J. Biol. Chem.*, 2010, **1**, 144.
24. R. Badiello, E. Gattavecchia, M. Mattii and M. Tamba, *Z Naturforsch., C: J. Biosci.*, 1974, **29C**, 647.
25. Z. Luo, *Sci. Rep.*, 2016, **6**, 37123.
26. S. K. Ha, M. Koketsu, K. Lee, S. Y. Choi, J. H. Park, H. Ishihara and S. Y. Kim, *Biol. Pharm. Bull.*, 2005, **28**, 838.
27. P. Merino-Montiel, S. Maza, S. Martos, O. Lopez, I. Maya and J. G. Fernandez-Bolanos, *Eur. J. Pharm. Sci.*, 2013, **48**, 582.
28. R. A. Hussain, A. Badshah, M. N. Tahir, H. Tamoor ul and A. Bano, *J. Biochem. Mol. Toxicol.*, 2014, **28**, 60.
29. R. A. Hussain, A. Badshah, J. M. Pezzuto, N. Ahmed, T. P. Kondratyuk and E. J. Park, *J. Photochem. Photobiol. B*, 2015, **148**, 197.
30. L. L. Romero-Hernandez, P. Merino-Montiel, S. Montiel-Smith, S. Meza-Reyes, J. L. Vega-Baez, I. Abasolo, S. Schwartz Jr., O. Lopez and J. G. Fernandez-Bolanos, *Eur. J. Med. Chem.*, 2015, **99**, 67.
31. M. E. Neganova, A. N. Proshin, O. M. Redkozubova, I. V. Serkov, T. P. Serkova, L. G. Dubova and E. F. Shevtsova, *Bull. Exp. Biol. Med.*, 2016, **160**, 340.
32. V. Alcolea, D. Plano, D. N. Karelia, J. A. Palop, S. Amin, C. Sanmartin and A. K. Sharma, *Eur J Med Chem.*, 2016, **113**, 134.
33. E. A. Cowan, C. D. Oldham and S. W. May, *Arch. Biochem. Biophys.*, 2011, **506**, 201.
34. R. Yu, J. P. Coffman, V. van Fleet-Stalder and T. G. Chasteen, *Environ. Toxicol. Chem.*, 1997, **16**, 140.
35. Y. Tan, Q. Feng, X. Sun, M. Xue, N. Jiang and X. Deng, *Zhong Nan Da Xue Xue Bao Yi Xue Ban*, 2016, **41**, 1305.
36. J. M. Park, D. H. Kim, H. K. Na and Y. J. Surh, *Oncotarget*, 2016, DOI: 10.18632/oncotarget.10289.
37. L. Wang, X. Guo, J. Wang, C. Jiang, M. C. Bosland, J. Lu and Y. Deng, *Cancer Prev. Res.*, 2016, **9**, 35.
38. M. Tarrado-Castellarnau, R. Cortes, M. Zanuy, J. Tarrago-Celada, I. H. Polat, R. Hill, T. W. Fan, W. Link and M. Cascante, *Pharmacol. Res.*, 2015, **102**, 218.
39. M. Liu, C. Hu, Q. Xu, L. Chen, K. Ma, N. Xu and H. Zhu, *Biosci. Rep.*, 2015, **35**.

40. I. Sinha, J. E. Allen, J. T. Pinto and R. Sinha, *Cancer Med.*, 2014, **3**, 252.
41. T. Okuno, K. Miura, F. Sakazaki, K. Nakamuro and H. Ueno, *Biomed. Res.*, 2012, **33**, 201.
42. S. Kassam, H. Goenaga-Infante, L. Maharaj, C. T. Hiley, S. Juliger and S. P. Joel, *Cancer Chemother. Pharmacol.*, 2011, **68**, 815.
43. M. S. Kumar, K. E. Pollok and M. L. Smith, *Anticancer Res.*, 2010, **30**, 291.
44. U. Gundimeda, J. E. Schiffman, D. Chhabra, J. Wong, A. Wu and R. Gopalakrishna, *J. Biol. Chem.*, 2008, **283**, 34519.
45. H. Hu, C. Jiang, G. Li and J. Lu, *Carcinogenesis*, 2005, **26**, 1374.
46. H. Zhao, M. L. Whitfield, T. Xu, D. Botstein and J. D. Brooks, *Mol. Biol. Cell*, 2004, **15**, 506.
47. H. M. Shen, W. X. Ding and C. N. Ong, *Free Radical Biol. Med.*, 2002, **33**, 552.
48. Z. Zhu, W. Jiang, H. E. Ganther and H. J. Thompson, *Cancer Res.*, 2002, **62**, 156.
49. P. Du, U. M. Viswanathan, Z. Xu, H. Ebrahimnejad, B. Hanf, T. Burkholz, M. Schneider, I. Bernhardt, G. Kirsch and C. Jacob, *J. Hazard. Mater.*, 2014, **269**, 74.
50. M. Iman, H. B. Kaboutaraki, R. Jafari, S. A. Hosseini, A. Moghimi, A. Khamesipour, A. B. Harchegani and A. Davood, *Comb. Chem. High Throughput Screening*, 2016, **19**, 847–854.
51. N. D. Facompre, I. Sinha, K. El-Bayoumy, J. T. Pinto and R. Sinha, *Int. J. Cancer*, 2012, **131**, 2134.
52. N. D. Facompre, K. El-Bayoumy, Y. W. Sun, J. T. Pinto and R. Sinha, *Cancer Prev Res.*, 2010, **3**, 975.
53. R. K. Das and S. Bhattacharya, *Asian Pac. J. Cancer Prev*, 2005, **6**, 181.
54. D. Su, Y. Li and V. N. Gladyshev, *Nucleic Acids Res.*, 2005, **33**, 2486.
55. H. Mix, A. V. Lobanov and V. N. Gladyshev, *Nucleic Acids Res.*, 2007, **35**, 414.
56. A. S. Gribling-Burrer, M. Leichter, L. Wurth, A. Huttin, F. Schlotter, N. Troffer-Charlier, V. Cura, M. Barkats, J. Cavarelli, S. Massenet and C. Allmang, *Nucleic Acids Res.*, 2017, **45**, 5399.
57. P. Arsenyan, J. Vasiljeva, S. Belyakov, E. Liepinsh and M. Petrova, *Eur. J. Org. Chem.*, 2015, **2015**, 5842.
58. J. Yan, Y. Guo, Y. Wang, F. Mao, L. Huang and X. Li, *Eur. J. Med. Chem.*, 2015, **95**, 220–229.
59. D. L. Lucas, R. K. Robins, R. D. Knight and D. G. Wright, *Biochem. Biophys. Res. Commun.*, 1983, **115**, 971.
60. M. Koketsu, H. Ishihara, W. Wu, K. Murakami and I. Saiki, *Eur. J. Pharm. Sci.*, 1999, **9**, 157.
61. E. Dominguez-Alvarez, M. Gajdacs, G. Spengler, J. A. Palop, M. A. Marc, K. Kiec-Kononowicz, L. Amaral, J. Molnar, C. Jacob, J. Handzlik and C. Sanmartin, *Bioorg. Med. Chem. Lett.*, 2016, **26**, 2821.
62. M. Gajdacs, G. Spengler, C. Sanmartin, M. A. Marc, J. Handzlik and E. Dominguez-Alvarez, *Bioorg. Med. Chem. Lett.*, 2017, **27**, 797.

63. E. Dominguez-Alvarez, D. Plano, M. Font, A. Calvo, C. Prior, C. Jacob, J. A. Palop and C. Sanmartin, *Eur. J. Med. Chem.*, 2014, **73**, 153.
64. C. Sanmartín, D. Plano, E. Domínguez, M. Font, A. Calvo, C. Prior, I. Encío and J. A. Palop, *Molecules*, 2009, **14**, 3313–3338.
65. G. I. Giles, K. M. Tasker, R. J. Johnson, C. Jacob, C. Peers and K. N. Green, *Chem. Commun.*, 2001, 2490.
66. N. M. Giles, G. I. Giles, J. E. Holley, N. J. Gutowski and C. Jacob, *Biochem. Pharmacol.*, 2003, **66**, 2021.
67. N. M. Giles, N. J. Gutowski, G. I. Giles and C. Jacob, *FEBS Lett*, 2003, **535**, 179.
68. F. H. Fry and C. Jacob, *Curr. Pharm. Des.*, 2006, **12**, 4479.
69. C. Jacob, *Biochem. Soc. Trans.*, 2011, **39**, 1247.
70. G. I. Giles, N. M. Giles, C. A. Collins, K. Holt, F. H. Fry, P. A. Lowden, N. J. Gutowski and C. Jacob, *Chem. Commun.*, 2003, 2030.
71. F. H. Fry, A. L. Holme, N. M. Giles, G. I. Giles, C. Collins, K. Holt, S. Pariagh, T. Gelbrich, M. B. Hursthouse, N. J. Gutowski and C. Jacob, *Org. Biomol. Chem.*, 2005, **3**, 2579.
72. S. Mecklenburg, S. Shaaban, L. A. Ba, T. Burkholz, T. Schneider, B. Diesel, A. K. Kiemer, A. Roseler, K. Becker, J. Reichrath, A. Stark, W. Tilgen, M. Abbas, L. A. Wessjohann, F. Sasse and C. Jacob, *Org. Biomol. Chem.*, 2009, **7**, 4753.
73. M. Doering, L. A. Ba, N. Lilienthal, C. Nicco, C. Scherer, M. Abbas, A. A. Zada, R. Coriat, T. Burkholz, L. Wessjohann, M. Diederich, F. Batteux, M. Herling and C. Jacob, *J. Med. Chem.*, 2010, **53**, 6954.
74. S. E. Denmark and Y. Fan, *J. Am. Chem. Soc.*, 2003, **125**, 7825.
75. J. Zhang, A. Jacobson, J. R. Rusche and W. Herlihy, *J. Org. Chem.*, 1999, **64**, 1074.
76. H. C. Kolb, M. G. Finn and K. B. Sharpless, *Angew. Chem., Int. Ed.*, 2001, **40**, 2004.
77. C. A. Collins, F. H. Fry, A. L. Holme, A. Yiakouvaki, A. Al-Qenaei, C. Pourzand and C. Jacob, *Org. Biomol. Chem.*, 2005, **3**, 1541.
78. S. Pariagh, K. M. Tasker, F. H. Fry, A. L. Holme, C. A. Collins, N. Okarter, N. Gutowski and C. Jacob, *Org. Biomol. Chem.*, 2005, **3**, 975.
79. C. W. Tornoe, C. Christensen and M. Meldal, *J. Org. Chem.*, 2002, **67**, 3057.
80. V. V. Rostovtsev, L. G. Green, V. V. Fokin and K. B. Sharpless, *Angew. Chem., Int. Ed.*, 2002, **41**, 2596.
81. G. C. Tron, T. Pirali, R. A. Billington, P. L. Canonico, G. Sorba and A. A. Genazzani, *Med. Res. Rev.*, 2008, **28**, 278.
82. J. F. Lutz and Z. Zarafshani, *Adv. Drug Delivery Rev.*, 2008, **60**, 958.
83. J. E. Moses and A. D. Moorhouse, *Chem. Soc. Rev.*, 2007, **36**, 1249.
84. S. L. de Castro, F. S. Emery and E. N. da Silva Junior, *Eur. J. Med. Chem.*, 2013, **69**, 678.
85. E. H. da Cruz, M. A. Silvers, G. A. Jardim, J. M. Resende, B. C. Cavalcanti, I. S. Bomfim, C. Pessoa, C. A. de Simone, G. V. Botteselle, A. L. Braga,

D. K. Nair, I. N. Namboothiri, D. A. Boothman and E. N. da Silva Junior, *Eur. J. Med. Chem.*, 2016, **122**, 1.
86. G. I. Giles, F. H. Fry, K. M. Tasker, A. L. Holme, C. Peers, K. N. Green, L. O. Klotz, H. Sies and C. Jacob, *Org. Biomol. Chem.*, 2003, **1**, 4317.
87. G. I. Giles, K. M. Tasker and C. Jacob, *Free Radical Biol. Med.*, 2001, **31**, 1279.
88. C. Jacob, G. I. Giles, N. M. Giles and H. Sies, *Angew. Chem., Int. Ed.*, 2003, **42**, 4742.
89. *Evolution,* https://en.wikipedia.org/wiki/Evolution_(2001_film).

CHAPTER 11

Ebselen and Thioredoxin Systems in Human Health, Disease and Therapeutic Potential

JUN LU,*[a,b] XIAOYUAN REN,[b] LILI ZOU[b] AND ARNE HOLMGREN*[b]

[a] School of Pharmaceutical Sciences, Southwest University, 400715 Chongqing, China; [b] Division of Biochemistry, Department of Medical Biochemistry and Biophysics, Karolinska Institutet, SE-171 77 Stockholm, Sweden
*Email: jun.lu@ki.se; junlu@swu.edu.cn; arne.holmgren@ki.se

11.1 Introduction

11.1.1 Mammalian and Bacterial Thioredoxin System

Thioredoxin systems, comprising thioredoxin (Trx), thioredoxin reductase (TrxR) and nicotinamide adenine dinucleotide phosphate (NADPH), are major disulfide reductase systems which play critical roles in maintaining cellular redox homeostasis and DNA synthesis and repair.[1–3] Oxidized Trx is reduced by TrxR to yield reduced Trx and oxidized TrxR, which can obtain electrons from NADPH. The reduced Trx thus provides electrons to its substrate proteins, such as ribonucleotide reductase (RNR), peroxiredoxin, methionine sulfoxide reductase (Msr), protein disulfide isomerase, and so

Organoselenium Compounds in Biology and Medicine: Synthesis, Biological and Therapeutic Treatments
Edited by Vimal Kumar Jain and K. Indira Priyadarsini

Published by the Royal Society of Chemistry, www.rsc.org

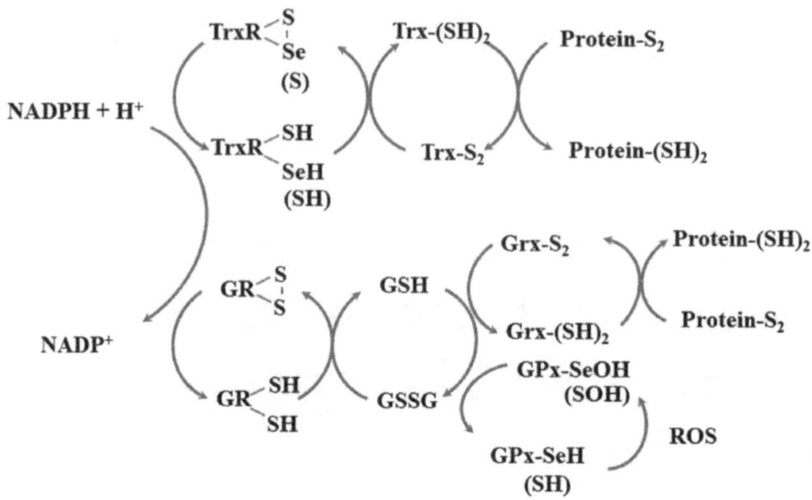

Figure 11.1 The thioredoxin system, comprising NADPH, thioredoxin (Trx) and thioredoxin reductase (TrxR), and the glutathione system, comprising NADPH, glutathione reductase (GR), glutathione (GSH) and glutaredoxin (Grx), are the two major disulfide reductase systems in the cell.

on, which are involved in various cellular events, *via* a dithiol–disulfide exchanging reaction, forming redox cycling processes (Figure 11.1).

The thioredoxin system is extensively present in all organisms,[3] but there is quite a distinct difference between mammalian and bacterial Trx systems, especially the TrxRs.[4] The mammalian TrxR is a dimeric selenoprotein with a molecular weight (MW) of ~115 kDa,[5] whereas bacterial TrxR is also a dimeric flavoprotein, but has a MW of ~70 kDa and does not contain selenocysteine (Sec, U) (Figure 11.1). Mammalian TrxR, which is a head-to-tail dimer, contains two active sites, an N-terminal active site containing a glutathione reductase CXXXXC motif, and a C-terminal active site containing a GCUG motif. The C-terminal -GCUG motif where U is a selenocysteine becomes a selenosulfide upon reduction of oxidized Trx. The mammalian TrxR with the essential Sec in the open active site[6] has remarkably broad substrate specificity, including reduction of hydrogen peroxide (H_2O_2).[7] TrxR1 is a cytosolic and nuclear enzyme, whereas mitochondria have their own thioredoxin system with TrxR2 and Trx2. TrxR2 is also a selenoenzyme with an essential Sec residue.[8] Bacterial TrxR has only one active site containing a CXXC motif. The reaction mechanism is different from mammalian TrxR. In mammalian cells, TrxR catalyzes electron transfer from NADPH to the N-terminal active site, then to the C-terminal active site in the other subunit of the head-to-tail dimer, and finally to its substrates.[9] In this process, no big conformational change is involved for mammalian TrxR.[10,11] In contrast, the reaction mechanism of bacterial TrxR is involved in the electron transfer from NADPH to its single active site and then to its substrates with a big conformational change in the protein.[12]

Trx is a ubiquitous small protein with a MW of 12 kDa[1] and a highly conserved active site -Trp-Cys-Gly-Pro-Cys-. Trx is an electron donor for

RNR, which is an essential enzyme to supply building blocks for DNA synthesis and repair by reducing ribonucleoside diphosphates to deoxyribonucleoside diphosphate. Trx is thus important for DNA synthesis and repair *via* RNR.[1] Besides RNR, methionine sulfoxide reductase reduce free or protein bond methionine sulfoxide using reduced Trx as an electron donor. Peroxiredoxins, which utilize its cysteine (Cys) residues to decompose H_2O_2, are also substrates of Trxs. *Via* these antioxidant enzymes, Trx systems therefore become a critical factor in defense against oxidative stress. In addition, mammalian Trxs regulate the activity of several transcription factors and thus mediate redox signaling processes.[3]

As mentioned earlier, there are two major types of Trx in mammalian cells, Trx1 in cytosol and Trx2 in mitochondria, both belonging to thioredoxin fold protein family. Human Trx2 has only two Cys residues in its active sites, while human Trx1 has three additional Cys residues besides the two Cys in the active site, which can form a second Cys62-Cys69 disulfide and regulate the reductive activity of Trx1.[3,13] Generally, bacterial thioredoxins, such as *Escherichia coli* Trx1, has only two Cys in its active site and the same as human Trx2.[4]

11.1.2 The Relationship of the Thioredoxin System with the Glutathione System

The glutathione (GSH)-glutaredoxin (Grx) system, comprising NADPH, glutathione reductase (GR), GSH and Grx, is another disulfide-reducing system. Although GSH-Grx has many overlapping functions with the thioredoxin system, the two systems are regarded as two parallel pathways in which Grx can also reduce RNR.[2] The coupled antioxidant enzymes in the GSH system are glutathione peroxidases (GPx), which are distributed in various tissues and are highly efficient at removing reactive oxygen species (ROS), particularly H_2O_2 and lipid peroxides (Figure 11.1). The presence of Sec in the active site of these enzymes is responsible for the high catalytic activity. A conserved catalytic triad containing Sec, Gln and Trp exerts a redox catalytic cycle by successive oxidation and reduction of Sec (Figure 11.1).[10]

Grx performs its disulfide reduction activity *via* a dithiol or monothiol reaction mechanism.[14] The major function of Grx is to catalyze the deglutathionylation from glutathionylated proteins. The GSH-Grx system and the Trx system exhibit cross-talk between the pathways. Trx and Grx can serve as a backup system when the other system is not working.[15] Mitochondrial Grx2 is reduced by the GSH system and TrxR. Moreover, a high concentration of GSH showed a similar efficiency to reduce oxidized Trx1 as TrxR.[15]

11.1.3 Ebselen

Ebselen (2-phenyl-1,2-benzisoselenazol-3(2*H*)-one) has been reported to exhibit anti-oxidant, anti-inflammatory, anti-atherosclerotic and cytoprotective effects *in vitro* and *in vivo*.[16] Ebselen is a seleno-organic compound, originally developed as a GPx mimic, which consumes electrons from reduced GSH, and its selenol form has a fast reaction rate with ROS (Figure 11.2). In recent years, ebselen was shown to have direct reactions with thioredoxin systems and to play a critical role in modulating ROS levels in cells, which may contribute to its antioxidant and antibacterial activities (Figures 11.2 and 11.3).

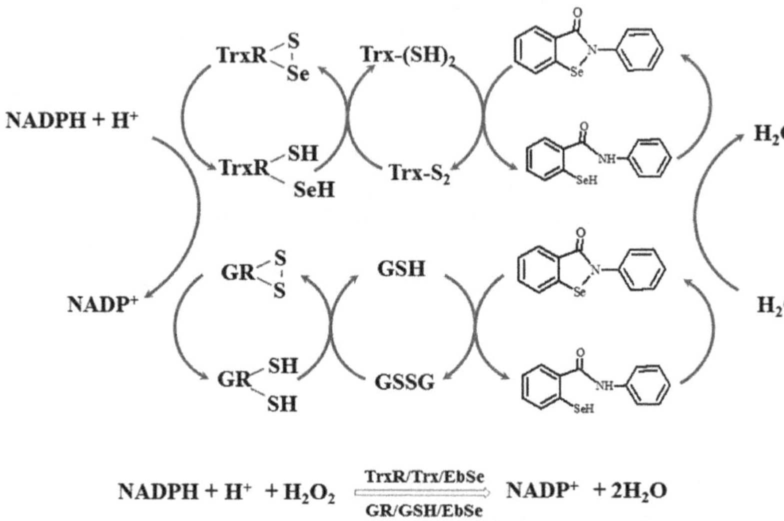

Figure 11.2 Ebselen acts as a glutathione peroxidase mimic or a thioredoxin-dependent peroxidase mimic to remove reactive oxygen species.

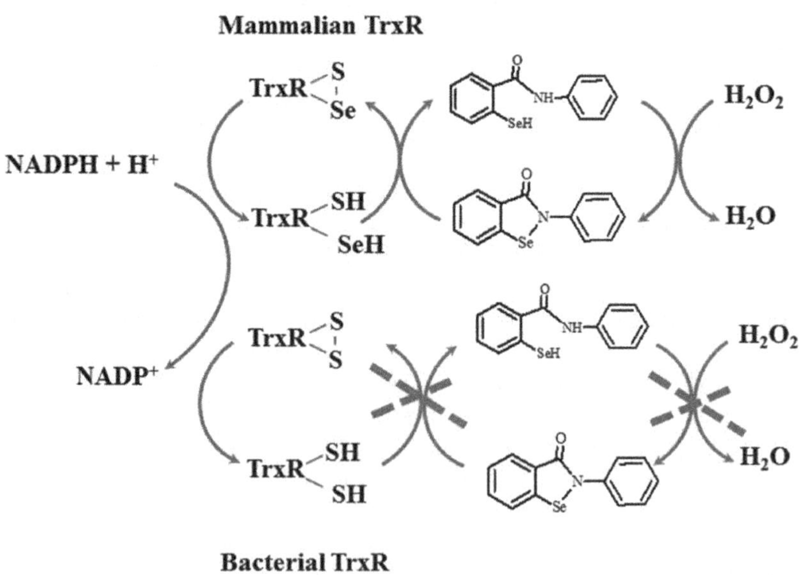

Figure 11.3 Ebselen is a substrate of mammalian thioredoxin reductase (TrxR), but an inhibitor of bacterial TrxR.

11.2 Interaction of Ebselen with Thioredoxin Systems

11.2.1 Reaction of Ebselen with Mammalian Thioredoxin Systems

Ebselen is a substrate for mammalian TrxR and can be converted to ebselen selenol by consuming NADPH (Figure 11.3).[1] The reduction reaction acts with an apparent K_M value of 2.5 μM and a k_{cat} of 588 min^{-1}. Oxidants such as H_2O_2 react with ebselen selenol to produce ebselen selenenic acid (EbSeOH), and spontaneously produce H_2O and ebselen to form a redox cycling mechanism for ebselen.

Upon the addition of Trx to TrxR, NADPH and ebselen, the reaction rate increases several-fold (Figure 11.2). This may be achieved by a very fast oxidation of reduced Trx by ebselen at a reaction rate constant $>2\times10^7$ M^{-1} s^{-1}.[17] This rate is orders of magnitude faster than the reaction of reduced Trx with a substrate such as insulin disulfides. Thus, ebselen can compete with disulfide substrates for the reduction by Trx. These results indicate that ebselen can serve as a mediator in the redox environment *via* the Trx system.

11.2.2 Ebselen as an Antioxidant to Remove ROS Together with the Mammalian Trx System

In combination with mammalian TrxR and Trx, ebselen can serve as a peroxidase to remove H_2O_2. Ebselen may react with selenol in TrxR and form diselenides, following the release of the ebselen selenenic form, which can directly react with H_2O_2. 2 μM ebselen enhanced the H_2O_2 reductase activity of mammalian TrxR 10-fold and 25-fold, each in the presence of additional 5 μM Trx (Figures 11.2 and 11.3). Therefore, ebselen acts not only as a GPx mimic, but also as an efficient TrxR (Trx)-related peroxiredoxin to scavenging ROS, particularly H_2O_2. In fact, TrxR at 50 nM and 2 μM ebselen provide an eight-fold higher activity compared with 1 mM GSH and 1 unit of GR to reduce H_2O_2, and the addition of 5 μM Trx in TrxR increases the difference to 13-fold.[18] This indicates that the Trx system is a major electron donor to ebselen to remove ROS.

Moreover, ebselen can form ebselen diselenide, which is a reservoir of ebselen and a substrate of TrxR/Trx.[17] Another example of ebselen serving as a catalyst is the reduction of the alpha-tocopherol-quinone (TQ) by TrxR,[19] in which TQ is a product of alpha-tocopherol oxidation. Thus, the generation of ebselen can work as a reactant to H_2O_2 and peroxynitrite and can also participate in the reduction of vitamin E, vitamin C or dehydroascorbate.[19]

11.2.3 Inhibition of the Bacterial Thioredoxin System by Ebselen

Ebselen is an excellent substrate for mammalian TrxR and a superfast oxidant of reduced thioredoxin.[17] In contrast, it acts as a reversible competitive inhibitor of *E. coli* TrxR, with $K_i = 0.52 \pm 0.13$ μM. Ebselen inhibits *E. coli* TrxR *in vitro* by binding to its active site Cys, forming a selenosulfide bond, which is difficult to reduce. Such a selenosulfide bond may be also formed between mammalian TrxR and ebsulfur, an analog of ebselen. However, ebsulfur is a substrate of *E. coli* TrxR.

That ebselen is a substrate of mammalian TrxR, but an inhibitor of *E. coli* TrxR may be due to the different reaction mechanisms between bacterial and mammalian TrxRs (Figure 11.3). *E. coli* TrxR is known to be highly specific for its *E. coli* Trx with a CATC active motif. The attacking Cys in *E. coli* TrxR forms an intermolecular disulfide with a Cys in its substrates like *E. coli* Trx or ebsulfur, and subsequently the resolving thiolate in *E. coli* TrxR reacts with the disulfide and releases the substrate. When ebselen reacts with *E. coli* TrxR, it forms a stable selenosulfide bond with the attacking Cys in its active site of *E. coli* TrxR, which cannot be attacked with the resolving thiolate from the other Cys residue, and thus blocks the enzyme activity.

For mammalian TrxR with ebselen or ebsulfur, the reaction occurs in the C-terminal active site GCUG motif; in this case, the resolving Cys is adjacent to the attacking Sec,[20] and the enzyme has a high capacity to reduce different substrates,[10] indicating that the C-terminal active site is not easily blocked by forming selenosulfide or diselenide bonds.

11.2.4 Reaction of Ebselen/Ebsulfur with other Disulfide Reductases

Ebselen is not an inhibitor of GR, in fact, in the activity assay of ebselen as a GPx, GR is always present in the assay system. However, some flavoproteins are potential targets of ebselen. For example, the unique trypanothione system mediator trypanothione reductase (TryR) can react with ebselen. All seven cysteines in TryR can be bound with ebselen, indicating that ebselen reacts with thiols unspecifically. In contrast, ebsulfur inhibits TryR with low efficiency, but only one Cys in TryR can bind with ebsulfur, indicating that ebsulfur inhibits TryR with a lower capacity but a higher specificity (Figure 11.4). This specificity is quite important for the drug, implying that other thiols, including the active thiol in bovine serum albumin, may have lower reactivity compared with the Cys in TryR. Thus, ebsulfur was shown to be a more potent killer of *Trypanosoma brucei* compared to ebselen, which may be the result of selective inhibition of ebsulfur for TryR.[21]

11.3 Therapeutic Potential of Ebselen in Disease

11.3.1 Ebselen use in Human Infections

11.3.1.1 Use of Ebselen Against GSH-negative Bacterial Pathogens

As discussed earlier, there is a distinct difference between mammalian and bacterial TrxRs, both in structure and reaction mechanism. More

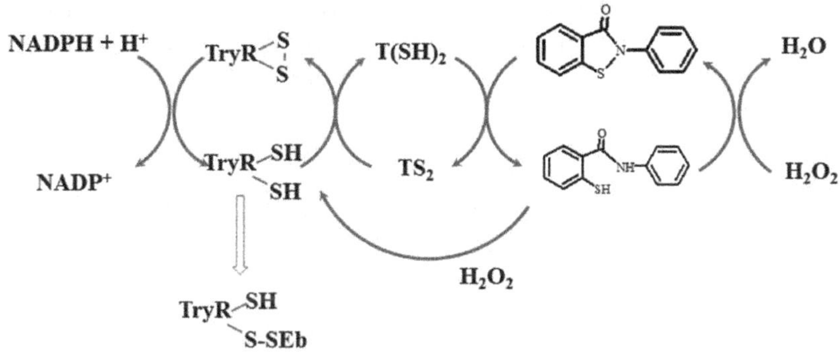

Figure 11.4 Ebselen sulfur analog ebsulfur works as a substrate of trypanothione reductase (TryR), and also as an inhibitor of TryR.

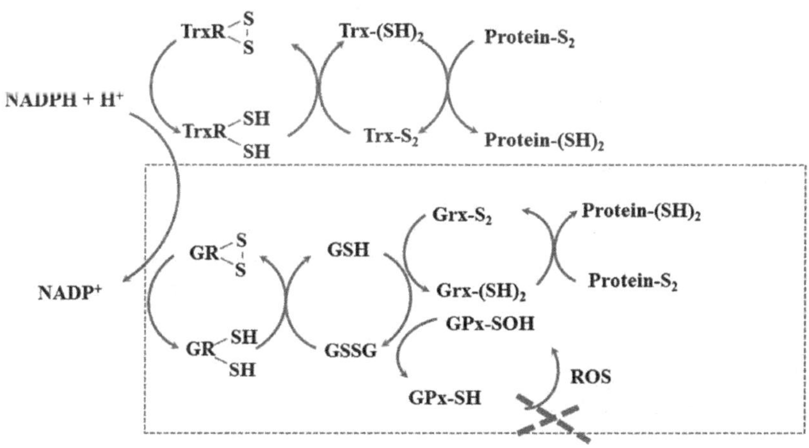

Figure 11.5 Bacterial thioredoxin reductase (TrxR) emerges as a novel antibiotic target against some bacterial pathogens in which the glutathione system is missing.

importantly, the roles of the thioredoxin system for cell viability in some pathogenic bacteria and mammalian cells are different. The presence of GSH in the mammalian cells makes TrxR dispensable in mammals. Even under conditions in which both GR and TrxR1 are deficient, mice can maintain sufficient hepatic redox homeostasis and intermediary metabolism for long-term survival.[22] But in some infectious pathogens, such as *Staphylococcus aureus*, *Helicobacter pylori* and *Mycobacterium tuberculosis*, GSH, the most abundant small thiol molecule in mammalian cells, is absent (Figure 11.5).[4] Therefore, the TrxR is essential for these pathogenic bacteria, and this fact can be used as a target for rational new antibiotic therapies.

Ebselen displays different activity against mammalian and bacterial TrxR and therefore emerges as a lead antibacterial drug. The concept that a drug can inhibit the bacterial thioredoxin system and kill the bacteria has been exploited.[23] *E. coli* DHB4 mutant strains with dysfunctional GSH systems, such as the *E. coli gor*$^-$ strain (lacking glutathione reductase) and the null mutant for gshA (*gshA*$^-$, lacking GSH) were more susceptible to ebselen than the wild-type strains. Moreover, in the growth-inhibited bacteria, the Trxs, including Trx1 and Trx2, were fully oxidized, suggesting that electron transfer *via* the Trx pathway is blocked.[23] This is a strong evidence that ebselen reacts with both the GSH and Trx systems in bacterial cells.

The sensitivity of bacterial pathogens lacking GSH confirmed that ebselen is a good potential antibacterial drug against GSH-negative bacterial pathogens, including *S. aureus*,[24] *H. pylori* and *M. tuberculosis.*[23] Interestingly, many multidrug-resistant bacterial strains are still sensitive to ebselen. To test the potential of ebselen as an antibiotic against clinically important GSH-negative pathogens, we examined whether the compound could inhibit the growth of these bacteria. Ebselen exhibited a marked inhibitory effect on these bacteria, including *H. pylori* strains resistant to macrolides and multidrug-resistant *M. tuberculosis* strains.[23] The therapeutic efficacy of ebselen was confirmed in a mouse model of staphylococcal skin infections. In a project to repurpose already-approved drugs for antibiotic development, ebselen was screened to be a potential efficient antibiotic against methicillin-resistant *S. aureus* (MRSA) in comparison to vancomycin and linezolid.[25,26] That bacterial TrxR can be used as a potential antibiotic target was verified by Harbut *et al.*[27] They found that auranofin, a known mammalian TrxR inhibitor, is an efficient inhibitor against bacterial TrxR, and demonstrates broad-spectrum bactericidal effects by disrupting bacterial thiol-redox homeostasis.[27]

11.3.1.2 Use of Ebselen Against Bacterial Drug Resistance

Ebselen and its analogs have also been shown to be inhibitors of *Bacillus anthracis* TrxR. The ebselen analogs displayed antibacterial activity on *Bacillus subtilis*, *S. aureus*, *Bacillus cereus* and *M. tuberculosis*. There existed a high resistance barrier for the bacteria to develop ebselen-resistant strains and isolating resistant mutants was found to be difficult and, ultimately, unsuccessful.[28] The reason for the difficulty of developing ebselen-resistant bacteria may be because ebselen has exerted a multiple drug-target mechanism, although the exact mechanism is not yet known. In addition to the thioredoxin and GSH systems, ebselen has been reported to have some other potential targets in bacteria, which may help it to overcome the bacterial drug resistance. Ebselen was identified as an inhibitor of diguanylate cyclases by a high-throughput screening in another study by covalent binding to Cys residues.[29] This may block the cyclic-di-GMP signaling pathway, regulating biofilm formation, flagella-mediated motility in *Pseudomonas aeruginosa*, and thus its pathogenesis. Ebselen was identified as an inhibitor of the cysteine protease domain (CPD) in the *Clostridium difficile* major virulence factor toxin B (TcdB) by another high-throughput screening study.

A CPD can bind with one or two ebselen molecules with a covalent modification probably occurring at the active site Cys. The treatment of ebselen blocked the toxic effects of TcdB and the pathology of *C. difficile* infection in mice.[30] Another study also indicated that ebselen is a potent inhibitor of the *M. tuberculosis* Ag85 complex by binding covalently to a Cys residue (C209) located near the Ag85C active site, which is central to the synthesis of major components of the inner and outer leaflets of the mycobacterial outer membrane.

11.3.1.3 Use of Ebselen in Combination with other Antibiotics

Ebselen exhibited a synergistic relationship with some conventional antimicrobials such as clindamycin, erythromycin and rifampicin, and the combination treatment significantly enhanced the killing effect on MRSA when compared to monotherapy.[25] Most recently, we discovered a new antibiotic strategy targeting bacterial thiol-dependent redox systems *via* strong synergistic effects of silver with ebselen against multidrug-resistant Gram-negative bacteria. Treatment with both silver and ebselen caused a fast depletion of GSH, inhibition of the thioredoxin system and elevation of cellular ROS, which accounted for bactericidal effect. The combination of the two drugs was efficient in the treatment of mild and acute multidrug-resistant *E. coli* peritonitis in mice.[56] Five of the most difficult-to-treat MDR Gram-negative pathogens, *Klebsiella pneumoniae*, *Acinetobacter baumannii*, *P. aeruginosa*, *Enterobacter cloacae* and *E.coli*, were sensitive to the combination of Ag^{+} and ebselen.[56]

11.3.1.4 Use of Ebselen Analogs in Parasitic Infections

Trypanosoma brucei, the major causative agent for African trypanosomiasis, including a human sleeping sickness, possess a unique thiol redox metabolism system.[31,32] The functional thiol-dependent redox system in the parasites contains TryR and trypanothione, but the antioxidant enzymes including TrxR, GR and catalase, which normally exist in other eukaryotes and prokaryotes are missing. This property makes the TryR a critical antioxidant enzyme to protect the parasites against oxidative stress.[33,34] As described earlier, ebselen has a non-specific binding with TryR, while its sulfur analog ebsulfur is a specific irreversible inhibitor for TryR. Thus, ebsulfur was shown to be an efficient trypanocidal compound with a high selectivity index over mammalian cells. The trypanocidal effect is dependent on the elevation of the parasitic intracellular ROS level. Some soluble ebsulfur analogs were synthesized and used to treat *T. b. brucei* infection in mice; these compounds demonstrated significant synergistic effects with nifurtimox and cured some of the infected mice.[21] Interestingly, besides TryR, ebselen and ebsulfur have been identified to be potent inhibitors of *T. brucei* hexokinase 1 (TbHK1) based on high-throughput screening.[35]

11.3.1.5 Use of Ebselen in other Microbial Infections

Ebselen has also been reported to have potent antifungal activity, including against clinically relevant isolates of both *Candida* and *Cryptococcus*. Similar to its antibacterial activity, the antifungal mechanism of ebselen involves the depletion of GSH and the elevation of ROS, although the effects on the thioredoxin system are not clear. As shown in two *Caenorhabditis elegans* models of infection, ebselen is better than the conventional antifungal drugs (fluconazole, flucytosine and amphotericin) at removing *Candida* and *Cryptococcus* fungi.[36]

Ebselen also exhibited a potential use in antiviral treatment. ROS play an important role during viral infection.[37] The protective effect of ebselen during virus infection largely depends on its antioxidant property as a GPx mimic which scavenges reactive oxygen/nitrogen species efficiently. *In vitro* studies showed that ROS are indispensable for virus binding and growth, especially during the initial phase of infection, and ebselen treatment lowered the virus titers in a cellular model.[38] In combination with apocynin, a NADPH oxidase inhibitor, ebselen ameliorated lung inflammation in a mouse model of influenza A virus-induced acute exacerbations of chronic obstructive pulmonary disease.[39] Beyond the general antioxidant effect, ebselen also targets other proteins important for viral replication. For example, the hepatitis C virus non-structural protein 3 (NS3) contains a serine protease, which cleaves proteins essential for innate immunity, and an RNA helicase, which is indispensable for viral genome replication and assembly.[40] Ebselen inhibited NS3 helicase activity, but had no effect on its protease activity, suggesting that a specific mechanism is involved in the process.[41] the human immunodeficiency virus type 1 capsid plays an important role during the viral life cycle. Ebselen was found to be a novel capsid inhibitor in a high-throughput screening study and was suggested as a promising drug for retroviral infection.[42]

11.3.2 Ebselen use in other Diseases

Ebselen is a lipid-soluble organoselenium compound. Numerous *in vivo* experiments have demonstrated that ebselen is able to inhibit both vasospasm and tissue damage in cerebral stroke/ischemia animal models (*i.e.* rat, mice, guinea pig and dog).[43,44] For example, ebselen has been shown to provide significant protection against ischemic damage in both gray and white matter,[45] and in the ventral posterior nucleus of rodent brains.[46] Additionally, administration of ebselen in gerbils significantly reduces neuronal death induced by ischemia and reperfusion in the hippocampal CA1 region.[47] It has been reported that ebselen inhibits inositol monophosphatase and acts as a safe treatment for bipolar disorder.[48]

The optimal GPx-like activity of ebselen was observed at neutral or near-physiological pH (7.4), and the activity of ebselen was barely detected in acidic medium.[49] During ischemia, brain pH falls rapidly within the first

5 min from 7.0 to 6.2, which is not optimal for ebselen's GPx-like activity.[50] Meanwhile, since ebselen has shown promising activity in the treatment of stroke, it seems that GPx-like activity of ebselen in the brain may not occur.[51,52] Thus, the enzymatic reduction of ebselen by TrxR might be the main contribution to its antioxidant property during ischemia. Unlike inorganic and aliphatic selenium compounds, ebselen does not liberate the selenium moiety, which remains within the ring structure. Subsequent metabolism involves methylation, glucuronidation and hydroxylation. Additionally, ebselen could also inhibit enzymes such as lipoxygenases, nitric oxide synthases, nitrogen oxides, protein kinase C and H^+/K^+-ATPase.[53]

Besides the relationship with redox systems, another neuroprotective mechanism of ebselen is involved in controlling the expression of γ-aminobutyric acid shunt enzymes to supply the tricarboxylic acid cycle,[54] and significantly inhibiting acetylcholinesterase activity,[55] demonstrating its engagement in the metabolic system.

11.4 Concluding Remarks

The thiol-dependent antioxidant systems including glutathione, thioredoxin and trypanothione systems are critical for thiol metabolism and cell viability, but roles of the thioredoxin and glutathione systems in regulating redox states of key proteins involved in bacterial cell death and drug resistance are poorly understood. The organic selenium compound ebselen targets thiol and primarily disrupts thiol metabolism, serving as a potent lead antimicrobial drug. This property provides an excellent probe for us to detect the role of thiol metabolism in the bacterial cell death pathway and the regulation of the development of drug resistance in bacteria.

References

1. A. Holmgren, *Annu. Rev. Biochem.*, 1985, **54**, 237.
2. A. Holmgren, *J. Biol. Chem.*, 1989, **264**, 13963.
3. C. H. Lillig and A. Holmgren, *Antioxid. Redox Signaling*, 2007, **9**, 25.
4. J. Lu and A. Holmgren, *Free Radical Biol. Med.*, 2014, **66**, 75.
5. L. Zhong, E. S. Arner, J. Ljung, F. Aslund and A. Holmgren, *J. Biol. Chem.*, 1998, **273**, 8581.
6. L. Zhong and A. Holmgren, *J. Biol. Chem.*, 2000, **275**, 18121.
7. M. Bjornstedt, M. Hamberg, S. Kumar, J. Xue and A. Holmgren, *J. Biol. Chem.*, 1995, **270**, 11761.
8. E. S. J. Arner, *Biochim. Biophys. Acta, Gen. Subj.*, 2009, **1790**, 495.
9. L. Zhong, E. S. Arner and A. Holmgren, *Proc. Natl. Acad. Sci. U. S. A.*, 2000, **97**, 5854.
10. L. V. Papp, J. Lu, A. Holmgren and K. K. Khanna, *Antioxid. Redox Signaling*, 2007, **9**, 775.
11. J. Lu and A. Holmgren, *J. Biol. Chem.*, 2009, **284**, 723.

12. B. W. Lennon Jr., C. H. Williams and M. L. Ludwig, *Science*, 2000, **289**, 1190.
13. X. Zhang, J. Lu, X. Ren, Y. Du, Y. Zheng, P. V. Ioannou and A. Holmgren, *Free Radical Biol. Med.*, 2015, **89**, 192.
14. C. H. Lillig, C. Berndt and A. Holmgren, *Biochim. Biophys. Acta*, 2008, **1780**, 1304.
15. Y. Du, H. Zhang, J. Lu and A. Holmgren, *J. Biol. Chem.*, 2012, **287**, 38210.
16. H. Sies and H. Masumoto, *Adv. Pharmacol.*, 1997, **38**, 229.
17. R. Zhao, H. Masayasu and A. Holmgren, *Proc. Natl. Acad. Sci. U. S. A.*, 2002, **99**, 8579.
18. R. Zhao and A. Holmgren, *J. Biol. Chem.*, 2002, **277**, 39456.
19. J. Fang, L. Zhong, R. Zhao and A. Holmgren, *Toxicol. Appl. Pharmacol*, 2005, **207**, 103.
20. Q. Cheng, T. Sandalova, Y. Lindqvist and E. S. Arnér, *J. Biol. Chem.*, 2009, **284**, 3998.
21. J. Lu, S. K. Vodnala, A. L. Gustavsson, T. N. Gustafsson, B. Sjoberg, H. A. Johansson, S. Kumar, A. Tjernberg, L. Engman, M. E. Rottenberg and A. Holmgren, *J. Biol. Chem.*, 2013, **288**, 27456.
22. S. Eriksson, J. R. Prigge, E. A. Talago, E. S. Arner and E. E. Schmidt, *Nat. Commun.*, 2015, **6**, 6479.
23. J. Lu, A. Vlamis-Gardikas, K. Kandasamy, R. Zhao, T. N. Gustafsson, L. Engstrand, S. Hoffner, L. Engman and A. Holmgren, *FASEB J.*, 2013, **27**, 1394.
24. R. Nozawa, T. Yokota and T. Fujimoto, *Antimicrob. Agents Chemother.*, 1989, **33**, 1388.
25. S. Thangamani, W. Younis and M. N. Seleem, *PLoS One*, 2015, **10**, e0133877.
26. S. Thangamani, W. Younisand, and M. N. Seleem, *Sci. Rep.*, 2015, **5**, 11596.
27. M. B. Harbut, C. Vilcheze, X. Luo, M. E. Hensler, H. Guo, B. Yang, A. K. Chatterjee, V. Nizet, W. R. Jacobs Jr., P. G. Schultz and F. Wang, *Proc. Natl. Acad. Sci. U. S. A.*, 2015, **112**, 4453.
28. T. N. Gustafsson, H. Osman, J. Werngren, S. Hoffner, L. Engman and A. Holmgren, *Biochim. Bophys. Acta*, 2016, **1860**, 1265.
29. O. J. Lieberman, M. W. Orr, Y. Wang and V. T. Lee, *ACS Chem. Biol.*, 2014, **9**, 183–192.
30. K. O. Bender, M. Garland, J. A. Ferreyra, A. J. Hryckowian, M. A. Child, A. W. Puri, D. E. Solow-Cordero, S. K. Higginbottom, E. Segal, N. Banaei, A. Shen, J. L. Sonnenburg and M. Bogyo, *Sci. Transl. Med.*, 2015, **7**, 306ra148.
31. A. H. Fairlamb, P. Blackburn, P. Ulrich, B. T. Chait and A. Cerami, *Science*, 1985, **227**, 1485–1487.
32. R. L. Krauth-Siegel and A. E. Leroux, *Antioxid. Redox Signal.*, 2012, **7**, 583–607.
33. S. Krieger, W. Schwarz, M. R. Ariyanayagam, A. H. Fairlamb, R. L. Krauth-Siegel and C. Clayton, *Mol. Microbiol.*, 2000, **35**, 542–552.

34. M. A. Comini, R. L. Krauth-Siegel and L. Flohe, *Biochem. J.*, 2007, **402**, 43–49.
35. E. R. Sharlow, T. A. Lyda, H. C. Dodson, G. Mustata, M. T. Morris, S. S. Leimgruber, K. H. Lee, Y. Kashiwada, D. Close, J. S. Lazo and J. C. Morris, *PLoS Negl. Trop. Dis.*, 2010, **4**, e659.
36. S. Thangamani, H. E. Eldesouky, H. Mohammad, P. E. Pascuzzi, L. Avramova, T. R. Hazbun and M. N. Seleem, *Biochim. Biophys. Acta*, 2017, **1861**, 3002–3010.
37. K. B. Schwarz, *Free Radic. Biol. Med.*, 1996, **21**, 641–649.
38. R. D. Michalek, S. T. Pellom, B. C. Holbrook and J. M. Grayson, *Virology*, 2008, **379**, 205–212.
39. L. C. Oostwoud, P. Gunasinghe, H. J. Seow, J. M. Ye, S. Selemidis, S. Bozinovski and R. Vlahos, *Sci. Rep.*, 2016, **6**, 20983.
40. K. D. Raney, S. D. Sharma, I. M. Moustafa and C. E. Cameron, *J. Biol. Chem.*, 2010, **285**, 22725–22731.
41. S. Mukherjee, W. S. Weiner, C. E. Schroeder, D. S. Simpson, A. M. Hanson, N. L. Sweeney, R. K. Marvin, J. Ndjomou, R. Kolli, D. Isailovic, F. J. Schoenen and D. N. Frick, *ACS Chem. Biol.*, 2014, **9**, 2393–2403.
42. S. Thenin-Houssier, I. M. de Vera, L. Pedro-Rosa, A. Brady, A. Richard, B. Konnick, S. Opp, C. Buffone, J. Fuhrmann, S. Kota, B. Billack, M. Pietka-Ottlik, T. Tellinghuisen, H. Choe, T. Spicer, L. Scampavia, F. Diaz-Griffero, D. J. Kojetin and S. T. Valente, *Antimicrob. Agents Chemother.*, 2016, **60**, 2195–2208.
43. M. Salazar, J. A. Pariente, G. M. Salido and A. Gonzalez, *Toxicology*, 2008, **244**, 280–291.
44. C. L. Dalla Corte, L. L. Bastos, F. Dobrachinski, J. B. T. Rocha and F. A. A. Soares, *Brain Res.*, 2012, **1430**, 101–111.
45. H. Imai, H. Masayasu, D. Dewar, D. I. Graham and I. M. Macrae, *Stroke*, 2001, **32**, 2149.
46. M. X. He, S. H. Xing, B. Yang, L. Q. Zhao, H. Y. Hua, Z. J. Liang, W. L. Zhou, J. S. Zeng and Z. Pei, *Brain Res.*, 2007, **1181**, 83.
47. H. Koizumi, H. Fujisawa, E. Suehiro, S. Shirao and M. Suzuki, *Neurol. Med. Chir*, 2011, **51**, 337.
48. N. Singh, A. C. Halliday, J. M. Thomas, O. V. Kuznetsova, R. Baldwin, E. C. Y. Woon, P. K. Aley, I. Antoniadou, T. Sharp, S. R. Vasudevan and G. C. Churchill, *Nat. Commun.*, 2013, **4**, 1332.
49. W. Hassan, M. Ibrahim, C. W. Nogueira, A. L. Braga, I. U. Mohammadzai, P. S. Taube and J. B. T. Rocha, *Brain Res.*, 2009, **1258**, 71.
50. E. M. Nemoto and S. Frinak, *Stroke*, 1981, **12**, 77.
51. I. J. Kade, B. D. Balogun and J. B. T. Rocha, *Chem.-Biol. Interact.*, 2013, **206**, 27.
52. W. S. Hassan, C. Oliveira, H. Noreen, J. P. Kamdem, C. W. Nogueira and J. B. T. Rocha, *Cur. Org. Chem.*, 2016, **20**, 218.
53. C. W. Nogueira and J. B. T. Rocha, *Arch. Toxicol.*, 2011, **85**, 1313.

54. J. Y. Seo, C. H. Lee, J. H. Cho, J. H. Choi, K. Y. Yoo, D. W. Kim, O. K. Park, H. Li, S. Y. Choi, I. K. Hwang and M. H. Won, *J. Neurol. Sci.*, 2009, **285**, 88.
55. C. M. Mazzanti, R. Spanevello, M. Ahmed, L. B. Pereira, J. F. Goncalves, M. Correa, R. Schmatz, N. Stefanello, D. B. R. Leal, A. Mazzanti, A. T. Ramos, T. B. Martins, C. C. Danesi, D. L. Graca, V. M. Morsch and M. R. C. Schetinger, *Int. J. Dev. Neurosci.*, 2009, **27**, 73.
56. L. Zou, J. Lu, J. Wang, X. Ren, L. Zhang, Y. Gao, M. E. Rottenberg and A. Holmgren, *EMBO Mol. Med.*, 2017, DOI: 10.15252/emmm.201707661.

Therapeutic Applications of Organoselenium Compounds

CHAPTER 12

History and Development of Selenium-based Radioprotectors: Distinctions between the Inorganic and Organic Forms

A. KUNWAR*[a,c] AND K. I. PRIYADARSINI*[b,c]

[a] Radiation & Photochemistry Division, Bhabha Atomic Research Centre, Mumbai – 400085, India; [b] Chemistry Division, Bhabha Atomic Research Centre, Mumbai – 400085, India; [c] Homi Bhabha National Institute, Bhabha Atomic Research Centre, Mumbai – 400085, India
Email: kamit@barc.gov.in; kindira@barc.gov.in

12.1 Radiation Exposure and Radioprotection

The exposure of living beings to ionising radiation such as high-energy γ-rays, X-rays and charged particles, depending on the extent of the exposure, can induce several deleterious effects, including mutations, radiation syndromes, cancer, *etc*. When radiation interacts with cells, chemical changes are initiated through radiolysis of cellular water to produce reactive oxygen species (ROS) which in turn induce biological changes.[1–5] Depending on the exposure and linear energy transfer of the

Organoselenium Compounds in Biology and Medicine: Synthesis, Biological and Therapeutic Treatments
Edited by Vimal Kumar Jain and K. Indira Priyadarsini

Published by the Royal Society of Chemistry, www.rsc.org

radiation, the physico-chemical events are initiated in an extremely short span of time, from femtoseconds to milliseconds, while the biological alterations start later, over seconds to hours.[1–5] Delayed and chronic effects are sometimes observed, after many months of exposure. The extent of radiation injury depends on the type and deposition pattern of radiation (track structure) and the total absorbed dose.[5,6] The radiation dose absorbed is expressed in Gray (Gy) units, one of which is equal to one joule of energy deposition in one kilogram ($1\ J\,kg^{-1}$) of the material.[7] Additionally, the sensitivity of different organs contributes to the varied manifestation of radiation injury. For example, organs such as the brain, bone, muscle, thyroid and liver are radio-resistant, whereas others, such as the lymphoid organs, reproductive organs, stem cells and intestinal crypts are radiosensitive.[4,6,8,9]

A radioprotector is a chemical substance, or a mixture of compounds, capable of minimizing the damaging effects of ionizing radiation on normal tissue.[10] Development of radioprotectors has been an area of active research from the beginning of the nuclear era. Most of the initial research was mainly focussed on radiotherapy, as it was realised that the protection of normal tissue is as important as the destruction of cancer cells.[11] Ideally, a radioprotector should be able to protect against the deleterious effect of every type of radiation, both during therapeutic and diagnostic procedures (planned exposure), and also during accidental radiation exposure (unplanned exposure).[11,12] The most commonly compared parameter with respect to the efficacy of a radioprotector is the dose modification factor (DMF).[12,13] It is defined as the ratio of radiation dose required to produce an effect in the presence and absence of a radioprotector, which varies with the type of radiation, dose rate, administered dosage of radioprotector, time and schedule of treatment, biological end point and animal type. Radioprotection by small molecules is offered in three ways: (i) as prophylactic agents administered before radiation exposure, like in radiotherapy and diagnostic exposures, (ii) as radiation mitigators, given immediately after exposure; and (iii) as therapeutic agents supplemented much after the induction of radiation damage.[9–13]

Extensive research employing several natural and synthetic small molecules has led to the identification of sulfhydryl compounds as promising radioprotectors. After screening thousands of sulfur compounds, only one compound, amifostine, also called WR-2721, was approved for clinical use to protect normal tissues in patients undergoing head and neck radiotherapy.[14] It is a pro-drug, having an active thiol group attached to a phosphate moiety to reduce toxicity.[14,15] Its action as radioprotector is attributed to free-radical scavenging, DNA repair and the reduction of radiation-induced oxidative stress and apoptosis. Clinically, it is administered during head and neck therapy to reduce early and late radiation effects, and is also carried by astronauts in space. In spite of its use in treatment, its considerable toxicity at radioprotective doses and limited utility post-irradiation has warranted the search for more effective and nontoxic alternative drugs.[14,15]

12.2 Selenium as a Radioprotector

Having understood that sulfur compounds act as good radioprotectors, it was anticipated that selenium, which shares same group as sulfur could be a better radioprotector.[12,16] Selenium is a stronger nucleophile than sulfur and is a micronutrient. Its role in cellular defense against oxidative stress occurs through its incorporation into antioxidant selenoenzymes such as glutathione peroxidase (GPx), thioredoxin reductase (TrxR) and selenoprotein P (SelP).[17] All these factors encouraged researchers to undertake extensive evaluations of selenium as a radioprotector in different chemical forms (inorganic and organic selenium) over the past two decades. Additionally, inorganic selenium has been tested clinically for reducing radiotherapy side effects, while the majority of studies on organic selenium are still at pre-clinical stages. The current status of all these studies is discussed in this chapter.

12.2.1 Inorganic Selenium

12.2.1.1 *Effect on Animal Survival*

The most commonly used forms of inorganic selenium are sodium selenite and sodium selenate. The first report on the radioprotective effect of selenium, by Hollo and Zlatarov in 1960 aimed to prevent death caused by X-ray radiation in a rat model.[18,19] In these experiments, animals were subjected to whole-body irradiation (WBI) using X-rays (8 Gy), and selenium in the form of sodium selenate or sodium selenite or selenium salt (sodium selenate containing 1.2% selenite) was administered after irradiation, at doses ranging from 4 to 10 $mg\,kg^{-1}$ by injection. The survival of the animals, monitored for a period of 2 months showed that neither sodium selenate nor sodium selenite protected rats from X-ray-induced mortality, but the selenium salt offered a 20–80% improvement in survival in the higher dose range. In these studies, the dosage of selenium salt needed to offer radioprotection was dependent on factors such as absorbed dose of radiation, mode (intraperitoneal (i.p.) or subcutaneous (s.c.) or combined) and time of administration. Much later, Weiss and co-workers at the Armed Forces Radiobiology Research Institute (Bethesda, MD, USA), performed a series of experiments to evaluate the radioprotective activity of selenium and its ability to modulate the efficacy of WR-2721 in terms of DMF in a mouse model.[20,21] The toxicity studies revealed that injection of 1.6 $mg\,kg^{-1}$ sodium selenite 24 h before WR-2721 (400–1200 $mg\,kg^{-1}$, i.p.) not only decreased its toxicity significantly ($P < 0.01$), but also improved the DMF. For example, at 400 $mg\,kg^{-1}$ the DMF for WR-2721 was 2.2, which increased to 2.6 when combined with sodium selenite. The same group has also evaluated the effect of selenium on survival and hematopoietic-regenerating effects in irradiated mice by administering it in combination with glucan and WR-2721, 30 min before the exposure to a lethal dose (6–13 Gy) of γ-radiation.[21,22] Their studies concluded that unlike the single-agent treatment, combination treatment was more effective in accelerating the bone marrow and splenic granulocyte-macrophage colony-forming cell regeneration. Recently, our group has evaluated the effect of low-dose (4 $\mu g\,kg^{-1}$) sodium selenite for improving 30 day survival and preventing radiation-induced genotoxicity in mice. When administered as a supplement 3 days a week starting from pre-irradiation to post-irradiation days, it prevented radiation (WBI ~8 Gy)-induced DNA damage in peripheral leukocytes, but did not offer a significant survival advantage.[23] All these studies established the radioprotective potential of sodium selenite, especially in combination with other agents.

12.2.1.2 Effect on Organ Specific Radiation Toxicities

Having recognised the radioprotective potential of sodium selenite, researchers have explored it for reducing the organ-specific toxicities incurred during WBI. For example, a single-dose injection of sodium selenite (0.5 $mg\,Se\,kg^{-1}$) into mice on day 9 of pregnancy, either 30 min or 2 h before 1.75 Gy WBI, significantly ($P<0.005$) reduced the number of foeti born with malformations such as short or kinked tails, rib and vertebral malformations, coloboma and deformation of the retina and iris.[24] In addition, selenium pre-treatment prevented the radiation-induced retardation of the sternum in newborn foeti.[24] Similarly, several studies have demonstrated radioprotection by sodium selenite in the head and neck region. Injection of sodium selenite (0.5 $mg\,kg^{-1}$) in rats 24 h before irradiation (15 Gy γ-radiation) to the head and neck region protected the secretary cells of submandibular glands from undergoing vacuolization, lysis of cytoplasmic inclusions and nuclear alterations.[25] Its pre-administration (15 $\mu g\,kg^{-1}$) also showed cytoprotective potency during a conventional fractioned irradiation (60 Gy/30 fractions/6 weeks) of the salivary glands in rats by delaying intra- and intercellular edema, vacuolization, degranulation and necrosis of the acinar cells.[26] On similar lines, Gehrisch and Dörr investigated the effects of systemic or topical administration of sodium selenite during clinically relevant fractionated irradiation protocols and observed it to prevent or significantly delay the radiation-induced mucositis in the mouse model.[27]

Not many reports are available on the radioprotective effect of selenium in other organs, except for a few observations in the intestine and kidney. Sodium selenite supplementation through drinking water (4 ppm) alone or in combination with vitamin E (30 $mg\,kg^{-1}$, i.p.) for 4 weeks reduced (by ~70%) oxidative damage and histological abnormalities in the intestine of rats exposed to γ-rays (10 Gy) in the abdominal region.[28] In another study, dietary supplementation with sodium selenite (100 $\mu g\,day^{-1}$) initiated immediately after WBI (10 Gy) and maintained for 21 weeks showed mitigation of radiation nephropathy in a rat model,[29] as evidenced by a decrease in blood urea nitrogen levels by 50%. Subsequently, this group also performed a follow-up study to explore the risks and benefits of delaying or shortening the periods of supplementation, and escalating the dosage beyond 100 μg day^{-1}. The results indicated that supplementation up to 200 μg sodium selenite per day substantially improved the mitigation of radiation nephropathy.[30] Shortening the period of supplementation to 3 or 2 months did not compromise efficacy and delaying the supplementation for 1 week reduced, but did not abrogate the mitigation of radiation nephropathy. From these studies it can be proposed that dietary supplementation with sodium selenite provides a safe, effective and practical way to mitigate radiation injury to the kidneys.

12.2.1.3 Radioprotection Studies in Cells

Unlike *in vivo* studies, cellular investigations have shown diverse effects with respect to the radioprotective activity of sodium selenite. For example, some reports indicated that sodium selenite pre-treatment protected cells from radiation-induced death, whereas others suggested that it did not affect their radiosensitivity.[31–39] Literature reports also suggest that sodium selenite showed differential activity of radioprotection and radiosensitization in normal and tumor cells, respectively. For example, sodium selenite pre-treatment protected cultured normal human skin fibroblasts, but sensitized the radiation-induced killing of squamous cell carcinoma and C6 rat glioma cells from single-dose and multiple fractionated dose of irradiation.[40] In another study, sodium selenite treatment radiosensitized both hormone dependent (LAPC-4) and hormone independent (DU-145) prostate cancer cells, but did not affect the sensitivity of intestinal epithelial cells to radiation.[41] Apart from these studies, sodium selenite has also been evaluated for anti-genotoxic effects against radiation exposure using cellular models. For example, Borek *et al.* showed that incubation of C3H/10T-1/2 cells with sodium selenite (2.5 μM) 24 h before exposure to X-rays resulted in inhibition of mutagenic transformation.[42] In another study, Diamond and colleagues reported that although sodium selenite treatment (30 nM) did not protect Chinese hamster ovary (CHO) and AA8 cells from γ-radiation (8 Gy)-induced cell death, it significantly ($P<0.05$) reduced the mutation frequency at the *hprt* locus in these cells.[38] Subsequently, the lower (30 nM) dosage of sodium selenite has been shown to protect against UV-induced DNA damage in human derived MCF-7 and LNCaP cells and mouse fibroblasts by augmenting DNA repair.[39,43,44] On similar lines, Seo *et al.* reported that sodium selenite augmented DNA repair following UV exposure in human fibroblasts through stimulation of DNA repair complexes.[45] Jerome-Morais *et al.* showed that sodium selenite pre-treatment in human MCF-7, HCT116 and NCM460 cell lines attenuated H2AX phosphorylation (a marker of DNA damage) following exposure to ionising radiation (γ-ray, 4 Gy) while the viability of these cells was unaffected.[39] Collectively, all these results established that sodium selenite pre-treatment prevented radiation-induced genotoxic stress and mutagenic transformation.

12.2.2 Organoselenium Compounds

12.2.2.1 *Naturally Occurring Organoselenium Compounds as Radioprotectors*

In general, it is believed that selenium in organic forms exhibits less toxicity than in inorganic forms. Over the years, organoselenium compounds from natural as well as synthetic origins have been evaluated for radioprotection. The naturally occurring organoselenium compounds are mainly selenoamino acids such as selenomethionine, selenocysteine and methylselenocystine. These are present in a variety of foods such as garlic, Brazil nuts and broccoli and consumed regularly as part of the diet. Shimazu and Tappel showed for the first time in 1964 that selenomethionine and selenocysteine protect amino acid and protein systems from radiation damage.[46,47] This study was performed in a cell-free system consisting of aqueous amino acids and proteins. The samples were exposed to γ-radiation, with dose ranges up to 1×10^5 Gy. The radiation dose that degrades 50% of amino acids or inhibits protein activity to half its original value ($D_{1/2}$), in the presence and absence of protectors was determined by monitoring the amount of the amino acids remaining after irradiation using chromatography and by measuring the enzymatic activity of the protein used in the study. Selenoamino acids showed two to five times greater $D_{1/2}$ values compared to analogous sulfur amino acids. Weiss and co-workers reported that selenomethionine (0.8–4 mg kg^{-1}) injected either before (−24 h and −1 h) or shortly after (+15 min) radiation exposure increased the 30 day survival of mice by 20–90% depending on the absorbed dose (γ-ray, 9 Gy and 10 Gy).[20,21] At equitoxic doses, selenomethionine (4.0 mg kg^{-1}) and sodium selenite (0.8 mg kg^{-1}) showed similar protection (80%) against WBI at 9 Gy.

In addition, selenomethionine has been evaluated as a dietary supplement to reduce radiation abnormalities. For example, Cekan *et al.* demonstrated a decrease in the incidence of radiation-induced teratogenic effects in offspring, from 62% in the irradiated controls to 47% when diets of female mice were supplemented with selenomethionine (0.8–3.4 ppm) during pregnancy prior to radiation exposure (1.75 Gy).[48] In another study, Knizhnikov *et al.* showed that a selenomethionine-enriched diet (10–100 μg day^{-1}) started after whole-body exposure (1.4 Gy) increased the average lifespan by a month and half and decreased the incidence of leukaemia and other malignancies by 1.5–3.5-fold in a rat model.[49] Sieber and co-workers compared the radio-mitigating activity of dietary selenomethionine (100–200 μg day^{-1}) with that of sodium selenite in a rat model to prevent radiation injuries in the kidneys and observed that selenomethionine was moderately (~10%) less effective than sodium selenite.[29,30] Another study reported that selenomethionine supplementation (0.5 ppm) in the diet of rats exposed to chronic irradiation (0.094 Gy h^{-1} for 20 h daily) neither protected them nor affected their survival.[50–52] Taken together selenomethionine appears to be a less toxic alternative, but yielded mixed results as a radioprotector.

12.2.2.2 Synthetic Selenium Compounds as Radioprotectors

With regard to the use of synthetic organoselenium compounds for radioprotection, the initial thrust was to examine the selenium analogues of sulfur compounds that had been examined for radioprotection earlier. Accordingly, compounds such as 2-aminoethaneselenol (selenocysteamine) hydrochloride, bis(2-aminoethyl) diselenide (selenocystamine) hydrochloride, 2-aminoethaneselenosulfuric acid, 2-aminoethylisoselenouronium hydrobromide, 2-aminoselenazoline and 2-selenoethylguanidine were tested.[53–55] The results from these studies did not show any radioprotection; the selenium compounds were found to be more toxic than their sulfur counterparts. The first successful synthetic organic selenium radioprotector was reported by Badiello *et al.* who studied the radioprotective action of selenourea in rats exposed to semi-lethal and lethal doses (6–9 Gy) of γ-radiation.[56] They compared selenourea at a dose of 2 mg per animal with cysteine administered at a dose of 100 mg per animal. The results suggested that selenourea showed significantly ($P < 0.05$) better protection with regard to blood changes, mortality and the general clinical condition of the animals compared to cysteine. Subsequently, a number of other biologically important organoselenium compounds such as selenocystine, selenomethionine, colloidal selenium, selenourea, selenoxanthene, selenoxanthone and selenochromone were compared for radioprotection in rats.[57] The compounds were injected at a dose of 2 mg of each selenium derivative per animal just 5 min prior to irradiation of absorbed doses of 6, 7.5 and 9 Gy. The results showed that all the above compounds improved survival by >50% up to a radiation dose of 7.5 Gy, while at 9 Gy, only selenourea was effective. From the clinical point of view, disorders such as leucopoenia were less pronounced in the animals treated with all these compounds, compared to the irradiated control animals. Despite showing promising results, selenourea did not receive much attention as a radioprotector.[58] We proposed that selenourea, upon reaction with ROS generated redox-active colloidal selenium, which may influence its activity.[59]

Based on previous reports on the radioprotective effects of sulfur-containing five-membered heterocyclic ring compounds, several five-membered selenoheterocycles such as 2-amino-selenazoles, 4-aryl-1,2,3-selenadiazoles, 2-dialkylamino-1,3,4-selenadiazolines, 2-benzylidene-1,3,4-selenadiazolines and 2-arylidene-1,3-diselenols were synthesized and evaluated for radioprotection.[60] Toxicity studies revealed that the introduction of a selenium atom into the ring in place of the sulfur atom led to an increase in toxicity. Furthermore, except for a few 2-amino-selenazole derivatives, these compounds did not exhibit radioprotective activities at the doses employed. For example,

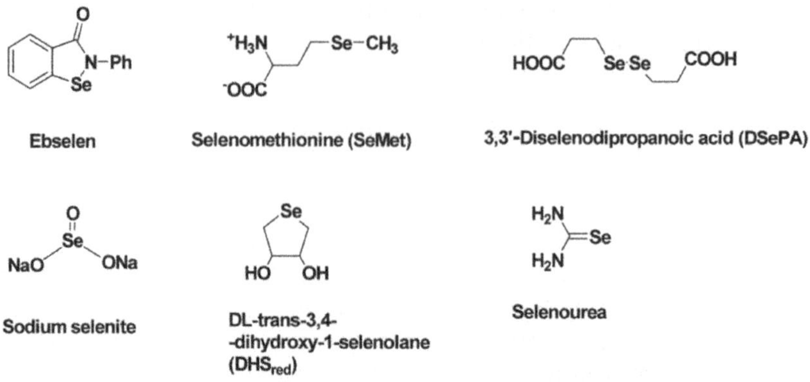

Figure 12.1 Chemical structures of selenium compounds studied for radioprotection.

Brucker and Bulka showed that treatment with 2-amino-4,5-dimethyl-selenazole hydrochloride reduced the radiation-induced degradation of ATP in a cell-free system.[61] Similarly, Kozák *et al.* demonstrated the protective effect of 2-amino-selenazoline in mice against WBI (5 Gy min^{-1}).[62]

With the understanding of GPx-like activity in organoselnium compounds and GPx being an important cellular antioxidant defence system, a number of GPx-active organoselenium compounds were evaluated for radio-protection. Ebselen is one such efficient GPx-mimicking compound which showed promising activity.[63] Radioprotection studies indicated that ebselen administration for 14 days at a daily dosage of 10 mg kg^{-1} body weight in mice before WBI (8 Gy) provided substantial protection (~60%) against killing and oxidative damage. In U937 monocytic cells, ebselen prevented radiation-induced oxidative stress and apoptosis significantly ($P<0.05$).[63] Chemical structures of important selenium compounds with promising radioprotection are given in Figure 12.1.

12.2.3 Organoselenium Radioprotectors Developed by our Group

With an aim to develop water-soluble and less toxic selenium radio-protectors, a number of organoselenium compounds such as selenoethers, diselenides and cyclic selenolanes have been synthesized in our laboratory and screened for radioprotection using *in vitro* and *in vivo* model systems, and the chemical structures are shown schematically in Figure 12.1. The *in vitro* studies indicated that aliphatic diselenides offer greater protection to biomolecules from the γ-radiation-induced degradations.[64,65] After screening several compounds, the most effective diselenide, 3,3′-diselenodipropionic (DSePA) was identified and evaluated for radio-protection in detail. DSePA is an effective free-radical scavenger and exhibits anti-hemolytic and GPx-mimicking activity.[64] The compound is non-toxic to normal cells and its maximum tolerable dose in mice was ~88 mg kg^{-1} body weight (i.p.). The radioprotective ability of DSePA was evaluated in BALB/c mice, wherein the administration of DSePA (2 mg kg^{-1} day^{-1} for five consecutive days) prior to WBI (10 Gy) led to a 35% increase in the 30 day survival of mice.[66,67] In irradiated mice, DSePA treatment prevented oxidative damage (such as lipid peroxidation and DNA damage), apoptosis and inflammatory response in radiosensitive organs such as the hematopoietic and gastrointestinal systems.[66,67] DSePA treatment prevented the depletion of endogenous antioxidants in the hepatic tissue of irradiated mice. Detailed toxicological investigations in CHO cells indicated that DSePA treatment did not induce any toxicity and prevented radiation-induced genotoxicity and subsequent cytotoxicity.[68] Late lung tissue responses such as pneumonitis and fibrosis are the most serious dose-limiting side effects of thoracic radiotherapy for several kinds of malignancies affecting organs in the thorax area.[69] Accordingly, DSePA treatment to prevent radiation-induced lung toxicities in C3H/HeJ mice was examined. Our results indicated that administration of DSePA during the post-irradiation period at a similar dosage (2 mg kg^{-1} 3 days a week) was capable of delaying the thoracic radiation (18 Gy)-induced pneumonitis response in mice by more than a month.[69] In this study, DSePA-treated mice had significantly reduced levels of lipid peroxidation and inflammatory cells influx in the lungs and increased GPx, compared to the irradiated mice.[69] Further pharmacokinetic studies of orally administered DSePA in different organ systems of tumor-bearing mice showed maximum bioavailability of DSePA in the lung, followed by other organs such as kidney, liver and intestine, while that in the tumor was significantly low.[70] This study also reiterated that DSePA, being a water soluble compound, is cleared from the body very fast. Encouraged by these results, our investigations are being extended to develop DSePA as an orally supplemented lung radioprotector for thoracic radiotherapy. All these are summarised in Figure 12.2.

Another organoselenium compound tested in our group is dihydroxyselenolane (DHS), a water-soluble compound, synthesised by Iwaoka and colleagues as a GPx mimic.[71] DHS was injected into mice at a dosage of 2 mg kg^{-1} for five consecutive days before 8 Gy WBI, and subsequently continued at the same dosage three times per week during the post-irradiation period. DHS-treated mice showed 30 day post-irradiation survival improvement of 40% compared to the radiation control group. All these studies clearly indicated that water-soluble organoselenium compounds exhibiting GPx-like activity have the potential to be explored as a new class of radioprotectors.

12.2.4 Molecular Mechanisms Contributing to the Radioprotective Effect of Selenium

Mechanistic insights into the radioprotective action of selenium have been understood mainly from the studies of sodium selenite and selenomethionine. The results so far have confirmed that most of the selenium

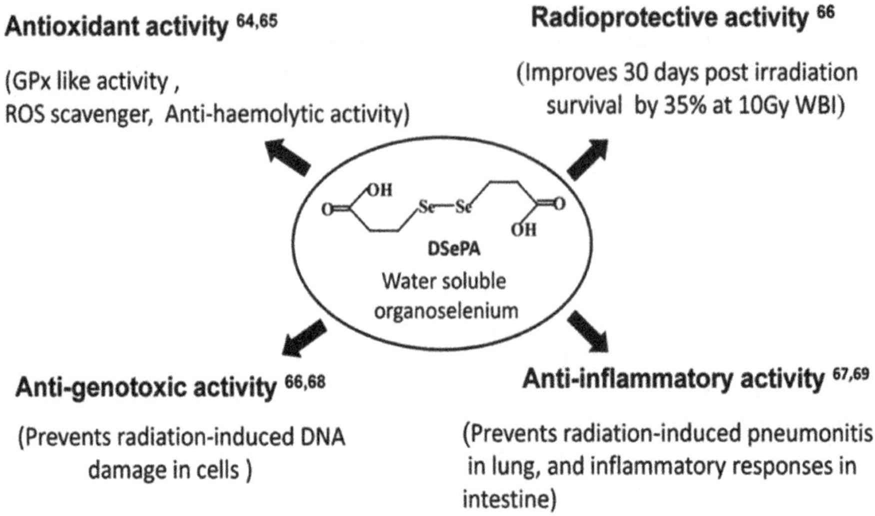

Figure 12.2 Summary of radioprotection studies of 3,3′-diselenodipropanoic acid (DSePA). GPx: glutathione peroxidase; ROS: reactive oxygen species; WBI: whole-body irradiation.

compounds generate hydrogen selenide (H_2Se) as a key metabolic intermediate, which serves as a precursor for the synthesis of selenocysteine, which is incorporated into the antioxidant selenoproteins including GPx.[21] However, the metabolism of sodium selenite is much faster compared to selenomethionine, leading to ROS generation and increased oxidative stress.[21] In the case of selenomethionine, while a fraction of it goes into the selenoprotein pool, most of it is converted in to easily eliminated smaller organic species such as methylated selnides, leading to lesser toxicity compared to sodium selenite.[21] The exact role of GPx in radioprotection is still not very clear. Earlier reports suggested that GPx is an important factor in cellular radiosensitivity;[31,33] however subsequent studies have failed to prove this. For example, Sandström *et al.* demonstrated that adding selenium in the form of selenite to cultured cells did not result in radioprotection of mammalian cells (HT29, P31 and N-18), despite inducing GPx.[32] Similarly, mammalian cell lines (Sup-T1 and CHO-AA8) engineered to over-express cytosolic GPx (~8–30-fold increase in activity) did not confer any protection with respect to survival in either cell type.[72] It is important to note that GPx activity in tissues reaches saturation upon administration of various selenium-containing compounds.[73–75] According to Sunde *et al.* a dietary selenium content of ~0.1 $\mu g\,g^{-1}$ is sufficient to maximally induce GPx expression.[74] However, the radioprotective effects of selenium compounds have been observed at a considerably higher doses (100 $\mu g\,day^{-1}$ to 4 $mg\,kg^{-1}$).[20–22] From these reports it appears that the radioprotective action of selenium may involve mechanisms beyond the induction of GPx. There are many other selenoproteins, which play key roles in antioxidant defence mechanisms and thus can contribute to the radioprotective effect of

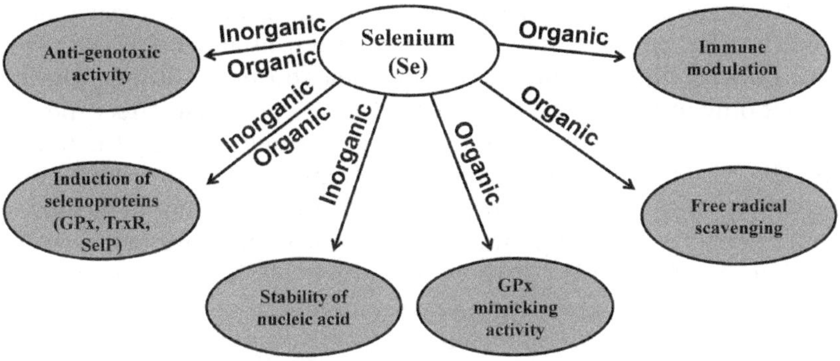

Figure 12.3 Probable mechanisms involved in the radioprotective activity of selenium compounds. GPx: glutathione peroxidase; TrxR: thioredoxin reductase; SelP: selenoprotein P.

selenium.[34,35] Supporting this assumption, Rafferty *et al.* have shown that pre-treatment with sodium selenite or selenomethionine protected cultured human keratinocytes and melanocytes from UVB-induced cell death by inducing TrxR and GPx levels.[35] In another study, the over-expression of SelP in normal human fibroblasts was shown to suppress the radiation-induced late ROS accumulation and thereby protected against radiation toxicity.[34] It has also been demonstrated that pre-treatment with sodium selenite prevented radiation-induced DNA damage, which can contribute to radioprotection.[38,39,43–45,76,77] The organoselenium compounds in addition to GPx-mimicking activity may exhibit free-radical scavenging and immunomodulatory activity, and all these can also contribute to their radioprotective activity.[63–71] For example, selenoamino acids such as selenomethione, selenocysteine, methyselenocysteine and selenourea have been reported to be efficient scavengers of hydroxyl and peroxyl radicals generated during radiolysis and prevented radiation-induced DNA damage in cellular and mice models.[46,47,78,79] Compounds such as DSePA and DHS offer protection by scavenging ROS, suppressing inflammatory responses and inducing GPx.[64–71] Ebselen, a mimic of GPx in which the selenium moiety is not available for metabolism, is shown to protect cells from radiation-induced apoptosis by modulating the redox state through its peroxidase activity using thioredoxin or glutathione present within the cells.[63,80] Thus, selenium compounds in various chemical forms (organic and inorganic) appear to act through multiple mechanisms, as shown in Figure 12.3, and understanding the relative importance of each of these mechanisms in contributing to the radioprotective effect requires further study.

12.3 Selenium in Clinical Radiotherapy

Pre-clinical evaluation of selenium as a radioprotective agent has prompted clinicians to undertake trials in human subjects to examine whether selenium administration can reduce radiotherapy-associated side effects. During the treatment of cancerous tissues with radiation, normal tissues are also exposed, leading to unwanted side effects such as mucositis, xerostomia, fibrosis, pneumonitis, lymphedema, diarrhea and many others. These side effects not only influence the treatment of cancer, but also affect the quality of life and sometimes even add to mortality in patients during the treatment phase. The ultimate goal of radiation therapy is thus to attain maximal tumor killing with minimal normal tissue damage, and to achieve this several radioprotective compounds including selenium are being evaluated in the clinic. As of the end of 2016, as many as 18 clinical studies are reported in the literature (Table 12.1) citing selenium and radiotherapy together.[81–100] These studies have been conducted across the world. Of these, seven are interventional studies demonstrating the therapeutic benefit of selenium administration in reducing the radiation-associated toxicities and the other 11 are observation studies investigating the effect of radiotherapy on patients' selenium levels.[82–100] Important results pertaining to these studies are discussed briefly in the following sections; for more detailed information, readers should refer to a recent review by Puspitasari *et al.*[81]

12.3.1 Effect of Selenium Supplementation on Radiation Toxicities

The types of radiation toxicities evaluated with selenium intervention are mucositis, xerostomia, lymphedema and diarrhoea manifested during the radiotherapy of cancers of larynx, head and neck, and gynecological origins. Sodium selenite has been used for selenium supplementation in these studies, except for one study wherein selenomethionine was used as a supplement. The effectiveness of selenium supplementation is accessed by monitoring parameters such as selenium levels in whole blood/plasma/serum, immune function, quality of life, levels of enzymatic and non-enzymatic antioxidants and pathological scores of side effects.

In the first clinical study, completed in 1998, Pakdaman reported that sodium selenite infusion in patients (16 females and 16 males) with brain tumors undergoing radiotherapy resulted in improvement in the general condition and a decrease in symptoms, such as nausea, emesis, headache, vertigo, unsteady gait, speech disorders and Jacksonian seizures in 76% of the patients.[82] In addition, the patients receiving selenium supplementation showed improvements in erythrocyte, hemoglobin and thrombocyte counts.

In another randomized double-blind placebo-controlled study, patients receiving radiation therapy for squamous cell carcinoma of the head and neck were administered oral sodium selenite daily for 8 weeks, beginning on the day of their first radiotherapy session, and their immune functions were monitored. Supplementation with selenium during therapy resulted in a significantly enhanced cell-mediated immune responsiveness ($P<0.05$).[83] This has been attributed to the ability of selenium to increase the expression of α-(p55) and β-(p70/75) subunits of the interleukin-2 receptor, which results in the generation of cytotoxic lymphocytes that finally destroy tumor cells.[89]

Micke *et al.* conducted a clinical study to evaluate the impact of selenium in the treatment of lymphedema after radiotherapy.[84] A total of 12 patients with edema of the arm and 36 patients with edema of the head and neck region previously treated with radiation received sodium selenite orally. The results indicated that sodium selenite was well tolerated by all the patients and a significant reduction in secondary lymphedema caused by radiation therapy was reported in 65% of the selenium-treated patients.

Elango *et al.* studied the effect of selenium supplementation on the antioxidant status of the plasma of radiation-treated oral cancer patients.[85] The control group ($n = 27$) were given radiotherapy alone and the selenium group ($n = 36$) were given radiotherapy and supplemented with sodium selenite daily. Both the groups were followed-up for 6 months and the results showed a marked increase in the concentrations of enzymatic antioxidants such as superoxide dismutase, catalase, GPx, glutathione reductase, glucose-6-phosphate dehydrogenase and non-enzymatic antioxidants such as glutathione, vitamin E, vitamin C, vitamin A and ceruloplasmin in the selenium-treated group compared to the radiation group.

Muecke *et al.* conducted a multicentre phase III trial to assess whether oral sodium selenite supplementation reduces the side effects of patients treated with radiotherapy for cervical and uterine cancer.[86] Patients with cervical cancer and uterine cancer who were undergoing radiotherapy were randomized to receive either sodium selenite or placebo. The selenium-treated group showed a decreased incidence of diarrhea of grade 2, attributed to the increase in the level of intestinal GPx and the concurrent neutralization of radiation-induced ROS in the small intestinal mucosa.

Büntzel *et al.* investigated the clinical impact of sodium selenite in reducing the side effects of patients undergoing radiotherapy for advanced head and neck cancer.[87] 39 patients (8 female, 31 male) were included in a randomised phase II study. 22 patients received sodium selenite orally during radiotherapy, while the remaining group received only radiation. The radiation toxicity grading indicated that although selenium supplementation did not affect xerostomia and stomatitis, it significantly reduced radiation-induced ageusia (loss of taste) and dysphagia in patients ($P < 0.05$).

Recently, in a multi-institutional, randomized, double-blind phase II trial, selenomethionine was evaluated to reduce the incidence of mucositis in patients with head and neck squamous cell cancer undergoing concurrent chemo-radiation therapy.[88] A total of 18 patients with stage III or IV head and neck squamous cell cancer were randomized to receive selenomethionine or placebo staring from 7 days before radiotherapy to 3 weeks after radiotherapy. The results showed that although selenomethionine was well tolerated, there was no difference in mucositis or patient-reported side effects between the control and selenium groups.

Taken together, the clinical studies undertaken so far suggest that selenium supplementation neither reduced the effectiveness of radiotherapy

Table 12.1 List of clinical trials with selenium in different chemical species in patients undergoing radiotherapy.

Study	Study objective	Study sites	Number of patients enrolled	Types of cancer/disease subjected to radiotherapy	Form and dosage of selenium administered	Remarks/conclusions	Study ref.
1	Effect of selenium supplementation on radiotherapy side effects and selenium level	Germany	$n=32$	Brain tumor	Sodium selenite (1000 μg day^{-1}) by infusion for 4–8 weeks	1) Reduction in intracranial pressure in patients 2) Increased selenium level at the end of radiotherapy	82
2	Effect of selenium supplementation on radiotherapy side effects and selenium level	USA	$n=33$	Head and neck cancer	Sodium selenite (200 μg day^{-1}) orally for 8 weeks	1) Significant improvement in cell-mediated immune response 2) Increased selenium level at the end of radiotherapy	83
3	Effect of selenium supplementation on radiotherapy side effects	Germany	$n=48$	Secondary lymphedema	Sodium selenite (500 μg day^{-1}) orally for 4–6 weeks	1) Significant improvement inquality of life	84
4	Effect of selenium supplementation on radiotherapy side effects	India	$n=126$	Oral cancer	Sodium selenite (400 μg day^{-1}) orally for 6 months	1) Improvement in the levels of enzymatic and non-enzymatic antioxidants in serum	85

5	Effect of selenium supplementation on radiotherapy side effects and selenium level	Germany	$n=81$	Cervical cancer ($n=11$), uterine cancer ($n=70$)	Sodium selenite (500 or 300 μg day^{-1}) orally during radiotherapy	1) Significant reduction in the occurrence of radiation-induced diarrhea 2) Increased selenium level at the end of radiotherapy	86
6	Effect of selenium supplementation on radiotherapy side effects	Germany	$n=39$	Head and neck cancer	Sodium selenite (500 or 300 μg day^{-1}) orally during radiotherapy	1) Reduction in symptoms of radiation-induced dysphagia	87
7	Effect of selenium supplementation on radiotherapy side effects	USA, New Zealand	$n=18$	Head and neck cancer	Selenomethionine (3600 μg m^{-2} day^{-1}) orally for 7 days prior to radiotherapy and continued during and up to 3 weeks after the radiotherapy	1) No reduction in the incidence of severe mucositis 2) No improvement in quality of life or survival outcomes	88
8	Effect of radiotherapy on selenium level	USA	$n=71$	Upper gastrointestinal cancer		1) Significant decrease in selenium level during radiotherapy	90
9	Effect of radiotherapy on selenium level	Finland	$n=24$	Breast cancer		1) No change in selenium level before, during and after the radiotherapy	91

Table 12.1 (*Continued*)

Study	Study objective	Study sites	Number of patients enrolled	Types of cancer/ disease subjected to radiotherapy	Form and dosage of selenium administered	Remarks/ conclusions	Study ref.
10	Effect of radiotherapy on selenium level	Italy	$n=66$	Breast ($n=38$) and lung cancer ($n=28$)		Not determined	92
11	Effect of radiotherapy on selenium level	Poland	$n=78$	Larynx carcinoma		1) No change in selenium level before and after the radiotherapy	93
12	Effect of radiotherapy on selenium level	India	$n=30$	Head and neck cancer		1) No change in selenium level before and after the radiotherapy	94
13	Effect of radiotherapy on selenium level	UK	$n=100$	Non-Hodgkin's lymphoma		Not determined	95
14	Effect of radiotherapy on selenium level	Germany	$n=224$	Breast cancer ($n=94$), cervical cancer ($n=25$), head and neck cancer ($n=23$), lung cancer ($n=19$), prostate cancer ($n=13$), other ($n=50$)		1) No change in selenium level before and after the radiotherapy	96

15	Effect of radiotherapy on selenium level	Brazil	$n = 209$	Breast cancer	1) Significant decrease in selenium level at the end of radiotherapy	97
16	Effect of radiotherapy on selenium level	China	$n = 95$	Nonsmall cell lung cancer	1) Significant decrease in selenium level at the end of radiotherapy	98
17	Effect of radiotherapy on selenium level	Turkey	$n = 47$	Head and neck cancer	1) No change in selenium level before and after the radiotherapy	99
18	Effect of radiotherapy on selenium level	Iran	$n = 80$	Breast cancer	1) Significant decrease in selenium level at the end of radiotherapy	100

nor caused any toxicity at the administered doses. Additionally, selenium supplementation showed a positive effect on the general health condition of the patients and their quality of life.

12.3.2 Effect of Radiation Therapy on Selenium status in the Body

Previously, reduced blood selenium levels have been observed under various pathological conditions linked with oxidative stress. Radiation injuries showing similar pathologies are also expected to modulate the selenium level in blood. This hypothesis has been examined through clinical trials wherein selenium levels were estimated in plasma, serum or whole blood using techniques such as atomic absorption spectrometry, inductively coupled plasma mass spectrometry and fluorimetry before, during and at the end of radiotherapy of various malignancies.[90–100] The results are mostly incongruent with regard to the effect of radiotherapy on selenium status. For example, some studies found a significant decrease in selenium levels during radiotherapy,[90,97,98,100] while in other trials, no difference was observed between pre- and post-radiotherapy values.[91–96,99] Considering the fact that malignancies alone are known to reduce selenium status in blood, the results may not be conclusive.[81,101,102] Furthermore, reduced selenium status during radiotherapy may also be attributed to poor dietary intake due to anorexia, nausea, vomiting and diarrhea.[81,101,102] It is possible that radiation-induced functional impairments in the intestinal lumen may lead to poor absorption of dietary selenium.[30] More detailed studies are required to assess the status of selenium in cancer patients during radiotherapy.

12.4 Conclusions

Selenium, a micronutrient and an active constituent of important redox-regulating enzymes, has been explored for radioprotection in inorganic and organic forms. Sodium selenite, the inorganic and highly active selenium species has showed promising radioprotection, when administered as a single agent or in combination with other radioprotectors. Although it suffers from dose-limiting toxicity, clinical trials have confirmed that it can be used as a supplement at low doses ($<200\ \mu g\ day^{-1}$) to reduce radiotherapy-associated side effects. However, whether selenium supplementation is necessary for radiotherapy patients needs to be corroborated with the basal selenium levels in the population. Selenium status subsequent to radiotherapy must be assessed before undertaking any selenium supplementation.

It is clear that selenium in its organic form is less toxic than the inorganic selenium. Natural organoselenium compounds such as selenomethionine, although exhibiting radioprotection under *in vivo* systems, were not very effective in the clinical trial. With the advancement of synthetic chemistry methodologies, several new organoselenium compounds are being developed, which exhibit interesting pharmacological properties. Among these, GPx mimetics, ROS-scavenging and water-soluble selenium compounds have shown encouraging results with regard to radioprotection. However, to date, no synthetic organoselenium compound has been examined in a clinical trial for radiotherapy.

One of the important factors to be understood is the mechanism of action of a selenium compound as a radioprotector. Intracellular speciation of selenium compounds (including inorganic and organic forms) and their entry to the total selenium pool plays a very important role in determining their biological activities. Considering this, it would be interesting in future to integrate the speciation and metabolism of selenium compounds with the *in vivo* radioprotection. Additionally, it is important to understand the underlying mechanisms at molecular and redox level, and the role of selenoproteins in imparting radioprotection at individual organ level are some of the future goals of study for selenium biologists.

Another interesting observation emerging out of the recent research is that selenium compounds show differential activity, where selenium acts as a radioprotector for normal cells while sensitizing tumor cells. This has been proposed from studies in cellular models, and more need to be performed under *in vivo* conditions. The outcome of such studies will be useful in designing new organoselenium compounds by synthetic chemists, for exploration as radiotherapy and chemotherapy adjuvants.

Acknowledgements

The authors thank all the colleagues and students whose names appeared as co-authors in their publications. The contributions of Dr V. K. Jain, BARC and Prof. M. Iwaoka, Tokai University, Japan to the synthesis of important selenium compounds are gratefully acknowledged.

References

1. P. A. Riley, *Int. J. Radiat. Biol.*, 1994, **65**, 27.
2. H. D. Kogelnik, *Int. J. Radiat. Oncol. Biol. Phys.*, 1996, **35**, 219.
3. C. V. Sonntag, *The Chemical Basis of Radiation Biology*, Taylor & Francis, London, 1987.
4. K. E. Carr, S. P. Hume, R. R. Ettarh, E. A. Carr and J. S. McCollough, in *Radiation and the Gastrointestinal Tract* ed. A. Dubois, G. L. King and D. R. Livengood, CRC Press, Boca Raton, 1995.
5. E. J. Ainsworth, S. M. Afzal, A. D. Crouse, W. R. Hanson and R. J. Fry, *Adv. Space Res.*, 1989, **9**, 299.
6. A. V. Barabanova, The Medical Basis for Radiation Accident Preparedness, The Clinical Care of Victims, NewYork: Parthenon, 2002.
7. J. W. T. Spinks and R. J. Woods, *An Introduction to Radiation Chemistry*, Weily Interscience Publication, New York, 1990.
8. J. F. Weiss, *Environ. Health Perspect.*, 1997, **105**, 1473.

9. J. F. Weiss, M. R. Landauer, J. B. Hogan, P. J. Gunter-Smith, K. A. Benson, R. Neta and W. R. Hanson, *Adv. Exp. Med. Biol.*, 1997, **400B**, 865.
10. J. F. Weiss and M. R. Landauer, *Int. J. Radiat. Biol.*, 2009, **85**, 539.
11. P. G. Prasanna, D. Narayanan, K. Hallett, E. J. Bernhard, M. M. Ahmed, G. Evans, B. Vikram, M. Weingarten and C. N. Coleman, *Radiat. Res.*, 2015, **184**, 235.
12. J. F. Weiss and M. R. Landauer, *Toxicology*, 2003, **189**, 1.
13. E. A. Bump and K. Malaker, *Radioprotectors: Chemical, Biological, and Clinical Perspectives*, CRC Press, Boca Raton, 1998.
14. C. N. Andreassen, C. Grau and J. C. Lindegaard, *Semin. Radiat. Oncol.*, 2008, **13**, 62–72.
15. P. E. Brown, *Nature*, 1967, **213**, 363.
16. L. A. Wessjohann, A. Schneider, M. Abbas and W. Brandt, *Biol. Chem.*, 2007, **388**, 97.
17. L. V. Papp, J. Lu, A. Holmgren and K. K. Khanna, *Antioxid. Redox Signaling*, 2007, **9**, 775.
18. Z. M. Hollo and S. z. Zlatarov, *Naturwissenschaften*, 1960, **47**, 87.
19. Z. M. Hollo and S. z. Zlatarov, *Naturwissenschaften*, 1960, **47**, 328.
20. J. F. Weiss, R. L. Hoover and K. S. Kumar, *Free Radical Res. Commun.*, 1987, **3**, 33.
21. J. F. Weiss, V. Srinivasan, K. S. Kumar and M. R. Landauer, *Adv. Space Res.*, 1992, **12**, 223.
22. M. L. Patchen, T. J. MacVittie and J. F. Weiss, *Int. J. Radiat. Oncol. Biol. Phys.*, 1990, **18**, 1069.
23. P. Verma, A. Kunwar and K. I. Priyadarsini, *Biol. Trace Elem. Res.*, 2017, DOI: 10.1007/s12011-017-0955-9.
24. E. Cekan, B. Tribukait and H. Vokal-Borek, *Acta Radiol. Oncol.*, 1985, **24**, 267.
25. M. L. Pontual, F. M. Tuji, S. P. Barros, F. N. Bóscolo, P. D. Novaes and S. M. de Almeida, *J. Appl. Oral. Sci.*, 2007, **15**, 162.
26. C. Sagowski, S. Wenzel, L. Jenicke, F. U. Metternich and M. Jaehne, *Eur. Arch. Otorhinolaryngol.*, 2005, **6**, 459.
27. A. Gehrisch and W. Dörr, *Strahlenther. Onkol.*, 2007, **183**, 36.
28. U. Mutlu-Türkoglu, Y. Erbil, S. Öztezcan, V. Olgac, G. Toker and M. Uysal, *Life Sci.*, 2000, **66**, 1905.
29. F. Sieber, S. A. Muir, E. P. Cohen, P. E. North, B. L. Fish, A. A. Irving, M. Mäder and J. E. Moulder, *Radiat. Res.*, 2009, **171**, 368.
30. F. Sieber, S. A. Muir, E. P. Cohen, B. L. Fish, M. Mäder, A. M. Schock, B. J. Althouse and J. E. Moulder, *Radiat. Res.*, 2011, **176**, 366.
31. S. L. Marklund, N. G. Westman, G. Roos and J. Carlsson, *Radiat. Res.*, 1984, **100**, 115.
32. B. E. Sandström, J. Carlsson and S. L. Marklund, *Radiat. Res.*, 1989, **117**, 318.
33. J. Sun, Y. Chen, M. Li and Z. Ge, *Free Radical Biol. Med.*, 1998, **24**, 586.
34. J. C. Eckers, A. L. Kalen, W. Xiao, E. H. Sarsour and P. C. Goswami, *Int. J. Radiat. Oncol. Biol. Phys.*, 2013, **87**, 619.

35. T. S. Rafferty, R. C. McKenzie, J. A. Hunter, A. F. Howie, J. R. Arthur, F. Nicol and G. J. Beckett, *Biochem. J.*, 1998, **332**, 231.
36. T. Hehr, M. Bamberg and H. P. Rodemann, *InFoOnkologie*, 1999, **2**, 25.
37. H. P. Rodemann, T. Hehr and M. Bamberg, *Med. Klin.*, 1999, **94**, 39.
38. A. M. Diamond, P. Dale, J. L. Murray and D. J. Grdina, *Mutat. Res.*, 1996, **356**, 147.
39. A. Jerome-Morais, S. Bera, W. Rachidi, P. H. Gann and A. M. Diamond, *Biochim. Biophys. Acta*, 2013, **1830**, 3399.
40. P. Schueller, S. Puettmann, O. Micke, V. Senner, U. Schaefer and N. Willich, *Anticancer Res.*, 2004, **24**, 2913.
41. B. Husbeck, D. M. Peehl and S. J. Knox, *Free Radical. Biol. Med.*, 2005, **38**, 50.
42. C. Borek, A. Ong, H. Mason, L. Donahue and J. E. Biaglow, *Proc. Natl. Acad. Sci. U. S. A.*, 1986, **83**, 1490.
43. M. S. Baliga, H. Wang, P. Zhuo, J. L. Schwartz and A. M. Diamond, *Biol. Trace Elem. Res.*, 2007, **115**, 227.
44. V. de Rosa, P. Erkekoğlu, A. Forestier, A. Favier, F. Hincal, A. M. Diamond, T. Douki and W. Rachidi, *Free Radical Res.*, 2012, **46**, 105.
45. Y. R. Seo, C. Sweeney and M. L. Smith, *Oncogene*, 2002, **21**, 3663.
46. F. Shimazu and A. L. Tappel, *Radiat. Res.*, 1964, **23**, 210.
47. F. Shimazu and A. L. Tappel, *Science*, 1964, **143**, 369.
48. E. Cekan, P. Slanina, K. Bergman and B. Tribukait, *Acta Radiol. Oncol.*, 1985, **24**, 459.
49. V. A. Knizhnikov, N. K. Shandala, M. Cand, V. A. Komleva, B. Cand, V. A. Knyazhev, M. Cand and V. A. Tutelyan, *Nutr. Res.*, 1996, **16**, 505.
50. H. D. Hurt, E. E. Cary, W. H. Allaway and W. J. Visek, *J. Nutr.*, 1971, **101**, 363.
51. G. Batist, A. Reynaud, A. G. Katki, E. L. Travis, M. C. Shoemaker, R. F. Greene and C. E. Myers, *Biochem. Pharmacol.*, 1986, **35**, 601.
52. C. E. Myers, A. Katki and E. Travis, *Ann. N. Y. Acad. Sci.*, 1982, **393**, 419.
53. W. H. H. Giinther and H. G. Mautner, *J. Am. Chem. Soc.*, 1960, **82**, 2762.
54. H. G. Mautner, S. H. Chu and C. M. Lee, *J. Org. Chem.*, 1962, **27**, 3671.
55. D. L. Klayman, *J. Org. Chem.*, 1965, **30**, 2454.
56. R. Badiello, A. Trenta, M. Mattii and S. Moretti, *Med. Nucl., Radiobiol. Lat.*, 1967, **10**, 57.
57. A. Breccia, R. Badiello, A. Trenta and M. Mattii, *Radiat. Res.*, 1969, **38**, 483.
58. R. Badiello, A. Trenta, M. Mattii and S. Moretti, *Ric. Sci.*, 1967, **37**, 452.
59. B. Mishra, P. A. Hassan, K. I. Priyadarsini and H. Mohan, *J. Phys. Chem. B*, 2005, **109**, 12718.
60. L. L. Petrova, L. V. Trufanova, M. L. Petrov, M. A. Abramov, N. I. Zmitrovich and N. A. Terenteva, *Pharm. Chem. J.*, 1994, 28.
61. W. Brucker and E. Bulka, *Stud. Biophys.*, 1966, **1**, 253.
62. I. Kozák, L. Kronrád and Z. Dienstbier, *Strahlentherapie*, 1976, **151**, 84.
63. J. K. Tak and J. W. Park, *Free Radical Biol. Med.*, 2009, **46**, 1177.
64. A. Kunwar, B. Mishra, A. Barik, L. B. Kumbhare, R. Pandey, V. K. Jain and K. I. Priyadarsini, *Chem. Res. Toxicol.*, 2007, **20**, 1482.

65. B. S. Kumar, A. Kunwar, A. Ahmad, L. B. Kumbhare, V. K. Jain and K. I. Priyadarsini, *Radiat. Environ. Biophys.*, 2009, **48**, 379.
66. A. Kunwar, P. Bansal, S. Jaya Kumar, P. P. Bag, P. Paul, N. D. Reddy, L. B. Kumbhare, V. K. Jain, R. C. Chaubay, M. K. Unnikrishnan and K. I. Priyadarsini, *Free. Radical Biol. Med.*, 2010, **48**, 399.
67. A. Kunwar, P. P. Bag, S. Chattopadhyay, V. K. Jain and K. I. Priyadarsini, *Arch. Toxicol.*, 2011, **85**, 1395.
68. R. K. Chaurasia, S. Balakrishnan, A. Kunwar, U. Yadav, N. Bhat, K. Anjaria, R. Nairy, B. K. Sapra, V. K. Jain and K. I. Priyadarsini, *Mutat. Res.*, 2014, **774**, 8.
69. A. Kunwar, V. K. Jain, K. I. Priyadarsini and C. K. Haston, *Am. J. Respir. Cell Mol. Biol.*, 2013, **49**, 654.
70. V. Gota, J. S. Goda, K. Doshi, A. Patil, S. Sunderajan, K. Kumar, M. Varne, A. Kunwar, V. K. Jain and K. I. Priyadarshini, *Eur. J. Drug Metab. Pharmacokinet.*, 2016, **41**, 839.
71. A. Kunwar, P. Verma, H. N. Bhilwade, M. Iwaoka and K. I. Priyadarsini, *Mutat. Res.*, 2016, **807**, 33.
72. D. B. Mansur, Y. Kataoka, D. J. Grdina and A. M. Diamond, *Radiat. Res.*, 2001, **155**, 536.
73. Y. Xia, K. E. Hill, D. W. Byrne, J. Xu and R. F. Burk, *Am. J. Clin. Nutr.*, 2005, **81**, 829.
74. R. A. Sunde, in *Selenium its Molecular Biology and Role in Human Health*, ed. D. L. Hatfield, M. J. Berry and V. N. Gladyshev, Springer, New York, 2006.
75. P. F. Surai, *Selenium in Nutrition and Health*, Nottingham University Press, Nottingham, 2006.
76. T. A. Kuliev, R. A. Babaev and K. A. Kuliev, *Radiobiologiia*, 1981, **21**, 451.
77. M. S. Baliga, V. Diwadkar-Navsariwala, T. Koh, R. Fayad, G. Fantuzzi and A. M. Diamond, *Mol. Nutr. Food Res.*, 2008, **52**, 1300.
78. R. Badiello and E. M. Fielden, *Int. J. Radiat. Biol. Relat. Stud. Phys., Chem. Med.*, 1970, **17**, 1.
79. A. Kunwar, S. Jayakumar, H. N. Bhilwade, P. P. Bag, H. Bhatt, R. C. Chaubey and K. I. Priyadarsini, *Radiat. Environ. Biophys.*, 2011, **50**, 271.
80. R. Zhao and A. Holmgren, *J. Biol. Chem.*, 2002, **277**, 39456.
81. I. M. Puspitasari, R. Abdulah, C. Yamazaki, S. Kameo, T. Nakano and H. Koyama, *Radiat. Oncol.*, 2014, **9**, 125.
82. A. Pakdaman, *Biol. Trace Elem. Res.*, 1998, **62**, 1.
83. L. Kiremidjian-Schumacher, M. Roy, R. Glickman, K. Schneider, S. Rothstein, J. Cooper, H. Hochster, M. Kim and R. Newman, *Biol. Trace Elem. Res.*, 2000, **73**, 97.
84. O. Micke, F. Bruns, R. Mücke, U. Schäfer, M. Glatzel, A. F. DeVries, K. Schönekaes, K. Kisters and J. Büntzel, *Int. J. Radiat. Oncol. Biol. Phys.*, 2003, **56**, 40.
85. N. Elango, S. Samuel and P. Chinnakkannu, *Clin. Chim. Acta*, 2006, **373**, 92.

86. R. Muecke, L. Schomburg, M. Glatzel, R. Berndt-Skorka, D. Baaske, B. Reichl, J. Buentzel, G. Kundt, F. J. Prott, A. Devries, G. Stoll, K. Kisters, F. Bruns, U. Schaefer, N. Willich and O. Micke, *Int. J. Radiat. Oncol. Biol. Phys.*, 2010, **78**, 828.
87. J. Büntzel, D. Riesenbeck, M. Glatzel, R. Berndt-Skorka, T. Riedel, R. Mücke, K. Kisters, K. G. Schönekaes, U. Schäfer, F. Bruns and O. Micke, *Anticancer Res.*, 2010, **30**, 1829.
88. M. Mix, A. K. Singh, M. Tills, S. Dibaj, A. Groman, W. Jaggernauth, Y. Rustum and M. B. Jameson, *World J. Clin. Oncol.*, 2015, **6**, 166.
89. M. Roy, L. Kiremidjian-Schumacher, H. I. Wishe, M. W. Cohen and G. Stotzky, *Proc. Soc. Exp. Biol. Med.*, 1992, **200**, 36.
90. L. Pothier, W. W. Lane, A. Bhargava, C. Michielson and H. O. Douglass Jr., *Cancer*, 1987, **60**, 2251.
91. H. M. Antila, M. S. Salo, V. Nanto, V. Nikkanen and O. Kirvela, *Acta Oncol.*, 1992, **31**, 569.
92. L. Piccinini, P. Borella, A. Bargellini, C. I. Medici and A. Zoboli, *Biol. Trace Elem. Res.*, 1996, **51**, 23.
93. B. Rostkowska-Nadolska, L. Pospiech and M. Bochnia, *Arch. Immunol. Ther. Exp.*, 1999, **47**, 321.
94. S. P. Yadav, A. Gera, I. Singh and R. Chanda, *J. Otolaryngol.*, 2002, **31**, 216.
95. K. W. Last, V. Cornelius, T. Delves, C. Sieniawska, J. Fitzgibbon, A. Norton, J. Amess, A. Wilson, A. Z. Rohatiner and T. A. Lister, *J. Clin. Oncol.*, 2003, **21**, 2335.
96. I. Fraunholz, K. Eberlein, B. Schopohl, H. D. Bottcher and C. Rodel, *Strahlenther. Onkol.*, 2008, **184**, 411.
97. C. A. Franca, C. R. Nogueira, A. Ramalho, A. C. Carvalho, S. L. Vieira and A. B. Penna, *Ann. Oncol.*, 2011, **22**, 1109.
98. Y. C. Zeng, M. Xue, F. Chi, Z. G. Xu, G. L. Fan, Y. C. Fan, M. H. Zheng, W. Z. Zhong, S. L. Wang, Z. Y. Zhang, X. D. Chen, L. N. Wu, X. Y. Jin, W. Chen, Q. Li, X. Y. Zhang, Y. P. Xiao, R. Wu and Q. Y. Guo, *Cancer Radiother.*, 2012, **3**, 179.
99. C. Eroglu, D. Unal, A. Cetin, O. Orhan, S. Sivgin and A. Ozturk, *Anticancer Res.*, 2012, **32**, 3587.
100. M. K. Arjmandi, D. Moslemi, A. Sadati Zarrini, M. G. Ebrahimnezhad, A. Mosapour, A. Haghhaghighi, S. Halalkhor, A. Bijani and H. Parsian, *Rep. Pract. Oncol. Radiother.*, 2016, **21**, 141.
101. R. Muecke, O. Micke, L. Schomburg, K. Kisters, J. Buentzel, J. Huebner and J. Kriz, *Radiat. Oncol.*, 2014, **9**, 289.
102. R. Muecke, L. Schomburg, J. Buentzel, K. Kisters and O. Micke, *Integr. Cancer Ther.*, 2010, **9**, 136.

CHAPTER 13

Toxicology and Anticancer Activity of Synthetic Organoselenium Compounds

JOÃO B. T. ROCHA,*[a] CLÁUDIA S. OLIVEIRA[b] AND PABLO A. NOGARA[b]

[a] Universidade Federal de Santa Maria, Departamento de Bioquímica e Biologia Molecular, Av Roraima, n° 1000, Santa Maria 97090-900, Brazil;
[b] Universidade Federal de Santa Maria, Programa de Pós-Graduação em Ciências Biológicas: Bioquímica Toxicológica, Av Roraima, n° 1000, Santa Maria 97090-900, Brazil
*Email: jbtrocha@yahoo.com.br

13.1 Introduction

Ever since the discovery of selenium in 1817, the chemical, medical, toxicological, technological, and pharmacological interest in selenium has been growing globally. Toxicity, and the potential pharmacological uses of synthetic organoselenium compounds in cancer are discussed in this chapter. Other well studied synthetic selenium compounds, for instance, ebselen, benzylselenocyanate (BSC) and 1,4-phenylenebis(methylene)selenocyanate (*p*-XSC) are discussed briefly, because there are several comprehensive reviews of their anti-carcinogenic effects.[1–4] Some new classes of promising organoselenium compounds such as 3′-azido-3′-deoxythymidine (AZT) are

Organoselenium Compounds in Biology and Medicine: Synthesis, Biological and Therapeutic Treatments
Edited by Vimal Kumar Jain and K. Indira Priyadarsini

Published by the Royal Society of Chemistry, www.rsc.org

also included.[5–7] Although the exact mechanisms involved in the anticancer activity of selenium compounds are not known, their anticancer effects are usually associated in part with selenium toxicity.

The question of the toxicity of organoselenium compounds is discussed in terms of the main bottlenecks in our limited biochemical and molecular knowledge of the subject. In fact, many phenomenological data about the toxicity of organoselenium compounds are available, but mechanistic studies establishing causal associations between molecular events and toxicological or pharmacological outcomes are scant. The identification of specific targets of specific selenium compounds are greatly needed, but the task is very hard due to the limited availability of analytical methods. These aspects are discussed in more detail later, because a better understanding of how selenium is toxic to normal and cancerous cells can help researchers to design ideal molecules that can be used to develop new therapeutic agents.

The physiological chemistry of selenium is discussed here, but readers can consult recent reviews.[8–10] In short, until now selenium has only one physiological functional group, *i.e.* the selenol group found in the active site of selenoenzymes or in the redox active moiety of selenoproteins.[8–11] The selenol group of selenocysteine is an analogue of the thiol group found in cysteine, and both groups can participate in redox reactions as nucleophile centers. The reactivity of the selenol group is much higher than that of thiol group to the point that we cannot find free selenocysteine in physiological media. In fact, the selenol group is found only in a small number of selenoproteins. For a detailed discussion about the differences between the use of selenium and sulfur in biology see Nogueira *et al.*[11] and Rocha *et al.*[12] The potential importance of the transitory formation of selenol groups during the metabolism of organoselenium compounds, both regarding their pharmacological and their toxicological effects are discussed in this chapter.

The methodological and theoretical limitations found in the field of organoselenium chemistry are also found in general when studying the toxicology of the majority of pharmaceutical agents and include lack of knowledge about the precise primary targets involved in triggering the cascade of events that result in cell toxicity or cell preservation. Organoselenium compounds have low selectivity for specific biological targets. They can interact with different classes of proteins.[11–13] Indeed, experimental uses of organoselenium compounds have been undertaken mainly on empirical grounds. The necessity of establishing more rational approaches to test the toxicity and potential therapeutic use of new and old organoselenium compounds have been timidly commented upon in the literature.[14,15] Thus, in addition to bioanalytical methodologies, the development of new *in silico* methodologies are needed to speed advances in the pharmacology and toxicology of organochalcogens. At the end of this chapter, we give a few examples on how *in silico* methods can facilitate the development of the field.

13.2 Selenium Toxicology

13.2.1 Inorganic Selenium

The toxicity of cationic salts of selenium was established in the earlier history of the element, both in experimental models and in humans.[11,12,16–21] The catalytic or stoichiometric reduction of Se^{+4} and Se^{+6} to Se^0 or hydrogen selenide or selenide (HSe^-/Se^{-2}) by thiol-containing molecules has been known for a long time and is considered important in the toxicity of selenium salts.[11,12,17,22] The chemical interactions between Se^{+4} and reduced glutathione (GSH) with the formation of seleno-diglutathione, Se^0, HSe^- and dimethylselenide was studied intensively by Ganther[23–25] who confirmed the early suggestions about the involvement of thiol group oxidation in the toxicity of Se^{+4}.[12,17,22]

The fate of selenide inside the living cell is determined by its incorporation into selenoproteins or its excretion either in volatile dimethylselenide (garlic-like odour) or in the urine as trimethylselenonium. Elemental selenium or selenium sulfide (S_2Se), which have red or orange colours, have been visually and chemically detected in rodents and humans heavily intoxicated with Se^{+4}; however, their formation has not been described after exposure to moderate levels of selenium salts.[12,26]

The garlic-like odour can also be detected in humans occupationally exposed to selenium.[16] Although the formation of dimethylselenide does not indicate selenium toxicity, the presence of garlic-like odour in the breath can indicate that –SH (thiol) groups have been oxidized during the reduction of Se^{+4} or Se^{+6} to HSe^- or Se^{-2}. The formation of Se^0 probably occurs only when the levels of Se^{+4} or Se^{+6} inside the tissues are too high (*i.e.* in the case of severe poisoning with selenium salts). In this case, the accumulation of seleno-diglutathione (selenopersulfide, as originally used by Ganther[25]) and glutathione-selenol intermediates will saturate the "normal pathway" (*i.e.* the route forming selenide, dimethylselenide and trimethylselenonium). Consequently, the decomposition of glutathione-selenol to Se^0 will occur, which, depending on its concentration, can precipitate in tissues. The transformation of selenide (Se^{-2}) into Se^0 is hypothetically possible, but the oxidizing agents have not been identified or proposed. In view of the instability of the –Se–S– bonds in R-S–Se–S-R or R-S–SeH, seleno-diglutathione can be decomposed to SeS_2.[24]

13.2.2 Naturally Occurring Organoselenium Compounds

The toxicity of naturally occurring organoselenium compounds has been investigated by Franke and colleagues.[17,18,20] In contrast to the cationic inorganic salts of selenium, naturally occurring selenium compounds (*e.g.* selenomethione, selenocystathionine, g-glutaminyl-semethylselenocysteine and methylselenocysteine) can form HSe^- or Se^{2-} and methylselenol, which are metabolized to volatile dimethylselenide (exhaled through the lungs) or trimethylselenonium (excreted in urine), without directly oxidizing thiol groups.[27–29] Thus, the molecular mechanisms involved in the toxicity of naturally occurring organoselenium compounds are not expected to be the same as the inorganic salts.

The control of selenomethione and methylselenocysteine metabolism after the exposure to toxic levels is little known. Methylation has been considered a route of detoxification, because the lethality of dimethylselenide and trimethylselenonium is lower than that of cationic selenium ions or the selenoamino acids. Indeed, the level of methylation is considerably increased after injection of toxic doses of selenomethionine or methylselenocysteine in rats.[27] The contribution of trimethylselenonium to the total selenium excreted through urine also increased considerably after exposure to high doses of the two selenoamino acids.[27]

Selenomethionine is as toxic as Se^{+4} or Se^{+6}, although the exact mechanism of selenomethionine toxicity is elusive.[30,31] As indicated in Figure 13.1, Se^{+4} and Se^{+6} can oxidize the thiol groups of GSH and target thiol-containing proteins.[32] In contrast, the selenomethionine cannot oxidize thiol groups directly and selenomethionine can be either non-specifically incorporated into the general body proteins of mammals or metabolized to methylselenol which is further methylated to dimethylselenide and trimethylselenonium.[28,29]

In contrast to the other organic forms of selenium, Se^{+4} and Se^{+6}, selenomethionine does not produce superoxide (O_2^-) in the presence of GSH or other reducing thiol molecules.[30] However, the metabolism of selenomethionine produces intermediate(s) that can impose O_2^- formation in vertebrates.[33] The redox cycle that leads to the formation of methylselenol and reactive oxygen species (ROS) is shown in Figure 13.1. The methylselenol/methylselenolate intermediate is common both to inorganic salts of selenium and naturally occurring organoselenium compounds, and consequently, parts of the mechanisms of their toxicities overlap. Here it is important to emphasize that little is known about the proportion that each specific pathway contributes to the toxicity of a specific selenium form (both organic and inorganic). For instance, the intermediate seleno-diglutathione, which is expected to be formed only after intoxication with cationic forms of selenium, can be metabolized as glutathione-selenol (GSSeH) (Figure 13.1). The GSSeH intermediate can react with ROS (for instance, O_2^-) to give a selenyl radical that can oxidize thiol groups. In short, the formation of intermediates containing the reactive –SeH/–Se^- (including $H_2Se \leftrightarrow HSe^-$, which are inorganic analogues of the organic R-SeH $\leftrightarrow$ R-Se^- groups) can form reactive oxygen, sulfur or selenium species.[34] These reactive species can attack different biomolecules and lead to the inactivation of proteins or other macromolecules. Thus, the formation of HSe^- or selenolate (RSe^-) intermediates can trigger the toxicity of different forms of selenium.

The experimental evidence that selenium compounds can induce oxidative stress (*i.e.*, an overproduction of reactive species) are still elusive. For instance, *in vitro*, different forms of selenium (both inorganic and organic)

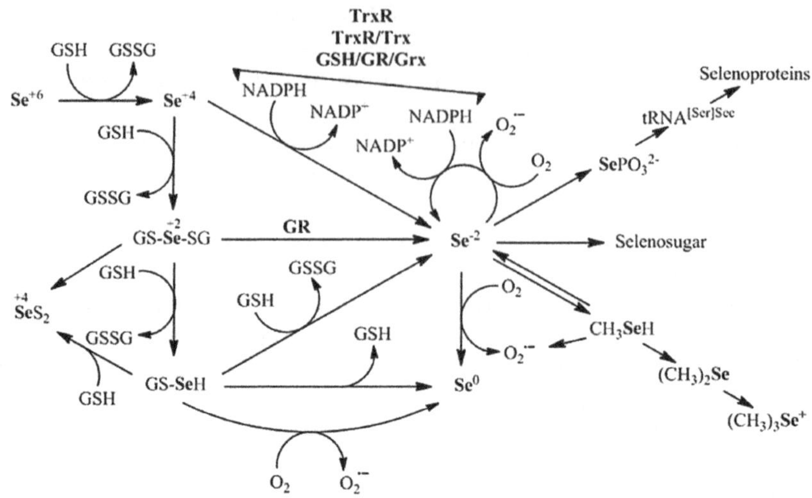

Figure 13.1 Metabolism of selenium in mammalian cells. The cationic ions of selenium, Se^{+6} and Se^{+4} (selenate and selenite, respectively), can be reduced to the "active selenide" (Se^{-2}) (the form that will be incorporated into the selenocysteinyl residue in the ^{SecSer}-t-RNA that is specifically incorporated in selenoproteins) by direct chemical interaction with thiol-containing molecules (in the scheme, reduced glutathione (GSH) is included, but it can be different proteins containing thiol groups, *e.g.* thioredoxin) or by enzymatic reactions catalyzed by glutathione reductase (GR) and thioredoxin reductase (TrxR). The ingestion of high quantities of Se^{+4} or Se^{+6} can be toxic, by either oxidizing thiol groups or by leading to the formation of selenol intermediates that may result in the formation of reactive oxygen intermediates. The scheme includes the potential formation of S_2Se and Se^0, which have been detected after intoxication with Se^{+4}. The formation of Se^0 and superoxide (O_2^-) from selenide is possibly not operant under physiologically relevant conditions. The same applies to the decomposition of the glutathione-selenol (GSSeH) intermediate. The increase in the formation of selenide can result in the formation of reactive species in several steps.

have been reported to cause excessive ROS/reactive nitrogen species production in different types of cells.[35–37]

The formation of selenyl radical and subsequent formation of O_2^- from synthetic organoseleno compounds (Figure 13.2) can be compared with the formation of O_2^- after the interaction of oxygen with thiyl radicals formed during the autoxidation of different types of thiol-containing molecules.[38] Furthermore, the reaction of O_2^- with selenol groups gives hydrogen peroxide and the selenyl radical, which has been implicated in the inactivation of glutathione peroxidase[38] and possibly of selenoproteins. Although the formation of selenyl radical intermediates during the interaction of thiols with diselenides and other organoselenium compounds have been demonstrated experimentally in chemical and *in vitro* models, the participation of selenyl radical in the toxicology of organoselenium compounds after *in vivo*

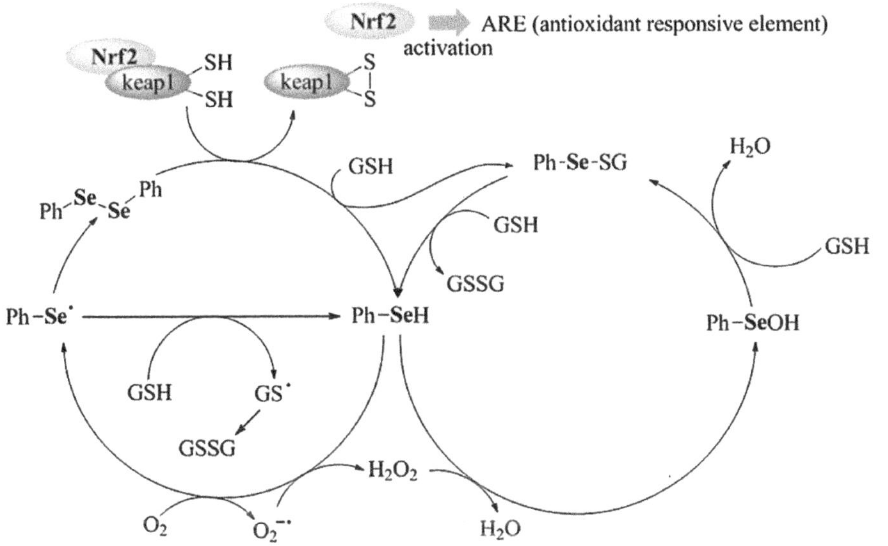

Figure 13.2 Metabolism of diphenyl diselenide inside the mammalian cell. Diphenyl diselenide can be metabolized to its selenol/selenolate intermediate which can either catalyze the decomposition of peroxide or can generate reactive selenium, sulfur or oxygen reactive species. In similar way to ebselen, diphenyl diselenide can also be reduced by thioredoxin reductase, but it is not presented in the figure (see Figure 13.3).

exposure still needs more convincing points of evidence. New experimental approaches will be needed to determine the formation and rate constant of reaction of selenyl radical intermediates with different biomolecules in order to predict potential toxicity and/or therapeutic properties of organoselenium compounds.

13.2.3 Synthetic Organoselenium Compounds

Like inorganic and natural organoselenium compounds, the different classes of synthetic organoselenium compounds (*e.g.* diselenides and ebselen) can be metabolized as selenol intermediates. Consequently, the selenol group can generate different radicals, such as selenyl, thiyl and ROS. Thus, excessive production of selenol intermediates can also be potentially toxic to living cells.

In relation to selenium toxicity, GSH, the most abundant low-mass thiol-containing tripeptide molecule found in mammalian cells, participates in the reduction of Se^{+4} and Se^{+6} and, in cases where they have caused poisoning, the levels of GSH can be decreased and S_2Se or Se^0 can be formed. Thus, the depletion of GSH can have an important role in selenium toxicity. GSH, as well as other dithiol-containing molecules (*e.g.* dithiothreitol and protein with vicinal thiol groups) can also interact with different types of

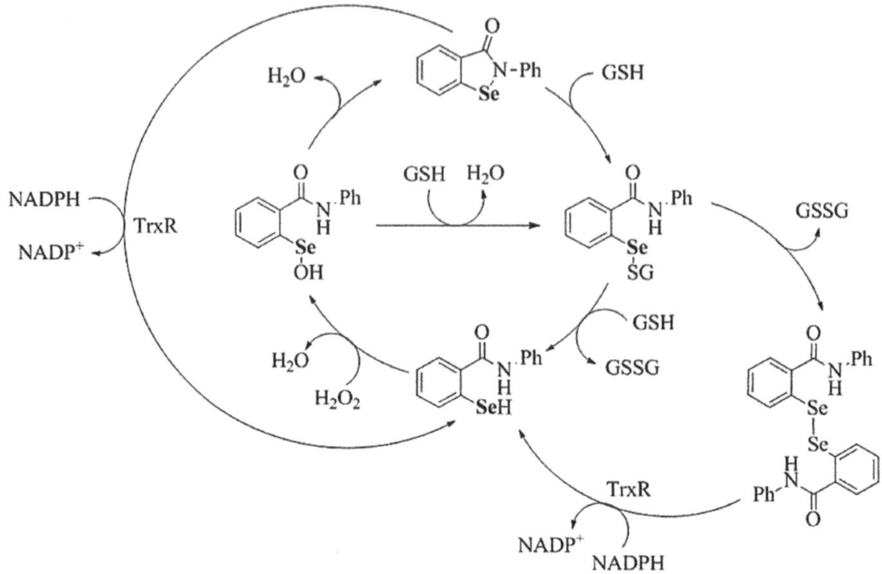

Figure 13.3 Glutathione (GSH) and ebselen interactions. Ebselen can be reduced by GSH or by thiol-containing proteins and forming the selenol intermediate that can decompose the toxic peroxides. In a similar way to diselenide (Figure 13.2), ebselen selenol can be metabolized to reactive selenium, sulfur or oxygen reactive species (this is not shown in the figure, see Figure 13.2).

synthetic organoselenium compounds.[32,39–47] The oxidation of thiol groups from GSH, and particularly of specific proteins can contribute to the toxicity of organoselenium compounds. Several studies have demonstrated inhibition of important physiological functions both at molecular, subcellular and cellular levels by high doses and/or concentrations of synthetic organoselenium compounds (Figures 13.2 and 13.3). For instance, diselenides and ebselen can target the mitochondrial respiratory chain and other proteins.[48–52] The exact molecular mechanism involved in their mitochondrial toxicity is still debatable.[45]

13.3 Environmental or Dietary Toxicity of Selenium: A Neglected Problem?

Since the establishment of selenium as a part of antioxidant proteins (selenoproteins) and the general belief that oxidative stress is central to the development of almost all types of chronic diseases, fears about the element have decreased considerably. Most importantly from the point of view of human health, inappropriate use of selenium in supplements has increased to dangerous levels.

Accordingly, recent literature indicates that it is necessary to be careful about selenium supplementation and have awareness of latent environmental exposure to selenium. For instance, the chronic use of dietary selenium supplementation or chronic ingestion of drinking water containing selenium slightly above normal or safe levels can facilitate the development of diabetes type 2 and neurodegenerative diseases.[53–58] These observations do not indicate that we have returned to the era of selenophobia (the fear of selenium observed in the 20th century[59]), but we have to be much more critical about the intentional or unintentional chronic use of selenium.

In short, selenium is not as safe as it seems to be and the toxicity of inorganic and organic forms needs to be studied using more nutritionally and environmentally relevant concentrations, particularly to determine the suitability or otherwise of exploiting organoselenium compounds for the development of drugs for chronic use in medicine. Although the mechanisms discussed earlier can operate under exposure to chronic supra-nutritional levels of selenium, it is certain that the targets will change considerably. However, we know very little about the molecular and cellular targets of chronic ingestion/exposure to low levels of selenium compounds and this type of study is needed to improve our understanding of the toxicity of this element at levels close to the physiological requirements.

13.4 Selenium in Cancer

Cancer is among the most important causes of morbidity and mortality throughout the world. The disease can be initiated by endogenous (hereditary, immunological and hormonal factors) or exogenous (dietary factors, cigarette smoking, radiation, specific parasites, bacteria and viruses and exogenous hormones) factors.[1]

The history of the use of selenium to treat cancer started in the 19th century, as cited by the British physician Watson-Williams,[60,61] and continues till today. Inorganic salts of selenium (*e.g.* selenite), naturally occurring organoselenium compounds (*e.g.* selenomethionine and methylselenocysteine) and Se-enriched yeast have been used as potential dietary agents to delay or cure cancer.[1,62–73] However, the outcome from clinical and epidemiological trials is contradictory and the use of selenium as a dietary supplement to decrease cancer incidence is questionable and debatable.[53,55,58,74,75] In contrast to naturally occurring selenium compounds, research into the potential use of synthetic organoselenium compounds as chemotherapeutic agents is still incipient and they are far from reaching the shelves of pharmacies. In fact, the first synthetic organoselenium compounds tested in cancer models were diselenide diacetic acid and diselenide dibutyric acid.[76] Although the diselenides did not exhibit antitumorogenic activity, selenium levels increased in the organs of the treated rats, whereas tumours showed lower levels of selenium than the surrounding tissues.

Over the years, investigations have been undertaken to understand the effects of selenium in cancer. An important study by Shamberger and Willis[77] demonstrated that the incidence of cancer in the USA was higher in states whose populations had lower intakes of selenium. Certainly, these findings contributed in motivating the experimental and clinical/epidemiological studies of selenium supplementation (in the form of selenomethionine, sodium selenite or selenized yeast) as a possible chemopreventive agent in the 1980s and 1990s.[1,62–71]

Subsequently, quite a number of experimental studies confirmed the importance of selenium status and the potential role of selenoproteins in the development of different types of cancer in animal models. For instance, Thompson[78] gave support to the epidemiological findings of Shamberger and Willis by demonstrating that rats fed a selenium-deficient diet were more susceptible to develop mammary carcinoma induced by 1-methyl-1-nitrosourea. The association between selenium status (determined by selenium content either in serum, plasma, toenail or hair) and cancer incidence was further confirmed by other studies (see Table 13.1).

The potential role of selenoproteins in cancer has been speculated upon for decades,[79] but our knowledge about the inhibitory or stimulatory function of selenoproteins in cancer is still incipient.[8,80] For instance, *in vitro* and *in vivo* models have demonstrated changes in the pattern of expression of selenoproteins in cancer cells.[81] The regulatory role of selenoproteins in cancer development has been indicated in the recent literature.[81–88] Accordingly, Diwadkar-Navsariwala *et al.*[89] observed that transgenic mice deficient in the expression of the selenoproteins presented a high incidence of prostate cancer.

In view of the molecular role of selenium in selenoproteins and the clinical points of evidence that good selenium status could be associated with low risk of development of chronic diseases associated with inflammation and disruption of the immune system, large human clinical trials were undertaken to confirm whether dietary selenium has a chemopreventive effect (Table 13.2).

Considering the inconclusive results obtained in the human clinical trials, the beneficial effect of selenium supplementation against cancer is highly debatable.[90] However, hypothetically selenium could be used to delay or treat specific cancers by improving selenium status (in the case of patients with low selenium status) or to shift selenium levels from below normal or to

Table 13.1 Correlation between selenium status and cancer incidence.

High Se levels in:	Low cancer incidence in:	Study type	Ref.
Serum	Thyroid	Case–control	224
Toenail	Prostate	Observational; prospective cohort	225, 226
Plasma	Prostate	Case–control	227
Serum	Prostate	Case–control	228
Serum	Fatal cancer	Case–control	229
Whole blood	Esophageal	Observational	230
Toenail	Head and neck	Observational	231
Serum	Gastric	Ecological	232
Toenail	Bladder	Prospective cohort	233

Table 13.2 Main selenium clinical trials.

Clinical trial	Number of participants	Dosage	Duration	Main results
Selenium clinical trial (China)	Se: 20 847 Placebo: 109 624	15 ppm of sodium selenite per day	5 years	Selenium-treated participants presented a decrease in primary liver cancer incidence[234]
Nutritional Prevention of Cancer trial (USA)	Se: 479 Placebo: 495	200 μg of selenium as selenized yeast per day	4.5 years	Selenium-treated participants presented a decrease in prostate cancer incidence[64]
Selenium and Vitamin E Cancer Prevention Trial (SELECT) (USA, Canada and Puerto Rico)	Se: 8752 Vitamin E: 8737 Vitamin E + Se: 8702 Placebo: 8696	Se: 200 μg of selenomethionine per day Vitamin E: 400 mg per day	12 years	Dietary supplementation with vitamin E significantly increased the risk of prostate cancer among healthy men and dietary selenium supplementation presented effects on prostate cancer incidence.[235]
Selenium and Bladder Cancer Trial (SELEBLAT) (Belgium)	Se: 151 Placebo: 141	200 μg of selenium as selenized yeast per day	3 years	Selenium supplementation did not reduce the probability of recurrence of bladder cancer[236]

the cytotoxic level. When the selenoprotein synthesis reaches a plateau, the dietary selenium is directed to the synthesis and excretion of metabolites, such as methylselenol. The overproduction of methylselenol in the cell milieu increases the formation of ROS (Figure 13.1), which in turn can have chemotherapeutic effects. In fact, it has been demonstrated that methylselenol is effective against several types of cancer cell lines[91–96] and in an animal model.[97]

13.4.1 Organoselenium Compounds and Cancer

Diselenide diacetic acid and diselenide dibutyric acid were used in 1941 in a rodent model of cancer, but the compounds were not effective.[76] In 1943, one study demonstrated that high levels of some organic compounds of selenium in the diet caused adenomatous hyperplasia in the liver and thyroid of rats, in particular the compound bis(4-aminophenyl)selenium hydroxide. The authors noticed high toxicity of Se^{4+}, but inorganic forms of selenium apparently did not cause the hyperplasic effects observed with the organoselenium compounds. The authors also observed the absence of toxicity of aromatic selenium compounds devoid of amino groups.[98] Although this study remained unnoticed by researchers working on organoselenium compounds, it is very informative and germane to our discussion, because it demonstrated that a small change in the organic moiety of selenium-containing molecules caused a profound change in the toxicological properties of the compounds. A myriad of organoselenium compounds have been synthetized and many of them have been tested against cancer using cell culture models.[5–7,99–133] In the next section the mechanism of some organoselenium molecules against cancer is briefly discussed.

13.5 The Mechanism of Action of Synthetic Organoselenium Compounds as Anti-cancer Agents

Cancer cells produce a large amount of ROS and possess a special antioxidant defence system.[134] Thus, molecules that act on the antioxidant capacity can be good candidates to act as new therapeutic anti-tumour agents. In this context, selenium compounds with pro-oxidant properties can be promising antitumor agents. Alterations in the activity of transcription factors that will modulate specific genes associated with tumorigenesis could also be used to predict potential chemotherapeutic effects of the synthetic organoselenium compounds.

Molecules that are metabolized to selenide can be incorporated into selenoproteins. Compounds that can be preferentially metabolized to intermediates containing selenol group can have their anticancer activity potentiated. In this context, some classes of synthetic organoselenium compounds, for instance, diselenides can be expected to be good chemotherapeutic candidates since they can be metabolized to selenol. In this section, we discuss the ebselen, diselenides, selenocyanides and AZT derivatives as potential chemotherapeutic agents. The inclusion of AZT derivatives is justified in view of their similarity to the natural existing nucleotides, which are constituents of nucleic acids.[5–7] Consequently, they can interfere with the process of cell division, which is much more intense in cancer cells than in normal cells.

13.5.1 Ebselen

The 2-phenyl-1,2-benzisoselenazol-3(2*H*)-one (ebselen) was synthesised in 1924 and rediscovered more recently as an antioxidant and anti-inflammatory agent.[12] Ebselen does not release Se to an inorganic or selenide pool; consequently, it is not a source of Se to selenoprotein synthesis. It has been demonstrated that ebselen can be reduced after reacting with thiol groups, forming a selenol intermediate.[11,12] Nowadays, it is being tested as a lithium mimetic agent both experimentally and clinically.[135] Interestingly, despite its several therapeutic effects, its potential against cancer has been little explored.

Like other selenium compounds, ebselen can trigger apoptosis in carcinogenic cells. Table 13.3 presents a summary of studies investigating ebselen in cancer. Briefly, the interaction of high levels of ebselen with the cellular thiol leads to depletion of important cellular defences and accumulation of ROS in the intracellular medium, which results in cell apoptosis.[50,136–139] In addition, ebselen alters the mitochondrial function, inducing the BAX protein activation, a pro-apoptotic protein (Figure 13.4).[138]

Table 13.3 Ebselen effects as a chemopreventive agent.

Effects	Lowest effective dose	Experimental model	Ref.
Inhibition of quiescin sulfhydryl oxidase 1	500 nM	Purified enzyme	136
58% reduction in tumor growth	160 μg day^{-1}	Mouse	136
Apoptosis preceded by intracellular glutathione and depletion of protein thiols	25 μM	Human hepatoma cell line	50
Decrease in cell viability	40 μM	Pancreatic rat tumorigenic cells	137
Increase in caspase-3 activity	10 μM	Pancreatic rat tumorigenic cells	137
Increase in ROS levels; decrease in cell viability; increase in apoptosis	40 μM	Human myeloma cell	138
Inhibits hepatocarcinogenesis induced by aflatoxin-B1	5 mg day^{-1}	Rats	139

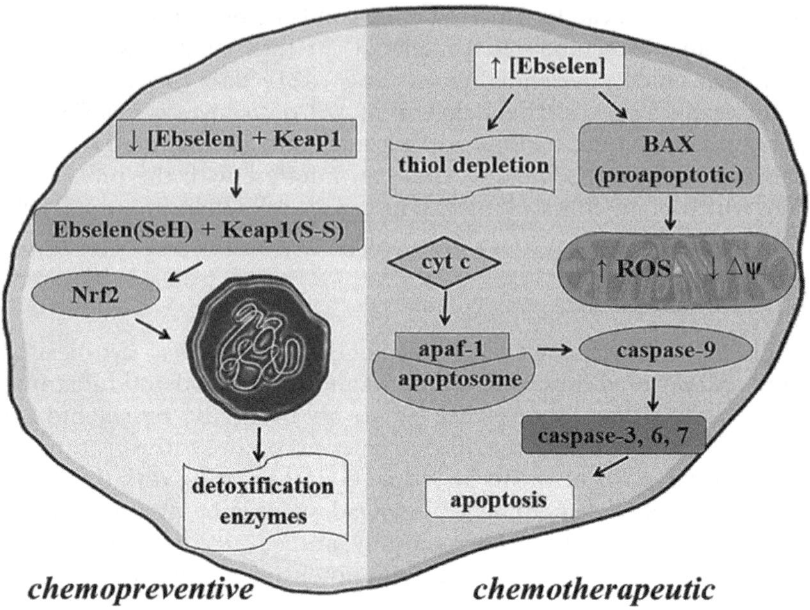

Figure 13.4 Schematic fate of ebselen inside the cell: chemopreventive and chemotherapeutic approach. The specific interaction of ebselen and other organoselenium compounds with specific proteins (keap-1) can be involved in their chemopreventive properties (left). The massive oxidation of thiols by high concentrations/doses of organoselenium compounds can be involved in their apoptotic mechanism (right).

13.5.2 Diselenides

Diselenides are toxic to cancer cells.[140–143] Diselenides that have been tested as chemopreventive agents are shown in Figure 13.5. Diselenides have been tested against breast cancer cells,[141] adenocarcinoma cells[142] and neuroblastoma cells.[143] The chemopreventive mechanism of the diselenides can be related to the reduction of the Se–Se bond, forming selenol intermediates. The selenol intermediates can promote the formation of selenyl and thiyl radicals or ROS, which can disrupt the redox balance of cancer cells.

Carcinogenic cells exposed to diphenyl diselenide result in cell apoptosis *via* an increase in Bax/Bacl2 ratio and the genes p53 and p21 cell cycle arrest.[142] Diphenyl diselenide can induce apoptosis *in vitro*, but this has not been investigated after *in vivo* administration. In contrast to ebselen,[12,144] diphenyl diselenide[145] and several other selenides and diselenides can release selenium from the organic moiety,[146] which feeds the selenide and methylselenol, methylselenide pool of selenium. The selenide can be used to up-regulate the synthesis of selenoproteins, which can improve the organism response against carcinogenic agents. Barbosa *et al.*[147] showed that supplementation with diphenyl diselenide delayed the development of mammary tumours induced by *N*-nitroso-*N*-methylurea in rats. The

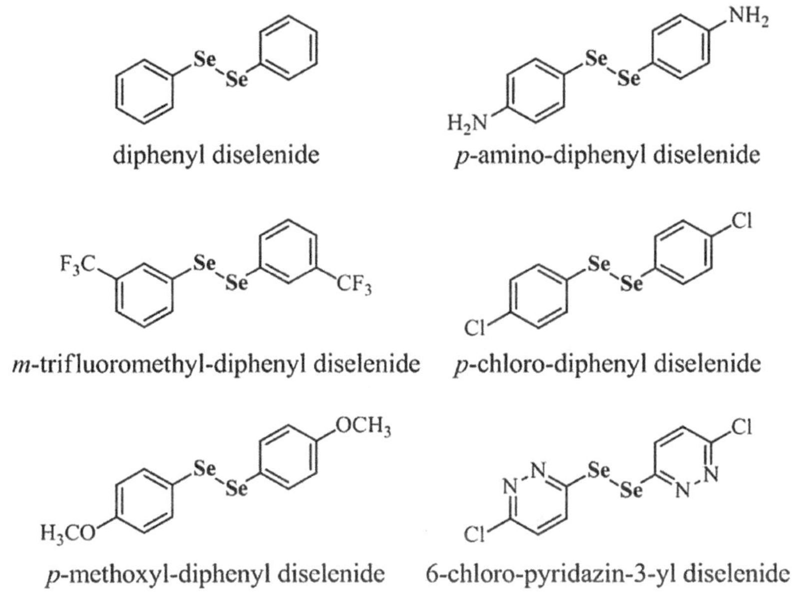

Figure 13.5 Molecular structure of some diselenides tested in tumorogenic cell culture.

methylselenol can promote the formation of reactive species that could be more toxic to cancer cells than normal cells.

13.5.3 Selenocyanates

Following the studies of Gusberg *et al.*,[76] one of the first tested molecules was benzylselenocyanate (BSC), which showed good activity against colon carcinogenesis,[148] forestomach tumor,[149] mammary carcinogenesis[150] and hepatocarcinogenesis.[151] Subsequently, the 1,4-phenylene-bis(methylene)-selenocyanate (*p*-XSC) was tested[152] and found to be less toxic than BSC.[153] Furthermore, in a tumorogenic cell culture study, *p*-XSC was proven to kill cells more efficiently than BSC and selenite.[154] For completeness, a brief discussion on the anticancer mechanism of selenocyanates is included here, as the subject matter has been reviewed earlier.[1–4]

El-Bayoumy and co-workers showed the effects of selenocyanates in pre- and post-cancer initiation induced by 7,12-dimethylbenz[*a*]anthracene (DMBA).[150,152,154–159] The *p*-XSC down-regulates the expression of cytochromes, oncogene activators (c-myc and cyclin D1/D2), cell proliferation activators (PCNA) and up-regulates the expression of phase II detoxification enzymes (for example, glutathione *S*-transferase) and pro-apoptotic proteins (for example, p21, p27, caspase3 and BAD).[157] All these molecular alterations can decrease the proliferation of cancer cells by triggering cellular apoptosis. In addition to apoptosis induction and interaction with GSH, *p*-XSC and its putative metabolite interacted with specific vicinal thiol-containing

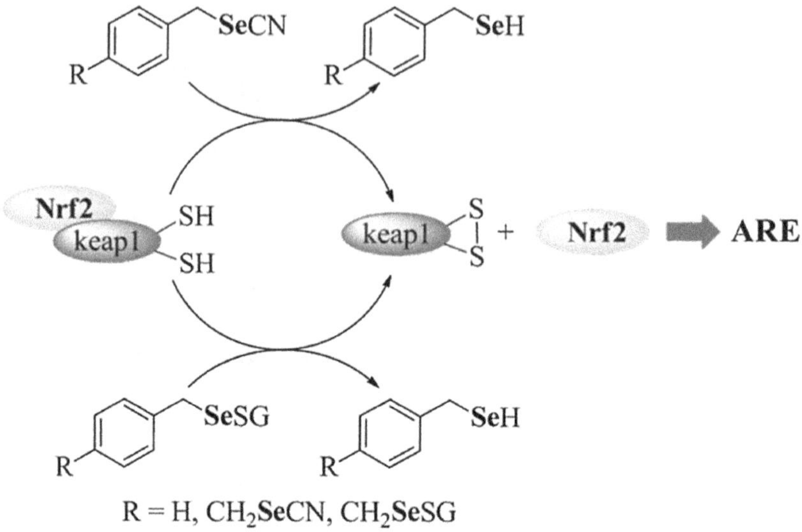

Figure 13.6 Benzylselenocyanate (BSC), 1,4-phenylene-bis(methylene)-selenocyanate (*p*-XSC) and 1,4-phenylene-bis(methylene)-selenoglutathione (*p*-SeSG) interaction with thiol groups. In a similar way to diselenides and ebselen, *p*-XSC and *p*-SeSG can interact with specific thiol-containing proteins (keap-1 in this case), modulating the transcription of antioxidant genes.

transcript factor inhibitor, *e.g.* keap-1. The oxidation of keap-1 by organoselenium compounds increases the transport of Nrf2 from the cytoplasm to the nucleus of the cell (Figure 13.6).[159] This stimulates the synthesis of phase II enzymes, for instance, glutathione *S*-transferase.

Contrary to ebselen, which does not release selenium from its structure,[12] *p*-XSC, similar to organoselenides and organodiselenides,[145,146] can release selenium to the inorganic pool.[158] A study by Ip *et al.*[158] showed that only rats fed with a high concentration of *p*-XSC (20 ppm) presented a recovery in glutathione peroxidase activity using a selenium-deficient diet model. Thus, the scission of the Se–C bond and the consequent release of inorganoselenium in the *p*-XSC molecule occur slowly, but it can furnish selenium-to-selenoprotein synthesis. However, to our knowledge the *p*-XSC, a promising synthetic organoselenium compound, has not been tested in humans.

13.5.4 3′-Azido-3′-deoxythymidine Derivatives

3′-Azido-3′-deoxythymidine (AZT) was first developed as potential anticancer agent. Nowadays, AZT has a crucial role in the treatment of acquired immunodeficiency syndrome, acting as an inhibitor of the viral reverse transcriptase enzyme.[160,161] Besides its antiretroviral activity, some studies have shown anticancer activity of AZT and its derivatives.[162–167] Among

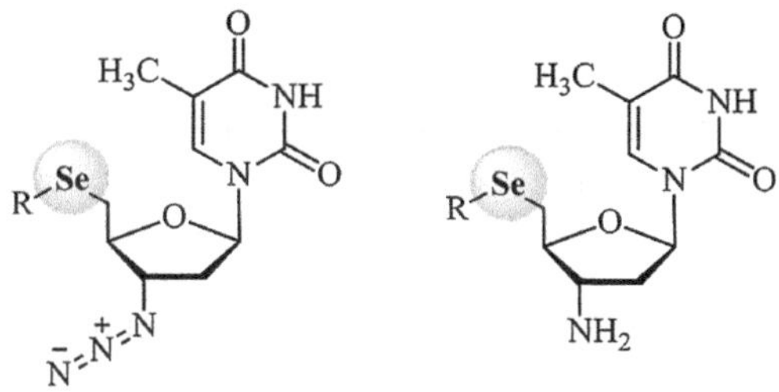

R = Ph, *p*Me-Ph, *p*MeO-Ph, *p*Cl-Ph, *o*Me-Ph, α-Np

Figure 13.7 3′-Azido-3′-deoxythymidine (AZT)-selenium derivatives tested in tumorigenic cells.

them, AZT-selenium derivatives tested against 5637 bladder cancer cells lines are quite promising (Figure 13.7).[5,7] The compounds seem to cause apoptosis in the cancer cells, *i.e.* increasing the mRNA expression of pro-apoptotic genes (BAX and caspase-9) and decreasing the mRNA expression of an anti-apoptotic gene (BCL-2). However, more studies are required to understand the exact mechanism by which the AZT derivatives produce anticancer effects.

13.6 *In silico* Study as a Virtual Screening of Potential Anticancer Drugs

As stated in the critical review of Orian and Toppo,[14] the future of organoselenium compounds as therapeutic agents will require the development of new *in silico* studies and techniques. Particularly, the intensification of *in silico* studies determining the interaction of organochalcogens with thiols (both low- and high-molecular mass compounds, *i.e.* thiol-containing proteins) and other targets will be needed. There are a few reports in the literature on *in silico* studies showing the interaction of organoselenium compounds with simple thiols.[15,168–174] However, studies dealing with the interaction of selenium compounds with specific proteins are still rare.

Density functional theory (DFT) studies have been applied to evaluate the structural and electronic properties, energetic aspects and reactivity of organoselenium compounds. These investigations facilitate drug design of new compounds that could present better properties than the originals.[175–177]

In silico is an expression used to mean "performed on computer", and the technique has the potential to speed the rate of drug discovery while reducing the need for expensive lab work and clinical trials. Virtual screening uses algorithms and models for the identification of novel bioactive compounds, where large libraries of molecules are screened computationally against specific targets, and those that are predicted to bind well are experimentally tested. This process has been shown to be more efficient than the traditional trial and error method.[178,179]

By using computational methods and the 3D structural information of the protein target, it is possible to investigate molecular interactions involved in ligand–protein binding at the atomic level. However, there are limited investigations involving virtual screening and organoselenium compounds.[14,15] However, the available studies can help us to understand the nature of interactions between the ligand and the target(s), the main residues involved, the nucleophile/electrophile centre(s) in specific moieties and the thermodynamics of the process. Hypothetically, these sets of parameters/data can facilitate design of new organoselenium compounds with higher activity and lower side effects. However, we are still far from having an organoselenium compound that interacts with a specific biological target.[12,14,15]

In addition, computational techniques help us to understand the most probable chemical mechanism involved in a particular reaction. For example, Pearson and Boyd[180] and Antony and Bayse[181] have explored the reaction mechanisms and thermodynamics of the reduction of hydrogen peroxide by ebselen. They also studied the effects of solvation and demonstrated that the reaction free energy barriers for the oxidation of ebselen are in agreement with experimental procedures. *In silico* studies involving organoselenium compounds have shown that there are at least five to six main potential targets in cancer, such as thioredoxin reductase (TrxR),[182] tubulin,[183] histone deacetylase (HDAC),[184] zinc finger proteins (ZFP)[185] and the DNA.[186]

13.6.1 Thioredoxin Reductase

The thioredoxin system regulates various cellular processes, including antioxidant defences, redox homeostasis, cell proliferation and apoptosis. Many mammalian tumour cell lines overexpress TrxR-1 and show enhanced TrxR-1 activity, which can be associated with tumor aggressiveness and apoptosis inhibition. It has been reported that reduction of TrxR-1 levels in cancer cells reverses many characteristics of malignancy and inhibits cancer cell growth. Thus, TrxR1 is considered to be an important anticancer drug target.[182,187,188]

The Trx system has been considered a target for cancer treatment because the reduction of the ribonucleotides to deoxyribonucleotides required in DNA synthesis depends indirectly on the activity of TrxR and Trx. In the case of cancer cells, these redox reactions are higher than in normal cells.[182,189]

In terms of the catalytic mechanism, fully oxidized TrxR accepts electrons from NADPH by reducing FAD to $FADH_2$, which passes the reducing equivalents to the *N*-terminal redox-active center (Cys59 and Cys64 in rat TrxR1) in one subunit. Finally, this reduced N-terminal redox center transfers two electrons to the C-terminal redox-active site (Cys497 and Sec498) in another subunit to generate the active enzyme. The TrxR passes the reducing equivalent to its native substrate Trx, which catalyzes thiol-disulfide exchange reactions and can scavenge ROS either directly or indirectly.[190,191]

Studies have shown that ethaselen (an ebselen analogue) significantly suppresses cell viability in several cultured human cancer cell lines in parallel with direct inhibition of TrxR1 activity, including lung (A549), stomach (BGC823), cervical (HeLa), nasopharyngeal (KB), prostate (PC-3, DU145) and leukemia (K562, HL-60) cell lines. As a downstream effect of TrxR1 inactivation, ethaselen caused a dose-dependent thioredoxin oxidation and enhanced the levels of cellular ROS in A549 cells.[188,192–195]

In silico molecular dynamics using SANDER (simulated annealing with NMR-derived energy restraints) and molecular docking using GoldScore fitness scoring function (which considers factors such as H-bonding and van der Waals energy, metal interaction and ligand torsion strain) demonstrated that ethaselen binds to the unique Sec–Cys reduced pair at the C-terminal active site of mammalian TrxR1, causing its inhibition. Selenol from Sec498, due its higher nucleophilicity than thiol from Cys497, is the first to react with ethaselen, forming a diselenide bond, after the thiol moiety reacts with another electrophilic selenium atom from ethaselen, forming a selenenylsulfide bond, inactivating the enzyme (Figure 13.8).[188,194]

Lan *et al.*[196] have reported that ethaselen induces apoptosis through inhibition on the TrxR-Trx-NF-κB pathway and down-regulates anti-apoptosis gene expression of Bcl-2, Bcl-xL, cIAP-2 and XIAP. These data indicate that ethaselen can act as both TrxR and tumor growth inhibitor. Ethaselen is undergoing further clinical research and evaluation.

As mentioned earlier, TrxR is also involved in the reductive metabolism of inorganoselenium compounds, particularly Se^{4+}. Zhao and Holmgren[197] have demonstrated that ebselen and its diselenide can be reduced by mammalian and bacterial enzymes. However, the concentration required to inhibit the enzyme is higher than the concentration required to give

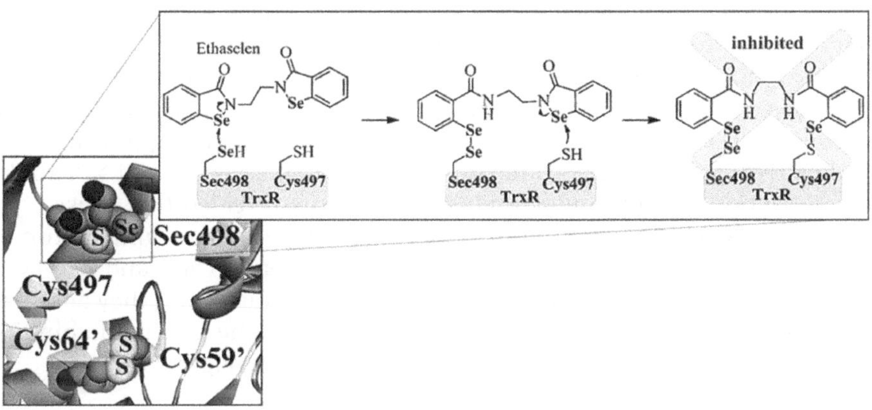

Figure 13.8 Representation of mechanism of thioredoxin reductase (TrxR)-1 inhibition by ethaselen. The formation of diselenide (Se–Se) and selenide-sulfide (Se–S) bonds between ethaselen and active residues of TrxR inactivates the enzyme and can contribute to its anticarcinogenic activity.

optimum reduction to its selenol intermediate. Similar findings have been reported with diphenyl diselenide and its analogues.[198,199]

Recently, it has been demonstrated that ebselen and its analogues can inhibit TrxR from different bacteria species; however, the structural basis for selective inhibitory effects of some analogues has not been investigated. The use of *in silico* methods could be used for better understanding interaction of ebselen and its analogues with different sources of TrxR.[200]

Shaaban *et al.*[201] synthetized novel tetrazole-based diselenides and selenoquinones and applied molecular docking into mammalian TrxR1, to predict the anticancer potential of new molecules. The compounds were evaluated for their anticancer activity against HepG2 (liver cancer) and MCF-7 (breast cancer) cell lines and compared with their cytotoxicity to normal WI-38 employing standard MTT assays. In general, diselenides were more cytotoxic than selenoquinones, and showed higher activity in HepG2 than MCF-7 cells.

13.6.2 Tubulin

Microtubules are key components of the cytoskeleton of eukaryotic cells and are involved in numerous cellular functions, such as maintenance of cell shape, cell signalling, cellular transport and cell division. The assembly of α- and β-tubulin heterodimers and the disassembly of the polymeric form are dynamic processes that lead to the formation of microtubules. The disruption of these dynamic processes blocks the cell division machinery at mitosis and results in cell death. Since cancer cells undergo cell division much more rapidly than normal cells, the cytoskeleton is affected more selectively in the cancer cells than normal cells.[183,202]

In studies by Guan *et al.*[203] and Wen *et al.*[204] involving organoselenium compounds, the combretastatin A-4 (CA-4) analogues were synthesised using a rigid 1,2,5-selenadiazole ring (I), to fix the *cis*-orientation between the ring-A and ring-B, and 3-(3,4,5-trimethoxyphenylselenyl)-1*H*-indole (II) (Figure 13.9A). Most of these compounds exhibited significant antiproliferative activity, with some showing nanomolar IC_{50} values. Tubulin polymerization and immunostaining experiments revealed that these organoselenium compounds inhibited tubulin polymerization and disrupted tubulin microtubule dynamics in a similar manner to CA-4 and cell cycle arrest in the G2/M phase. Docking studies demonstrated that they adopt an orientation similar to that of colchicine and CA-4 at the colchicine binding site on tubulin, probably due its structural similarity (Figure 13.9B).

13.6.3 Histone Deacetylase

HDACs are enzymes that remove acetyl groups from ε-*N*-acetyl lysine residues from histone (added by histone acetyltransferases) and are involved in the remodelling of chromatin (condensed or uncondensed), and have a key

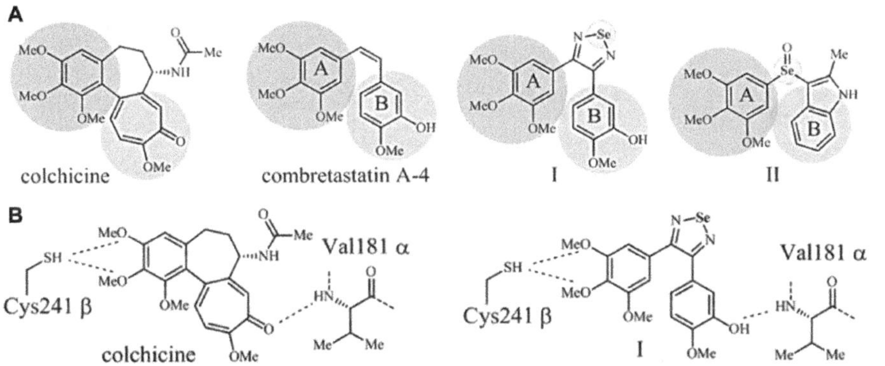

Figure 13.9 Inhibitors of tubulin polymerization. (A) The structures of two classical and two new organoselenium compounds that are inhibitors of tubulin polymerization are shown. The pharmacophore groups are highlighted. (B) Comparison of the binding modes between the colchicine and the compound I. Both compounds interact similarly with tubulin.

role in the epigenetic regulation of gene expression. In recent years, inhibition of HDACs has emerged as a potential strategy in cancer treatment. HDAC inhibitors can reactivate gene expression and inhibit the growth and survival of tumour cells. Silencing of tumor suppressor genes (such as p21) through hypoacetylation is a hallmark of many cancers, and switching these back on through HDAC inhibition has shown clinical benefit. However, such studies have also indicated that the effects of HDAC inhibitors could be considerably broader and more complicated than originally understood.[184,205–207] The proposed mechanism of deacetylation of histones demonstrated the importance of substrate coordination with zinc and H-bonds with tyrosine residue, from the catalytic site of HDAC.[208]

Studies by Nian *et al.*[209] and Lee *et al.*[210] demonstrated that β-methylselenopyruvate (MSP) and α-keto-γ-methylselenobutyrate (KMSB), α-keto acid metabolites from methylselenocysteine (MSC) and selenomethionine can alter HDAC activity and histone acetylation status. In a cell-free system, glutamine transaminase K and L-amino acid oxidase convert MSC to the corresponding α-keto acid MSP, and L-amino acid oxidase converts selenomethionine to its corresponding α-keto acid, KMSB.

These findings provide a potential new paradigm by which natural organoselenium metabolites might prevent the progression of human colon and prostate cancer. In addition, computational modelling studies by molecular docking using Monte Carlo-based global energy minimization provide a probable mechanism involving reversible interaction with the active-site zinc atom in the active site of HDAC, probing the entrance of substrate (acetylated lysine from histones). Both α-keto acid metabolites presented the same pattern of interactions, the coordination of keto and carboxyl moiety with the zinc, the H-bond with Tyr306 and the hydrophobic interactions between selenium aliphatic chain and the benzene ring from Phe152 (Figure 13.10).[211,212]

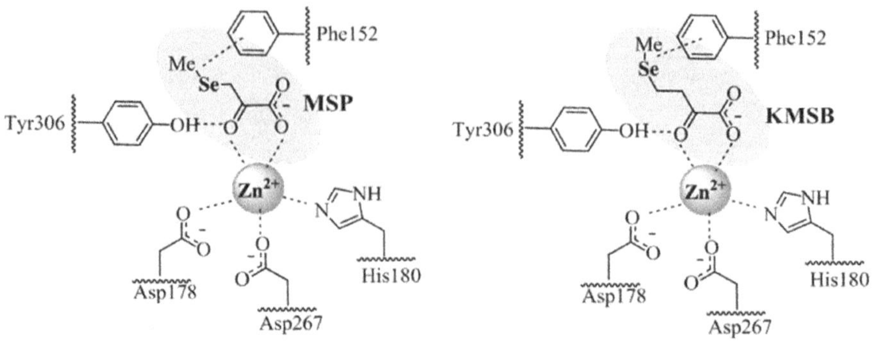

Figure 13.10 Interactions between methylselenopyruvate (MSP) and α-keto-γ-methylselenobutyrate (KMSP) in the human histone deacetylase (HDAC)8 enzyme. The α-keto acid metabolites interact in the active site of HDAC8, hindering the entrance of substrate.

13.6.4 Zinc Finger Proteins

Zinc finger proteins (ZFP) are a family of proteins where zinc is complexed through four invariant cysteine and/or histidine residues, forming a zinc finger domain that is mostly involved in DNA binding in transcription factors, and in protein–protein interactions. It is assumed that several of these proteins are regulated by oxidation of zinc-binding cysteine residues, which leads to a loss of DNA binding, zinc release and/or formation of disulfide bridges, with subsequent loss of the tertiary structure in ZFP_{ox} necessary for recognition.[185,213–215] Studies revealed that aberrant expression of ZFP contributes to tumorigenesis. For example, the ZNF139, which is overexpressed in gastric cancer cells, promotes proliferation and inhibit apoptosis through up-regulating the expression of survivin, x-IAP and Bcl-2, and down-regulating Caspase-3 and Bax.[216,217] Zinc finger transcription factors incorporate Zn^{2+} ions tetrahedrally coordinated to Cys and His residues (typically Cys_2His_2, Cys_3His or Cys_4) to ensure proper folding of the ZFP tertiary structure for biological recognition. Thus, oxidation of Cys molecules could be a therapeutic strategy against tumours, silencing genes important for the cancer cell's progression (Figure 13.11).

Studies by Bayse *et al.*[218] using DFT with the B3PW91 functional and solvent-assisted proton exchange, showed important insights into the thiol reduction mechanisms in ZFPs (and other S–Zn proteins), by different reducible organoselenium compounds (rSe).[219–221] According to their results, the S···Se chalcogen bonding interactions consist of the donation of electron density from the S lone pair of the ZFP model (highest occupied molecular orbital, HOMO) to a Se–X antibonding molecular orbital from the rSe compound (lowest unoccupied molecular orbital, LUMO). The strength of the interaction depends on the Lewis basicity of the zinc finger model, where increasing the number of thiolates (Cys) coordinated to Zn^{2+} ($Cys_4 > Cys_3His > Cys_2His_2$), increases the softness of the model, leading to

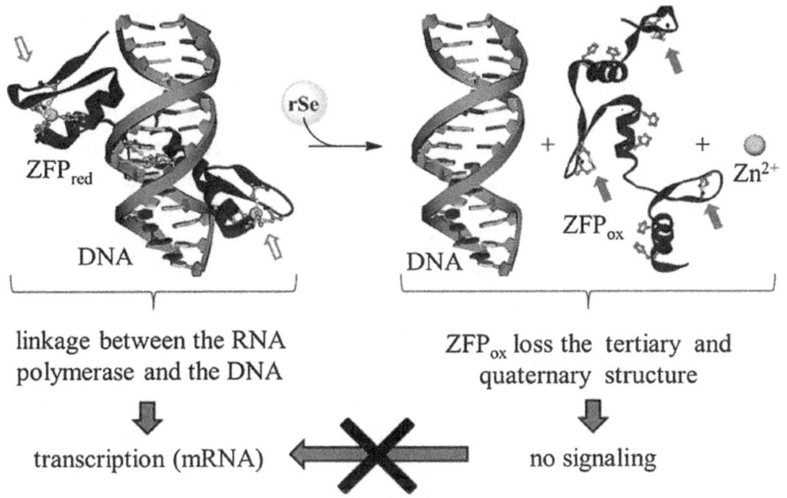

Figure 13.11 Representation of the zinc finger dynamics. The oxidation of cysteinyl residues (Cys) cause the loss of tertiary and quaternary structure of protein and consequently zinc release and gene silencing (adapted from PDB: 1TF3). Tetrahedral zinc complexes are indicated by white arrows and the oxidised cysteinyl residues by dark grey arrows.

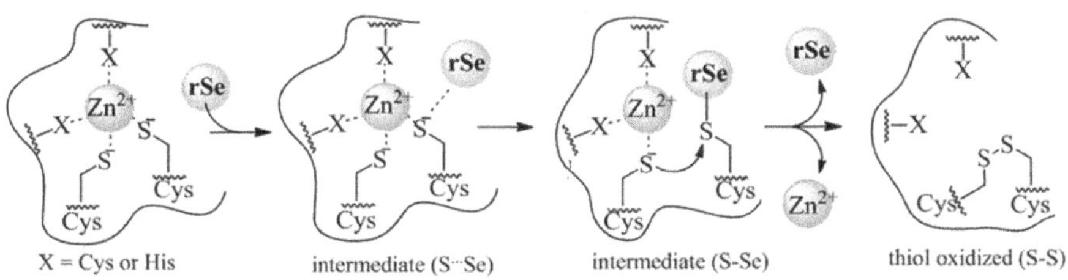

Figure 13.12 Possible mechanism of Cys oxidation and zinc release in zinc finger proteins in the presence of reducible selenium compounds (rSe).

stronger intermolecular S···Se interactions in a hypothetical intermediate. The Lewis acidity of the rSe compound measured by the LUMO energy, showed that rSe compounds are more reactive than their respective sulfur analogues, making them a better electrophile for the soft Lewis base ZFP models.

Transition states energy calculations revealed that the formation of a selenosulfide (S–Se) bond from the S···Se intermediate complex between ebselen and a Cys_2His_2 zinc finger model, which can occur through a moderate activation barrier that is consistent with experimental observations. In a subsequent step, the thiol oxidation (S–S) occurs with the release of zinc and the rSe (Figure 13.12). In addition, studies comparing the S···Se interactions of the zinc finger model and MeSH (methenethiol, simulating a endogenous thiol source) complexes, suggest that rSe reacts with ZFP even in the presence of GSH, because the thermodynamics of these processes is more favourable ($\Delta G \leq 0$).

Porphobilinogen synthase or δ-aminolevulinate dehydratase (δ-ALA-D) is an enzyme involved in the porphyrin and heme biosynthetic pathway, which catalyses asymmetric condensation of two aminolevulinic acid molecules to form the porphobilinogen. Like ZFP, δ-ALA-D presents three vicinal thiol groups in its active centre, which are involved in the co-ordination of essential Zn(II) ion. Barbosa *et al.*[32] showed that diphenyl diselenide and its analogues can inhibit the δ-ALA-D, by oxidizing enzyme cysteinyl residues. Later, Saraiva *et al.*,[220] through *in silico* protein–ligand molecular docking studies, demonstrated that the diphenyl diselenides can access the active site of δ-ALA-D, and the thiolate from Cys124 attacks one of the Se atoms from diselenides, releasing one selenophenol molecule. Subsequently, the thiolate from Cys132 attacks the sulfur atom of Cys124 (in the S–Se bond intermediate) oxidizing and inhibiting the enzyme. These results help us to understand the mechanism of action and reactivity of organoselenium compounds in a biological medium. The interaction of organoselenium compounds with proteins containing vicinal thiol groups can be exploited in the treatment of cancer; however, knowledge about this subject is still elusive and *in silico* approaches have to be applied to search for potential targets of organoselenium compounds. Moieties of the ZFP and ALA-D (containing vicinal thiols) types can guide the search for target proteins in protein databank exploration. In addition, the search for target proteins could be considered the first step in drug discovery, and offers an alternative pathway for the exploration of biochemical assays.

13.6.5 DNA

DNA as an anticancer drug target can make transcriptional machinery to drive the cancer cell toward apoptosis. The mode of action of the DNA-targeting ligands focuses either on the sequence-specificity by groove binding and strand cleavage, or by identifying the morphologically distinct higher order structures like that of the G-quadruplex DNA. Recognition and characterization of the interactions of small molecules with DNA is significant as they give effective information for designing new and more efficient therapeutic agents in controlling gene expression.[186]

In vitro and *in silico* DNA binding studies of organoselenium compounds were investigated by Rizvi *et al.*[222] and Mati *et al.*[223] using absorption and emission titration methods which revealed that these recognize the minor groove of DNA in accordance with the molecular docking studies with the DNA duplex. The studies of Rizvi *et al.*[222] involving the compound bis(3,5-bis(trifluoromethyl)phenyl)selenide (BTFPS), demonstrated the ability of this molecule to cleave pBR322 DNA through a hydrolytic process induced by the nucleophilic Se atom (Figure 13.13). *In silico* DFT calculations and HOMO energy analysis confirmed that Se atom is the nucleophile, and the compound[222] binds to DNA through external contact and hydrogen bonding, in the minor groove.[222]

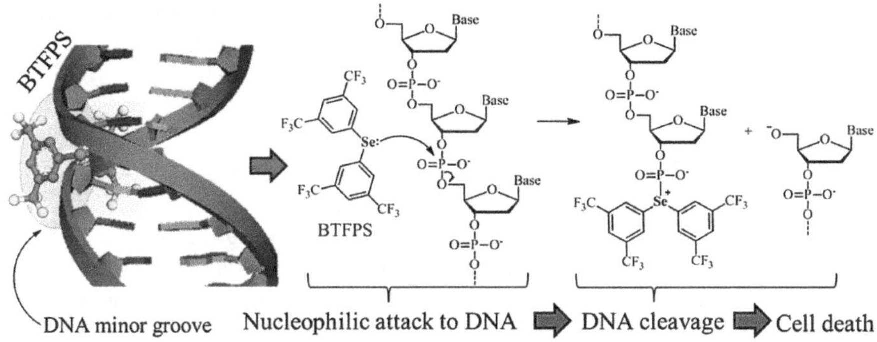

Figure 13.13 Bis(3,5-bis(trifluoromethyl)phenyl)selenide (BTFPS) interactions with DNA minor groove and a possible mechanism of DNA hydrolytic cleavage.

13.7 Conclusion

Although selenium has a potential to be used in cancer prevention and treatment, we are far from reaching a clear consensus. Regarding the potential preventive effects of selenium supplementation, it is inconclusive, as the clinical trials have indicated both positive and negative effects of selenium. Regarding the use of selenium as therapeutic agent, past clinical studies have indicated a positive effect of colloidal preparations of elemental selenium against some types of inoperable carcinomas, but the abrupt interruption of their use is enigmatic and difficult to grasp from the analysis of the literature. Concerning the main objective of this chapter, the potential use of synthetic organoselenium compounds as potential remedies against cancer, we can say that we are still in the beginning of a promising field of research. As we have discussed, our knowledge about the interaction of hundreds of organoselenium compounds with specific molecular targets is still negligible. The interaction of some compounds have been studied in a more detailed manner, but nonetheless, we have not yet been able to construct a theoretical framework to guide the chemical and biological approaches to accelerate the development of anticancer therapeutic agents containing organoselenium moieties. The studies conducted with a few molecules have been undertaken on empirical grounds, and although they have been informative in a broad sense, the future of the field of exploring organoselenium compounds in cancer treatment will require refinements.

There is a need to establish more rational approaches to test the toxicity and potential therapeutic use of new and old organoselenium compounds, although a few reviews have covered them briefly.[14,15] In short, it is time for the development of *in silico* and high-throughput *in vitro* methods to determine which selenium compounds can be considered worth testing *in vivo*, either as anticancer agents or for other pathologies. The partial success of ebselen has animated the field, but although ebselen is considered a safe drug, ironically it has no target specificity to exploit for treatment.

References

1. R. Sinha and K. El-Bayoumy, *Curr. Cancer Drug Targets*, 2004, **4**, 13.
2. K. El-Bayoumy, *Mutat. Res.*, 2001, **475**, 123.
3. K. El-Bayoumy and R. Sinha, *Mutat. Res.*, 2005, **591**, 224.
4. N. Facompre and K. El-Bayoumy, *Cancer Res.*, 2009, **69**, 2699.
5. D. de Souza, D. O. C. Mariano, F. Nedel, E. Schultze, V. F. Campos, F. Seixas, R. S. da Silva, T. S. Munchen, V. Ilha, L. Dornelles, A. L. Braga, J. B. T. Rocha, T. Collares and O. E. D. Rodrigues, *J. Med. Chem.*, 2015, **58**, 3329.
6. D. O. C. Mariano, D. de Souza, D. F. Meinerz, J. Allebrandt, A. F. de Bem, W. Hassan, O. E. D. Rodrigues and J. B. T. da Rocha, *Hum. Exp. Toxicol.*, 2017, DOI: 10.1177/0960327116674529.
7. R. M. da Rosa, B. C. Piccoli, F. D'A. da Silva, L. Dornelles, J. B. T. Rocha, M. S. Sonego, K. R. Begnini, T. Collares, F. K. Seixas and O. E. D. Rodrigues, *Med. Chem. Commun.*, 2017, DOI: 10.1039/C6MD00640J.
8. D. L. Hatfield, P. A. Tsuji, B. A. Carlson and V. N. Gladyshev, *Trends Biochem. Sci.*, 2014, **39**, 112.
9. E. G. Varlamova and I. V. Cheremushkina, *J. Trace Elem. Med. Biol.*, 2017, **39**, 76.
10. J. K. Wrobel, R. Power and M. Toborek, *IUBMB Life*, 2016, **68**, 97.
11. C. W. Nogueira, G. Zeni and J. B. T. Rocha, *Chem. Rev.*, 2004, **104**, 6255.
12. J. B. T. Rocha, B. C. Piccoli and C. S. Oliveira, *ARKIVOC*, 2017, **ii**, 457.
13. C. H. Schiesser, *ARKIVOC*, 2001, **7**, 49.
14. L. Orian and S. Toppo, *Free Radical Biol. Med.*, 2014, **66**, 65.
15. L. Wolters and L. Orian, *Curr. Org. Chem.*, 2016, **20**, 189.
16. A. Hamilton, *Industrial Poisons in the United States*, 1925, p. 303.
17. E. P. Painter, *Chem. Rev.*, 1941, **28**, 179.
18. A. L. Moxon and M. Rhian, *Phys. Rev.*, 1943, **23**, 305.
19. C. H. Walker and E. Klein, *Am. Med.*, 1915, 62832.
20. K. W. Franke and E. P. Painter, *Cereal Chem.*, 1938, **15**, 1.
21. H. C. Dudley, *Am. J. Hyg.*, 1936, **23**, 169.
22. C. C. Tsen and A. L. Tappel, *J. Biol. Chem.*, 1958, **233**, 1230.
23. H. E. Ganther, *Biochemistry*, 1966, **5**, 1089.
24. H. E. Ganther, *Biochemistry*, 1968, **7**, 2898.
25. H. E. Ganther, *Biochemistry*, 1971, **26**, 4089.
26. C. Köppel, H. Baudisch, K. H. Beyer, I. Klöppel and P. V. Schneider, *J. Toxicol., Clin. Toxicol.*, 1986, **24**, 21.
27. S. J. Foster, R. J. Kraus and H. E. Ganther, *Arch. Biochem. Biophys.*, 1986, **15**, 77.
28. G. N. Schrauzer, *Adv. Food Nutr. Res.*, 2003, **47**, 73.
29. G. N. Schrauzer, *J. Nutr.*, 2000, **130**, 1653.

30. J. E. Spallholz, *Free Radical Biol. Med.*, 1994, **17**, 45.
31. O. E. Olson, *J. Am. Coll. Toxicol.*, 1986, **5**, 45.
32. N. B. V. Barbosa, J. B. T. Rocha, G. Zeni, T. Emanuelli, M. M. Beque and A. L. Braga, *Toxicol. Appl. Pharmacol.*, 1998, **149**, 243.
33. V. P. Palace, J. E. Spallholz, J. Holm, K. Wautier, R. E. Evans and C. L. Baron, *Ecotoxicol. Environ. Saf.*, 2004, **58**, 17.
34. J. Chaudiere, O. Courtin and J. Leclaire, *Arch. Biochem. Biophys.*, 1992, **296**, 328.
35. T. Maraldi, M. Riccio, L. Zambonin, M. Vinceti, A. De Pol and G. Hakim, *Neurotoxicology*, 2011, **32**, 180.
36. M. Wallenberg, E. Olm, C. Hebert, M. Björnstedt and A. P. Fernandes, *Biochem. J.*, 2010, **429**, 85.
37. M. Lazard, M. Dauplais, S. Blanquet and P. Plateau, *J. Biol. Chem.*, 2015, **290**, 10741.
38. H. P. Misra, *J. Biol. Chem.*, 1974, **249**, 2151.
39. B. Zadehvakili, S. M. McNeill, J. P. Fawcett and G. I. Giles, *Biochem. Pharmacol.*, 2016, **15**, 19.
40. E. A. Cowan, C. D. Oldham and S. W. May, *Arch. Biochem. Biophys.*, 2011, **506**, 201.
41. V. De Silva, M. M. Woznichak, K. L. Burns, K. B. Grant and S. W. May, *J. Am. Chem. Soc.*, 2004, **126**, 2409.
42. M. Iwaoka and F. Kumakura, *Phosphorus, Sulfur Silicon Relat. Elem.*, 2008, **183**, 1009.
43. M. Farina, V. Folmer, R. C. Bolzan, L. H. Andrade, G. Zeni, A. L. Braga and J. B. T. Rocha, *Toxicol. Lett.*, 2001, **119**, 27.
44. K. Arai and M. Iwaoka, *Curr. Org. Chem.*, 2016, **20**, 155.
45. R. L. Puntel, D. S. Ávila, D. H. Roos and S. Pinton, *Curr. Org. Chem.*, 2016, **20**, 198.
46. R. L. Puntel, D. H. Roos, V. Folmer, C. W. Nogueira, A. Galina, M. Aschner and J. B. T. Rocha, *Toxicol. Sci.*, 2010, **117**, 133.
47. J. B. T. Rocha, R. A. Saraiva, S. C. Garcia, F. S. Gravina and C. W. Nogueira, *Toxicol. Res.*, 2012, **1**, 85.
48. R. L. Puntel, D. H. Roos, R. L. Seeger and J. B. T. Rocha, *Toxicol. In Vitro*, 2013, **27**, 59.
49. R. L. Puntel, D. H. Roos, R. L. Seeger, M. Aschner and J. B. T. Rocha, *Neurotoxic. Res.*, 2013, **24**, 109.
50. C. F. Yang, H. M. Shen and C. N. Ong, *Arch. Biochem. Biophys.*, 2000, **380**, 319.
51. D. Morin, R. Zini, H. Ligeret, W. Neckameyer, S. Labidalle and J. P. Tillement, *Biochem. Pharmacol.*, 2003, **65**, 1643.
52. P. Santofimia-Castaño, G. M. Salido and A. González, *DNA Cell Biol.*, 2013, **32**, 147.
53. M. Vinceti, T. Maraldi, M. Bergomi and C. Malagoli, *Rev. Environ. Health*, 2009, **24**, 231.
54. M. Vinceti, F. Bonvicini, K. J. Rothman, L. Vescovi and F. Wang, *Environ. Health*, 2010, **9**, 77.

55. M. Vinceti, C. M. Crespi, F. Bonvicini, C. Malagoli, M. Ferrante, S. Marmiroli and S. Stranges, *Sci. Total Environ.*, 2013, **443**, 633.
56. M. Vinceti, I. Bottecchi, A. Fan, Y. Finkelstein and J. Mandrioli, *Rev. Environ. Health*, 2012, **27**, 19.
57. M. Vinceti, N. Solovyev, J. Mandrioli, C. M. Crespi, F. Bonvicini, E. Arcolin, E. Georgoulopoulou and B. Michalke, *Neurotoxicology*, 2013, **38**, 25.
58. M. Vinceti, P. Ballotari, C. Steinmaus, C. Malagoli, F. Luberto, M. Malavolti and P. G. Rossi, *Environ. Res.*, 2016, **150**, 348.
59. D. V. Frost and O. E. Olson, *CRC Crit. Rev. Toxicol.*, 1972, **1**, 467–514.
60. E. Watson-Williams, *Br. Med. J.*, 1919, **2**, 463.
61. E. Watson-Williams, *Br. J. Surg.*, 1920, **8**, 50.
62. J. Lü, J. Zhang, C. Jiang, Y. Deng, N. Özten and M. C. Bosland, *Nutr. Cancer*, 2016, **68**, 1.
63. E. A. Klein, I. M. Thompson, S. M. Lippman, P. J. Goodman, D. Albanes, P. R. Taylor and C. Coltman, *Prostate Cancer Prostatic Dis.*, 2000, **3**, 145.
64. L. C. Clark, B. Dalkin, A. Krongrad, G. F. J. Combs, B. W. Turnbull, E. H. Slate, R. Witherington, J. H. Herlong, E. Janosko, D. Carpenter, C. Borosso, S. Falk and J. Rounder, *Br. J. Urol.*, 1998, **81**, 730.
65. R. J. Shamberger, *J. Natl. Cancer Inst.*, 1970, **4**, 931.
66. H. Shen, C. Yang, J. LIU and C. Ong, *Free Radical Biol. Med.*, 2000, **28**, 1115.
67. G. Nilsonne, X. Sun, C. Nyström, A. Rundlöf, A. P. Fernandes, M. Björnstedt and K. Dobra, *Free Radical Biol. Med.*, 2006, **41**, 874.
68. D. F. Birt, A. D. Julius, C. E. Runice and S. Salmasi, *J. Natl. Cancer Inst.*, 1986, **77**, 1281.
69. D. F. Birt, T. A. Lawson, A. D. Julius, C. E. Runice and S. Salmasi, *Cancer Res.*, 1982, **42**, 4455.
70. R. H. Thompson, L. D. Meeker and S. Kokoska, *Cancer Res.*, 1984, **44**, 2803.
71. L. Yan, J. A. Yee, D. Li, M. H. McGuire and G. L. Graef, *Anticancer Res.*, 1999, **19**, 1337.
72. A. S. Weisberger and L. G. Suhrland, *Blood*, 1956, **11**, 11.
73. A. S. Weisberger and L. G. Suhrland, *Blood*, 1956, **11**, 19.
74. L. Novotny, P. Rauko, S. B. Kombian and I. O. Edafiogho, *Neoplasma*, 2010, **57**, 383.
75. R. Muecke, L. Schomburg, J. Buentzel, K. Kisters and O. Micke, *Integr. Cancer Ther.*, 2010, **9**, 136.
76. S. B. Gusberg, P. Zamecnik and J. C. Aub, *J. Pharmacol. Exp. Ther.*, 1941, **71**, 239.
77. J. Shamberger and C. E. Willis, *Crit. Rev. Clin. Lab. Sci.*, 1971, **2**, 211.
78. H. J. Thompson, *Carcinogenesis*, 1991, **12**, 2175.
79. H. E. Ganther, *Carcinogenesis*, 1999, **20**, 1657.
80. D. L. Hatfield, M. H. Yoo, B. A. Carlson and V. N. Gladyshev, *Biochim. Biophys. Acta*, 2009, **1790**, 1541.

81. V. N. Gladyshev, V. M. Factor, F. Housseau and D. L. Hatfield, *Biochem. Biophys. Res. Commun.*, 1998, **251**, 488.
82. X. Lan, J. Xing, H. Gao, S. Li, L. Quan, Y. Jiang, S. Ding and Y. Xue, *Biol. Trace Elem. Res.*, 2016, 1.
83. A. G. Cox, A. Tsomides, A. J. Kim, D. Saunders, K. L. Hwang, K. J. Evason, J. Heidel, K. K. Brown, M. Yuan, E. C. Lien and B. C. Lee, *Proc. Natl. Acad. Sci. U. S. A.*, 2016, 201600204.
84. D. N. Ekoue, S. Zaichick, K. Valyi-Nagy, M. Picklo, C. Lacher, K. Hoskins, M. A. Warso, M. G. Bonini and A. M. Diamond, *J. Trace Elem. Med. Biol.*, 2017, **39**, 227.
85. D. N. Ekoue, C. He, A. M. Diamond and M. G. Bonini, *Biochim. Biophys. Acta*, 2017, **1858**, 628.
86. N. Potenza, F. Castiello, M. Panella, G. Colonna, G. Ciliberto, A. Russo and S. C. Costantini, *PLoS One*, 2016, **11**, 1.
87. R. B. Flohé, *Chem. Biodiversity*, 2008, **5**, 389.
88. R. Irons, B. A. Carlson, D. L. Hatfield and C. D. Davis, *J. Nutr.*, 2006, **136**, 1311.
89. V. Diwadkar-Navsariwala, G. S. Prins, S. M. Swanson, L. A. Birch, V. H. Ray, S. Hedayat, D. L. Lantvit and A. M. Diamond, *Proc. Natl. Acad. Sci. U. S. A.*, 2006, **103**, 8179.
90. M. Vinceti and K. J. Rothman, *Am. J. Clin. Nutr.*, 2016, **104**, 245.
91. Z. Wang, C. Jiang and J. Lü, *Mol. Carcinog.*, 2002, **34**, 113.
92. H. Zeng, M. Briske-Anderson, J. P. Idso and C. D. Hunt, *J. Nutr.*, 2006, **136**, 1528.
93. A. Kim, J. H. Oh, J. M. Park and A. S. Chung, *J. Cell. Physiol.*, 2007, **212**, 386.
94. H. Zeng, M. Wu and J. H. Botnen, *J. Nutr.*, 2009, **139**, 1613.
95. H. Zeng, M. Briske-Anderson, M. Wu and M. P. Moyer, *Nutr. Cancer*, 2012, **64**, 128.
96. Y. Zhan, B. Cao, Y. Qi, S. Liu, Q. Zhang, W. Zhou, D. Xu, H. Lu, O. Sartor, W. Kong, H. Zhang and Y. Dong, *Int. J. Cancer*, 2013, **133**, 2225.
97. H. Zeng, W. H. Cheng and L. K. Johnson, *J. Nutr. Biochem.*, 2013, **24**, 776.
98. A. A. Nelson, O. G. Fithugh and H. O. Calvery, *Cancer Res.*, 1943, **3**, 230.
99. P. Guo, P. Zhao, J. Liu, H. Ma, J. Bai, Y. Cao, Y. Liu, H. He and C. Qi, *Biol. Trace Elem. Res.*, 2013, **151**, 301.
100. D. Plano, Y. Baquedano, E. Ibáñez, I. Jiménez, J. A. Palop, J. E. Spallholz and C. Sanmartín, *Molecules*, 2010, **21**, 7292.
101. S. Ye, X. Zheng, T. Hu and H. Zeng, *Cell. Mol. Biol.*, 2016, **30**, 6.
102. F. Xing, S. Li, X. Ge, C. Wang, H. Zeng, D. Li and L. Dong, *Oral Oncol.*, 2008, **44**, 963.
103. K. Bijian, Z. Zhang, B. Xu, S. Jie, B. Chen, S. Wan, J. Wu, T. Jiang and M. A. Alaoui-Jamali, *Eur. J. Med. Chem.*, 2012, **48**, 143.
104. D. Plano, C. Sanmartín, E. Moreno, C. Prior, A. Calvo and J. A. Palop, *Bioorg. Med. Chem. Lett.*, 2007, **15**, 6853.

105. R. Terazawa, D. R. Garud, N. Hamada, Y. Fujita, T. Itoh, Y. Nozawa, K. Nakane, T. Deguchi, M. Koketsu and M. Ito, *Bioorg. Med. Chem.*, 2010, **1**, 7001.
106. J. He, D. Li, K. Xiong, Y. Ge, H. Jin, G. Zhang, M. Hong, Y. Tian, J. Yin and H. Zeng, *Bioorg. Med. Chem.*, 2012, **15**, 3816.
107. S. U. Hossain, S. Sengupta and S. Bhattacharya, *Bioorg. Med. Chem.*, 2005, **15**, 5750.
108. P. Guo, Q. Wang, J. Liu, L. Liu, P. Zhao, Y. Cao, Y. Liu, C. Qi and Y. Liu, *Biol. Trace Elem. Res.*, 2013, **154**, 304.
109. J. Yan, Y. Guo, Y. Wang, F. Mao, L. Huang and X. Li, *Eur. J. Med. Chem.*, 2015, **5**, 220.
110. S. Shaaban, A. Negm, M. A. Sobh and L. A. Wessjohann, *Eur. J. Med. Chem.*, 2015, **5**, 190.
111. F. Jing, X. Fu, S. Li, B. Li, J. Zhao, X. Wang, Y. Liu and B. Chen, *Chem. Pharm. Bull.*, 2015, **63**, 431.
112. E. Domínguez-Álvarez, D. Plano, M. Font, A. Calvo, C. Prior, C. Jacob, J. A. Palop and C. Sanmartín, *Eur. J. Med. Chem.*, 2014, **12**, 153.
113. I. L. Martins, J. P. Miranda, N. G. Oliveira, A. S. Fernandes, S. Gonçalves and A. M. Antunes, *Molecules*, 2013, **8**, 5251.
114. D. Plano, D. N. Karelia, M. K. Pandey, J. E. Spallholz, S. Amin and A. K. Sharma, *J. Med. Chem.*, 2016, **10**, 1946.
115. V. Alcolea, D. Plano, D. N. Karelia, J. A. Palop, S. Amin, C. Sanmartín and A. K. Sharma, *Eur. J. Med. Chem.*, 2016, **4**, 134.
116. H. C. Zhao, Y. P. Shi, Y. M. Liu, C. W. Li, L. N. Xuan, P. Wang, K. Zhang and B. Q. Chen, *Bioorg. Med. Chem. Lett.*, 2013, **15**, 6577.
117. C. Chen, M. H. Hsu, Y. Y. Cheng, C. Y. Liu, L. C. Chou, L. J. Huang, T. S. Wu, X. Yang, K. H. Lee and S. C. Kuo, *Eur. J. Med. Chem.*, 2011, **46**, 6046.
118. D. Zuo, D. Guo, X. Jiang, Q. Guan, H. Qi, J. Xu, Z. Li, F. Yang, W. Zhang and Wu, *Chem. -Biol. Interact.*, 2015, **5**, 7.
119. J. Xu, D. Zuo, H. Qi, Q. Shen, Z. Bai, M. Han, Z. Li, W. Zhang and Y. Wu, *Biomed. Pharmacother.*, 2016, **78**, 308.
120. M. A. Rizvi, S. Guru, T. Naqvi, M. Kumar, N. Kumbhar, S. Akhoon, S. Banday, S. K. Singh, S. Bhushan, G. Mustafa Peerzada and B. A. Shah, *Bioorg. Med. Chem. Lett.*, 2014, **24**, 3440.
121. D. Plano, E. Moreno, M. Font, I. Encío, J. A. Palop and C. Sanmartín, *Arch. Pharm.*, 2010, **343**, 680.
122. A. Das, J. Bortner, D. Desai, S. Amin and K. El-Bayoumy, *Chem. -Biol. Interact.*, 2009, **15**, 158.
123. B. L. Li, B. Li, R. L. Zhang, J. J. Zhao, X. F. Wang, Y. M. Liu, Y. P. Shi, J. B. Liu and B. Q. Chen, *Bioorg. Med. Chem. Lett.*, 2016, **15**, 1279.
124. M. A. Pino, M. Pietka-Ottlik and B. Billack, *Arh. Hig. Rada Toksikol.*, 2013, **64**, 77.
125. Y. W. Liang, J. Zheng, X. Li, W. Zheng and T. *Chen*, *Eur. J. Med. Chem.*, 2014, **12**, 335.

126. Y. J. Huang, Y. Y. Zhang, G. Liu, J. Tang, J. G. Hu, Z. Z. Feng, F. Liu, Q. Y. Wang and D. Li, *Asian Pac. J. Cancer Prev.*, 2014, **15**, 6301.
127. Z. Wen, J. Xu, Z. Wang, H. Qi, Q. Xu, Z. Bai, Q. Zhang, K. Bao, Y. Wu and W. Zhang, *Eur. J. Med. Chem.*, 2015, **27**, 184.
128. M. Zhou, S. Ji, Z. Wu, Y. Li, W. Zheng, H. Zhou and Chen, *Eur. J. Med. Chem.*, 2015, **96**, 92.
129. D. Desai, S. V. Madhunapantula, K. Gowdahalli, A. Sharma, R. Chandagaludoreswamy, K. El-Bayoumy, G. P. Robertson and S. Amin, *Bioorg. Med. Chem. Lett.*, 2010, **15**, 2038.
130. M. Doering, L. A. Ba, N. Lilienthal, C. Nicco, C. Scherer, M. Abbas, A. A. Zada, R. Coriat, T. Burkholz, L. Wessjohann, M. Diederich, F. Batteux, M. Herling and C. Jacob, *J. Med. Chem.*, 2010, **14**, 6954.
131. A. I. Gutiérrez-Hernández, J. G. López-Cortés, M. C. Ortega-Alfaro, M. T. Ramírez-Apan, J. J. Cázares-Marinero and R. A. Toscano, *J. Med. Chem.*, 2012, **24**, 4652.
132. F. Hayat, A. Salahuddin, J. Zargan and A. Azam, *Eur. J. Med. Chem.*, 2010, **45**, 6127.
133. Y. A. Ivanenkov, M. S. Veselov, I. G. Rezekin, D. A. Skvortsov, Y. B. Sandulenko, M. V. Polyakova, D. S. Bezrukov, S. V. Vasilevsky, M. E. Kukushkin, A. A. Moiseeva, A. V. Finko, V. E. Koteliansky, N. L. Klyachko, L. A. Filatova, E. K. Beloglazkina, N. V. Zyk and A. G. Majouga, *Bioorg. Med. Chem.*, 2016, **15**, 802.
134. S. Misra, M. Boylan, A. Selvam, J. E. Spallholz and M. Björnstedt, *Nutrients*, 2015, **7**, 3536.
135. C. Pisanu, C. Melis and A. Squassina, *Drug Dev. Res.*, 2016, **77**, 368.
136. P. D. Hanavan, C. R. Borges, B. A. Katchman, D. O. Faigel, T. H. Ho, C. T. Ma, E. A. Sergienko, N. Meurice, J. L. Petit and D. F. Lake, *Oncotarget*, 2015, **6**, 18418.
137. P. Santofimia-Castaño, L. Garcia-Sanchez, D. C. Ruy, M. Fernandez-Bermejo, G. M. Salido and A. Gonzalez, *Toxicol. Lett.*, 2014, **229**, 465.
138. L. Zhang, L. Zhou, J. Du, M. Li, C. Qian, Y. Cheng, Y. Peng, J. Xie and D. Wang, *BioMed Res. Int.*, 2014, **2014**, 696107.
139. C. F. Yang, J. Liu, S. Wasser, H. M. Shen, C. E. Tan and C. N. Ong, *Carcinogenesis*, 2000, **21**, 2237.
140. B. Romano, D. Plano, I. Encío, J. A. Palop and C. Sanmartín, *Bioorg. Med. Chem.*, 2015, **23**, 1716.
141. C. Kim, J. Lee and M. S. Park, *Arch. Pharmacal Res.*, 2015, **38**, 659.
142. F. Nedel, V. F. Campos, D. Alves, A. J. McBride, O. A. Dellagostin, T. Collares, L. Savegnago and F. K. Seixas, *Life Sci.*, 2012, **91**, 345.
143. T. Posser, M. T. de Paula, J. L. Franco, R. B. Leal and J. B. T. Rocha, *Arch. Toxicol.*, 2011, **85**, 645.
144. N. Noguchi, *Arch. Biochem. Biophys.*, 2016, **595**, 109.
145. W. J. Adams Jr., J. J. Kocsis and R. Snyder, *Toxicol. Lett.*, 1989, **48**, 301.
146. K. Schwarz and C. M. Foltz, *J. Biol. Chem.*, 1958, **233**, 245.

147. N. B. V. Barbosa, C. W. Nogueira, T. N. Guecheva, M. L. Bellinaso and J. B. T. Rocha, *Arch. Toxicol.*, 2008, **82**, 655.
148. B. S. Reddy, T. Tanaka and K. El-Bayoumy, *J. Natl. Cancer Inst.*, 1985, **74**, 1325.
149. K. El-Bayoumy, *Cancer Res.*, 1985, **45**, 3631.
150. J. Nayini, K. El-Bayoumy, S. Sugie, L. A. Cohen and B. S. Reddy, *Carcinogenesis*, 1989, **10**, 509.
151. S. Sugie, B. S. Reddy, K. El-Bayoumy and T. Tanaka, *Jpn. J. Cancer Res.*, 1989, **80**, 952.
152. K. El-Bayoumy, Y. H. Chae, P. Upadhyaya, C. Meschter, L. A. Cohen and B. S. Reddy, *Cancer Res.*, 1992, **52**, 2402.
153. O. S. Sohni, H. Li, A. Surace, K. El-Bayoumy, P. Upadhyaya and E. S. Fiala, *Anticancer Res.*, 1995, **15**, 1849.
154. Z. Ronai, J. K. Tillotson, F. Traganos, Z. Darzynkiewicz, C. C. Conaway, P. Upadhyaya and K. El-Bayoumy, *Int. J. Cancer*, 1995, **63**, 428.
155. K. El-Bayoumy, A. Das, T. Boyiri, D. Desai, R. Sinha, B. Pittman and S. Amin, *Chem. –Biol. Interact.*, 2003, **25**, 179.
156. K. El-Bayoumy, Y. H. Chae, J. G. Rosa, L. K. Williams, D. Desai, S. Amin and E. Fiala, *Cancer Lett.*, 2000, **3**, 7.
157. K. El-Bayoumy, B. A. Narayanan, D. H. Desai, N. K. Narayanan, B. Pittman, S. G. Amin, J. Schwartz and D. W. Nixon, *Carcinogenesis*, 2003, **24**, 1505.
158. C. Ip, K. El-Bayoumy, P. Upadhyaya, H. Ganther, S. Vadhanavikit and H. Thompson, *Carcinogenesis*, 1994, **15**, 187.
159. S. W. Emmert, K. El-Bayoumy, A. Das, Y.-W. Sun, S. Amin, D. Desai, C. Aliaga and J. P. Richie Jr, *Free Radical Biol. Med.*, 2012, **52**, 2064.
160. J. B. Weissman, *N. Engl. J. Med.*, 1987, **30**, 1158.
161. M. A. Fischl, D. D. Richman, M. H. Grieco, M. S. Gottlieb, P. A. Volberding, O. L. Laskin, J. M. Leedom, J. E. Groopman, D. Mildvan, R. T. Schooley, G. G. Jackson, D. T. Durack and D. King, *N. Engl. J. Med.*, 1987, **23**, 185.
162. S. Hassani, S. H. Ghaffari, F. Zaker, R. Mirzaee, H. Mardani, D. Bashash, A. Zekri, M. Yousefi, A. Zaghal, K. Alimoghaddam and A. Ghavamzadeh, *Ann. Hematol.*, 2013, **92**, 1207.
163. C. P. Souza Sobrinho, A. Gragnani, I. D. Santos, A. F. Oliveira, M. V. Lipay and L. M. Ferreira, *Acta Cir. Bras.*, 2012, **27**, 855.
164. J. Pereira, D. Levy, J. L. Ruiz, G. A. Brocardo, K. A. Ferreira, R. O. Costa, R. G. Queiroz, D. A. Maria, A. E. Neto, D. A. Chamone and S. P. Bydlowski, *Anticancer Agents Med. Chem.*, 2013, **13**, 186.
165. K. S. Jeng, I. S. Sheen and W. J. Jeng, *Hepatogastroenterology*, 2011, **58**, 2091.
166. J. Humer, B. Ferko, A. Waltenberger, R. Rapberger, H. Pehamberger and T. Muster, *Melanoma Res.*, 2008, **18**, 314.
167. G. S. Heverly-Coulson and R. J. Boyd, *J. Phys. Chem. A*, 2010, **114**, 1996.

168. M. Bortoli, L. P. Wolters, L. Orian and F. M. Bickelhaupt, *J. Chem. Theory Comput.*, 2016, **12**, 2752.
169. J. K. Pearson, F. Ban and R. J. Boyd, *J. Phys. Chem. A*, 2005, **109**, 10373.
170. R. Kheirabadi, M. Izadyar and M. R. Housiandokht, *J. Theor. Biol.*, 2016, **409**, 108.
171. J. K. Pearson and R. J. Boyd, *J. Phys. Chem. A*, 2006, **110**, 8979.
172. E. D. A. dos Santos, E. Hamel, R. Bai, J. C. Burnett, C. S. S. Tozatti, D. Bogo, R. T. Perdomo, A. M. Antunes, M. M. Marques, M. Matos and D. P. de Lima, *Bioorg. Med. Chem. Lett.*, 2013, **23**, 4669.
173. U. S. F. Tambunan, R. Bakri, A. A. Parikesit, T. Ariyani, R. D. Puspitasari and D. Kerami, *IOP Conf. Ser.: Mater. Sci. Eng.*, 2016, **107**, 12054.
174. M. S. Shahabuddin, M. Nambiar, B. Choudhary, G. M. Advirao and S. C. Raghavan, *New Drugs*, 2010, **28**, 35.
175. F. Zaccaria, L. P. Wolters, C. F. Guerra and L. Orian, *J. Comput. Chem.*, 2016, **37**, 1672.
176. M. Torsello, A. C. Pimenta, L. P. Wolters, I. S. Moreira, L. Orian and A. Polimeno, *J. Phys. Chem. A*, 2016, **120**, 4389.
177. G. Ribaudo, M. Bellanda, I. Menegazzo, L. P. Wolters, M. Bortoli, G. Ferrer-Sueta, G. Zagotto and L. Orian, *Chem. – Eur. J.*, 2017, **23**, 1.
178. E. Lionta, G. Spyrou, D. K. Vassilatis and Z. Cournia, *Curr. Top. Med. Chem.*, 2014, **14**, 1923.
179. G. Schneider, *Nat. Rev. Drug Discovery*, 2010, **9**, 273.
180. J. K. Pearson and R. J. Boyd, *J. Phys. Chem. A*, 2007, **111**, 3152.
181. S. Antony and C. A. Bayse, *Inorg. Chem.*, 2011, **50**, 12075.
182. E. S. J. Arner and A. Holmgren, *Semin. Cancer Biol.*, 2006, **16**, 420.
183. M. A. Jordan and L. Wilson, *Nat. Rev. Cancer*, 2004, **4**, 253.
184. B. E. Gryder, Q. H. Sodji and A. K. Oyelere, *Future Med. Chem.*, 2012, **4**, 505.
185. J. Jen and Y. C. Wang, *J. Biomed. Sci.*, 2016, **23**, 53.
186. A. Ali and S. Bhattacharya, *Bioorg. Med. Chem.*, 2014, **22**, 4506.
187. J. D. Pennington, K. M. Jacobs, L. Sun, G. Bar-Sela, M. Mishra and D. Gius, *Curr. Pharm. Des.*, 2007, **13**, 3368.
188. K. F. Tonissen and G. D. Trapani, *Mol. Nutr. Food Res.*, 2009, **53**, 87.
189. L. Wang, Z. Yang, J. Fu, H. Yin, K. Xiong, Q. Tan, H. Jin, J. Li, T. Wang, W. Tang, J. Yin, G. Cai, M. Liu, S. Kehr, K. Becker and H. Zeng, *Free Radical Biol. Med.*, 2012, **52**, 898.
190. R. Sengupta and A. Holmgren, *World J. Biol. Chem.*, 2014, **26**, 68.
191. D. Mustacich and G. Powis, *Biochem. J.*, 2000, **346**, 1.
192. W. Cai, L. Zhang, Y. Song, B. Wang, B. Zhang, X. Cui, G. Hu, Y. Liu, J. Wu and J. Fang, *Free Radical Biol. Med.*, 2012, **52**, 257.
193. C. Shi, L. Yu, F. Yang, J. Yan and H. Zeng, *Biochem. Biophys. Res. Commun.*, 2003, **309**, 578.

194. F. Zhao, J. Yan, S. Deng, L. Lan, F. He, B. Kuang and H. Zeng, *Cancer Lett.*, 2006, **236**, 46.
195. Z. F. Peng, L. X. Lan, F. Zhao, J. Li, Q. Tan, H. W. Yin and H. H. Zeng, *J. Zhejiang Univ., Sci., B*, 2008, **9**, 16.
196. L. Lan, F. Zhao, Y. Wang and H. Zeng, *Eur. J. Pharmacol.*, 2007, **555**, 83.
197. R. Zhao and A. Holmgren, *J. Biol. Chem.*, 2002, **277**, 39456.
198. A. S. Freitas, A. S. Prestes, C. Wagner, J. H. Sudati, D. Alves, L. O. Porciúncula, I. J. Kade and J. B. T. Rocha, *Molecules*, 2010, **15**, 7699.
199. A. S. Freitas and J. B. T. Rocha, *Neurosci. Lett.*, 2011, **503**, 1.
200. T. N. Gustafsson, H. Osman, J. Werngren, S. Hoffner, L. Engman and A. Holmgren, *Biochim. Biophys. Acta*, 2016, **1860**, 1265.
201. S. Shaaban, A. Negm, A. M. Ashmawy, D. M. Ahmed and L. A. Wessjohann, *Eur. J. Med. Chem.*, 2016, **122**, 55.
202. E. Pasquier and M. Kavallaris, *IUBMB Life*, 2008, **60**, 165.
203. Q. Guan, F. Yang, D. Guo, J. Xu, M. Jiang, C. Liu, K. Bao, Y. Wu and W. Zhang, *Eur. J. Med. Chem.*, 2014, **87**, 1.
204. Z. Wen, J. Xu, Z. Wang, H. Qi, Q. Xu, Z. Bai, Q. Zhang, K. Bao, Y. Wu and W. Zhang, *Eur. J. Med. Chem.*, 2015, **90**, 184.
205. S. Minucci and P. G. Pelicci, *Nat. Rev. Cancer*, 2006, **6**, 38.
206. J. E. Bolden, M. J. Peart and R. W. Johnstone, *Nat. Rev. Drug Discovery*, 2006, **5**, 769.
207. X. Qiu, X. Xiao, N. Li and Y. Li, *Prog. Neuropsychopharmacol. Biol. Psychiatry*, 2017, **72**, 60.
208. T. L. Newkirk, A. A. Bowersab and R. M. Williams, *Nat. Prod. Rep.*, 2009, **26**, 1293.
209. H. Nian, W. H. Bisson, W. M. Dashwood, J. T. Pinto and R. H. Dashwood, *Carcinogenesis*, 2009, **30**, 1416.
210. J. I. Lee, H. Nian, A. J. L. Cooper, R. Sinha, J. Dai, W. H. Bisson, R. H. Dashwood and J. T. Pinto, *Cancer Prev. Res.*, 2009, **2**, 683.
211. D. E. Wilcox, A. D. Schenk, B. M. Feldman and Y. Xu, *Antioxid. Redox Signaling*, 2001, **3**, 549.
212. H. Blessing, S. Kraus, P. Heindl, W. Bal and A. Hartwig, *Eur. J. Biochem.*, 2004, **271**, 3190.
213. P. I. Oteiza, *Free Radical Biol. Med.*, 2012, **53**, 1748.
214. K. D. Kröncke and L. O. Klotz, *Antioxid. Redox. Signaling*, 2009, **11**, 1015.
215. L. Fan, B. Tan, Y. Li, Q. Zhao, Y. Liu, D. Wang and Z. Zhang, *Int. J. Clin. Exp. Pathol.*, 2015, **8**, 12428.
216. M. P. Foster, D. S. Wuttke, I. Radhakrishnan, D. A. Case, J. M. Gottesfeld and P. E. Wright, *Nat. Struct. Biol.*, 1997, **4**, 605.
217. S. Quintal, A. Viegas, S. Erhardt, E. J. Cabrita and N. P. Farrell, *Biochemistry*, 2012, **51**, 1752.
218. C. A. Bayse, S. M. Whitty and S. Antony, *Curr. Chem. Biol.*, 2013, **7**, 57.

219. S. Antony and C. A. Bayse, *Inorg. Chem.*, 2013, **52**, 13803.
220. R. A. Saraiva, D. C. Bueno, P. A. Nogara and J. B. T. Rocha, *J. Toxicol. Environ. Health, Part A*, 2012, **75**, 1012.
221. A. Casas, H. Fukuda, P. Riley and A. M. del C Batlle, *Cancer Lett.*, 1997, **121**, 105.
222. M. A. Rizvi, M. Zaki, M. Afzal, M. Mane, M. Kumar, B. A. Shah, S. Srivastav, S. Srikrishna, G. M. Peerzada and S. Tabassum, *Eur. J. Med. Chem.*, 2015, **90**, 876.
223. S. S. Mati, S. S. Roy, S. Chall, S. Bhattacharya and S. C. Bhattacharya, *J. Phys. Chem. B*, 2013, **117**, 14655.
224. E. Glattre, Y. Thomassen, S. O. Thoresen, T. Haldorsen, P. G. Lund-Larsen, L. Theodorsen and J. Aaseth, *Int. J. Epidemiol.*, 1989, **18**, 45.
225. N. E. Allen, R. C. Travis, P. N. Appleby, D. Albanes, M. J. Barnett, A. Black, H. B. Bueno-de-Mesquita, M. Deschasaux, P. Galan, G. E. Goodman, P. J. Goodman, M. J. Gunter, M. Heliövaara, K. J. Helzlsouer, B. E. Henderson, S. Hercberg, P. Knekt, L. N. Kolonel, C. Lasheras, J. Linseisen, E. J. Metter, M. L. Neuhouser, A. Olsen, V. Pala, E. A. Platz, H. Rissanen, M. E. Reid, J. M. Schenk, M. J. Stampfer, P. Stattin, C. M. Tangen, M. Touvier, A. Trichopoulou, P. A. van den Brandt and T. J. Key, *J. Natl. Cancer Inst.*, 2016, DOI: 10.1093/jnci/djw153.
226. P. A. Van den Brandt, M. P. A. Zeegers, P. Bode and R. A. Goldbohm, *Cancer Epidemiol., Biomarkers Prev.*, 2003, **12**, 866.
227. J. D. Brooks, E. J. Metter, D. W. Chan, L. J. Sokoll, P. Landis, W. G. Nelson, D. Muller, R. Andres and H. B. Carter, *J. Urol.*, 2001, **166**, 2034.
228. T. M. Vogt, R. G. Ziegler, B. I. Graubard, C. A. Swanson, R. S. Greenberg, J. B. Schoenberg, G. M. Swanson, R. B. Hayes and S. T. Mayne, *Int. J. Cancer*, 2003, **103**, 664.
229. J. Ringstad, B. K. Jacobsen, S. Tretli and Y. Thomassen, *J. Clin. Pathol.*, 1988, **41**, 454.
230. K. Jaskiewicz, W. F. Marasas, J. E. Rossouw, F. E. Van Niekerk and E. W. Heine Tech, *Cancer*, 1988, **62**, 2635.
231. D. H. E. Maasland, L. J. Schouten, B. Kremer and P. A. Van Den Brandt, *Eur. J. Cancer*, 2016, **60**, 83.
232. M. Nouarie, A. Pourshams, F. Kamangar, M. Sotoudeh, M. H. Derakhshan, M. R. Akbari, H. Fakheri, M. J. Zahedi, K. Caldwell, C. C. Abnet, P. R. Taylor, R. Malekzadeh and S. M. Dawsey, *World J. Gastroenterol.*, 2004, **10**, 2544.
233. M. P. A. Zeegers, R. A. Goldbohm, P. Bode and P. A. Van den Brandt, *Cancer Epidemiol., Biomarkers Prev.*, 2002, **11**, 1292.
234. S. Y. Yu, Y. J. Zhu, W. G. Li, Q. S. Huang, C. Z. Huang, Q. N. Zhang and C. Hou, *Biol. Trace Elem. Res.*, 1991, **29**, 289.
235. E. A. Klein, I. M. Thompson Jr, C. M. Tangen, J. J. Crowley, M. S. Lucia, P. J. Goodman, L. M. Minasian, L. G. Ford, H. L. Parnes, J. M. Gaziano,

D. D. Karp, M. M. Lieber, P. J. Walther, L. Klotz, J. K. Parsons, J. L. Chin, A. K. Darke, S. M. Lippman, G. E. Goodman, F. L. Meyskens Jr and L. H. Baker, *JAMA, J. Am. Med. Assoc.*, 2011, **306**, 1549.
236. M. E. Goossens, M. P. Zeegers, H. V. Poppel, S. Joniau, K. Ackaert, F. Ameye, I. Billiet, J. Braeckman, A. Breugelmans, J. Darras, K. Dilen, L. Goeman, B. Tombal, S. V. Bruwaene, B. V. Cleyenbreugel, F. V. der Aa, K. Vekemans and F. Buntinx, *Eur. J. Cancer*, 2016, **69**, 9e18.

CHAPTER 14

Metabolism of Selenium/ Selenocystine and Their Roles in the Prevention and Treatment of Human Cancer

SOUGAT MISRA* AND MIKAEL BJÖRNSTEDT

Division of Pathology F46, Department of Laboratory Medicine, Karolinska Institutet, Karolinska University Hospital Huddinge, Stockholm, Sweden 141 86
*Email: Sougat.Misra@ki.se

14.1 Introduction

The history of treating cancer with selenium goes back to the early 20th century when "colloidal selenium" was used to treat patients with inoperable carcinoma. A number of case reports were summarized by E. Watson-Williams in 1919.[1] The original report also describes the treatment effects of intravenous administration of "colloidal suspension of erythro-selenium beta"—a dichroic coral red fluid, in 24 cancer patients at King's College Hospital (London, UK) and at the throat department of the Bristol Royal Infirmary (Bristol, UK). The highest administered dose was equivalent to 5 mg selenium per day. The author reported beneficial effects in 19 patients. The diminution or disappearance of cancer-related pain and improvement in anemia and cachexia have been described as the most striking effects. The author concludes that "the degree of relief already witnessed is sufficiently

Organoselenium Compounds in Biology and Medicine: Synthesis, Biological and Therapeutic Treatments
Edited by Vimal Kumar Jain and K. Indira Priyadarsini

Published by the Royal Society of Chemistry, www.rsc.org

considerable to justify that in the treatment of inoperable carcinoma, colloidal selenium deserves more attention than it has received in this country".

Shortly after the initial reports of E. Watson-Williams and others, Dr A. T. Todd, an assistant physician at Bristol Royal Infirmary initiated treatment of inoperable carcinoma with colloidal lead selenide in patients with different cancer diagnoses.[2] The treatment response differed among these patients. Later, he combined radiation therapy along with lead selenide administration, but without any positive results. He had further used colloidal sulfur selenium (SSe) in combination with low doses of radiation.[3] The report also describes the use of colloidal radioactive selenium, termed therein as RAS—selenium combined with the lead-like products of radium disintegration. The author described some remarkable results in patients with inoperable mammary metastases, at least 10 patients having "an apparent cure". However, it is important to note that such results might be limited by subjective assessment, with "no sign or symptom which suggests that any cancer remains". Nevertheless, Dr Todd's remarkable attempt to treat cancer with colloidal selenium still remains an authentic testimonial about the possibilities, limitations and uncertainties of developing selenium-based cancer chemotherapeutics.

Following these early attempts to treat cancer with different selenium preparations, no further clinical studies were undertaken until 1956, when Weisberger and Suhrland reported the anti-leukemic effects of oral administration of selenocystine in four patients.[4] The patients received a daily dose varying between 50 and 200 mg selenocystine for a duration spanning between 10 days and 57 days. The authors reported a dramatic reduction in the frequency of circulating tumour cells with a significant decrease in spleen size. A prominent anti-leukemic effect of selenocystine was evident in one patient with documented resistance to other cancer chemotherapeutic drugs. However, dose-limiting toxicities such as nausea and vomiting and the appearance of drug resistance in one patient precluded prolonged treatment in these patients. Two patients died shortly after the cessation of selenocystine therapy. Postmortem examinations in these patients revealed no signs of toxicity attributed to selenocystine administration. Nevertheless, the potential of this selenium compound was not further tested in follow-up human clinical studies on leukemia patients.

Shortly after the publication of the seminal work of Weisberger and Suhrland, Klaus Schwarz and Calvin M. Foltz discovered that selenium is an essential element.[5] Their discovery led to a plethora of research aiming at understanding the efficacy of different selenium compounds as nutritional supplements in the subsequent years. However, it was not known then how selenium exerts its essentiality until the discovery of glutathione peroxidase in 1973.[6,7] Subsequently, a great deal of effort led to the discovery of a number of selenoproteins and the characterization of their functions.[8] At the same time, there was a growing interest on cancer prevention studies. The concept of antioxidant supplementation in chemoprevention was one of the leading hypotheses that had been tested both in animal models and human clinical trials. Selenium incurred a special interest with its known function as an antioxidant. This led to the one of the biggest nutritional trials (SELECT) on the use of selenium (as selenomethionine) and vitamin E as cancer chemopreventive agents, either as single agents or in combination.[9] The study concluded that neither vitamin E nor selenomethionine supplementation prevented the incidence of prostate cancer. This study raised an important question on the perceived concept of antioxidant supplementation in cancer chemoprevention. The long-term supplementation of selenomethionine was perhaps sufficient to elevate the level of selenoproteins with antioxidant functions, including glutathione peroxidase, but without discernible effects on cancer prevention. This suggests that the antioxidant effects of selenium in the form of selenoproteins may be limited in their efficacy to prevent carcinogenesis. In contrast to the beneficial effects of antioxidants, it has recently been shown that supplementation of vitamin E and *N*-acetyl cysteine in fact accelerate the progression of lung cancer[10] and malignant melanoma[11] in mouse models.

In this chapter, we discuss the metabolism of different organic selenium compounds and their roles in cancer chemoprevention and possible chemotherapeutic applications. We also discuss why selenium compounds are unique in their prospective application as anticancer agents in the context of ever-increasing interest in harnessing the potential of redox-directed cancer chemotherapeutics.

14.2 Organic Selenium Compounds and Their Structures

The chemical structures of some of the relevant organic selenium compounds discussed in this chapter are presented in Figure 14.1.

14.3 Transport and Metabolism of Organic Selenium Compounds in Humans

A detailed knowledge of the pharmacokinetics (absorption, distribution, metabolism and excretion) of drugs is a prerequisite for their successful clinical application. In this section, we briefly discuss the transport and metabolism of selenium compounds outlined in Figure 14.1.

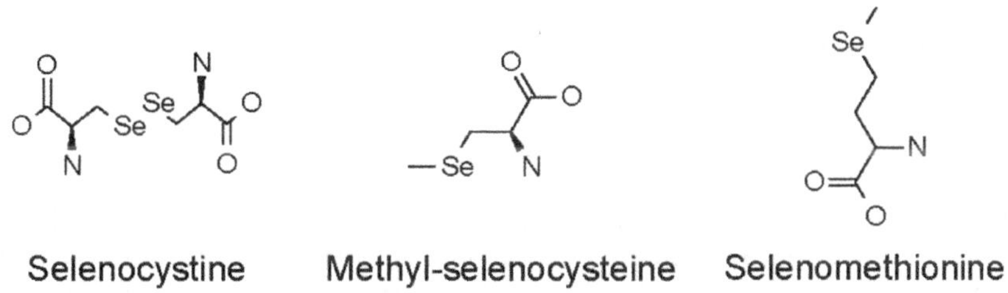

Figure 14.1 Structures of the main seleno compounds discussed in this chapter.

14.3.1 Transport

The structural similarity of selenocystine, methylselenocysteine and selenomethionine with their corresponding sulfur analogues suggests plausible common modalities in their respective transport mechanisms. Cognate transport systems for the sulfur analogues are discussed, along with the available information on the counterpart selenium compounds.

14.3.1.1 Cystine/Selenocystine

There are four known transporters involved in cellular cystine transport. The major cystine transport system (xCT, gene *SLC7A11*) was first described by Bannai and Kitamura in human diploid fibroblast cells.[12] It is a pH-dependent and electroneutral anionic amino acid transporter with an estimated K_m value of 43 μM for cystine. The selenium analogue of cystine, selenocystine, was not used as an inhibitor in this study. Therefore it was not known whether selenocystine could share the same transport pathway. In a later study, it was shown that D,L-selenocystine can inhibit cystine transport with an approximate K_i value of 66 μM in a rat hepatoma cell line.[13] Transport inhibition by structurally similar compounds indicates possible competitive inhibition at the uptake site. The calculated K_m value for cystine was measured to be 45 μM, comparable to that reported by Bannai and Kitamura.[12] These findings suggest similar transport properties between murine and human cystine uptake systems at whole-cell level.

The two other well-characterized cystine transporter are rBAT (gene *SLC3A1*) and $b^{0,+}$AT1 (gene *SLC7A9*). Their expression is mostly tissue-specific, unlike the ubiquitously expressed *SLC7A11*. *SLC3A1* is primarily expressed in the small intestine, kidney cortex, pancreas and transverse colon, while *SLC7A9* is expressed in the small intestine, kidney cortex and liver. Heterologous expression of human $b^{0,+}$ AT1 in *Xenopus laevis* oocytes showed that this transporter is involved in the influx of both cystine and selenocystine, with K_m values of 110 and 220 μM, respectively.[14] However, mouse slc6a19 didn't show any measurable cystine or selenocystine transport activity in this model system.

The asparate/glutamate transporter 1 (AGT1, gene name *SLC7A13*) is the fourth cystine transporter that has been characterized recently.[15] This transporter is expressed in the renal apical membrane with an estimated K_m value of 67.6 μM. However, its function as a selenocystine transporter is yet to be established.

14.3.1.2 Cysteine/Methylselenocysteine

Methylselenocysteine is not a direct selenium analogue of cysteine. It has an additional $-CH_3$ group attached to its selenium moiety. Cysteine is transported *via* five different known transporters (EAAT3, ASCT1, ASCT2, NTT4 and SNAT1). Among these, ASCT1 and ASCT2 are the two major transporters, with high substrate specificity towards cysteine. It has also been shown that $b^{0,+}$rBAT and $b^{0,+}$AT1 can transport both cysteine and methylselenocysteine.[14] Methylselenocysteine exhibits lower K_m values for both of these transporters compared to cysteine. In addition, SIT1 (*SLC6A20*) is involved in the influx of cysteine ($K_m = 2.64$ mM) and methylselenocysteine ($K_m = 0.47$ mM).

In a recent study, it was shown that selenocysteine (the Se analogue of cysteine) can be transported *via* EAAT1 (*SLC1A3*), EAAT2 (*SLC1A2*) and EAAT3 (*SLC1A1*).[16] The rates of selenocysteine transport are comparable to the uptake of glutamate, a cognate substrate of these transporters. This is in contrast with cysteine, which is a poor substrate for both EAAT1 and EAAT2.

14.3.1.3 Methionine/Selenomethionine

There are at least 13 different transporters that are implicated in the transport of methionine in humans (Figure 14.2).[17,18] Among these, B^0AT1 (*SLC6A19*) and $ATB^{0,+}$ (*SLC6A14*) are the key methionine transporters expressed in the luminal membrane. So far, it has been shown that selenomethionine influx can be carried out by $b^{0,+}$rBAT, B^0AT1 and SIT1. The $b^{0,+}$rBAT system has the highest affinity for selenomethionine ($K_m = 0.12$ mM), while SIT1 has the lowest affinity ($K_m = 3.69$ mM). Interestingly, SIT1 exhibits a higher affinity for selenomethionine than for methionine ($K_m = 6.93$ mM).[14]

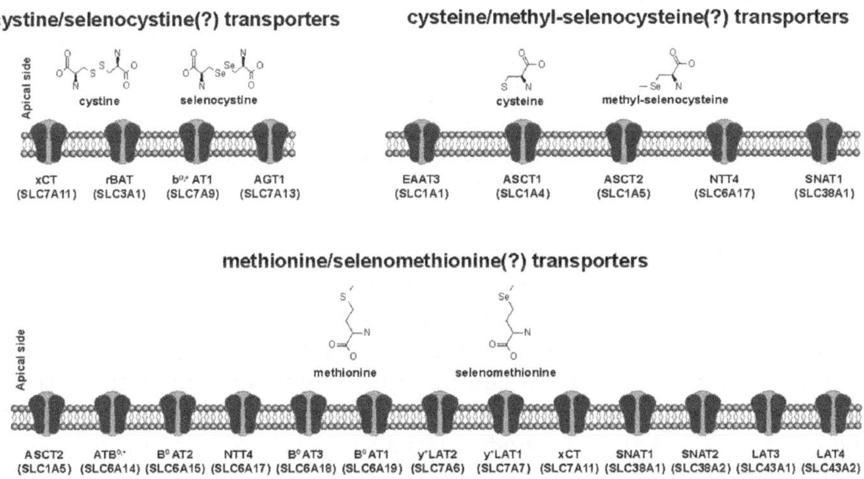

Figure 14.2 A schematic diagram showing the transporters/carriers involved in the cellular uptake of cystine, cysteine and methionine in humans. Some of these transporters/carriers are implicated in the uptake of structurally similar selenium compounds with different uptake properties. However, there is limited experimental evidence whether these transporters are indeed involved in the transport of corresponding seleno compounds.

It is still not known whether all of the aforementioned transporters are involved in the influx of Se analogues of sulfur amino acids. Further studies are required to identify the transport systems involved in the uptake of these selenium compounds, given their increasing importance both in human nutrition and therapeutic applications.

14.4 Metabolism

The biochemical pathways involving metabolic transformations of sulfur compounds are well characterized. Due to the similar physical and chemical properties of sulfur and selenium, the metabolic pathways overlap between analogous selenium and sulfur compounds. However, differences exist in the redox chemistry between selenium and sulfur.[19] Such differences often make selenium metabolites much more reactive compared to their sulfur counterparts, especially when the Se moiety is present as selenolate. This, together with a larger atomic radius, lower electronegativity and thus a higher nucleophilicity of the free selenium moiety of selenium compounds compared to their sulfur counterparts explains the higher reactivity of selenolates. These chemical properties together lead to lower basicity of selenolate compared to thiolate, as could be determined by low pK_a values of selenolates.[20,21] For example, at neutral pH, selenocysteine is ionized to selenolate ($pK_a = 5.24$), while cysteine ($pK_a = 8.25$) remains in the thiol state. Such chemical properties demand special consideration while discussing the metabolism of certain selenium compounds due to the inherent chemical reactivity of selenol. In such a state, spontaneous chemical transformations may override the enzyme-mediated biochemical catalysis for which the concentrations of substrates and metabolizing enzymes in the local microenvironment may be limiting factors.

The metabolic pathway involving incorporation of selenium into selenoproteins and their metabolic roles have often been extensively discussed in the context of selenium metabolism. A common pool of hydrogen selenide is utilized for the biosynthesis of selenoproteins *via* the intermediate metabolite selenophosphate. However, a comprehensive understanding of the metabolisms of different selenium compounds beyond their transformation into hydrogen selenide is necessary, specifically when selenium is administered at high doses. Such investigations will aid in prospective applications of different selenium compounds for chemotherapeutic purposes, specifically, in targeting the redox imbalances in cancer. Hence, both chemical and biochemical transformations of selected selenium compounds are discussed in this section.

14.4.1 Selenocystine/Selenomethionine

The metabolic transformations of selenocystine and selenomethionine are discussed together due to a perceived overlapping of their metabolic pathways. The availability of free selenocystine is perhaps limited to high selenium-containing dietary sources. However, this compound is of particular interest from the perspective of cancer chemotherapeutics. The first step in selenocystine metabolism involves cleavage of the diselenide bond, resulting in the formation of selenocysteine. It has been shown that human thioredoxin reductase (TXNRD) efficiently reduces selenocystine to selenocysteine in the presence of excess NADPH.[22] Selenocystine can also be reduced by glutaredoxin 1, another protein with disulfide reductase activity, albeit with low turnover capacity.[23] It is not known whether TRP14, a dedicated cystine reductase,[24] is capable of reducing selenocystine. Similarly, glutathione reductase may play an important role in reducing selenocystine in an identical mechanism to that by which it reduces oxidized glutathione (GSH).[25]

A reducing mileu (specifically GSH content) in the cell further determines the fate of the liberated selenocysteine. As indicated earlier, free selenocysteine remains in a deprotonated form and is redox reactive. It can either be oxidized back to selenocystine[26] or forms selenocysteine-glutathione selenenyl sulfide (CysSeSG) upon spontaneous reaction with GSH.[27] The latter intermediate is a substrate for selenocysteine β-lyase which liberates hydrogen selenide. Selenocysteine is also a catalytic substrate for the enzyme, with the same end product.[28] It has also been shown that CysSeSG is converted to hydrogen selenide by glutathione reductase in the presence of GSH and NADPH. Together, the combined actions of both enzymatic and non-enzymatic pathways are involved in the metabolism of selenocystine, with hydrogen selenide being a key intermediate metabolite. In the context of cancer chemotherapeutic applications, hydrogen selenide generation is one of the key aspects of selenocystine metabolism, since this metabolite plays a critical role in targeting redox dysregulation in cancer cells. A schematic diagram of selenocystine metabolism is presented in Figure 14.3.

Selenomethionine is randomly incorporated in the place of methionine during protein biosynthesis when co-administered. The relative cellular

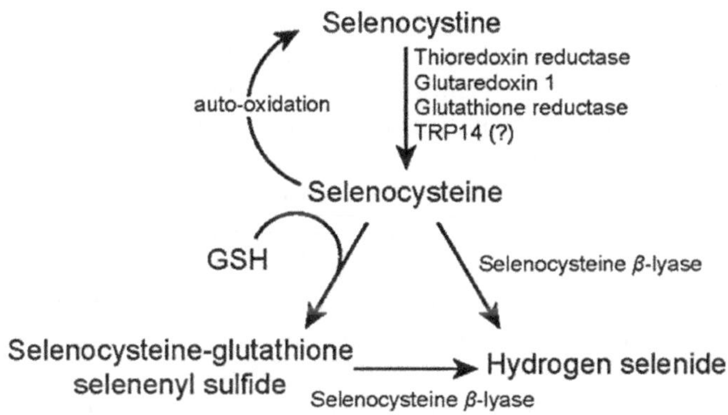

Figure 14.3 Enzymatic and non-enzymatic pathways involved in the metabolic transformations of selenocystine.

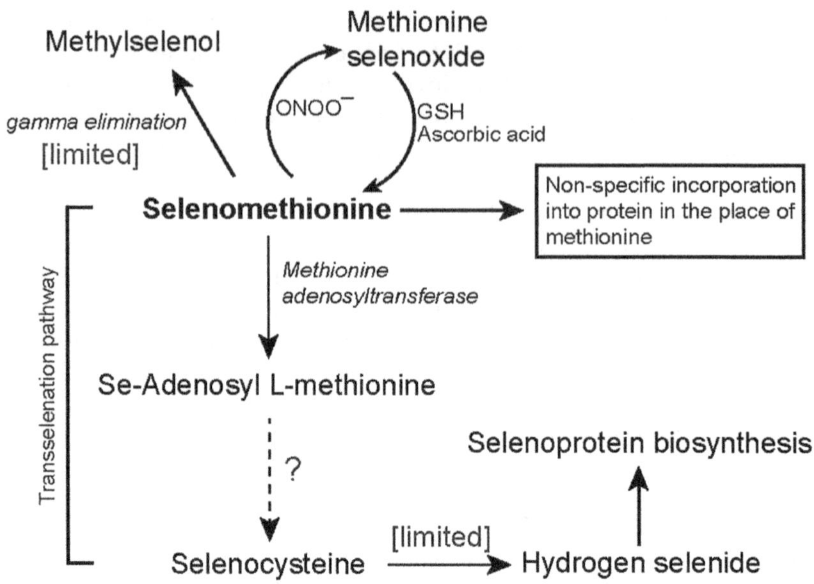

Figure 14.4 Human metabolism of selenomethionine. The *trans*-selenation pathway appears to play a key role in selenomethionine metabolism. However, the involvement of key enzymes of the *trans*-sulfuration pathway requires validation. Selenomethionine metabolism *via* γ-elimination reaction is limited in humans, but contributes a major pathway in certain bacteria. Selenomethionine can also be oxidized to methionine sulfoxide upon reaction with peroxynitrite. However, intracellular concentration of reductants (glutathione (GSH) or ascorbic acid) is sufficient to convert methionine sulfoxide to selenomethionine. The biosynthesis of selenoproteins from selenomethionine is relevant in nutritional supplementation. However, this component only partially accounts for the total intake of selenomethionine.

abundance of selenomethionine and methionine determines which of these structural analogues is incorporated into proteins (Figure 14.4). It is generally accepted that the metabolism of selenomethionine is similar to methionine metabolism. It implies that the *trans*-sulfuration pathway may play a key role in which selenomethionine can be converted into selenocysteine (termed herein as *trans*-selenation pathway). It has been shown that selenomethionine is a substrate for methionine adenosyltransferase purified from rat liver.[29] However, the enzyme exhibits a lower K_m value for selenomethionine compared to its cognate substrate L-methionine. Under the assumption of a similar mechanism of substrate catalysis, this would result in the formation of Se-adenosyl-L-methionine. There is no further information concerning the possible transformation of Se-adenosyl-L-methionine into selenocysteine, which remains to be investigated.

The possible involvement of *trans*-selenation pathway in selenomethionine metabolism needs to be elaborated despite a lack of substantial experimental evidence. Why is selenomethionine supplementation effective in increasing the expression of selenoproteins both *in vitro* and *in vivo*? Considering that selenoprotein biosynthesis requires the generation of selenophosphate, it is tempting to speculate that an intermediate metabolite of selenomethionine may in fact be selenocysteine following the *trans*-sulfuration pathway. Such a metabolic conversion would facilitate an easy route towards hydrogen selenide generation and subsequent synthesis of selenophosphate by selenophosphate synthetase. In contesting the presented hypothesis, it may be argued that if the *trans*-selenation pathway is involved in selenomethionine metabolism and selenocysteine is generated, then why is this compound less toxic in comparison to selenocystine, which is metabolized to hydrogen selenide upon conversion into selenocysteine? It is perhaps due to competition of structurally analogous compounds for the substrate binding sites of the enzymes of the *trans*-sulfuration pathway, and therefore limited turnover of selenocysteine from selenomethionine to exert appreciable cytotoxicity. Such a scenario could be validated in cytotoxicity experiments under methionine-restricted conditions. In line with this, it has been shown that selenomethionine is highly toxic to *Saccharomyces cerevisiae* when grown in methionine-free medium in comparison to methionine-supplemented medium.[30]

In relation to this discussion, selenomethionine metabolism was uniquely addressed in recent studies using X-ray absorption near-edge (XANE) spectroscopy. When human lung adenocarcinoma A549 cells were treated with selenomethionine for 24 h, selenocysteine was the key selenium metabolite, while a significant proportion of selenomethionine remained unaltered.[31] However, in the SH-SY5Y neuroblastoma cell line, selenocysteine was the major selenium metabolite.[32] The authors indicated that the identified selenocysteine component from spectral analyses perhaps belonged to the selenocysteine moiety in selenoproteins. Limitations in data analyses, including poor spectral resolution and a limited library of model selenium compounds have been addressed in these studies. The absence of a diselenide intermediate in selenomethionine-treated cells in both studies is indicative of the limited production of methylselenol from selenomethionine. Since methylselenol is implicated as one of the key intermediates in cancer prevention and treatment, its limited metabolic turnover from selenomethionine provides a plausible basis as to why large prospective case–control studies with selenomethionine supplementation have found no beneficial effects in cancer chemoprevention.

XANE spectroscopy studies may have been limited in capturing the volatile selenium compounds such as methylselenol or dimethyl-diselenide. Using head-space gas chromatography mass spectrometry (GCMS), it has been shown that treatment with selenomethionine in Jurkat cells for 24 h neither produces methylselenol nor dimethyl-diselenide.[33] In an identical setting, purified L-methionine-γ-lyase can convert selenomethionine to these metabolites *in vitro*. A similar observation has been made in an *in vivo* study in which transformation of volatile selenium metabolites has been minimal following subcutaneous injection of very high doses of selenomethionine.

All these observations bring forward a common question on the effectiveness of enzymatic γ-elimination of selenomethionine in producing methylselenol in humans. Under such limitations, the *trans*-selenation pathway of selenomethionine metabolism provides an important connection in relation to its cytotoxic effects.

The intrinsic differences between the utilization and excretion of methionine and selenomethionine in humans and rodents provide some key insights in their metabolic transformation. In human sulfur excretion studies, it has been shown that free sulfate (77–92%), ester sulfate (7–9%) and taurine (2–6%) are the key sulfur-containing urinary metabolites of the total sulfur excreted.[34,35] However, anaverage of ∼70% of total urinary selenium has been shown to be excreted as selenosugar 1 in a high-dose selenomethionine supplementation study.[36] It has also been shown that urinary excretion of selenium is faster following oral administration of selenomethionine compared to selenite.[37] Selenosugar 1 comprises the key urinary metabolite for both of these compounds. Of total ingested selenium, ∼30% is excreted as urinary metabolites within 48 h. The above findings suggest that there exists a common modality in their excretion pathway, perhaps *via* a common intermediate from which methylselenol group is transferred to the *N*-acetyl galactosamine to form selenosugar 1.

14.4.2 Methylselenocysteine

The metabolism of methylselenocysteine differs from other selenocompounds. The selenium moiety is covalently attached to a methyl group and thereby requires enzymatic catalysis. Kynurenine aminotransferases (KATs) are the key β-lyase enzymes that play key roles in metabolizing methylselenocysteine. These pyridoxal 5′-phosphate-dependent β-lyase enzymes use methylselenocysteine as a substrate to form methylselenol, pyruvate and ammonia. Notably, these enzymes also function as transaminases, resulting in the formation of β-methylselenopyruvate using methylselenocysteine as a substrate. It has been shown that purified rat renal KAT1 has 4.5 times higher transaminase activity compared to its β-elimination reaction using methylselenocysteine as a substrate (Figure 14.5).[38]

Cystathionine γ-lyase can also catalyze β-elimination reactions in addition to its known functions in γ-elimination reactions.[39] It has been shown that cystathionine γ-lyase can also catalyze methylselenocysteine into methylselenol, pyruvate and ammonia.[40] However, the rate of product formation (pyruvate) was much lower when methylselenocysteine was used as a substrate for the enzyme compared to its physiologically cognate substrate L-cystathionine.

Once methylselenol is formed, it is spontaneously converted to dimethyldiselenide. As it is highly volatile, it is difficult to quantify in conventional metabolomics analyses. However, this metabolite could not be detected by the head-space GC-MS technique when Jurkat cells were incubated with methylselenocysteine for 24 h.[33] Instead, the majority of the

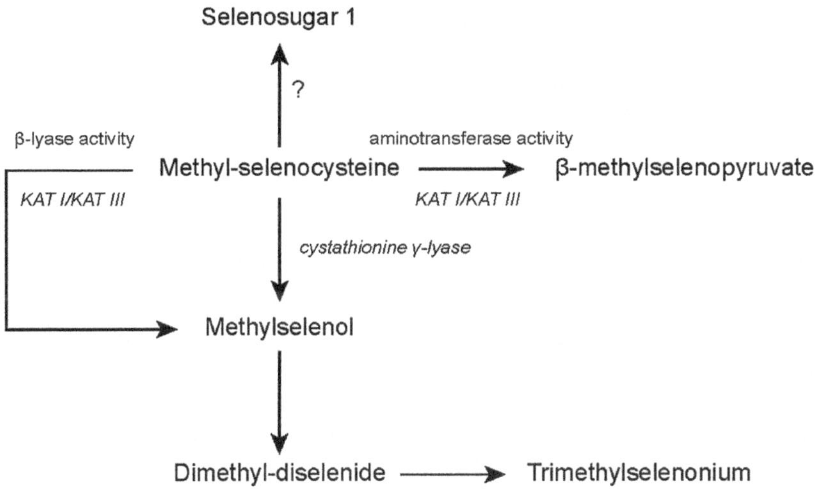

Figure 14.5 A route of metabolic transformation of methylselenocysteine in humans. A single step enzymatic conversion of methylselenocysteine by kynurenine aminotransferase (KAT)I and KATIII enzymes leads to the formation of methylselenol—a key selenium metabolite implicated both in cancer chemoprevention and prospective therapeutic applications. Both of these enzymes can convert methylselenocysteine into β-methylselenopyruvate by their dedicated aminotransferase activity. Methylselenocysteine is an uncharacteristic β-elimination substrate for cystathionine γ-lyase, leading to the formation of methylselenol. Methylselenol is further oxidized into dimethyl diselenide or trimethylselenonium and excreted. Selenosugar 1 is the key urinary metabolite of methylselenocysteine.

parent compound was found to be unchanged in the growth medium. One plausible explanation could be poor uptake of methylselenocysteine and/or low β-lyase activities in these cells, and therefore the transformation into methylselenol was probably below detection limit.

The metabolism of methylselenocysteine has been studied in A549 and SH-SY5Y cells using XANE spectroscopy.[31,32] From spectral fitting analyses, it has been shown that selenocystine and selenocysteine constitute the major fraction of selenium metabolites. A significant proportion of methylselenocysteine remained unchanged. The identification of selenocystine as a major metabolite opens up a new paradigm in understanding the metabolism of methylselenocysteine in view of the existing understanding of its metabolic pathway. In a similar way, identification of high proportion of selenocysteine in A549 cells following methylselenocysteine treatment is also contradictory to our current knowledge of methylselenocysteine metabolism. It is probably unlikely that the detected level of selenocysteine would all be incorporated into selenoproteins.

In order to compare the metabolism of methylselenocysteine and selenomethionine, Ohta *et al.*[41] treated rats with a mixture of ^{76}Se-labelled methylselenocysteine, ^{77}Se-labelled selenomethionine and, as a reference,

^{82}Se-labelled methylseleninic acid. Analysis revealed that methylseleninic acid and methylselenocysteine were both metabolized to trimethylselenonium and the volatile dimethylselenide, detected in urine and exhaled gas, respectively. These findings clearly indicate that both methylseleninic acid and methylselenocysteine are good sources of monomethylselenol and thus are highly interesting candidates for the prevention and treatment of cancer.

14.5 Selenium Compounds and Cancer Prevention

The hypothesis that the progression of preneoplastic cells into a state of malignancy can be prevented or that dysplastic cells can be eliminated with the right intervention assumes two key considerations. First, we need to understand the biology of these cells at large and the molecular drivers of the process. Second, we need to devise appropriate interventions that can effectively eliminate dysplastic cells.[42]

Since the discovery of glutathione peroxidase as a selenoprotein in 1973, the field of selenium and human health has largely focused on selenium as an antioxidant. To this end, many of the biological effects of selenium have been associated with the functions of selenoproteins. The diversity of selenoproteome in mammals and their roles *in vitro* and *in vivo* perhaps provides useful insights into their physiological functions. Many of these selenoproteins function as oxido-reductases and thus play key roles in redox biology. The antioxidant functions of these selenoproteins are often discussed in the context of cancer chemoprevention.

Dysregulation of redox homeostasis is one of the key features of many cancers. Increased levels of reactive oxygen species (ROS) have been implicated in strengthening pro-oncogenic signals, specifically altering the kinase signaling pathways. It has also been shown that expression of thioredoxin reductase 1 plays critical roles in tumour growth in mice.[43] Knockdown of thioredoxin reductase 1 results in smaller tumour size compared to xenoplanted wild-type thioredoxin reductase 1 expressing cells. A plethora of scientific articles have supported and disputed the roles of selenium and selenoproteins in the carcinogenesis process. Results from nutritional intervention studies are discussed here in order to comprehend the cancer chemopreventive effects of the tested selenium compounds.

There are five published major studies on the cancer chemopreventive effects of selenium.[44–48] The first of these is the Nutritional Prevention of Cancer (NPC) trial.[44] In this study, patients who had previous history of basal or squamous cell carcinoma were supplemented with 200 µg of selenium as 0.5 mg high-selenium brewer's yeast tablets daily. The primary end-points for the trial were the recurrent incidences of these carcinomas. No statistically different incidences of these carcinomas were found between the supplemented and the placebo group. Further analyses of the study results indicated that there were excess risks for non-melanoma skin cancer,[49] other cancers[50] and possible diagnostic bias for prostate cancer in the original trial.[51] In a subsequent randomized control study (the SELECT trial), the efficacy of selenium supplementation as selenomethionine (200 μg Se per day) in the prevention of prostate cancer was evaluated. The study findings demonstrated no significant differences in the pre-specified cancer endpoints, including prostate cancer. Further analyses revealed increased risks for high-grade prostate cancer in subjects with a high baseline selenium level following selenomethionine supplementation.

Discrepancies in the results from these two trials are often attributed to the baseline selenium level in these subjects. Participants in the NPC trial had lower plasma selenium levels (median 113.4 $\mu g L^{-1}$) than the participants (median 135.0 $\mu g L^{-1}$) in the SELECT trial. When the study findings were further analyzed, baseline plasma selenium levels failed to account for the perceived differences in the results.[52] If any key variable differed between these trials it was the selenium preparation and patient characteristics. A recent follow-up study on the SELECT trial indicated no beneficial effects of selenomethionine supplementation in the subjects belonging to the lowest quintiles of selenium status.[53]

Similar results were found in subjects who were at higher risk for high-grade prostatic intraepithelial neoplasia[46] and prostate cancer[47] following selenomethionine supplementation. Notably, the latter study also had an arm in which the study participants received 400 μg of Se equivalent daily without any beneficial effects. In another study, selenized yeast (200 μg Se equivalent) was supplemented in patients who underwent surgical interventions due to stage I non-small cell lung carcinoma.[48] There were no differences in the incidence of second primary tumours in both the supplemented and the placebo group.

It is often argued that low baseline plasma levels in study subjects may be beneficial in eliciting cancer chemoprevention of selenium. A recent report described the results from a randomized clinical trial (SELEBLAT) evaluating the effect of selenium supplementation (200 μg Se as high-selenium yeast, Seleno Precise) on the recurrence of non-invasive urothelial carcinoma in a Belgian study cohort.[54] The median baseline plasma selenium level in these subjects was 83.3 $\mu g L^{-1}$. With a median follow-up of 18 months, no beneficial effects of selenium supplementation were found in this group of patients. However, it is important to note that the enrolled study participants were previously diagnosed with cancer both in the NPC and SELEBLAT trials. Hence the prevention of cancer recurrence could describe these trials at best.

Unfortunately, the aforementioned studies have not led to any conclusive results, but rather merely created questions and doubts about the chemopreventive potential of selenium, when supplemented as selenomethionine or selenized yeast preparations, the composition of which vary in terms of selenium species. However, these studies do not preclude the possibilities that selenium, in other chemical forms, may be effective in eliciting cancer chemopreventive effects.

Beyond cancer chemoprevention, selenium plays critical and undisputable roles in human health and disease. In a recently published Swedish study, the mean serum selenium concentration in an elderly study population was measured to be 67.1 $\mu g L^{-1}$, a level well below the saturation level of selenoproteins.[55] The mortality from all causes was 43% higher in the lowest quartile (<57.2 $\mu g L^{-1}$) compared to the highest quartile (>76.1 $\mu g L^{-1}$), indicating a potential health hazard associated with low selenium status. In a 5 year prospective randomized double-blind placebo-controlled study, it was shown that long-term combined supplementation of selenium (200 μg Se as high-selenium yeast, Seleno Precise) and Q10 (Bio-Quinon 100 mg) together reduced cardiovascular dysfunction-related mortality in an elderly population. It is important to note that such positive effects may not be solely attributed to selenium.

The chemical form of selenium is apparently critical for any preventive or therapeutic effects of selenium in cancer. The active forms are limited to redox-active compounds and their precursors.[56] The published human trials along with substantial bodies of preclinical data clearly show that selenomethionine is largely inefficient in the prevention or treatment of cancer.[57–60] Contrary to selenomethionine, redox-active selenium compounds are eventually metabolized to selenide or methylated species, including the highly reactive monomethylselenol. Lü *et al.*[61] have defined precursors of monomethylselenol as next-generation selenium compounds in cancer prevention. We extend this definition to cancer therapy. These classes of selenium compounds comprise methylselenocysteine, methylseleninic acid and methylselenocyanate, as shown in the preclinical studies.[62–64] A limiting factor for further studies is the lack of conclusive and systematic human safety and pharmacokinetic data on these compounds. In line with this, examining the chemopreventive effects of methylselenocysteine would be interesting, specifically in a population with a low baseline plasma selenium concentration.

14.6 Key Aspects of Selenium Compounds as Prospective Cancer Therapeutics

Before pondering the field of selenium compounds as cancer chemotherapeutics, it would be appropriate to define what is meant by cancer chemotherapeutics. In its strictest definition, it refers to a class of drugs that are used for the systemic treatment of cancer with the goals to cure, control or palliate of the disease (American Cancer Society). The major classes of cancer chemotherapeutics belong to alkylating agents, antimetabolites, anti-tumour antibiotics, topoisomerase inhibitors, mitotic inhibitors, corticosteroids and other miscellaneous drugs. A majority of these classes of drugs target cell division and their modes of action are based on the supposition that cancer cells divide rapidly. Furthermore, there are other classes of drugs that target the unique physiological or molecular adaptations of cancer cells not common to normal cells. These types of treatment comprise targeted therapies, differentiating agents, hormone therapies and immunotherapy.

Can selenium compounds be categorized into one of the above classifications based on their modes and mechanisms of actions? The pharmacodynamics of redox-active selenium compounds span cell cycle arrest,[65] topoisomerase II inhibition,[66] specific cytotoxic effects on cancer cells compared to their normal counterparts *in vitro*[67,68] and enhanced anti-tumour immune functions,[69,70] among others. It is obvious that these diverse anti-tumour effects render redox-active selenium compounds into a class of "unique" cancer therapeutics with polypharmacological properties; we have partially addressed this in an earlierpublication.[71]

An understanding of key adaptive physiological processes in cancer cells may provide some important clues towards the development of new therapeutic strategies, including use of redox-active selenium compounds. Rapidly dividing cells, including cancer cells, reprogram their metabolism to support the sustained growth. Such reprogramming of metabolism consists of (i) increased ATP generation to satisfy the energy requirement of dividing cells; (ii) increased synthesis of macromolecules; and (iii) tight regulation of cellular redox homeostasis under an altered metabolic state.[72] Similar metabolic adaptations are also a prerequisite for rapidly proliferating normal cells under the influence of physiological growth signals as opposed to constitutive autonomous adaptations in cancer cells.[73] In the context of therapeutic interventions, selective targeting of energy metabolism and macromolecular biosynthesis in cancer cells can be challenging, given that these metabolic pathways are ubiquitous to all cell types for their renewal and survival, and therefore require targeted drug delivery. However, there are unique differences in redox regulation between cancer cells and normal cells. This is of key interest concerning applications of redox-active selenium compounds for cancer therapeutic purposes.

The anti-proliferative effects of selenium compounds are often attributed to their capacity in targeting the dysregulated redox homeostasis of cancer cells at pharmacological doses. In Section 14.4, we discussed the metabolic transformations of selenocystine and methylselenocysteine into hydrogen selenide and methylselenol, respectively. Since these selenol intermediates are highly redox reactive, they can indiscriminately oxidize intracellular pools of thiols (both protein and non-protein). The manifestation of effects could range from loss of cellular redox balance, inactivation of proteins with functional –SH groups and protein misfolding to eventual cell death. An over-reliance on redox regulatory mechanisms in cancer cells can make these vulnerable to persistent oxidative stress, as has been the therapeutic basis for many alkylating agents currently used for chemotherapeutic interventions.

There exists a uniqueness about selenol-generating selenium compounds in the context of redox-directed cancer therapy. The intermediate metabolites hydrogen selenide and methylselenol not only oxidize thiols in redox reactions, but also produce ROS at the same time. The cumulative effect is manifested as induction of ROS-mediated cellular damage under a gradual loss of cellular reductive milieu with concomitant oxidation of thiols. The latter events are highly effective in potentiating oxidative stress. We have also shown that methylselenol is a superior substrate in redox cycling with the thioredoxin and glutaredoxin systems, suggesting the roles of these key redox proteins in eliciting oxidative stress by consuming NADPH.[74] These properties together partly explain the superior nature of methylselenol precursors in cancer therapy.

The generation of redox-reactive selenols have broader implications on cellular physiology beyond their recognized roles in ROS-induced oxidative stress. Alterations of numbers of cellular signaling pathways have been reported. Methyl selenocysteine is a potent inhibitor of protein kinase C activity, cdk kinase activity and DNA synthesis.[75] An inhibition of protein kinase C is associated with the activation of sphingomyelinase, leading to ceramide overproduction.[76] This latter event triggers the release of cytochrome C from mitochondria and subsequently culminates in the activation of apoptotic signaling. The cell death mechanisms are complex, and depending on cell types and selenium species apoptosis, necroptosis or necrosis may occur.[77] Methylselenol also induces cell cycle arrest at the G1 phase.[65] In contrast to hydrogen selenide, it does not induce DNA single-strand breaks as shown in both adenocarcinoma and leukemic cells.[78]

Methylated selenium compounds are of special therapeutic interest as they negatively regulate key signaling pathways in androgen-sensitive prostate cancer. At pharmacological concentrations, these compounds down-regulate androgen receptor protein expression, transcriptional inhibition of prostate-specific antigen (PSA) promoter in response to androgen stimulation and degradation of PSA, as described in detail elsewhere.[79] These observations are in line with the down-regulation of a number of androgen receptor-regulated genes including ABCC4, GUCY1A3 and DHCR24.[80] These findings highlight a direct inhibition of oncogenic signals by methylselenol-generating compounds.

Methylselenol alters the tumour cell surface expression of ligands (MICA/B) that are cognate targets of NKG2D receptor of different immune cells including natural killer cells, $CD8^+$ T cells, $\gamma\delta$ T cells, and some activated $CD4^+$ T cells.[70] However, the hydrogen selenide-generating compound selenite does not affect MICA/B.[69] In contrast, selenite at pharmacological concentration abrogates HLA-E expression in different cancer cells, thereby making these vulnerable to CD94/NK group 2A-positive NK cells. Such immuno-modulatory effects have direct implications on therapeutic outcome.

It has been shown that methylselenol exerts reduced tumour cell migration and invasion along with inhibition of angiogenesis. In human fibrosarcoma cells, methylselenol inhibits their invasion and migration.[81] In the FaDu xenograft model, it has been shown that methylselenocysteine down-regulates irinotecan-induced cyclo-oxygenase (COX)2, inducible nitric oxide synthase (iNOS) and hypoxia inducible factor (HIF) 1α.[82] A similar effect to methylselenocysteine treatment on iNOS expression has been made in RAW 264.7 murine macrophages upon challenge with lipopolysaccharide. Reduced expression of these key neovascularization regulators indicates possible anti-angiogenic effects of methylselenocysteine. In another study with the FaDu xenograft model, methylselenocysteine treatment (0.2 mg per day for 14 days) resulted in a marked reduction in microvessel density and a concomitant increase in vascular maturation index.[83] At the same time, it was shown that treatment resulted in improved vessel functionality along with a significant reduction in vascular permeability. A manifestation of such changes ensures enhanced drug delivery to tumours, which is highly desirable for treatment efficacy.

Animal models of cancer provide some valuable information concerning the anti-tumour efficacy of methylselenocysteine. Dietary supplementation of methylselenocysteine (1 $mg\,kg^{-1}$ diet) for 18 weeks following subcutaneous transplantation of Dunning R3327-H rat prostate adenocarcinoma cell in Copenhagen rats resulted in decreased tumour weight compared to either lycopene or vitamin E supplementation.[84] However, the most dramatic anticancer effects of this compound have been reported in combination with established anticancer agents in preclinical animal models.[85] Apart from its high efficiency in incurring complete remission in these models, methylselenocysteine exerts remarkable protection against myelosuppression induced by oxaliplatin when used in combination.

14.6.1 Pharmacokinetics

Given that the cancer therapeutic/chemopreventive applications of these potent selenium compounds require systemic administration, understanding their pharmacology is a key consideration.

Methylselenocysteine is stable, has a high oral bioavailability and is suggested to be taken up efficiently.[86] Its metabolic conversion to methylselenol makes it a highly interesting prospective cancer therapeutic selenium compound. To pave the way for human trials using methylselenocysteine, a pharmacokinetic and safety study of methylselenocysteine was performed using rats and dogs as non-primate model systems.[87] The no observed adverse effects level (NOAEL) for methylselenocysteine in rats was estimated to be <0.5 $mg\,Se\,kg^{-1}$ per day and <0.15 $mg\,Se\,kg^{-1}$ per day in dogs. However, the NOAELs were not "clean", since histological signs of liver damage were observed in the absence of serum transaminase elevations. This is an important consideration in designing human clinical trials with methylselenocysteine. We believe that dose escalation studies in human must be performed carefully with careful monitoring of the appearance of any side effects.

As with other selenium compounds, the lack of systematic dose escalation phase I clinical trials limits the possibilities of exploring the full potential of this compound. Marshall *et al.* have reported the pharmacokinetics of methylselenocysteine in humans after single oral doses of methylselenocysteine up to 1.2 mg. The t_{max} ranged between 3 and 5 h and the median C_{max} ranged from 21.0 $ng\,mL^{-1}$ (dose: 0.4 mg Se) to 57 $ng\,mL^{-1}$ (dose: 1.2 mg Se).[88] A non-linear increase in C_{max} and area under the curve to the administered dose indicates a high potential of accumulation. It was also found that urinary selenium concentration returned back to the pre-dose level within 48 h and was independent of the administered dose. In a follow-up study, the same research group reported multiple-dose pharmacokinetics of methylselenocysteine in selenium-replete subjects.[89] The study reports no treatment-associated adverse events in the study subjects following daily administration of methylselenocysteine for 84 days.

To the best of our knowledge, there are no reported pharmacokinetics data on selenocystine, either in rodents or non-primates. In ICR male mice, it was shown that daily oral administration of 15, 10 and 5 mg of selenocystine for 30, 60 and 90 days resulted in no significant changes in plasma aspartate transaminase and alanine transaminase.[90]

The mode of administration and the chemical nature of the selenium compounds are critical for their anti-tumour effects. A golden standard for studying the therapeutic potential of selenium is the use of the inorganic salt selenite. Selenite must be administered intravenously, since this compound will react with thiols present in the gut mucosa and per oral administration will likely not led to any detectable level of the active compound, selenite, in plasma. We have recently published the first-in-man phase I clinical trial in humans using intravenously administered selenite. Our data showed a half-life of 18.2 h of total selenium and a surprisingly high tolerance, with a maximum tolerated dose of 10.2 $mg\,m^{-2}$ body surface.[91] Preliminary data indicate that the half-life of selenite is very short, and therefore we will next perform a modified study with prolonged infusions in order to achieve a longer exposure of selenite in plasma.

14.7 Conclusion

There is a major gap in our understanding of the transporters involved in the uptake of different selenium compounds. Existing knowledge of the transport mechanisms of the cognate sulfur analogues could greatly facilitate in deciphering the transport systems involved. Such investigations warrant better understanding of the pharmacological aspects that drive target tissue accumulation.

With the advent of next-generation instrumentation in analytical chemistry, immense progress has been made in identifying the metabolites of different selenium compounds in diverse model systems, both *in vitro* and *in vivo*. However, many important questions remain to be investigated. For example, how is selenosugar 1 formed as one of the key excretory metabolites of both organic and inorganic selenium compounds? On the same note, how important is the *trans*-selenation pathway in the metabolic transformation of selenomethionine?

In the context of chemoprevention, selenomethionine continues to be evaluated in prevention trials. Most of the case–control nutritional intervention studies have failed to harness its beneficial effects, if there are any. Should the interest be focused on the right choice of selenium compounds, as outlined by Marshall and colleagues,[88,89] or the supplemented dose based on the available pharmacokinetics data? We believe that it can be both, as the founding studies on the cancer chemopreventive effects of selenium have been performed with other redox-active selenium compounds and at higher doses.

Nevertheless, we must acknowledge the immense undertakings and scientific contributions made by the NPC and SELECT trials in our understanding on the roles of certain chemical forms of selenium in cancer chemoprevention. These studies highlight the complexity of selenium functions in biology. A colossal body of successful preclinical studies provides important clues in transforming these earlier experiences of failure into plausible success stories. Perhaps the undesirable outcome of these studies cannot negate the high level of anticipation of success. A quote by Max Beerbohm, a humorist, is worth mentioning:

> "There is much to be said for failure. It is much more interesting than success."
>
> *Mainly on the Air*, 1946

The cancer preventive mechanisms of selenium compounds are complex. Possible prevention mechanisms include inhibition of pre-neoplastic transformation of normal cells among others. However, chemotherapeutic application relies on multiple modes and mechanisms of actions at therapeutic doses. Such properties are of great advantage in cancer therapy, since a multitarget approach may render therapeutic efficacy and prevent the development of drug resistance.[71] The pro-oxidant/antioxidant effects are thus likely to comprise only part of the preventive and anti-proliferative effects of selenium. In fact, several authors have suggested that selenium-mediated chemoprevention of cancer in nutritionally adequate subjects are independent of antioxidant selenoproteins.[56,92]

Acknowledgements

The contributing authors would like to thank Cancerfonden, Cancer-och Allergifonden, KI Fonder, Jochnick Foundation, Radiumhemmetsforsknings Fonder and the County Council of Stockholm for funding.

References

1. E. Watson-Williams, *Br. Med. J.*, 1919, **2**, 463–464.
2. A. T. Todd, *Lancet*, 1927, **209**, 575.
3. F. Hernaman-Johnson, *Proc. R. Soc. Med.*, 1935, **28**, 681.
4. A. S. Weisberger and L. G. Suhrland, *Blood*, 1956, **11**, 19.
5. K. Schwarz and C. M. Foltz, *J. Am. Chem. Soc.*, 1957, **79**, 3292.
6. J. T. Rotruck, A. L. Pope, H. E. Ganther, A. B. Swanson, D. G. Hafeman and W. G. Hoekstra, *Science*, 1973, **179**, 588.

7. L. Flohe, W. A. Günzler and H. H. Schock, *FEBS Lett.*, 1973, **32**, 132.
8. G. V. Kryukov, S. Castellano, S. V. Novoselov, A. V. Lobanov, O. Zehtab, R. Guigó and V. N. Gladyshev, *Science*, 2003, **300**, 1439.
9. S. M. Lippman, E. A. Klein, P. J. Goodman, M. S. Lucia, I. M. Thompson, L. G. Ford, H. L. Parnes, L. M. Minasian, J. M. Gaziano, J. A. Hartline, J. K. Parsons, J. D. Bearden, E. D. Crawford, G. E. Goodman, J. Claudio, E. Winquist, E. D. Cook, D. D. Karp, P. Walther, M. M. Lieber, A. R. Kristal, A. K. Darke, K. B. Arnold, P. A. Ganz, R. M. Santella, D. Albanes, P. R. Taylor, J. L. Probstfield, T. J. Jagpal, J. J. Crowley, F. L. Meyskens Jr, L. H. Baker and C. A. Coltman Jr., *JAMA, J. Am. Chem. Soc.*, 2009, **301**, 39.
10. V. I. Sayin, M. X. Ibrahim, E. Larsson, J. A. Nilsson, P. Lindahl and M. O. Bergo, *Sci. Transl. Med.*, 2014, **6**, 221ra215.
11. K. Le Gal, M. X. Ibrahim, C. Wiel, V. I. Sayin, M. K. Akula, C. Karlsson, M. G. Dalin, L. M. Akyürek, P. Lindahl, J. Nilsson and M. O. Bergo, *Sci. Transl. Med.*, 2015, **7**, 308re308.
12. S. Bannai and E. Kitamura, *J. Biol. Chem.*, 1980, **255**, 2372.
13. M. Makowske and H. N. Christensen, *J. Biol. Chem.*, 1982, **257**, 5663.
14. A. Nickel, G. Kottra, G. Schmidt, J. Danier, T. Hofmann and H. Daniel, *Chem. -Biol. Interact.*, 2009, **177**, 234.
15. S. Nagamori, P. Wiriyasermkul, M. E. Guarch, H. Okuyama, S. Nakagomi, K. Tadagaki, Y. Nishinaka, S. Bodoy, K. Takafuji, S. Okuda, J. Kurokawa, R. Ohgaki, V. Nunes, M. Palacín and Y. Kanai, *Proc. Natl. Acad. Sci. U. S. A.*, 2016, **113**, 775.
16. S. D. Watts, D. Torres-Salazar, C. B. Divito and S. G. Amara, *PLoS One*, 2014, **9**, e109245.
17. S. P. Alexander, E. Kelly, N. Marrion, J. A. Peters, H. E. Benson, E. Faccenda, A. J. Pawson, J. L. Sharman, C. Southan and J. A. Davies, *Br. J. Pharmacol.*, 2015, **172**, 6110.
18. L. Mastrototaro, G. Sponder, B. Saremi and J. R. Aschenbach, *IUBMB Life*, 2016, **68**, 924.
19. H. J. Reich and R. J. Hondal, *ACS Chem. Biol.*, 2016, **11**, 821.
20. Y. Sugiura, Y. Hojo, Y. Tamai and H. Tanaka, *J. Am. Chem. Soc.*, 1976, **98**, 2339.
21. R. E. Huber and R. S. Criddle, *Arch. Biochem. Biophys.*, 1967, **122**, 164.
22. M. Björnstedt, M. Hamberg, S. Kumar, J. Xue and A. Holmgren, *J. Biol. Chem.*, 1995, **270**, 11761.
23. M. Wallenberg, E. Olm, C. Hebert, M. Björnstedt and A. P. Fernandes, *Biochem. J.*, 2010, **429**, 85.
24. I. Pader, R. Sengupta, M. Cebula, J. Xu, J. O. Lundberg, A. Holmgren, K. Johansson and E. S. J. Arnér, *Proc. Natl. Acad. Sci. U. S. A.*, 2014, **111**, 6964–6969.
25. G. Batist, A. G. Katki, R. W. Klecker and C. E. Myers, *Cancer Res.*, 1986, **46**, 5482.
26. C. Jacob, G. I. Giles, N. M. Giles and H. Sies, *Angew. Chem., Int. Ed.*, 2003, **42**, 4742.

27. T. Hasegawa, T. Okuno, K. Nakamuro and Y. Sayato, *Arch. Toxicol.*, 1996, **71**, 39.
28. N. Esaki, T. Nakamura, H. Tanaka and K. Soda, *J. Biol. Chem.*, 1982, **257**, 4386.
29. F. Pan and H. Tarver, *Arch. Biochem. Biophys.*, 1967, **119**, 429.
30. M. Lazard, M. Dauplais, S. Blanquet and P. Plateau, *J. Biol. Chem.*, 2015, **290**, 10741.
31. C. M. Weekley, J. B. Aitken, S. Vogt, L. A. Finney, D. J. Paterson, M. D. de Jonge, D. L. Howard, I. F. Musgrave and H. H. Harris, *Biochemistry*, 2011, **50**, 1641.
32. C. M. Weekley, J. B. Aitken, I. F. Musgrave and H. H. Harris, *Biochemistry*, 2012, **51**, 736.
33. C. Gabel-Jensen, K. Lunoe and B. Gammelgaard, *Metallomics*, 2010, **2**, 167.
34. J. Martensson, *Metabolism*, 1982, **31**, 487.
35. H. Nakamura, R. Kajikawa and T. Ubuka, *Amino Acids*, 2002, **23**, 427.
36. D. Kuehnelt, D. Juresa, K. A. Francesconi, M. Fakih and M. E. Reid, *Toxicol. Appl. Pharmacol.*, 2007, **220**, 211.
37. D. Kuehnelt, N. Kienzl, P. Traar, N. H. Le, K. A. Francesconi and T. Ochi, *Anal. Bioanal. Chem.*, 2005, **383**, 235.
38. J. N. M. Commandeur, I. Andreadou, M. Rooseboom, M. Out, L. J. De Leur, E. Groot and N. P. E. Vermeulen, *J. Pharmacol. Exp. Ther.*, 2000, **294**, 753.
39. D. Cavallini, C. De Marco, B. Mondovi and B. G. Mori, *Enzymologia*, 1960, **22**, 161.
40. J. T. Pinto, B. F. Krasnikov, S. Alcutt, M. E. Jones, T. Dorai, M. T. Villar, A. Artigues, J. Li and A. J. Cooper, *J. Biol. Chem.*, 2014, **289**, 30950.
41. Y. Ohta, Y. Kobayashi, S. Konishi and S. Hirano, *Chem. Res. Toxicol.*, 2009, **22**, 1795.
42. B. M. Ryan and J. M. Faupel-Badger, *Semin. Oncol.*, 2016, **43**, 22.
43. M. H. Yoo, X. M. Xu, B. A. Carlson, A. D. Patterson, V. N. Gladyshev and D. L. Hatfield, *PLoS One*, 2007, **2**, e1112.
44. L. C. Clark, G. F. Combs, Jr., B. W. Turnbull, E. H. Slate, D. K. Chalker, J. Chow, L. S. Davis, R. A. Glover, G. F. Graham, E. G. Gross, A. Krongrad, J. L. Lesher, Jr., H. K. Park, B. B. Sanders, Jr., C. L. Smith and J. R. Taylor, *JAMA, J. Am. Chem. Soc.*, 1996, **276**, 1957.
45. S. M. Lippman, E. A. Klein, P. J. Goodman, M. S. Lucia, I. M. Thompson, L. G. Ford, H. L. Parnes, L. M. Minasian, J. M. Gaziano, J. A. Hartline, J. K. Parsons, J. D. Bearden, 3rd, E. D. Crawford, G. E. Goodman, J. Claudio, E. Winquist, E. D. Cook, D. D. Karp, P. Walther, M. M. Lieber, A. R. Kristal, A. K. Darke, K. B. Arnold, P. A. Ganz, R. M. Santella, D. Albanes, P. R. Taylor, J. L. Probstfield, T. J. Jagpal, J. J. Crowley, F. L. Meyskens, Jr., L. H. Baker and C. A. Coltman, Jr., *JAMA, J. Am. Chem. Soc.*, 2009, **301**, 39.
46. J. R. Marshall, C. M. Tangen, W. A. Sakr, D. P. Wood, Jr., D. L. Berry, E. A. Klein, S. M. Lippman, H. L. Parnes, D. S. Alberts, D. F. Jarrard,

W. R. Lee, J. M. Gaziano, E. D. Crawford, B. Ely, M. Ray, W. Davis, L. M. Minasian and I. M. Thompson, Jr., *Cancer Prev. Res.*, 2011, **4**, 1761.
47. A. M. Algotar, M. S. Stratton, F. R. Ahmann, J. Ranger-Moore, R. B. Nagle, P. A. Thompson, E. Slate, C. H. Hsu, B. L. Dalkin, P. Sindhwani, M. A. Holmes, J. A. Tuckey, D. L. Graham, H. L. Parnes, L. C. Clark and S. P. Stratton, *Prostate*, 2013, **73**, 328.
48. D. D. Karp, S. J. Lee, S. M. Keller, G. S. Wright, S. Aisner, S. A. Belinsky, D. H. Johnson, M. R. Johnston, G. Goodman, G. Clamon, G. Okawara, R. Marks, E. Frechette, W. McCaskill-Stevens, S. M. Lippman, J. Ruckdeschel and F. R. Khuri, *J. Clin. Oncol.*, 2013, **31**, 4179.
49. A. J. Duffield-Lillico, E. H. Slate, M. E. Reid, B. W. Turnbull, P. A. Wilkins, G. F. Combs, Jr., H. K. Park, E. G. Gross, G. F. Graham, M. S. Stratton, J. R. Marshall and L. C. Clark, *J. Natl. Cancer Inst.*, 2003, **95**, 1477.
50. A. J. Duffield-Lillico, M. E. Reid, B. W. Turnbull, G. F. Combs, Jr., E. H. Slate, L. A. Fischbach, J. R. Marshall and L. C. Clark, *Cancer Epidemiol., Biomarkers Prev.*, 2002, **11**, 630.
51. A. J. Duffield-Lillico, B. L. Dalkin, M. E. Reid, B. W. Turnbull, E. H. Slate, E. T. Jacobs, J. R. Marshall and L. C. Clark, *BJU Int.*, 2003, **91**, 608.
52. M. Vinceti, G. Dennert, C. M. Crespi, M. Zwahlen, M. Brinkman, M. P. Zeegers, M. Horneber, R. D'Amico and C. Del Giovane, *Cochrane Database Syst. Rev.*, 2014, CD005195.
53. A. R. Kristal, A. K. Darke, J. S. Morris, C. M. Tangen, P. J. Goodman, I. M. Thompson, J. F. L. Meyskens, G. E. Goodman, L. M. Minasian, H. L. Parnes, S. M. Lippman and E. A. Klein, *J. Natl. Cancer Inst.*, 2014, **106**, djt456.
54. M. E. Goossens, M. P. Zeegers, H. van Poppel, S. Joniau, K. Ackaert, F. Ameye, I. Billiet, J. Braeckman, A. Breugelmans, J. Darras, K. Dilen, L. Goeman, B. Tombal, S. Van Bruwaene, B. Van Cleyenbreugel, F. Van der Aa, K. Vekemans and F. Buntinx, *Eur. J. Cancer*, 2016, **69**, 9.
55. U. Alehagen, P. Johansson, M. Bjornstedt, A. Rosen, C. Post and J. Aaseth, *Eur. J. Clin. Nutr.*, 2016, **70**, 91.
56. G. F. Combs, Jr. and W. P. Gray, *Pharmacol. Ther.*, 1998, **79**, 179.
57. D. G. Menter, A. L. Sabichi and S. M. Lippman, *Cancer Epidemiol., Biomarkers Prev.*, 2000, **9**, 1171.
58. N. M. Corcoran, M. Najdovska and A. J. Costello, *J. Urol.*, 2004, **171**, 907.
59. G. X. Li, H. J. Lee, Z. Wang, H. Hu, J. D. Liao, J. C. Watts, G. F. Combs, Jr. and J. Lu, *Carcinogenesis*, 2008, **29**, 1005.
60. D. L. McCormick, K. V. Rao, W. D. Johnson, M. C. Bosland, R. A. Lubet and V. E. Steele, *Cancer Prev. Res.*, 2010, **3**, 381.
61. J. Lü, J. Zhang, C. Jiang, Y. Deng, N. Ozten and M. C. Bosland, *Nutr. Cancer*, 2016, **68**, 1.
62. J. Lu, C. Jiang, M. Kaeck, H. Ganther, S. Vadhanavikit, C. Ip and H. Thompson, *Biochem. Pharmacol.*, 1995, **50**, 213.
63. J. Lu, H. Pei, C. Ip, D. J. Lisk, H. Ganther and H. J. Thompson, *Carcinogenesis*, 1996, **17**, 1903.

64. M. Kaeck, J. Lu, R. Strange, C. Ip, H. E. Ganther and H. J. Thompson, *Biochem. Pharmacol.*, 1997, **53**, 921.
65. R. Sinha, T. K. Said and D. Medina, *Cancer Lett.*, 1996, **107**, 277.
66. N. Zhou, H. Xiao, T.-K. Li, A. Nur-E-Kamal and L. F. Liu, *J. Biol. Chem.*, 2003, **278**, 29532.
67. B. Husbeck, L. Nonn, D. M. Peehl and S. J. Knox, *Prostate*, 2006, **66**, 218.
68. B. Husbeck, D. M. Peehl and S. J. Knox, *Free Radical Biol. Med.*, 2005, **38**, 50.
69. M. Enqvist, G. Nilsonne, O. Hammarfjord, R. P. Wallin, N. K. Bjorkstrom, M. Bjornstedt, A. Hjerpe, H. G. Ljunggren, K. Dobra, K. J. Malmberg and M. Carlsten, *J. Immunol.*, 2011, **187**, 3546.
70. M. Hagemann-Jensen, F. Uhlenbrock, S. Kehlet, L. Andresen, C. Gabel-Jensen, L. Ellgaard, B. Gammelgaard and S. Skov, *J. Biol. Chem.*, 2014, **289**, 31576.
71. S. Misra, M. Wallenberg, O. Brodin and M. Björnstedt, *Diversity of Selenium Functions in Health and Disease*, CRC Press, 2015, pp. 109–136.
72. R. A. Cairns, I. S. Harris and T. W. Mak, *Nat. Rev. Cancer*, 2011, **11**, 85.
73. M. G. Vander Heiden, L. C. Cantley and C. B. Thompson, *Science*, 2009, **324**, 1029.
74. A. P. Fernandes, M. Wallenberg, V. Gandin, S. Misra, F. Tisato, C. Marzano, M. P. Rigobello, S. Kumar and M. Bjornstedt, *PLoS One*, 2012, **7**, e50727.
75. R. Sinha, S. C. Kiley, J. X. Lu, H. J. Thompson, R. Moraes, S. Jaken and D. Medina, *Cancer Lett.*, 1999, **146**, 135.
76. S. J. Chmura, E. Nodzenski, R. R. Weichselbaum and J. Quintans, *Cancer Res.*, 1996, **56**, 2711.
77. M. Wallenberg, S. Misra, A. M. Wasik, C. Marzano, M. Bjornstedt, V. Gandin and A. P. Fernandes, *J. Cell. Mol. Med.*, 2014, **18**, 671.
78. J. Lu, C. Jiang, M. Kaeck, H. Ganther, S. Vadhanavikit, I. P. Clement and H. Thompson, *Biochem. Pharmacol.*, 1995, **50**, 213.
79. G. F. Combs and J. Lü, in *Selenium: Its Molecular Biology and Role in Human Health*, ed. D. L. Hatfield, M. J. Berry and V. N. Gladyshev, Springer, US, Boston, MA, 2006, pp. 249–264.
80. Y. Dong, H. Zhang, A. C. Gao, J. R. Marshall and C. Ip, *Mol. Cancer Ther.*, 2005, **4**, 1047.
81. H. Zeng, M. Briske-Anderson, J. P. Idso and C. D. Hunt, *J. Nutr.*, 2006, **136**, 1528.
82. M. B. Yin, Z. R. Li, K. Toth, S. Cao, F. A. Durrani, G. Hapke, A. Bhattacharya, R. G. Azrak, C. Frank and Y. M. Rustum, *Oncogene*, 2006, **25**, 2509.
83. A. Bhattacharya, M. Seshadri, S. D. Oven, K. Toth, M. M. Vaughan and Y. M. Rustum, *Clin. Cancer Res.*, 2008, **14**, 3926.
84. B. L. Lindshield, N. A. Ford, K. Canene-Adams, A. M. Diamond, M. A. Wallig and J. W. Erdman, Jr., *PLoS One*, 2010, **5**, e10423.
85. S. Cao, F. A. Durrani, K. Toth and Y. M. Rustum, *Br. J. Cancer*, 2014, **110**, 1733.

86. K. T. Suzuki, Y. Ohta and N. Suzuki, *Toxicol. Appl. Pharmacol.*, 2006, **217**, 51.
87. W. D. Johnson, R. L. Morrissey, I. Kapetanovic, J. A. Crowell and D. L. McCormick, *Food Chem. Toxicol.*, 2008, **46**, 1068.
88. J. R. Marshall, C. Ip, K. Romano, G. Fetterly, M. Fakih, B. Jovanovic, M. Perloff, J. Crowell, W. Davis, R. French-Christy, A. Dew, M. Coomes and R. Bergan, *Cancer Prev. Res.*, 2011, **4**, 1938.
89. J. R. Marshall, R. F. Burk, R. Payne Ondracek, K. E. Hill, M. Perloff, W. Davis, R. Pili, S. George and R. Bergan, *Oncotarget*, 2017, **8**, 26312.
90. T. Hasegawa, S. Taniguchi, M. Mihara, K. Nakamuro and Y. Sayato, *Arch. Toxicol.*, 1994, **68**, 91.
91. O. Brodin, S. Eksborg, M. Wallenberg, C. Asker-Hagelberg, E. H. Larsen, D. Mohlkert, C. Lenneby-Helleday, H. Jacobsson, S. Linder, S. Misra and M. Bjornstedt, *Nutrients*, 2015, **7**, 4978.
92. C. Ip, *J. Nutr.*, 1998, **128**, 1845.

CHAPTER 15

Organoselenium Compounds as Cancer Therapeutic Agents

VALENTINA GANDIN*[a] AND ARISTI P. FERNANDES*[b]

[a] Department of Pharmaceutical and Pharmacological Sciences, University of Padova, Via Marzolo 5, 35131 Padova, Italy; [b] Division of Biochemistry, Department of Medical Biochemistry and Biophysics (MBB), Karolinska Institutet, SE-171 77 Stockholm, Sweden
*Email: valentina.gandin@unipd.it; Aristi.Fernandes@ki.se

15.1 Introduction

During recent decades, the connection between selenium and cancer has been studied widely and from many different aspects, ranging from the use as a trace element to a cancer preventive agent and more recently as a cancer therapeutic agent. Epidemiological studies performed over the past four decades have shown lower death rates for cancer in regions characterized by higher levels of Se in the soil, and a correlation between the occurrence of certain cancers and Se levels in plasma.[1–3] While the cancer-preventing activity of Se derivatives has been pursued since the 1980s, and has been extensively studied, only recent studies have focused on Se derivatives as potential and attractive therapeutic agents for the treatment of aggressive/advanced neoplasias. In addition, Se supplementation in patients receiving conventional anticancer therapies has been shown to enhance efficacy, limit side effects, and improve general patient condition.[4–6]

In recent decades, selenium-containing compounds have attracted a growing interest, with mounting reports demonstrating their high efficacy and selectivity against cancer cells.[7] However, the cancer therapeutic

Organoselenium Compounds in Biology and Medicine: Synthesis, Biological and Therapeutic Treatments
Edited by Vimal Kumar Jain and K. Indira Priyadarsini

Published by the Royal Society of Chemistry, www.rsc.org

properties of selenium compounds seem to be many, and with inorganic and organic selenium derivatives being metabolized differently *in vivo*, they activate distinct molecular mechanisms and exhibit varying degrees of toxicity and anticancer activity.[8] Inorganic selenium forms generally exhibit genotoxic effects, which in turn may account for a lower therapeutic window, a higher systemic toxicity, and an increased risk of second malignancy in treated patients. Conversely, organic Se compounds, besides retaining a significant antitumor activity with an increased ability to prevent metastasis, have fewer side effects and lack the genotoxic potential.[9,10]

With organic compounds showing the most promising results and with many new compounds being synthesized and characterized, this chapter focuses on listing different types of organoselenium compounds proposed as potential anticancer agents and describes the known antitumor mechanisms underlying this group of compounds.

15.2 Selenium and Cancer Treatment

15.2.1 Increased Oxidative Stress as a Selective Anticancer Strategy

Redox homeostasis is fundamental for maintaining normal cellular functions and ensuring cell survival. The regulation of intracellular reactive oxygen species (ROS) levels is ultimately crucial for upholding the cellular homeostasis. At low to moderate levels, ROS have a physiological role as signaling molecules that sustain cellular proliferation, differentiation, and survival. For instance, hydrogen peroxide (H_2O_2) can serve as a signal for proliferation and differentiation as well as migration, but can also act as a signal transduction molecule that can induce pro-inflammatory cytokines and the nuclear factor (NF)-kB. At higher levels, ROS can impair the cellular redox homeostasis generating oxidative stress, which, among other things can give rise to DNA damage, irreversible protein oxidation, and lipid peroxidation, which can be detrimental to the cell. To control ROS levels, cells use scavenging molecules including antioxidants, such as the glutathione system (comprising glutathione peroxidases (GPx), glutathione-*s*-transferases (GST), and glutaredoxins (Grx)) and the thioredoxin (Trx) system with peroxiredoxins,[11,12] and also low molecular-weight compounds such as ubiquitin and vitamin E.

Although cellular transformation during carcinogenesis arises from many different pathways, the metabolic requirements of the different tumor cells are generally believed to be very similar.[13] One central issue is to counteract the increase in oxidizing effects arising from their aberrant metabolism and increasing ROS production, through induction of antioxidant scavenging molecules. Consequently, while healthy cells are characterized by a low steady-state level of ROS, and in some way constant levels of reducing equivalents, cancer cells are endowed with increased levels of ROS and reducing equivalents. It is the uncontrolled growth in cancer cells that mainly accounts for the excess production of ROS, associated with accelerated rates of glycolysis (the Warburg effect) and pentose phosphate cycle. To counteract oxidative stress, cancer cells must perform concomitant up-regulation of the ROS scavenging mechanisms, leaving little or no room for further adaptation to additional induction of oxidative stress. If the redox balance is impaired, oxidative stress will arise, causing detrimental effects to the cell. Hence, additional inhibition or loss of antioxidant systems may tip the redox balance specifically in cancer cells, generating oxidative stress and ultimately cell death. Alternatively, further induction of oxidative stress may also lead to the preferential killing of cancer cells, as cancer cells will be more sensitive than normal cells to the accumulation of ROS, which offers an interesting therapeutic window.[14,15] Consequently, inducing ROS to reach a threshold that is incompatible with cell viability and targeting the enhanced antioxidant mechanisms may selectively kill cancer cells, without affecting normal cells.[16,17] Comprehensive reviews on redox cancer metabolism and therapeutic strategies have been published.[13,15,18,19]

With respect to the induction of oxidative stress as an anticancer strategy using pro-oxidants, Se compounds have been shown to be interesting candidates. Among redox modulators generating oxidative stress as a cancer therapeutic approach, Se compounds have shown to have many beneficial properties and many new compounds have been designed, synthesized, and evaluated for their efficacy and specificity as anticancer agents. At present, organoselenium compounds have proven to be the most promising, and while more studies of their effects on cancer cell metabolism are required, the demonstrated mechanisms and antiproliferative properties are discussed herein.

15.2.1.1 *Selenium Cytotoxicity: Selective Uptake and Oxidative Stress Induction*

As tumor cells generally are more vulnerable to induce oxidative stress, they have also proven to be more susceptible to Se compounds.[20–22] The cytotoxic effects are realized at pharmacologically achievable doses, offering a narrow therapeutic window for the use of Se compounds as anticancer agents. The mechanisms behind the anticancer properties of selenium compounds have been shown to be diverse, and highly dependent on the dose and specific chemical, as well as on the redox state and experimental model.[23] Once inside the cells, Se compounds are metabolized and can undergo redox reactions and change form and oxidation state, producing several metabolites, such as hydrogen selenide (HSe^-) and methyl selenol (CH_3Se^-) in response to the cell environment.[24] These redox-active Se metabolites are highly nucleophilic and therefore very reactive and have proven to be the most efficient anticancer agents. These compounds have the ability to generate ROS, thereby generating oxidative stress. The mechanism is through redox cycling of selenolates with glutathione (GSH) or the Trx/Grx systems, which together with oxygen will produce superoxide and hydrogen peroxide. In addition, Se compounds may, by direct interaction with free thiols and selenols, also cause oxidation of a large number of proteins and low molecular-weight compounds. These modifications result in the formation of intra- or inter-molecular bonds, including selenotrisulfide bonds (S–Se–S), selenenylsulfide bonds (Se–S), and diselenide bonds (Se–Se) with protein selenols.[25] The redox-active Se compounds may also catalyze the formation of disulfide bonds (S–S) and/or mixed disulfide bonds with glutathione (S–SG) or nitric oxide (S–NO). These oxidations may result in numerous biological downstream effects, and may directly affect the structure, biological function, or enzyme activity of proteins. Direct modification and regulation of signaling proteins through thiol oxidation include protein kinases, phosphatases, and transcription factors.[26] Among those that have been best characterized are caspases, p53, Jun, AP-1, APE-1/Ref-1, Sp1, NF-κB, ASK-1, and JNK.[27–30] In addition, the redox modifications of specific thiol residues may also result in an altered iron–sulfur cluster biogenesis,[31] as well as changes in calcium and iron homeostasis.[32–34] Redox modification of thiol/disulfide exchange in proteins could ultimately lead to protein unfolding and loss of protein biological functions/activities.[35,36]

Redox-active Se compounds do not only affect proteins, but can, by generation of ROS as well as through direct interaction and binding, cause DNA damage, and alter the DNA response.[37–41] These redox-active metabolites are known to cause both single- and double-strand breaks.[42]

Selenium is known to have a remarkable selectivity for accumulation in cancer cells at nutritional doses. This was first proposed and demonstrated in human studies in the 1960s when ^{75}Se-sodium selenite and ^{75}Se-selenomethionine (SeMet) were assessed as scanning agents in the diagnosis of tumors. By using ^{75}Se as a tumor radiotracer, a surprisingly high accuracy was observed in localizing intracranial tumors as well as thoracic and abdominal neoplasms.[43–46] This selective uptake and accumulation of Se in tumor cells is believed to be one of the mechanisms behind the observed tumor selectivity and specificity. However, the uptake mechanism varies between selenium compounds, and is not clearly elucidated. Selenide has been suggested to be transported *via* ATPases,[47] nevertheless it remains a challenge to determine the exact transport for each Se compound, as they are very reactive molecules. The uptake of selenite is less efficient, but has been shown to be facilitated by the presence of extracellular reducing thiols, indicating that the reduced form, presumably selenide, is more readily taken up by the cells.[47] It was later shown that the accumulation in tumors could be partly explained by the overexpression of the cystine/glutamate antiporter xCT observed in several tumors,[48] generating a more reducing extracellular microenvironment, and thus facilitating the uptake by reducing selenite to selenide.[49] The selective uptake of selenium by cancer cells has also been attributed to their aberrant metabolism, which requires an increased amount of selenium, generally being rate limiting for the incorporation of central selenoproteins such as GPx and thioredoxinreductase (TrxR), key antioxidants, and enzymes important in upholding the redox homeostasis in the cell.[50]

15.3 Organoselenium Compounds as Potential Anticancer Agents

The interest in synthetic organoselenium chemistry has increased persistently since the 1970s, due to the great variety of applications of organoselenium compounds in modern organic synthesis,[51] but also due to the many biological properties discovered in recent years, including the anticancer activity of organoselenium compounds.[52] Accordingly, several groups of organoselenium-containing compounds have been described in the literature, and were evaluated for their antitumor properties (Table 15.1). We discuss the major and most widely studied, based on their chemical properties and mode of action.

15.3.1 Selenoaminoacid Derivatives

Selenocystine is a diselenide oxidation product of the selenocysteine (Sec) amino acid. Selenocystine, which is highly redox active, is efficiently reduced by low molecular-weight thiols and disulfide reductases, such as thioredoxin reductase to Sec. Due to the low pK_a value, Sec retains a highly nucleophilic property which makes it very unstable and prone to reaction. In *in vitro* assays, it has been shown to be effective against human lung and cervical cancers,[53,54] as well as against melanoma, both alone or in combination with 5-fluoro uracil.[55,56] From a mechanistic point of view, selenocystine treatment seems to give rise to both paraptotic-like cell death with endoplasmic reticulum stress and unfolded protein response induction, and apoptotic cell death, depending on the cell cycle stage at the time of treatment.[53] The apoptotic cell death has been shown to be p53- and caspase-dependent.[55] In human melanoma cells, it was further shown that selenocystine induced PARP cleavage, activation of multiple caspases (-3, -7, -9, -8, -10), release of cytochrome c, apoptosis-inducing factor and Smac/Diablo from mitochondria to the cytosol, thus indicating the activation of both intrinsic and extrinsic apoptosis.

In vivo, selenocystine was shown to be highly effective in a melanoma xenograft model, with no signs of systemic toxicity.[55] Interestingly, selenocystine has been administered in humans for the treatment of acute and chronic myeloid leukemia. Selenocystine elicited a noteworthy activity, being particularly effective against immature leucocytes with respect to mature leucocytes, without inducing any significant effect on the bone marrow.[57] However, even though selenocystine is endowed with a promising antitumor activity, its poor stability and low solubility has impeded its further development as an anticancer drug.

Se-methylselenocysteine (MSC) is a monomethylated seleno-amino acid derivative. MSC can be metabolized to $MeSe^-$ by selenocysteine Se-conjugate β-lyases.[58] Like HSe^-, it has not been possible to directly identify $MeSe^-$ in cell or animal models due to its high volatility and reactivity; nevertheless, the cleavage of MSC to $MeSe^-$ has been indirectly monitored in rat liver supernatant.[59] In the same study, γ-lyase cleavage of SeMet to $MeSe^-$ was not

Table 15.1 Organoselenium compunds.

Compound	Structure	Biological models	Ref.
Selenoaminoacid derivatives			
Selenocystine (SeC)	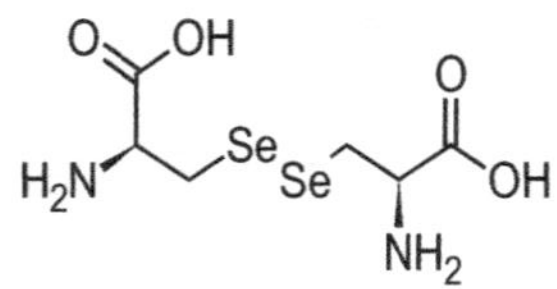	*In vitro*	
		Human melanoma cells	55, 56
		Human cervical cancer cells	54
		Human lung cancer cells	53
		In vivo	
		Human melanoma	55
Se-methylselenocysteine (MSC)	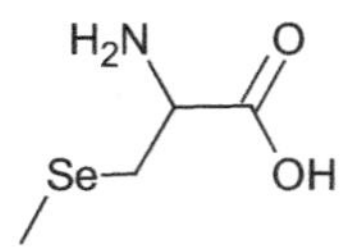	*In vitro*	
		Human lung cancer cells	60
		Human colon cancer cells	61
		Human ovarian cancer cells	62
		Human leukemia cells	64
		In vivo (combination therapy)	66
		Human colorectal carcinoma and head and neck squamous cell carcinoma	66
		Human breast carcinoma	65
		Human colon carcinoma	68, 69
Selenides and diselenides			
Methylseleninic acid (MSA)	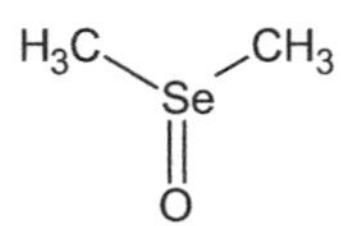	*In vitro*	
		Human lung cancer cells	71
		Human prostate cancer cells	72–75
		Human breast cancer cells	23
		In vivo	
		Mouse mammary epithelial cells	76
		Human prostate carcinoma	72, 77
		In vivo (combination therapy)	
		Triple-negative breast cancer	78
		Human prostate cancer	79

Selenodiglutathione (SDG)	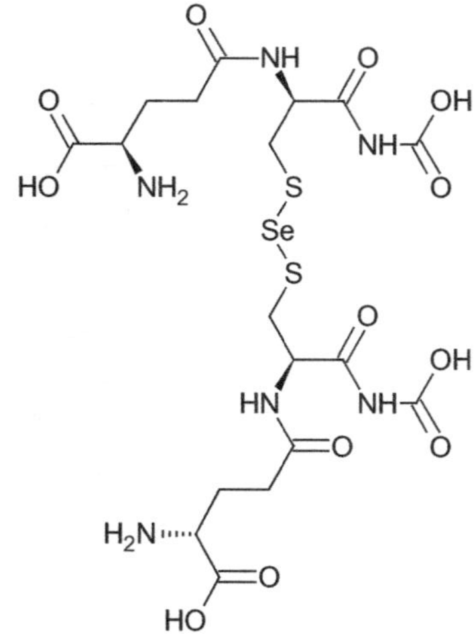	*In vitro* Human promyelocytic leukemia cells Human ovarian cancer cells Human oral carcinoma cells Human cervical cancer cells	 86 87 88 53
Quinazoline and pyrido[2,3-*d*]pyrimidine selenium compounds	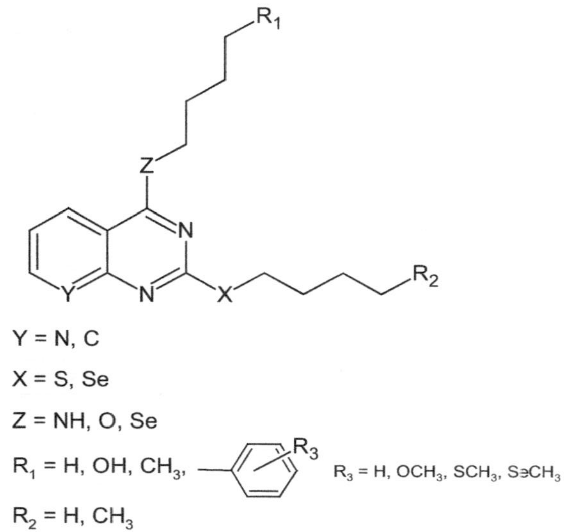	*In vitro* Human leukemia cells Human colon, lung, and breast cancer cells	 91

Table 15.1 (*Continued*)

Compound	Structure	Biological models	Ref.
Methylimidoselenocarbamates	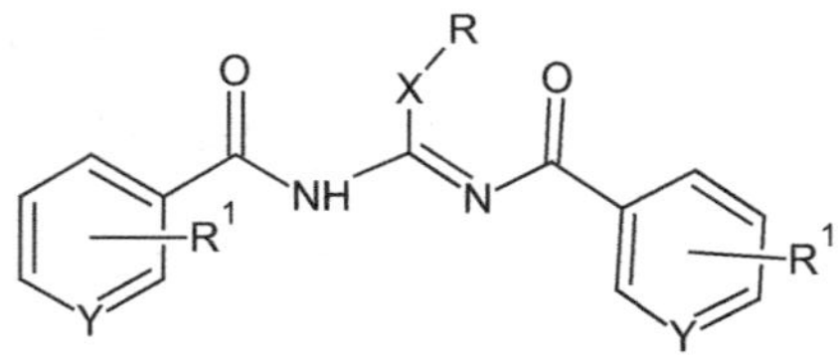 X = S, Se Y = C, N R = alkyl R^1 = H, alkyl, OCH_3, Cl, CF_3, CN, NO_2	*In vitro* Human prostate cancer cells Human breast and colon cancer cells *In vivo* Human prostate cancer	 89, 90 90 90
Diphenyl diselenides	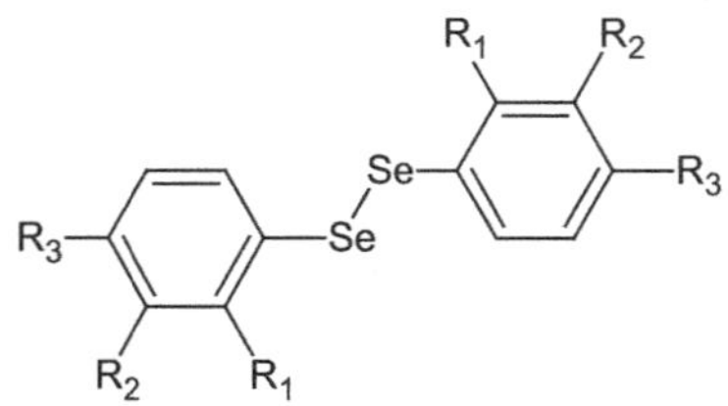 R_1 = H, OCH_3, NH_2 R_2 = H, CF_3 R_3 = H, OCH_3	*In vitro* Human neuroblastoma cells Human colon carcinoma cells	 93 94

Peptoid selenobenzoic acid derivatives	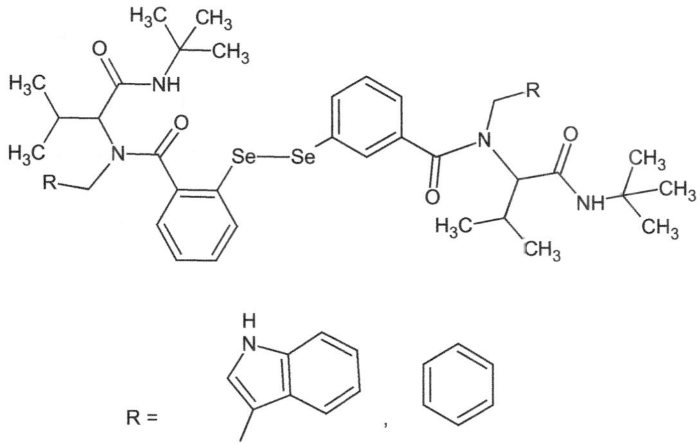	*In vitro* Human hepatocellular carcinoma cells	97
Tetrazole-based symmetric diselenides	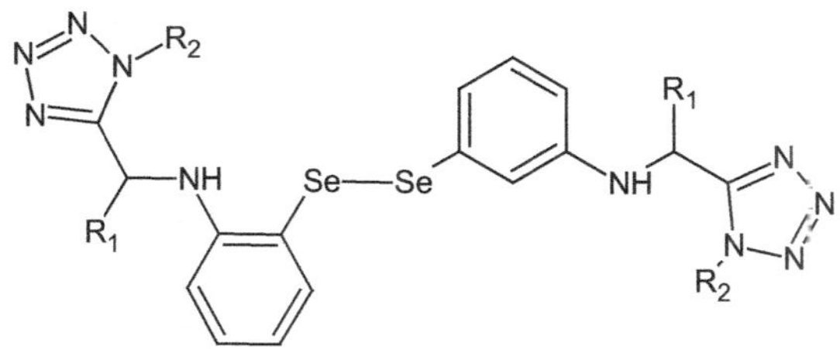 R_1 = H, $CH(CH_3)_2$, furan, *p*-toluene R_2 = *tert*-buthyl-, *p*-CH_3O-Ph-, 2,4-CH_3O-Ph-, 1,1,2-trimethoxy-, cyclohexane, Ph,	*In vitro* Human hepatocellular carcinoma cells	101

Table 15.1 (*Continued*)

Compound	Structure	Biological models	Ref.
Diseleno- and selenium-based quinones	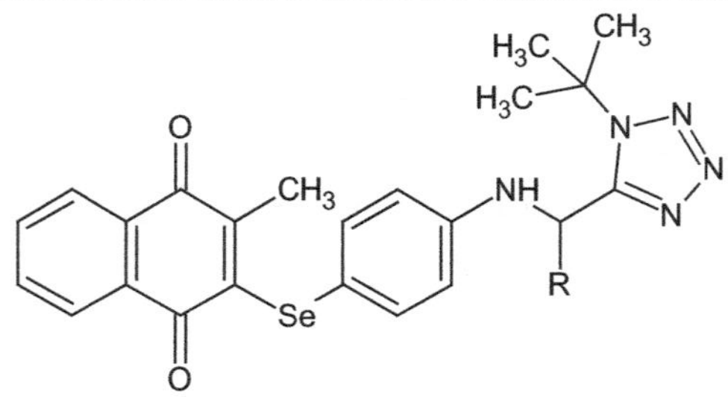 R = H, *p*-toluene, furan	*In vitro* Human hepatocellular carcinoma cells	 101
Selenocyantes			
Isoselenocyanates	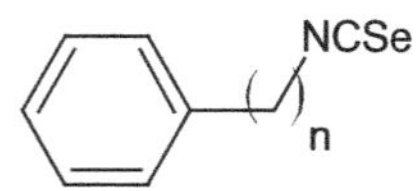	*In vitro* Human prostate, breast, and colon cancer cells Melanoma, glioblastoma and sarcoma cells *In vivo* Human melanoma	 102 102 102, 103
Naphthalimide-based selenocyanates	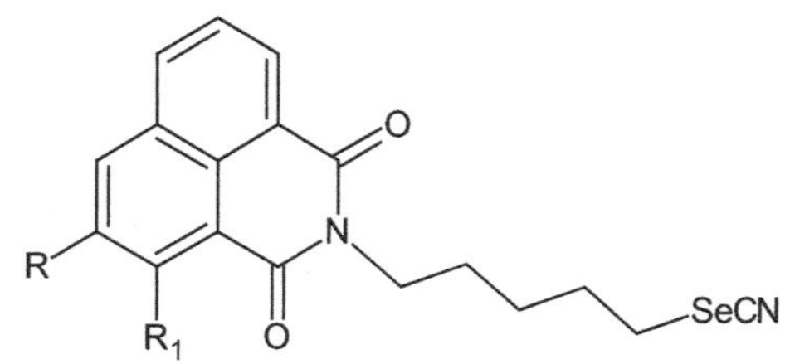 R = H, Cl, NH_2 R_1 = H, CH_3, NO_2	*In vivo* Ehrlich ascites mammary carcinoma	 108, 109

1,4-Phenylenebis(methylene)-selenocyanate, (p-XSC)	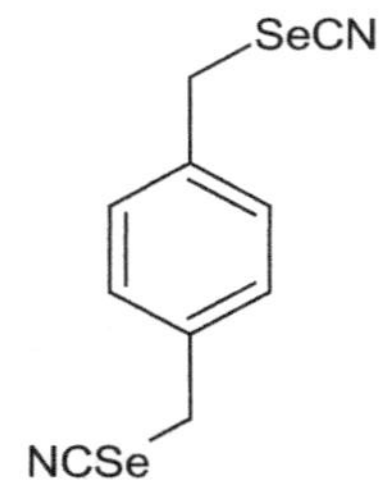	*In vitro* Human prostate cancer cells Human colon cancer cells	 88 110
5,7-Dibromoisatin selenocyanates	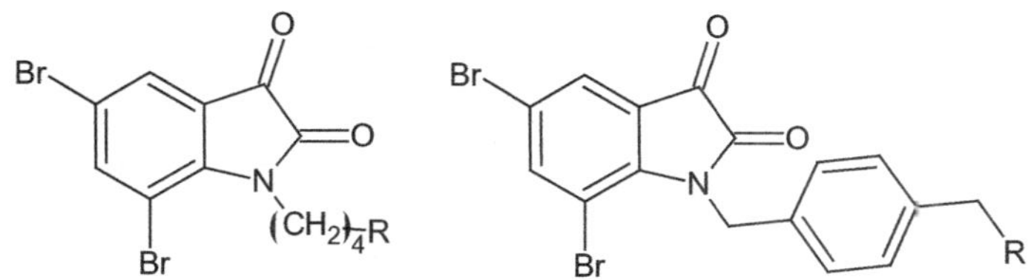 R = SCN, SeCN	*In vitro* Human colon, breast, and lung cancers and melanoma	 113
Selenoesters			
Alkyl and aryl selenoesters	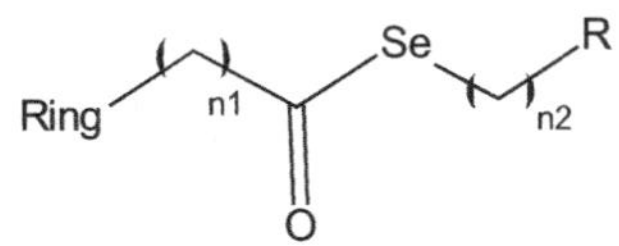 n1 = 0, 1 n2 = 0, 1, 2 R = CH_3, $COC(CH_3)_3$, $COCH_3$, $CONH_2$, $COOCH_3$, $COOC(CH_3)_3$, COOPh Ring = 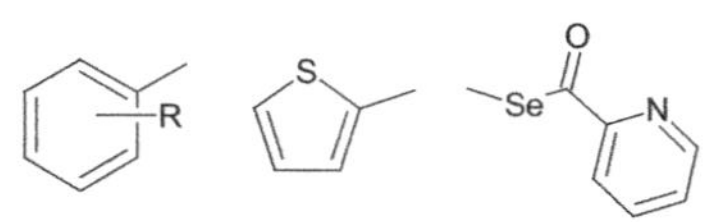	*In vitro* Human prostate, colon, breast, and lung cancers	 114, 115

Table 15.1 (*Continued*)

Compound	Structure	Biological models	Ref.
Xylitol selenious ester	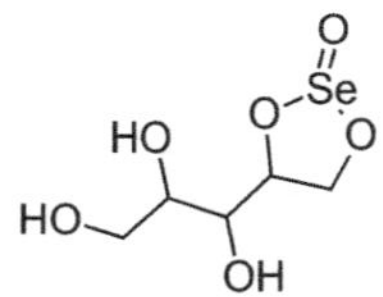	*In vitro* Human hepatoma cells	 117–119
Sucrose selenious ester	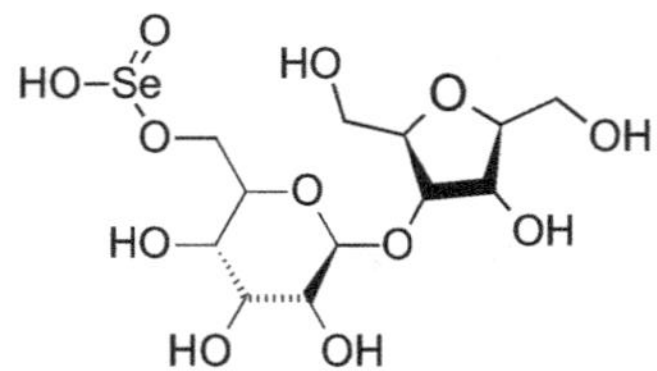	*In vitro* Human hepatoma cells Human cervical, bladder, gastric cancer cells, and melanoma cells	 117 120
Ethaselen, ebselen and their derivatives			
1,2-[Bis(1,2-benzisoselenazolone-3(2*H*)-ketone)]ethane (Ethaselen or BBSKE)	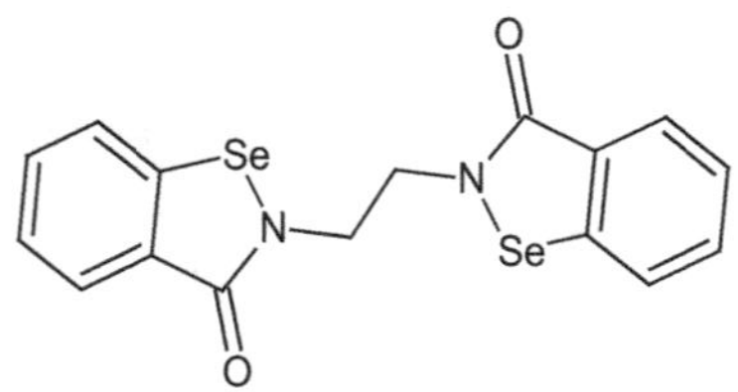	*In vitro* Human lung cancer cells Human prostate cancer cells Human cervical and gastric cancer cells and hepatoma cells *In vivo* Human prostate carcinoma Human tongue carcinoma Human liver carcinoma *In vivo* (combination therapy) Human lung carcinoma	 121, 124 123 121 125 126 127 127, 131

Benzisoselenazolone derivatives	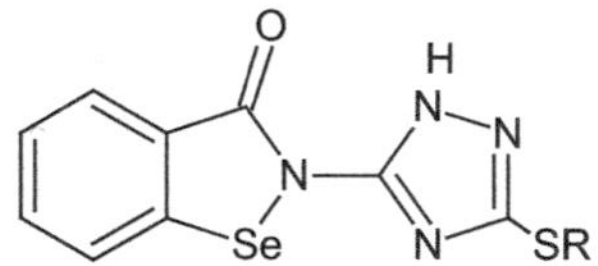 R = CH_2CH_3, $CH_2CH_2CH_3$, $CH_2(CH_3)_2$, butyl-, isobutyl-, pentyl-, hexyl-, benzyl-, *p*-F-benzyl, *p*-Cl-benzyl-, *o*-CH_3-benzyl-, *m*-CH_3-benzyl-, *p*-NO_2-benzyl-	*In vitro* Human cervical, lung cancer cells, and hepatocellular carcinoma Human breast, lung cancer cells, and hepatocellular carcinoma	 134 133, 135
2-Phenyl-1,2-benzisoselenazol-3(2*H*)-one (ebselen)	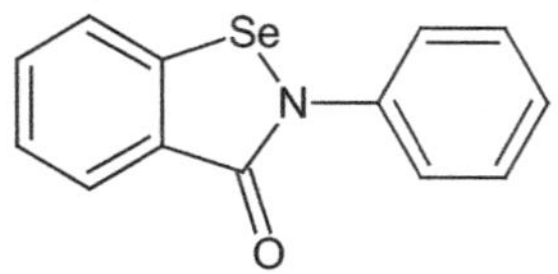	*In vitro* Human myeloma cells Human pancreatic and renal cancer cells *In vivo* Human pancreatic carcinoma	 138 139 139
Benzoselenazole-stilbene hybrids	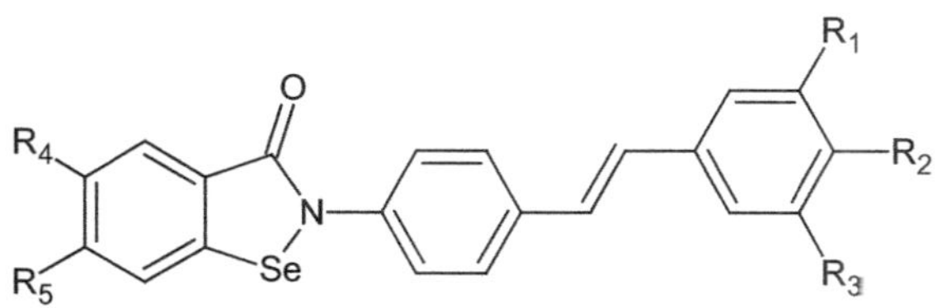 R_1, R_2, R_3 = H, OCH_3 R_4 = H, F, Cl R_5 = H, Cl	*In vitro* Human breast, lung, cervical cancer cells, and hepatoma cells	 141

Table 15.1 *(Continued)*

Compound	Structure	Biological models	Ref.
Other Se heterocycles			
2,5-Bis(5-hydroxymethyl-2-selenienyl)-3-hydroxymethyl-*N*-methylpyrrole (D-501036)	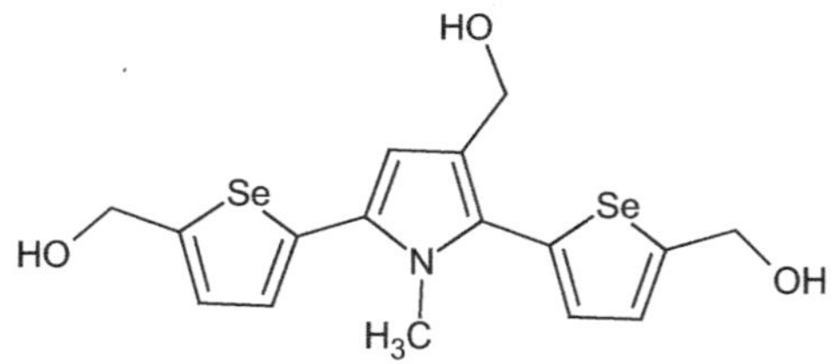	*In vitro* Human cervical cancer cells and hepatoma cells Human renal, breast, lung, prostate, colorectal, and nasopharyngeal cancer cells *In vivo* Human renal carcinoma	143, 144 142 142
1,2,5-Selenadiazolo[3,4-*d*]pyrimidine-5,7(4*H*,6*H*)-dione	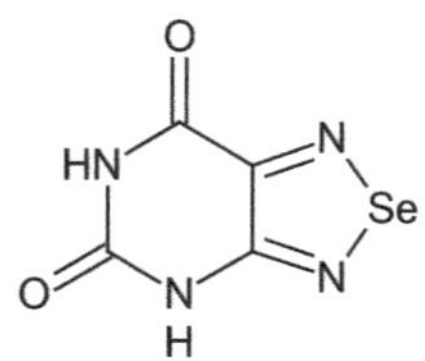	*In vitro* Human breast cancer cells Human hepatoma and melanoma cell	146
Anthrax[1,2-*c*][1,2,5]selenadiazolo-6,11-dione	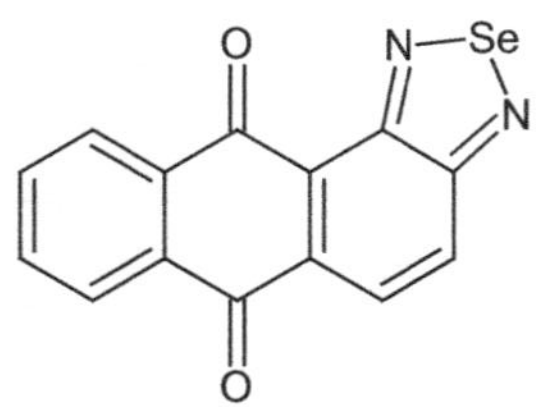	*In vitro* Human breast cancer cells	147
4-(Benzo[*c*][1,2,5]selenadiazol-6-yl)benzene-1,2-diamine (BSBD)	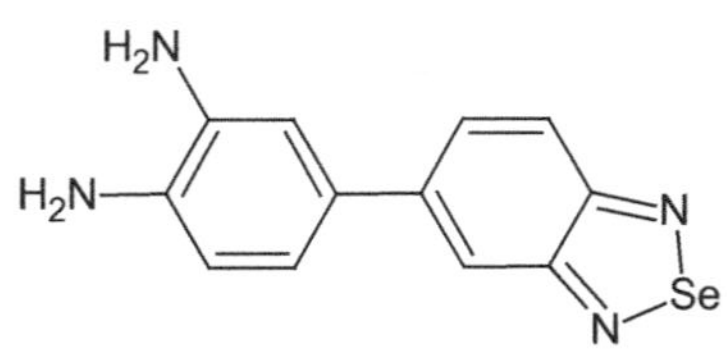	*In vitro* Human melanoma, hepatoma, neuroblastoma, and glioma cells Human breast and prostate cancer cells Rat pheochromocytoma cells	148

2-β-*N*-Ribofuranosylselenazole-4-carboxamide (selenazofurin) and 5-β-D-Ribofuranosylselenophene-3-carboxamide (selenophenfurin)	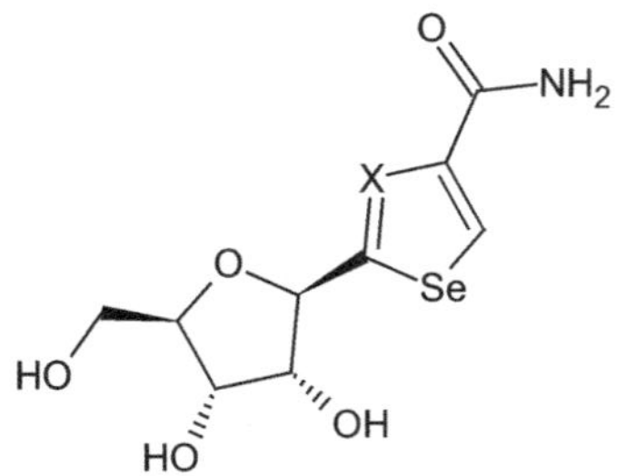 X = N (Selenazofurin) X = CH (Selenophenfurin)	*In vitro* Murine leukemia cells	149–151
		In vitro Human colon, cervical, renal, bladder cancer cells, and lymphoma cells	153
1,3-Selenazine compounds	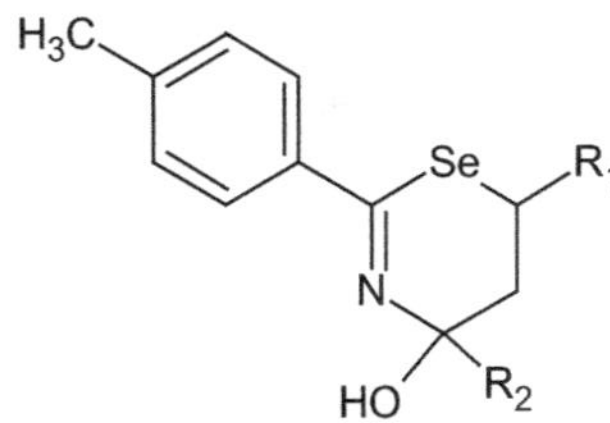 R_1 = H, $CH_2CH_2CH_3$, $CH(CH_3)_2$ R_2 = CH_2CH_3, CH_3	*In vitro* Human fibrosarcoma cells and human gastric cancer cells	154, 155

Table 15.1 (*Continued*)

Compound	Structure	Biological models	Ref.
1,4-Oxaselenins	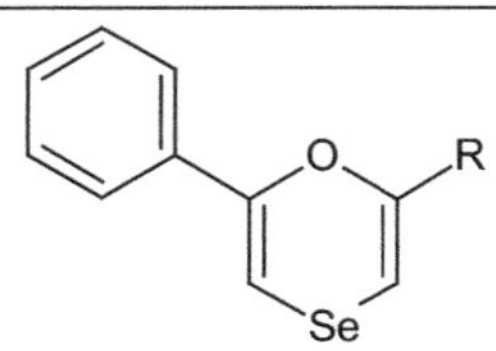 R = Ph, *p*-CH_3O-Ph- , *p*-Cl-Ph-	*In vitro* Human ovarian and cervical cancer cells	 156
Selenium containing non-steroidal anti-inflammatory drugs			
Selenocoxib-1-GSH	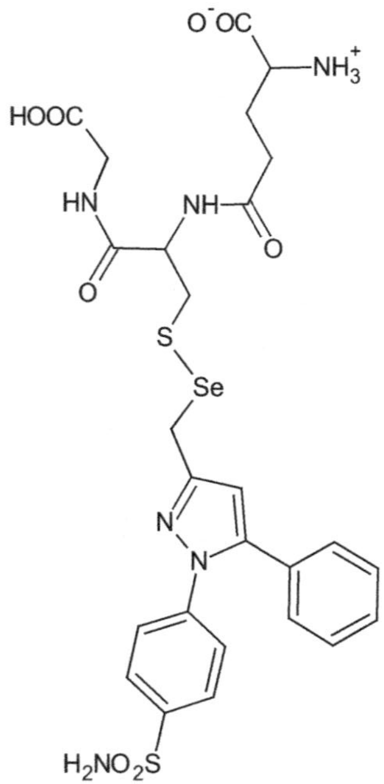	*In vitro* Human melanoma *In vivo* Human melanoma	 158 158

Selenium NSAIDs	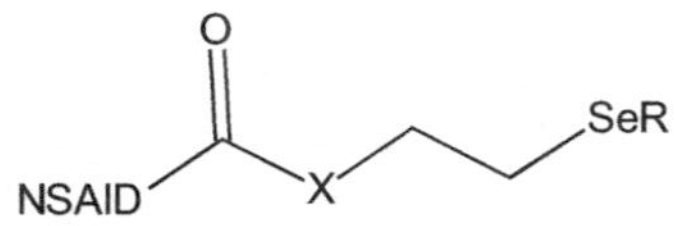 X = NH, O, Se R = H, CH_3, CN NSAID = Aspirin, Ibuprofen, Naproxen	*In vitro* Human colon, breast, pancreatic cancer cells, and melanoma	159
Se-containing histone deacetylase inhibitor			
Suberoylanilide hydroxamic acid (SAHA) selenoderivatives	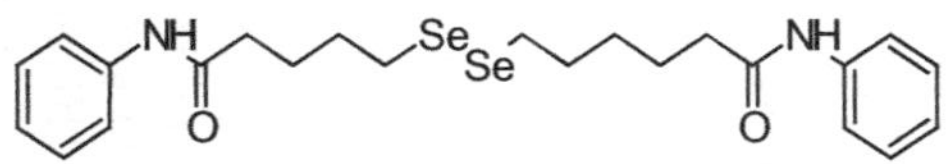 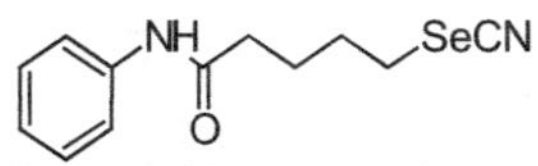	*In vitro* Human lung cancer cells	160–162

detectable, and together with its high K_m, the biological significance of this activity, with subsequent demethylation to produce selenide, is questionable for SeMet. Conversely, MSC has been shown to be effective in inhibiting cancer cell viability and proliferation at micromolar concentrations in numerous human cancer cell lines.[60,61] MSC has been shown to kill cancer cells by triggering apoptosis in a caspase-dependent manner,[62–64] but the mitochondrial involvement in MSC-induced cell death has not been unеqivocally demonstrated.

Similar to the *in vitro* studies, the *in vivo* antitumor potential of MSC has been demonstrated in different tumor models, and apart from being dose- and schedule-dependent when used as a single agent, MSC has been recognized for its ability to significantly enhance the anticancer potential of conventional chemotherapeutic drugs, leading to an excellent therapeutic synergy *in vivo*. For instance, MSC has increased the antitumor efficacy of irinotecan and tamoxifen in a dose-dependent manner and at the same time managed to protect against their toxicity.[65,66] Similar effects have been seen in combination with cisplatin and oxaliplatin, in a variety of drug-sensitive and -resistant human tumor xenografts.[66,67] In addition, MSC seems to significantly protect against organ-specific toxicity induced by lethal doses of cisplatin and oxaliplatin, including diarrhoea, stomatitis, alopecia, bladder, kidney, and bone marrow toxicities. In addition, the reduced toxicity has been associated with an enhanced antitumor activity in rats bearing advanced Ward colorectal carcinoma, and in xenograft mice of human squamous cell carcinoma of the head and neck. Recently, MSC has attracted the attention of researchers owing to its ability to modulate metastatic processes. Animal studies have emphasized its antiangiogenic and antitumor efficacy in human colon carcinoma.[68,69] Treatment with MSC in the current study resulted in tumor growth inhibition and in a significant reduction in blood volume and microvessel density. Furthermore, a reduction of tumor interstitial fluid pressure was also observed, thus allowing a higher intra-tumoral drug delivery, and subsequently, a higher therapeutic response of combined chemotherapy. However, this chemo-modulating effect was more pronounced and significant in tumors with uniform histopathological features, characterized by little or no hypoxic regions. Conversely, it is important to take into consideration that the MSC efficacy is fully dependent on the cell/organ/tissue β-lyase activity to generate the active methylselenol metabolite.[70] This could explain biologically why *in vitro* studies with MSC do not display such prominent results as *in vivo* studies, even though in theory at least the same active metabolite should be generated, namely $MeSe^-$. Overall, the *in vivo* data provide the rationale for clinically evaluation of MSC in polychemotherapy.

15.3.2 Selenides and Diselenides

Methaneseleninic acid, also known as methylseleninic acid (MSA), is an oxoselenium compound which has been found to be effective *in vitro* against a large variety of tumors, including human lung,[71] prostate[72–75] and breast[76] tumor cell lines. *In vivo* MSA has been administered in a mouse model of mammary epithelial cells,[76] as well as in two prostate tumor xenograft models, where MSA was found to considerably reduce tumor growth without inducing substantial animal weight loss or other signs of systemic toxicity, nor any evidence of genotoxic side effects.[72,77] In combination therapy, MSA has resulted in an enhancement of paclitaxel efficacy for the treatment of triple-negative breast cancer[78] and has potentiated cell death induced by various chemotherapeutic drugs in prostate cancer cells.[79] MSA is readily reduced by intracellular thiols including GSH and TrxR to $MeSe^-$, and is often used as a $MeSe^-$ precursor,[80] but has also been proposed to be further metabolized to selenide through demethylation. Despite *in vitro* studies that show a higher antiproliferative activity of MSA compared to MSC, it retains a similar efficacy profile to MSC *in vivo*.[81] Most likely, this can be attributed to the presence of β-lyase in many tissues, especially liver, giving animals an ample capacity to metabolize MSC systemically. Mechanistically, even though MSA shows similar patterns as selenite with regards to deactivation of Akt and ERK1/2,[82,83] many differences have been observed when studying cell death pathways. Notably, the dephosphorylation induced by MSA of Akt and ERK1/2 in prostate cancer cells was not mediated by p38MAPK and JNK1/2.[75] Additionally, when comparing their effects on androgen receptor expression in prostate cancer, even though both selenite and MSA disrupt androgen receptor signaling, they have different mechanisms of action. Selenite acts by decreasing Sp1 levels, whereas MSA does not affect Sp1, but under hypoxic conditions, affected mTOR signaling through the induction of REDD1 and Akt.[30,84] Furthermore, MSA has been shown to induce apoptotic cancer cell death in different cell lines by the activation of multiple caspases (caspase-3, -7, -8, and -9), cytochrome c release and PARP cleavage.[73–75] In particular, MSA has been shown to induce apoptosis in p53 wild-type, p53-mutant,[74] and p53-null cells,[85] thus demonstrating a p53-independent cancer cell death pathway.

Selenodiglutathione (SDG) is the primary cellular metabolite of selenite. It has been evaluated for its anticancer potential *in vitro* against a wide range of cancer cells since the 1990s and has been found to be a more effective antiproliferative agent than selenite.[86–88] Interestingly, it retains a greater activity against cancer cells with respect to non-transformed cells, thus attesting a certain selectivity against neoplastic cells. Unfortunately, SDG has not been completely explored for its antitumor potential at present, probably because the believed mechanism of action is similar to that of selenite, and hence would retain similar adverse side effects. Conversely, we have recently demonstrated that selenite and SDG elicit diverse cytotoxic effects. Although both compounds are metabolized to selenide, their mechanism of action is profoundly different, both in terms of intracellular target compartments, induced cancer cell expression patterns, and specific cell death mechanism (necroptosis-like cell death for selenite and apoptosis-like cell death for SDG).[53] The mechanism accounting for these different effects was further shown to be attributed to the ability of SDG to glutathionylate free protein thiols.

Other selenides have also been evaluated for their antitumor potential. Plano *et al.* showed that methylimidoselenocarbamates act as cytotoxic agents at micromolar doses against human prostate cancer cells.[89] Later, the same authors showed that these compounds act as multi-kinase inhibitors, being able to inhibit the PI3K/AKT/mTOR pathway and several of its downstream regulators (p70S6K and 4E-BP1) and ERK1/2 in prostate, colon, and breast cancer cells. This ultimately led to the induction of both autophagy and apoptosis cell death. In addition, the most effective quinolone-based imidoselenocarbamate derivative was able to suppress prostate cancer cell tumor growth *in vivo*, in a mouse xenograft model.[90] In the same year, Moreno *et al.* synthesized and tested a series of quinazoline and pyrido[2,3-*d*]pyrimidine selenium compounds, some of them demonstrating a promising cytotoxicity against several human cancer cell lines (*i.e.* lymphocytic leukemia cells, as well as colon, lung, and breast cancer cells) at low micromolar concentrations.[91] The evaluation of the structure–activity relationships has highlighted that the position of the Se moiety in the molecule can affect the antiproliferative profile. It was further suggested that they induce cell death without affecting cell cycle phases.

Among selenides, symmetrical diselenides have become very attractive since they are quite stable, easy to prepare, and depending on the substituents, hardly toxic.[92] Recently, Posser *et al.* showed, for the first time, that diphenyldiselenide possesses a significant *in vitro* antitumor potential against neuoblastoma cells, and induces cancer cell death through apoptosis *via* the ERK1/2 pathway.[93] Subsequently, Nedel *et al.* demonstrated that substituted diaryldiselenides (3-(trifluoromethyl)-diphenyldiselenide and 4-methoxydiphenyl diselenide) have the potential to induce apoptosis in HT-29 cells through the activation of caspase-dependent and independent pathways and through cell-cycle arrest.[94]

Recent evidence has also suggested that diphenyldiselenide is a good substrate for TrxR, thus stimulating the research groups working in this field and laying down the basis for the development of novel diselenide derivatives as antitumor agents.[95] On this basis, Shaaban *et al.* synthesized different series of diselenides containing bioactive pharmacophores (*e.g.* quinones, naphthalene, cyclic imides) or pharmacologically relevant heterocycles (*e.g.* thiazolidinone, pyrazole and thiazolopyrimidine).[96–100] Some of these compounds exhibit cytotoxicity at sub-micromolar concentrations against various types of cancer cells, being particularly effective against hepatocellular and breast carcinomas. Later, Shaaban *et al.* showed that tetrazole-based diselenides and selenoquinones were effective against hepatocarcinoma cells by downregulating Bcl-2 and Ki-67 expression levels and activatingcaspase-8.[101]

15.3.3 Selenocyantes

Selenocyanate has proven its effectiveness in the prevention and treatment of a variety of cancers, both *in vitro* and *in vivo*. Sharma *et al.* synthesized and evaluated a series of naturally occurring and synthetic phenylalkyl isothiocyanates (ITCs; R-N=C=S), by isosterically replacing sulfur (S) with selenium (Se) to make corresponding isoselenocyanates (ISCs; R-N=C=Se). The replacement of S in ITCs with Se was an attempt to add the properties of selenium to the electrophilic isothiocyanate (–N=C=S) group to generate isoselenocyanates with an increased electrophilicity (–N=C=Se) group, thus resulting in a panel of more potent compounds with similar reactivity pattern towards GSH.[102] These ISCs were effective both *in vitro*, against glioblastoma, melanoma, sarcoma, prostate, breast, and colon cancer cell lines, as well as *in vivo*, towards a preclinical melanoma tumor xenograft model with no evidence of systemic toxicity. Remarkably, the structure–activity relationship studies concluded that tumor inhibitory effect increased with increasing chain length (probably due to an increase in lipophilicity), where $n = 4$ was found to be optimal. In addition, the authors formulated a hypothesis concerning their metabolism, suggesting that the *in vivo* activity of ISCs may be mediated by metabolic transformation to GSH conjugates. Later, the same authors showed that ISCs were effective in cultured melanoma cells and tumors.[103] Notably, *in vivo*, in a melanoma xenograft model, the most promising ISC was able to decrease tumor development by ~60% compared to corresponding ITCs, without inducing a detectable change in body weight or in blood parameters indicative of liver, kidney, or cardiac related toxicity. Mechanistically, the ISCs selectively induced apoptosis in melanoma cells by inhibiting the Akt3 signaling cascade. Moreover, Crampsie *et al.* assessed and confirmed their ability to redox cycle in the presence of GSH to form nitric oxide, superoxide, and other ROS, thus affecting cancer cell redox state, and through this inducing apoptotic cell death.[104] Modulation of antioxidant enzymes such as GPx, GST, TrxR, and catalase has also been reported.[104–106]

Accordingly, in combination treatment, the diphenylmethylselenocyanate (DMSE) has been shown to be effective by decreasing the tumor burden significantly through ROS generation and modulation of antioxidant and detoxifying enzyme systems in tumor cells. This, in turn, was associated with a caspase-dependent apoptotic cell death in tumor cells through the induction of p53 and Bax, and suppression of Bcl-2. It is noteworthy that a decrease in cisplatin-induced nephrotoxicity and hematopoietic toxicity in DMSE combination treatment was observed, suggesting that the combination may offer potential therapeutic benefit, and overcome some of the limitations associated with the chemotherapeutic use of cisplatin.[107]

Similar results were obtained by Roy *et al.* with a naphthalimide-based selenocyanate compound, 2-[5-selenocyanato-pentyl]-7-amino benzo[*de*]isoquinoline-1,3-dione (ANOS), which was able to significantly potentiate the therapeutic efficacy of cyclophosphamide by reducing the tumor burden and chemotherapy-induced toxicity in the host in Ehrlich ascites carcinoma-bearing Swiss albino mice.[108] In addition, Gosh *et al.* assessed the chemoenhancing efficacy of ANOS during cisplatin chemotherapy in the same cancer cell model. Results clearly attested that the tested naphthalimide-based selenocyanate compound not only reduced the toxicity of cisplatin, but also enhanced its antitumor efficacy by inducing apoptosis in tumor cells through mitochondria- and DNA-mediated damage, consequently increasing the lifespan of the tumor-bearing host.[109]

The synthetic compound 1,4-phenylenebis(methylene)selenocyanate (p-XSC) has also been proven to be effective in the low micromolar range *in vitro*, against both androgen-responsive and androgen-independent human prostate cancer cells. The mechanisms accounting for its antiproliferative activity include its ability to induce JNK and p38 kinase and activate ERKs 1 and 2 and Akt.[88] Furthermore, p-XSC-mediated apoptosis has proven to be independent of p53 expression in human colon cancer cells.[110] It induces cell cycle arrest in the G1 phase and inhibits Akt[111] and m-TOR signaling.[112]

Following the cutting-edge multi-target approach for developing new putative anticancer drugs, Krishnegowda *et al.* developed a series of ISC derivatives by combining an indoleheterocycle with a selenocyanate moiety, with the idea of yielding novel dual-targeted inhibitors. These compounds were shown to exhibit very promising *in vitro* antitumor activity (in the low micromolar range) against colon, breast, and lung cancers and against melanoma, being able to induce cancer cell death through a mechanism encompassing both microtubule and Akt inhibition.[113]

15.3.4 Selenoesters

Several selenoesters have been synthesized with the rationale that they may act as scaffolds that could undergo hydrolysis or enzymatic reduction, generating redox-active compounds.[114] In addition, they can be modulated by the inclusion of different functional groups in the moiety bound to the selenium atom of the selenoester. These redox-active metabolites, being ionic species of selenium (such as selenols), could readily participate in redox processes, and hence encompass an enhanced cytotoxicity against cancer cells. *In vitro* evaluation of the cytotoxic properties of these compounds has shown that several of them exhibited significant cytotoxicity against all the cancer cell lines investigated (PC-3, MCF-7, A-549, and HT-29), with some being even more potent than etoposide and cisplatin.[114] In addition, selenoesters have been evaluated for their ability to reverse multidrug resistance (MDR) in mouse MDR T-lymphoma cells.[115] Results showed that the selenoesters with ketone terminal fragments exerted 1.7–3.6-fold stronger efflux pump inhibitory activity compared to the established ABCB1 inhibitor verapamil. Moreover, three of the ketone selenoesters showed strong and selective cytotoxic effects with IC_{50} values in the nanomolar range in the resistant colon cancer cells, while in the non-tumor fibroblasts the IC_{50} increased to the micromolar range.[116]

Among sugar-based esters, both a xylitolselenious ester (xylitol-Se) and a sucrose selenious ester (sucrose-Se) have been synthesized. While barely characterized, the xylitol-Se compound has shown cytostatic effects on human hepatocellular carcinoma cells in a dose-dependent manner, while having no negative influences on the proliferation of normal human hepatic cells.[117,118] Xylitol-Se has also been reported to deplete cells from GSH,[119] and hence partially explains its underlying cytotoxic properties. In a similar manner, sucrose-Se has effectively inhibited the proliferation of several cancer cell lines in a dose-dependent manner without killing the normal human liver cell line HL-7702. In addition, sucrose-Se has a very high median lethal dose compared to inorganic sodium selenite (290.0 and 13.1 ppm, respectively), in an acute toxicity test on mice,[120] significantly increasing the therapeutic window for this compound.

15.3.5 Ethaselen and Ebselen

Ethaselen (1,2-[bis(1,2-benzisoselenazolone-3(2*H*)-ketone)]ethane), also known as BBSKE, has been synthesized and extensively studied by Zeng and co-workers in a number of cell lines.[121] They found that it targets and inhibits TrxR, a central antioxidant and redox regulatory enzyme, which has been proposed as a promising target for anticancer drugs.[122] Ethaselen has also been shown to inhibit the proliferation of tumor cells through S phase arrest, *via* its TrxR-inhibitory effect.[123] More detailed studies revealed that TrxR inhibition by ethaselen occurs through targeting of the C-terminal, but not N-terminal active site, and that it does not inhibit thioredoxin or glutathione reductase. The same study showed that downstream effects of ethaselen caused oxidation of thioredoxin and increased the levels of ROS.[124] Concomitantly, *in vivo* experiments have also shown promising results of ethaselen, with tumor inhibitory effects in the range of 40–80% in prostate, tongue, and liver cancer models.[125–127] The metabolism and secretion of ethaselen has further been suggested to occur *via* oxidation, methylation, and glucoronidation.[128] Ethaselen has reached phase I clinical trials in patients with non-small cell carcinoma. In the Ia/b phase, 1200 mg day^{-1} was found to be a safe and tolerable dose (ClinicalTrials.gov identifier: NCT02166242). Moreover, the effects of ethaselen have been tested *in vivo* in several combination therapies. One study used ethaselen in combination with cisplatin (*cis*-diaminedichloroplatinum II) in a lung xenograft mouse model. Compared to single drug administration, the combination treatment showed significant synergistic reduction of tumor size, with no evident signs of systemic or organ toxicity.[5] Co-treatment of ethaselen with sunitinib, a multitargeted tyrosine kinase inhibitor, has similarly displayed synergistic effects against proliferation of colorectal cancer cells.[129] In accordance with other studies[130] showing that inhibition of TrxR sensitizes cancer cells to radiotherapy, Wang *et al.* have shown that ethaselen enhances the efficacy of radiation therapy *in vitro* and *in vivo* without causing any signs of toxicity.[131] Despite the promising activity of ethaselen, its solubility in physiological media is not optimal, thus precluding its application without a suitable formulation. The formulation as copolymer micelles recently reported by Liu *et al.* allowed for an increased water solubility, and ultimately led to an enhanced antitumor activity, which was attributed to a massive accumulation into the tumor site.[132]

More recently, novel benzisoselenazolone derivatives have been synthesized and evaluated for their antiproliferative properties compared to ethaselen. Among the new compounds, some showed a superior antitumor activity compared to ethaselen, and they retained a good selectivity toward cancer cells, with only weak cytotoxic effects against non-cancerous cells.[133–135] Notably, Ye *et al.* have synthesized some ethaselen derivatives by linking it to a carmustine-based moiety. These combination molecules were shown to possess an improved solubility, a low toxicity profile and a very efficient antitumor activity in several cancer cell lines.[136] The authors also established a structure–activity relationship model relevant to further structural modification. Overall, their results have highlighted the potential of ethaselen-based combination molecules for development of novel antitumor agents.

Ebselen, a compound structurally related to ethaselen, has mainly been considered as a GPx mimic and hence as a strong antioxidant. In comparison to ethaselen, ebselen is an efficient substrate for TrxR, and the reaction is known to be stimulated by Trx, leading to a rapid oxidation of Trx.[137] The antioxidative properties of ebselen have been widely studied, but a few studies have also suggested that ebselen might possess antiproliferative and anticancer properties *via* ROS production.[138] These anticancer properties have also been suggested to be *via* the inhibition of quiescin sulfhydryl oxidase 1, an enzyme shown to promote the growth and invasion of tumor cells and to alter the extracellular matrix composition. In the same study, the authors described that daily oral treatment with ebselen resulted in a 58% reduction in tumor growth in mice bearing human pancreatic tumor xenografts.[139] New chiral ebselen analogues have been synthesized by Pacula *et al.*, but they only exhibited moderate cytotoxic properties.[140] A series of benzoselenazole-stilebene hybrids have been synthesized by combining reservatrol and ebselen, with the aim to create compounds that possess the pharmacological activity of both compounds simultaneously. These compounds were found to possess antiproliferative properties, causing cell cycle arrest and programed cell death in the low micromolar range, and they significantly induced ROS formation.[141]

15.3.6 Other Seleno-heterocycles

15.3.6.1 Selenium-containing Five-membered Rings

Different classes of Se-containing five-membered ring compounds (selenophenes, selenazoles, and selenadiazoles) have been prepared and evaluated for their pharmacological potential. Among compounds endowed with antitumor potential, one of the most promising derivatives is selenophene, D-501036, 2,5-bis(5-hydroxymethyl-2-selenienyl)-3-hydroxymethyl-*N*-methylpyrrole, which was identified as a potent antineoplastic agent with a broad spectrum of activity against human cancer cell lines. It elicited IC_{50} values in the low micromolar range, being highly selective against cancer cells compared to normal cells and highly effective against MDR cancer cells. Furthermore, D-501036 exhibited strong antitumor activity *in vivo*, in a mouse xenograft model of human renal carcinoma. Concerning its mechanism of action, it was found to induce cell death through a p53- and mitochondrial-mediated apoptotic pathway.[142–144]

Selenadiazoles have also attracted attention in the past owing to their potential as anticancer agents. As first example, a series of 4-methyl-1,2,3-selenadiazole-5-carboxylic acid amides were synthesized and characterized for their antitumor potential *in vitro* and *in vivo* against murine cancers.[145] Later, a series of 20 selenadiazole derivatives was designed and synthesized, with 1,2,5-selenadiazolo-[3,4-*d*]pyrimidine-5,7-(4*H*,6*H*)-dione being the most promising derivative. This compound was found to be cytotoxic toward different human cancer cells, namely melanoma and hepatoma, as well as breast cancer cells.[146] This compound was able to kill cells through the induction of both extrinsic and intrinsic caspase-dependent apoptosis. Mechanistically, an overproduction of ROS and depletion of mitochondrial membrane potential through regulation of the expression of pro-survival and pro-apoptotic Bcl-2 family members contribute to cancer cell death induction. The same authors later developed the anthrax[1,2-*c*][1,2,5]selenadiazolo-6,11-dione and highlighted its efficacy in inducing a time- and dose-dependent cell death of human breast carcinoma cells.[147] Similarly to the previous selenadiazole derivative, the authors showed that the cell death pathway was driven by depletion of mitochondrial membrane potential, up-regulation of Bax, Bad, and PUMA expression and down-regulation of Bcl-xl expression. However, the caspase-dependent cell killing effect was not mediated by ROS production in this case.

More recently, Zhang *et al.* developed a new selenodiazole compound, 4-(benzo[*c*][1,2,5]selenadiazol-6-yl) benzene-1,2-diamine, and tested it against different human cancer cell lines.[148] Remarkably, this compound was effective (at micromolar doses) against all tested cell lines, while barely affecting non-tumor cells. From mechanistic investigation in glioma cells, the authors showed that the selenodiazole derivative induced a caspase-dependent mitochondrial-mediated apoptotic cell death as a consequence of AKT dephosphorylation and p53 activation.

Selenazofurin (2-β-D-ribofuranosylselenazole-4-carboxamide), the selenium analogue of tiazofurin synthesized by Srivastava and Robins,[149] was initially shown to possess a pronounced antitumor activity against murine melanoma cancers both *in vitro* and in animal models, as well as against human promyelocytic leukemia cells *in vitro* at nanomolar levels.[150,151] Similar results were reported by Gebeyehu *et al.* with a series of selenazofurin dinucleotides; these latter dinucleotide analogues were more effective than the mononucleotide precursor.[152] Mechanistically, selenazofurin and its derivatives are metabolized to the corresponding selenazole-4-carboxamide-adenine dinucleotides and act as potent non-competitive inhibitors of inosine monophosphate dehydrogenase, a rate-limiting enzyme of *de novo* guanine nucleotides biosynthesis. Based on these very interesting and promising results, Franchetti *et al.* developed a new analogue in which the selenazole ring was placed into a selenophene heterocycle, the selenophenfurin.[153] This compound was effective toward leukemia, lymphoma, and solid tumor cell lines at concentrations similar to those of selenazofurin.

15.3.6.2 Selenium-containing Six-membered Rings

The development of Se-containing six-membered rings has increased over the past 20 years. Among six-membered ring-based compounds, Koketsu and coworkers developed a series of 1,3-selenazine-based compounds and demonstrated their antiproliferative effects against fibrosarcoma and gastric cancer cells *via* the induction of apoptotic cell death.[154,155] Later, they also developed some 1,4-oxaselenins and showed their inhibitory effect on the proliferation of human ovarian and cervical cancer cells by inducing apoptosis.[156]

15.3.7 Selenium-non-steroidal Anti-inflammatory Drugs

A study performed by Bi *et al.* presented evidence for a synergistic effect between a non-steroidal anti-inflammatory drug (NSAID) sulindac, and selenium.[157] In this study, a non-toxic dose of sulindac was used in combination with selenium, which significantly inhibited intestinal tumorigenesis, determined by the reduction in tumor incidence (52%) and tumor multiplicities (80%). The results were linked to significant induction of the expression of p27 and p53 as well as JNK1 phosphorylation. Recently, a selenium-containing selenocoxib derivative was developed and authors highlighted the ability of this compound to decrease melanoma cell growth by arresting cells in the G0–G1 phase of the cell cycle, to promote melanoma cell apoptosis and inhibit cellular proliferation.[158] Mechanistically, the selenium-containing selenocoxib derivative retains COX-2 inhibitory activity and PI3K/Akt inhibitory activity. Similarly, Plano *et al.* later incorporated selenium into NSAIDs, generating Se-NSAID hybrids.[159] The most effective was a SeCN-aspirin analogue, which was shown to be extremely selective toward cancer cells and was found to cause cell cycle arrest in G1 and G2/M phases and to induce apoptosis by activating caspase 3/7 and PARP cleavage in colorectal cancer cells.

15.3.8 Selenium-containing Histone Deacetylase Inhibitors

Histone deacetylases (HDACs) are involved in the epigenetic regulation of gene expression by remodeling of chromatin through their deacetylase activity. Inhibition of HDACs has emerged as a target for specific epigenetic changes associated with cancer and other diseases. The use of HDAC inhibitors is therefore believed to be an attractive therapeutic strategy against cancer. Selenium-containing derivatives of the United States Food and Drug Administration-approved HDAC inhibitor suberoylanilidehydroxamic acid (SAHA) have been synthesized and studied in melanoma and lung cancer cell lines. Both of the reported compounds, bis(5-phenylcarbamoylpentyl) diselenide and 5-phenylcarbamoylpentyl selenocyanide, were able to inhibit HDAC activity, and were also found to possess PI3 kinase pathway inhibitory properties. More importantly, both were significantly more effective in inducing cytotoxicity towards different cancer cell lines compared to SAHA.[160–162]

HDAC inhibition has also been reported for MSC and SeMet, which are transaminase substrates of glutamine transaminase K (GTK) and L-amino acid oxidase.[163] GTK and L-amino acid oxidase are able to convert MSC to the corresponding α-keto acid, β-methylselenopyruvate (MSP), and L-amino acid oxidase converts selenomethionine to its corresponding α-keto acid, α-keto-γ-methylselenobutyrate (KMSB). These metabolites structurally resemble the known HDAC inhibitor butyrate, and both MSP and KMSB have been shown to inhibit HDAC activity.[164]

15.4 Conclusion

Many different classes of organoselenium compounds have been explored as antiproliferative agents so far, several of them manifesting a pronounced cytotoxic activity against cancer cells compared with non-transformed ones. Coherently, *in vivo* testing of some derivatives clearly confirm their scarce systemic toxicity and great tolerability, both when used in monotherapy and, interestingly, when administered in combination with classical chemotherapeutic agents. In the latter case, they not only increase the therapeutic potential of well-known chemotherapeutic agents, but they also seem to protect against unwanted side effects.

The precise mechanism by which the antitumor activity is mediated remains unclear, and more profound *in vivo* experiments are warranted to elucidate their anticancer properties. However, what is clear is that organoselenium derivatives are more efficient, selective, and well-tolerated than the corresponding non-chalcogenated molecules, thus illustrating the importance of further exploiting the potential of novel orgnoselenium derivatives for the development of cancer-specific anticancer agents.

References

1. H. J. Jung and Y. R. Seo, *Biofactors*, 2010, **36**, 153.
2. M. P. Rayman, *Lancet*, 2012, **379**, 1256.
3. D. J. Hughes, V. Fedirko, M. Jenab, L. Schomburg, C. Meplan, H. Freisling, H. B. Bueno-De-Mesquita, S. Hybsier, N. P. Becker, M. Czuban, A. Tjonneland, M. Outzen, M. C. Boutron-Ruault, A. Racine, N. Bastide, T. Kuhn, R. Kaaks, D. Trichopoulos, A. Trichopoulou, P. Lagiou, S. Panico, P. H. Peeters, E. Weiderpass, G. Skeie, E. Dagrun,

M. D. Chirlaque, M. J. Sanchez, E. Ardanaz, I. Ljuslinder, M. Wennberg, K. E. Bradbury, P. Vineis, A. Naccarati, D. Palli, H. Boeing, K. Overvad, M. Dorronsoro, P. Jakszyn, A. J. Cross, J. R. Quiros, M. Stepien, S. Y. Kong, T. Duarte-Salles, E. Riboli and J. E. Hesketh, *Int. J. Cancer*, 2015, **136**, 1149.
4. I. M. Puspitasari, R. Abdulah, C. Yamazaki, S. Kameo, T. Nakano and H. Koyama, *Radiat. Oncol.*, 2014, **9**, 125.
5. Q. Tan, J. Li, H. W. Yin, L. H. Wang, W. C. Tang, F. Zhao, X. M. Liu and H. H. Zeng, *Invest. New Drugs*, 2010, **28**, 205.
6. E. Yakubov, M. Buchfelder, I. Y. Eyupoglu and N. E. Savaskan, *Biol. Trace Elem. Res.*, 2014, **161**, 246.
7. A. P. Fernandes and V. Gandin, *Biochim. Biophys. Acta*, 2015, **1850**, 1642.
8. C. M. Weekley and H. H. Harris, *Chem. Soc. Rev.*, 2013, **42**, 8870.
9. Y. C. Chen, K. S. Prabhu, A. Das and A. M. Mastro, *Int. J. Cancer*, 2013, **133**, 2054.
10. V. Valdiglesias, E. Pasaro, J. Mendez and B. Laffon, *Arch. Toxicol.*, 2010, **84**, 337.
11. A. P. Fernandes and A. Holmgren, *Antioxid. Redox Signaling*, 2004, **6**, 63.
12. A. Holmgren and J. Lu, *Biochem. Biophys. Res. Commun.*, 2010, **396**, 120.
13. R. A. Cairns, I. Harris, S. Mccracken and T. W. Mak, *Cold Spring Harbor Symp. Quant. Biol.*, 2011, **76**, 299.
14. M. Diehn, R. W. Cho, N. A. Lobo, T. Kalisky, M. J. Dorie, A. N. Kulp, D. Qian, J. S. Lam, L. E. Ailles, M. Wong, B. Joshua, M. J. Kaplan, I. Wapnir, F. M. Dirbas, G. Somlo, C. Garberoglio, B. Paz, J. Shen, S. K. Lau, S. R. Quake, J. M. Brown, I. L. Weissman and M. F. Clarke, *Nature*, 2009, **458**, 780.
15. C. Gorrini, I. S. Harris and T. W. Mak, *Nat. Rev. Drug Discovery*, 2013, **12**, 931.
16. R. A. Cairns, I. S. Harris and T. W. Mak, *Nat. Rev. Cancer*, 2011, **11**, 85.
17. V. Sosa, T. Moline, R. Somoza, R. Paciucci, H. Kondoh and L. L. Me, *Ageing Res. Rev.*, 2013, **12**, 376.
18. D. Trachootham, J. Alexandre and P. Huang, *Nat. Rev. Drug Discovery*, 2009, **8**, 579.
19. M. Liu, J. Fu, J. Li, L. Wang, Q. Tan, X. Ren, Z. Peng and H. Zeng, *Int. J. Pharm.*, 2010, **391**, 292.
20. G. Nilsonne, X. Sun, C. Nyström, A. K. Rundlöf, A. P. Fernandes, M. Björnstedt and K. Dobra, *Free Radical Biol. Med.*, 2006, **41**, 874.
21. B. Husbeck, L. Nonn, D. M. Peehl and S. J. Knox, *Prostate*, 2006, **66**, 218.
22. H. K. Rooprai, I. Kyriazis, R. K. Nuttall, D. R. Edwards, D. Zicha, D. Aubyn, D. Davies, R. Gullan and G. J. Pilkington, *Int. J. Oncol.*, 2007, **30**, 1263.
23. R. J. Jariwalla, B. Gangapurkar and D. Nakamura, *Br. J. Nutr.*, 2009, **101**, 182.
24. B. Gammelgaard, M. I. Jackson and C. Gabel-Jensen, *Anal. Bioanal. Chem.*, 2011, **399**, 1743.
25. H. E. Ganther, *Carcinogenesis*, 1999, **20**, 1657.

26. C. Pantano, N. L. Reynaert, A. Van Der Vliet and Y. M. Janssen-Heininger, *Antioxid. Redox Signaling*, 2006, **8**, 1791.
27. L. Flohe, R. Brigelius-Flohe, C. Saliou, M. G. Traber and L. Packer, *Free Radical Biol. Med.*, 1997, **22**, 1115.
28. Y. Zou, P. Niu, J. Yang, J. Yuan, T. Wu and X. Chen, *Cancer Biol. Ther.*, 2008, **7**, 689.
29. P. Ranawat and M. P. Bansal, *Mol. Cell. Biochem.*, 2008, **309**, 21.
30. B. Husbeck, R. S. Bhattacharyya, D. Feldman and S. J. Knox, *Mol. Cancer Ther.*, 2006, **5**, 2078.
31. R. Lill, B. Hoffmann, S. Molik, A. J. Pierik, N. Rietzschel, O. Stehling, M. A. Uzarska, H. Webert, C. Wilbrecht and U. Muhlenhoff, *Biochim. Biophys. Acta*, 2012, **1823**, 1491.
32. U. Muhlenhoff, S. Molik, J. R. Godoy, M. A. Uzarska, N. Richter, A. Seubert, Y. Zhang, J. Stubbe, F. Pierrel, E. Herrero, C. H. Lillig and R. Lill, *Cell Metab.*, 2010, **12**, 373.
33. A. Sheftel, O. Stehling and R. Lill, *Trends Endocrinol. Metab.*, 2010, **21**, 302.
34. C. M. Sag, S. Wagner and L. S. Maier, *Free Radical Biol. Med.*, 2013, **63**, 338.
35. S. Muller, J. Heider and A. Bock, *Arch. Microbiol.*, 1997, **168**, 421.
36. M. Zorn, C. H. Ihling, R. Golbik, R. G. Sawers and A. Sinz, Selective Selc-Independent Selenocysteine Incorporation Into Formate Dehydrogenases, *PLos One*, 2013, **8**, E61913.
37. T. Tobe, K. Ueda, M. Ando, Y. Okamoto and N. Kojima, *J. Biol. Inorg. Chem.*, 2015, **20**, 687.
38. K. El-Bayoumy, Y. H. Chae, P. Upadhyaya, C. Meschter, L. A. Cohen and B. S. Reddy, *Cancer Res.*, 1992, **52**, 2402.
39. Y. Qi, N. W. Schoene, F. M. Lartey and W. H. Cheng, *J. Biol. Chem.*, 2010, **285**, 33010.
40. B. J. Wycherly, M. A. Moak and M. J. Christensen, *Nutr. Cancer*, 2004, **48**, 78.
41. N. Zhou, H. Xiao, T. K. Li, E. K. A. Nur and L. F. Liu, *J. Biol. Chem.*, 2003, **278**, 29532.
42. J. Lu, M. Kaeck, C. Jiang, A. C. Wilson and H. J. Thompson, *Biochem. Pharmacol.*, 1994, **47**, 1531.
43. R. R. Cavalieri and K. G. Scott, *JAMA, J. Am. Med. Assoc.*, 1968, **206**, 591.
44. R. R. Cavalieri, K. G. Scott and E. Sairenji, *J. Nucl. Med.*, 1966, **7**, 197.
45. J. Esteban, D. Lasa and S. Perez-Modrego, *Radiology*, 1965, **85**, 149.
46. R. P. Spencer, G. Montana, G. T. Scanlon and O. R. Evans, *J. Nucl. Med.*, 1967, **8**, 197.
47. D. Ganyc and W. T. Self, *FEBS Lett.*, 2008, **582**, 299.
48. J. Lewerenz, S. J. Hewett, Y. Huang, M. Lambros, P. W. Gout, P. W. Kalivas, A. Massie, I. Smolders, A. Methner, M. Pergande, S. B. Smith, V. Ganapathy and P. Maher, *Antioxid. Redox Signaling*, 2013, **18**, 522.

49. E. Olm, A. P. Fernandes, C. Hebert, A. K. Rundlöf, E. H. Larsen, O. Danielsson and M. Björnstedt, *Proc. Natl. Acad. Sci. U. S. A.*, 2009, **106**, 11400.
50. V. M. Labunskyy, D. L. Hatfield and V. N. Gladyshev, *Physiol. Rev.*, 2014, **94**, 739.
51. D. M. Freudendahl, S. Santoro, S. A. Shahzad, C. Santi and T. Wirth, *Angew. Chem.*, 2009, **48**, 8409.
52. E. R. T. Tiekink, *Dalton Trans.*, 2012, **41**, 6390.
53. M. Wallenberg, S. Misra, A. M. Wasik, C. Marzano, M. Bjornstedt, V. Gandin and A. P. Fernandes, *J. Cell. Mol. Med.*, 2014, **18**, 671.
54. R. L. Poerschke and P. J. Moos, *Biochem. Pharmacol.*, 2011, **81**, 211.
55. T. Chen and Y. S. Wong, *Cell. Mol. Life Sci.*, 2008, **65**, 2763.
56. C. Fan, J. Chen, Y. Wang, Y. S. Wong, Y. Zhang, W. Zheng, W. Cao and T. Chen, *Free Radical Biol. Med.*, 2013, **65**, 305.
57. A. S. Weisberger and L. G. Suhrland, *Blood*, 1956, **11**, 19.
58. I. Andreadou, B. Van De Water, J. N. Commandeur, F. J. Nagelkerke and N. P. Vermeulen, *Toxicol. Appl. Pharmacol.*, 1996, **141**, 278.
59. K. T. Suzuki, K. Kurasaki and N. Suzuki, *Biochim. Biophys. Acta*, 2007, **1770**, 1053.
60. M. Suzuki, M. Endo, F. Shinohara, S. Echigo and H. Rikiishi, *Cancer Chemother. Pharmacol.*, 2011, **67**, 1129.
61. L. Schroterova, V. Kralova, A. Voracova, P. Haskova, E. Rudolf and M. Cervinka, *Toxicol. In Vitro*, 2009, **23**, 1406.
62. J. K. Yeo, S. D. Cha, C. H. Cho, S. P. Kim, J. W. Cho, W. K. Baek, M. H. Suh, T. K. Kwon, J. W. Park and S. I. Suh, *Cancer Lett.*, 2002, **182**, 83.
63. B. C. Jang, E. S. Choi, K. J. Im, W. K. Baek, T. K. Kwon, M. H. Suh, S. P. Kim, J. W. Park and S. I. Suh, *Int. J. Mol. Med.*, 2003, **12**, 733.
64. T. Kim, U. Jung, D. Y. Cho and A. S. Chung, *Carcinogenesis*, 2001, **22**, 559.
65. Z. Li, L. Carrier, A. Belame, A. Thiyagarajah, V. A. Salvo, M. E. Burow and B. G. Rowan, *Breast Cancer Res. Treat.*, 2009, **118**, 33.
66. S. Cao, F. A. Durrani and Y. M. Rustum, *Clin. Cancer Res.*, 2004, **10**, 2561.
67. S. Cao, F. A. Durrani, K. Toth and Y. M. Rustum, *Br. J. Cancer*, 2014, **110**, 1733.
68. A. Bhattacharya, S. G. Turowski, I. D. San Martin, A. Rajput, Y. M. Rustum, R. M. Hoffman and M. Seshadri, *Anticancer Res.*, 2011, **31**, 387.
69. A. Bhattacharya, K. Toth, A. Sen, M. Seshadri, S. Cao, F. A. Durrani, E. Faber, E. A. Repasky and Y. M. Rustum, *Clin. Colorectal Cancer*, 2009, **8**, 155.
70. K. T. Suzuki, Y. Tsuji, Y. Ohta and N. Suzuki, *Toxicol. Appl. Pharmacol.*, 2008, **227**, 76.
71. R. L. Poerschke, M. R. Franklin and P. J. Moos, *Toxicol. In Vitro*, 2008, **22**, 1761.

72. S. O. Lee, J. Yeon Chun, N. Nadiminty, D. L. Trump, C. Ip, Y. Dong and A. C. Gao, *Prostate*, 2006, **66**, 1070.
73. G. X. Li, H. Hu, C. Jiang, T. Schuster and J. Lu, *Int. J. Cancer*, 2007, **120**, 2034.
74. C. Jiang, Z. Wang, H. Ganther and J. Lu, *Cancer Res.*, 2001, **61**, 3062.
75. C. Jiang, Z. Wang, H. Ganther and J. Lu, *Mol. Cancer Ther.*, 2002, **1**, 1059.
76. U. Singh, K. Null and R. Sinha, *Mol. Nutr. Food Res.*, 2008, **52**, 1281.
77. L. Wang, M. J. Bonorden, G. X. Li, H. J. Lee, H. Hu, Y. Zhang, J. D. Liao, M. P. Cleary and J. Lu, *Cancer Prev. Res.*, 2009, **2**, 484.
78. Y. Qi, X. Fu, Z. Xiong, H. Zhang, S. M. Hill, B. G. Rowan and Y. Dong, *PLos One*, 2012, **7**, E31539.
79. H. Hu, C. Jiang, C. Ip, Y. M. Rustum and J. Lu, *Clin. Cancer Res.*, 2005, **11**, 2379.
80. S. Gromer and J. H. Gross, *J. Biol. Chem.*, 2002, **277**, 9701.
81. C. Ip, H. J. Thompson, Z. Zhu and H. E. Ganther, *Cancer Res.*, 2000, **60**, 2882.
82. H. Hu, C. Jiang, G. Li and J. Lu, *Carcinogenesis*, 2005, **26**, 1374.
83. Y. Wu, K. Zu, M. A. Warren, P. K. Wallace and C. Ip, *Mol. Cancer Ther.*, 2006, **5**, 246.
84. I. Sinha, J. E. Allen, J. T. Pinto and R. Sinha, *Cancer Med.*, 2014, **3**, 252.
85. K. Zu, T. Bihani, A. Lin, Y. M. Park, K. Mori and C. Ip, *Oncogene*, 2006, **25**, 546.
86. D. Y. Cho, U. Jung and A. S. Chung, *Biochem. Mol. Biol. Int.*, 1999, **47**, 781.
87. J. Lanfear, J. Fleming, L. Wu, G. Webster and P. R. Harrison, *Carcinogenesis*, 1994, **15**, 1387.
88. A. Ghose, J. Fleming, K. El-Bayoumy and P. R. Harrison, *Cancer Res.*, 2001, **61**, 7479.
89. D. Plano, C. Sanmartin, E. Moreno, C. Prior, A. Calvo and J. A. Palop, *Bioorg. Med. Chem. Lett.*, 2007, **17**, 6853.
90. E. Ibanez, A. Agliano, C. Prior, P. Nguewa, M. Redrado, I. Gonzalez-Zubeldia, D. Plano, J. A. Palop, C. Sanmartin and A. Calvo, *Curr. Med. Chem.*, 2012, **19**, 3031.
91. E. Moreno, D. Plano, I. Lamberto, M. Font, I. Encio, J. A. Palop and C. Sanmartin, *Eur. J. Med. Chem.*, 2012, **47**, 283.
92. D. Caeran Bueno, D. F. Meinerz, J. Allebrandt, E. P. Waczuk, D. B. Dos Santos, D. O. Mariano and J. B. Rocha, *BioMed Res. Int.*, 2013, 537279.
93. T. Posser, M. T. De Paula, J. L. Franco, R. B. Leal and J. B. Da Rocha, *Arch. Toxicol.*, 2011, **85**, 645.
94. F. Nedel, V. F. Campos, D. Alves, A. J. Mcbride, O. A. Dellagostin, T. Collares, L. Savegnago and F. K. Seixas, *Life Sci.*, 2012, **91**, 345.
95. A. S. De Freitas and J. B. Rocha, *Neurosci. Lett.*, 2011, **503**, 1.
96. S. Shabaan, L. A. Ba, M. Abbas, T. Burkholz, A. Denkert, A. Gohr, L. A. Wessjohann, F. Sasse, W. Weber and C. Jacob, *Chem. Commun.*, 2009, 4702.

97. S. Shaaban, A. Negm, M. A. Sobh and L. A. Wessjohann, *Eur. J. Med. Chem.*, 2015, **97**, 190.
98. N. Abbas, S. Shaaban and E. Abdel-Latif, *Res. J. Pharm., Biol. Chem. Sci.*, 2015, **6**, 1655.
99. S. Shaaban, E. H. Gaffer, M. Alshahd and S. S. Elmorsy, *Int. J. Res. Dev. Pharm. Life Sci.*, 2015, **4**, 1654.
100. S. Shaaban, E. H. Gaffer, Y. Jabar and S. S. Elmorsy, *Int. J. Pharm.*, 2015, **5**, 738.
101. S. Shaaban, A. Negm, A. M. Ashmawy, D. M. Ahmed and L. A. Wessjohann, *Eur. J. Med. Chem.*, 2016, **122**, 55.
102. A. K. Sharma, A. Sharma, D. Desai, S. V. Madhunapantula, S. J. Huh, G. P. Robertson and S. Amin, *J. Med. Chem.*, 2008, **51**, 7820.
103. A. Sharma, A. K. Sharma, S. V. Madhunapantula, D. Desai, S. J. Huh, P. Mosca, S. Amin and G. P. Robertson, *Clin. Cancer Res.*, 2009, **15**, 1674.
104. M. A. Crampsie, M. K. Pandey, D. Desai, J. Spallholz, S. Amin and A. K. Sharma, *Chem. -Biol. Interact.*, 2012, **200**, 28.
105. P. Chakraborty, S. S. Roy, U. H. Sk and S. Bhattacharya, *Free Radical Res.*, 2011, **45**, 177.
106. J. K. Das, S. Sarkar, S. U. Hossain, P. Chakraborty, R. K. Das and S. Bhattacharya, *Indian J. Med. Res.*, 2013, **137**, 1163.
107. P. Chakraborty, S. S. Roy and S. Bhattacharya, *Anti-Cancer Agents Med. Chem.*, 2015, **15**, 501.
108. S. S. Roy, P. Chakraborty, J. Biswas and S. Bhattacharya, *Biochimie*, 2014, **105**, 137.
109. P. Ghosh, S. Singha Roy, A. Basu, A. Bhattacharjee and S. Bhattacharya, *Free Radical Res.*, 2015, **49**, 453.
110. S. K. Lee, Y. H. Heo, V. E. Steele and J. M. Pezzuto, *Anticancer Res.*, 2002, **22**, 97.
111. N. D. Facompre, K. El-Bayoumy, Y. W. Sun, J. T. Pinto and R. Sinha, *Cancer Prev. Res.*, 2010, **3**, 975.
112. N. D. Facompre, I. Sinha, K. El-Bayoumy, J. T. Pinto and R. Sinha, *Int. J. Cancer*, 2012, **131**, 2134.
113. G. Krishnegowda, A. S. Prakasha Gowda, H. R. Tagaram, K. F. Carroll, R. B. Irby, A. K. Sharma and S. Amin, *Bioorg. Med. Chem.*, 2011, **19**, 6006.
114. E. Dominguez-Alvarez, D. Plano, M. Font, A. Calvo, C. Prior, C. Jacob, J. A. Palop and C. Sanmartin, *Eur. J. Med. Chem.*, 2014, **73**, 153.
115. E. Dominguez-Alvarez, M. Gajdacs, G. Spengler, J. A. Palop, Marc, K. Kiec-Kononowicz, L. Amaral, J. Molnar, C. Jacob, J. Handzlik and C. Sanmartin, *Bioorg. Med. Chem. Lett.*, 2016, **26**, 2821.
116. M. Gajdacs, G. Spengler, C. Sanmartin, M. A. Marc, J. Handzlik and E. Dominguez-Alvarez, *Bioorg. Med. Chem. Lett.*, 2017, **27**, 797.
117. P. Guo, Q. Wang, J. Liu, L. Liu, P. Zhao, Y. Cao, Y. Liu and C. Qi, *Biol. Trace Elem. Res.*, 2013, **154**, 304.

118. X. Wu, Y. Cao, J. Zhang, M. Lei, X. Deng, K. R. Zahid, Y. Liu, K. Liu, J. Yang, G. Xiong, H. Yao and C. Qi, *Biotechnol. Lett.*, 2016, **38**, 761.
119. M. Lei, D. Chen, X. Deng, J. Liu, L. Chen, Y. Liu, B. Li, H. Yao, G. Xiong, Y. Cao, J. Yang and C. Qi, *Biotechnol. Lett.*, 2012, **34**, 1617.
120. P. Guo, P. Zhao, J. Liu, H. Ma, J. Bai, Y. Cao, Y. Liu, H. He and C. Qi, *Biol. Trace Elem. Res.*, 2013, **151**, 301.
121. F. Zhao, J. Yan, S. Deng, L. Lan, F. He, B. Kuang and H. Zeng, A Thioredoxin Reductase Inhibitor Induces Growth Inhibition And Apoptosis In Five Cultured Human Carcinoma Cell Lines, *Cancer Lett.*, 2006, **236**, 46.
122. S. Urig and K. Becker, *Semin. Cancer Biol.*, 2006, **16**, 452.
123. C. Shi, L. Yu, F. Yang, J. Yan and H. Zeng, *Biochem. Biophys. Res. Commun.*, 2003, **309**, 578.
124. L. Wang, Z. Yang, J. Fu, H. Yin, K. Xiong, Q. Tan, H. Jin, J. Li, T. Wang, W. Tang, J. Yin, G. Cai, M. Liu, S. Kehr, K. Becker and H. Zeng, *Free Radical Biol. Med.*, 2012, **52**, 898.
125. C. J. Shi, H. H. Zeng, H. W. Li, F. G. Yang, X. Q. Wu and L. Z. Yu, *Zhonghua Yi Xue Za Zhi*, 2003, **83**, 1984.
126. F. Xing, S. Li, X. Ge, C. Wang, H. Zeng, D. Li and L. Dong, *Oral Oncol.*, 2008, **44**, 963.
127. Q. Tan, X. Y. Ren, J. Li, F. Wang, S. J. Deng, H. H. Zeng and Ai Zheng, *Chin. J. Cancer*, 2009, **28**, 472.
128. H. Y. Zhou, Z. Y. Meng, G. F. Dou, J. L. Ma, Y. Q. Lou and G. L. Zhang, *Yao Xue Xue Bao*, 2010, **45**, 627.
129. X. Zheng, Y. Zhang, L. Zhang, W. Xu, W. Ma, R. Sun and H. Zeng, *Biomed. Pharmacother.*, 2016, **83**, 212.
130. M. Selenius, M. Hedman, D. Brodin, V. Gandin, M. P. Rigobello, J. Flygare, C. Marzano, A. Bindoli, O. Brodin, M. Bjornstedt and A. P. Fernandes, *J. Cell. Mol. Med.*, 2012, **16**, 1593.
131. L. Wang, J. N. Fu, J. Y. Wang, C. J. Jin, X. Y. Ren, Q. Tan, J. Li, H. W. Yin, K. Xiong, T. Y. Wang, X. M. Liu and H. H. Zeng, Selenium-Containing Thioredoxin Reductase Inhibitor Ethaselen Sensitizes Non-Small Cell Lung Cancer To Radiotherapy, *Anti-Cancer Drugs*, 2011, **22**, 732.
132. J. Liu and Z. Wang, *Oxid. Med. Cell. Longevity*, 2015, **2015**, 294303.
133. X. Fu, S. Li, F. Jing, X. Wang, B. Li, J. Zhao, Y. Liu and B. Chen, *Med. Chem.*, 2016, **12**, 631.
134. B. L. Li, B. Li, R. L. Zhang, J. J. Zhao, X. F. Wang, Y. M. Liu, Y. P. Shi, J. B. Liu and B. Q. Chen, *Bioorg. Med. Chem. Lett.*, 2016, **26**, 1279.
135. F. Jing, X. Fu, S. Li, B. Li, J. Zhao, X. Wang, Y. Liu and B. Chen, *Chem. Pharm. Bull.*, 2015, **63**, 431.
136. S. Ye, X. Zheng, T. Hu and H. Zeng, *Cell. Mol. Biol.*, 2016, **62**, 6.
137. R. Zhao, H. Masayasu and A. Holmgren, *Proc. Natl. Acad. Sci. U. S. A.*, 2002, **99**, 8579.
138. L. Zhang, L. Zhou, J. Du, M. Li, C. Qian, Y. Cheng, Y. Peng, J. Xie and D. Wang, Induction Of Apoptosis In Human Multiple Myeloma Cell

Lines By Ebselen Via Enhancing The Endogenous Reactive Oxygen Species Production, *BioMed Res. Int.*, 2014, **2014**, 696107.
139. P. D. Hanavan, C. R. Borges, B. A. Katchman, D. O. Faigel, T. H. Ho, C. T. Ma, E. A. Sergienko, N. Meurice, J. L. Petit and D. F. Lake, *Oncotarget*, 2015, **6**, 18418.
140. A. J. Pacula, K. B. Kaczor, J. Antosiewicz, A. Janecka, A. Dlugosz, T. Janecki, A. Wojtczak and J. Scianowski, *Molecules*, 2017, 22.
141. J. Yan, Y. Guo, Y. Wang, F. Mao, L. Huang and X. Li, *Eur. J. Med. Chem.*, 2015, **95**, 220.
142. S. H. Juang, C. C. Lung, P. C. Hsu, K. S. Hsu, Y. C. Li, P. C. Hong, H. S. Shiah, C. C. Kuo, C. W. Huang, Y. C. Wang, L. Huang, T. S. Chen, S. F. Chen, K. C. Fu, C. L. Hsu, M. J. Lin, C. J. Chang, C. L. Ashendel, T. C. Chan, K. M. Chou and J. Y. Chang, *Mol. Cancer Ther.*, 2007, **6**, 193.
143. H. S. Shiah, W. S. Lee, S. H. Juang, P. C. Hong, C. C. Lung, C. J. Chang, K. M. Chou and J. Y. Chang, *Biochem. Pharmacol.*, 2007, **73**, 610.
144. Y. N. Yang, K. M. Chou, W. Y. Pan, Y. W. Chen, T. C. Tsou, S. C. Yeh, C. H. Cheung, L. T. Chen and J. Y. Chang, *Cancer Lett.*, 2011, **309**, 110.
145. P. Arsenyan, K. Rubina, I. Shestakova and I. Domracheva, *Eur. J. Med. Chem.*, 2007, **42**, 635.
146. T. Chen, W. Zheng, Y. S. Wong and F. Yang, *Biomed. Pharmacother.*, 2008, **62**, 77.
147. T. Chen, Y. S. Wong, W. Zheng and J. Liu, *Chem. -Biol. Interact.*, 2009, **180**, 54–60.
148. Y. Zhang, S. Zheng, S. M. Ngai, W. Zheng, J. Li, T. Chen and X. Zhong, *Chem. Pharm. Bull.*, 2014, **62**, 994.
149. P. C. Srivastava and R. K. Robins, *J. Med. Chem.*, 1983, **26**, 445.
150. T. J. Boritzki, D. A. Berry, J. A. Besserer, P. D. Cook, D. W. Fry, W. R. Leopold and R. C. Jackson, *Biochem. Pharmacol.*, 1985, **34**, 1109.
151. D. L. Lucas, R. K. Robins, R. D. Knight and D. G. Wright, *Biochem. Biophys. Res. Commun.*, 1983, **115**, 971.
152. G. Gebeyehu, V. E. Marquez, A. Van Cott, D. A. Cooney, J. A. Kelley, H. N. Jayaram, G. S. Ahluwalia, R. L. Dion, Y. A. Wilson and D. G. Johns, *J. Med. Chem.*, 1985, **28**, 99.
153. P. Franchetti, L. Cappellacci, G. A. Sheikha, H. N. Jayaram, V. V. Gurudutt, T. Sint, B. P. Schneider, W. D. Jones, B. M. Goldstein, G. Perra, A. De Montis, A. G. Loi, P. La Colla and M. Grifantini, *J. Med. Chem.*, 1997, **40**, 1731.
154. M. Koketsu, H. Ishihara, W. Wu, K. Murakami and I. Saiki, *Eur. J. Pharm. Sci.*, 1999, **9**, 157.
155. W. Wu, K. Murakami, M. Koketsu, Y. Yamada and I. Saiki, *Anticancer Res.*, 1999, **19**, 5375.
156. M. Koketsu, H. O. Yang, Y. M. Kim, M. Ichihashi and H. Ishihara, *Org. Lett.*, 2001, **3**, 1705.
157. X. Bi, N. Pohl, H. Dong and W. Yang, *J. Hematol. Oncol.*, 2013, **6**, 8.

158. R. Gowda, S. V. Madhunapantula, D. Desai, S. Amin and G. P. Robertson, *Mol. Cancer Ther.*, 2013, **12**, 3.
159. D. Plano, D. N. Karelia, M. K. Pandey, J. E. Spallholz, S. Amin and A. K. Sharma, *J. Med. Chem.*, 2016, **59**, 1946.
160. D. Desai, U. Salli, K. E. Vrana and S. Amin, *Bioorg. Med. Chem. Lett.*, 2010, **20**, 2044.
161. N. Karelia, D. Desai, J. A. Hengst, S. Amin, S. V. Rudrabhatla and J. Yun, *Bioorg. Med. Chem. Lett.*, 2010, **20**, 6816.
162. R. Gowda, S. V. Madhunapantula, D. Desai, S. Amin and G. P. Robertson, *Cancer Biol. Ther.*, 2012, **13**, 756.
163. J. N. Commandeur, I. Andreadou, M. Rooseboom, M. Out, L. J. De Leur, E. Groot and N. P. Vermeulen, *J. Pharmacol. Exp. Ther.*, 2000, **294**, 753.
164. J. I. Lee, H. Nian, A. J. Cooper, R. Sinha, J. Dai, W. H. Bisson, R. H. Dashwood and J. T. Pinto, *Cancer Prev. Res.*, 2009, **2**, 683.

CHAPTER 16

Cancer Prevention by Different Forms of Selenium

A. M. DIAMOND

University of Illinois at Chicago, Department of Pathology, 840 South Wood Street, Chicago, Illinois 60612, USA
Email: adiamond@uic.edu

16.1 Introduction

Selenium has long been considered for its potential in the prevention or treatment of cancer. The role of selenium as an essential component of the diet stemmed from the work of Klaus Schwartz in the 1950s when he obtained data indicating that rats maintained on a casein-only diet and consequently developing liver cirrhosis could be protected by the inclusion of a new factor that was determined to be selenium. However, the focus on selenium as an agent to prevent cancer did not follow until many years later, when an epidemiological study reported that there was an inverse association between selenium dietary intake and cancer mortality in different countries, as well as between regions within individual nations.[1] Among the lines of research that followed was a large effort to investigate the efficacy of selenium in cancer prevention using animal models. An impressive number of studies have now been published indicating that low-level, non-toxic levels of selenium provided in the diet of laboratory animals was capable of reducing both spontaneous and carcinogen-induced cancers, and did so in a wide variety of organs and tissue types. During this same time period, selenium was discovered to be an essential component of dozens of proteins in species across the phylogenetic tree, being inserted co-translationally as

Organoselenium Compounds in Biology and Medicine: Synthesis, Biological and Therapeutic Treatments
Edited by Vimal Kumar Jain and K. Indira Priyadarsini

Published by the Royal Society of Chemistry, www.rsc.org

the amino acid selenocysteine in response to UGA codons that would otherwise signal the termination of translation. The mechanisms used by various species to direct the insertion of selenocysteine in response to certain UGA codons is not discussed here, but readers are referred to several reviews on this topic.[2–4] As the list of proteins that fell into this category was established, including 25 human and 24 mouse selenoproteins,[5] it became apparent that many selenoproteins possess anti-oxidant enzyme activities that might account for the cancer-reducing properties of selenium.[6,7] Enthusiasm for the development of the use of selenium in cancer prevention was further sparked by the results of a selenium supplementation trial in which patients who were at risk of developing skin cancers were provided either a low dose of selenium or placebo, and subsequently monitored for the recurrence of their disease.[8] While the results of this trial did not indicate any benefit to the selenium-supplemented group in preventing skin cancer, there appeared to be a reduction in the risk of other common forms of cancer, those of the colon, lung and prostate, in the selenium-supplemented arm of that study. With promising results from cell culture, animal and human research, the National Institutes of Health embarked on the largest prostate cancer prevention trial ever conducted, the Selenium and Vitamin E Cancer Prevention Trial (SELECT). SELECT was terminated early due to the increased risk of prostate cancer in the vitamin E arm of the study and an apparent lack of benefit of selenium supplementation.[9] There have been several commentaries providing opinions as to why selenium was not effective in SELECT, with several of these suggesting that the form of selenium used may have been a factor: the previous study from 1997 used a selenized yeast supplement, while SELECT used selenomethionine. A timeline ranging from the discovery of selenium 100 years ago to the current efforts to evaluate whether selenium supplements can prevent cancer is presented in Figure 16.1. In this chapter,

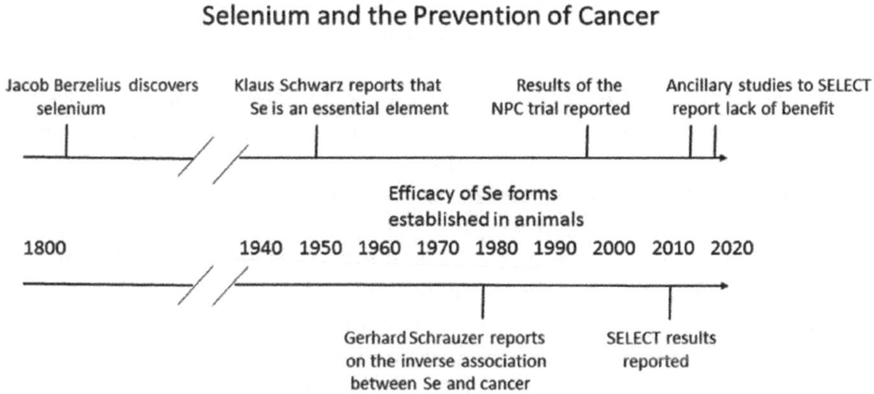

Figure 16.1 Timeline from the discovery of selenium through results from human clinical studies assessing its efficacy in preventing cancer. NPC: Nutritional Prevention of Cancer; SELECT: Selenium and Vitamin E Cancer Prevention Trial.

the differences in the forms of selenium used for cancer prevention research are discussed, as this relates to likely molecular and clinical outcomes.

16.2 Does Selenium Prevent Cancer?

If we are to consider the efficacy of different forms of selenium with regard to cancer prevention, it is necessary to first address whether selenium can in fact prevent the disease. This requires some consideration of what the term "prevention" means in the context of this chapter. The term "prevention" is often used in reference to different processes that can be difficult to distinguish. Cancer progression can refer to the conversion of a normal cell to one that is evolving along the path to malignancy. In this context, prevention is focused upon the elimination or reduction of those molecular events that initiate the accumulation of changes that cause the transformation to occur. Prevention may also refer to a slowing of the process by which a premalignant cell, one that has already accumulated some phenotypic changes, progresses to a more clinically significant state. For example, colonic polyps are considered a preneoplastic lesion, which, if arrested in that condition or stimulated to either regress or be eliminated, would fall into this category. The prevention of a tumor cell to acquire the necessary characteristics to disseminate throughout the body, avoid the surveillance of the host immune system and seed and grow in a distant site could also fit in this classification. The last category to consider is the circumstance where an agent causes the regression or elimination of the cancer itself, as would be the case when the administration of a substance or treatment causes an existing fully malignant cancer to regress, for example when a compound is either specifically toxic to cancer cells or more generally toxic to more rapidly dividing cells that often comprise the tumor. There has been extensive research demonstrating the toxicity of organic and inorganic forms of selenium towards a wide variety of cancer cells[10] and the therapeutic potential of selenium is not be considered here in any detail.

The interest in selenium for the prevention of cancer was sparked by the publication of Schrauzer *et al.* in 1977 indicating an inverse association between the dietary intake of selenium and cancer incidence.[1] Almost 40 years later, a comprehensive meta-analysis was published by Vinceti *et al.*, who compiled the data from 55 prospective observational studies that included data from more than one million study participants.[11] The results of their analysis indicated that there was indeed an inverse association between selenium status and both cancer incidence and death from the disease, with this association being strongest for cancers of the prostate, stomach and bladder. Further support for the ability of selenium to prevent cancer has come from a large number of studies using animal models where rodents were protected to varying degrees from developing cancers following carcinogenic exposures such as hormones, chemicals or radiation due to supplementation with selenium. Most of these studies examined the ability of selenium to prevent the formation of tumors following the injection of tumor cells in xenograft models where human cancer cells are injected into immunocompromised mice. Enthusiasm for the use of selenium as a supplement for preventing cancer was reduced given the negative results of SELECT, which indicated no benefit from selenium supplementation for the reduction of prostate cancer incidence, or other cancers evaluated in ancillary analyses examining whether providing selenium reduced the incidence of other cancer types. Vinceti *et al.* also examined the results from eight randomized controlled trials using cancer as an endpoint and concluded that there was no consistent benefit from selenium supplementation, and in some cases there may have been an increased risk.[11] The apparently conflicting results from observational/animal studies and supplementation trials raises the possibility that the chemical form of selenium accounts for the different effects obtained.

16.3 Mechanisms by which Selenium can Prevent Cancer

16.3.1 *In Vitro* Studies

Examining the biological effects of various selenium compounds using cultured cells offers several advantages, including (1) the rapid detection of molecular changes, (2) limited variables such as the impact of absorption and excretion, (3) a cost-efficient examination of the consequences of cell type specificity and (4) the ability to examine a variety of forms of selenium on the same cell population. The vast majority of such studies have used immortal and transformed cells derived from human cancers,[10] and cytotoxic or cytostatic effects on these cells can be attributed to the pro-oxidant capacity of these compounds or direct effects of signaling pathways that are altered in cancer cells to establish their transformed phenotype. Examples abound of such effects of selenium, and many have been characterized as pro-oxidant effects resulting in the production of reactive oxygen species (ROS) or thiol modification which impacts signal transduction or chromatin binding and modification. There have been some reports of common effects on transformed and non-transformed cells exposed to selenium compounds. For example, selenite can attenuate the expression of the anti-apoptotic protein BCl-2 while increasing the levels of the p53 tumor suppressor in immortal human hepatic cells and a hepatoma cell line.[12] However, work done with tumor cells may have limited utility in identifying selenium-influenced pathways and processes that are also common to "normal cells" and contribute to cancer prevention.

Experimental evidence from studies using mammalian cells in culture have identified many possible target proteins and molecular pathways that might account for the chemopreventive properties of selenium, as well as differential effects depending on the chemical form administered. The most common forms of selenium used for culture studies are the inorganic forms selenite and selenate, and the less toxic selenomethionine (SeMet), the major chemical form in foods. Some studies have also investigated the efficacy of selenomethylselenocysteine (MSC), a major constituent of selenium-enriched garlic,[13] a form shown to be chemopreventive in animals.[14]

16.3.2 Effects of Chemical Forms of Selenium on Selenoprotein Levels

A likely major effector of the biological effects of selenium is the consequences of selenium availability on the levels and activity of selenoproteins. There are 25 human selenium-containing proteins encoded by mRNAs where the UGA codon directs the translational machinery to insert a selenocysteine into the elongating peptide.[5] Many of the biological consequences of selenium consumption can be attributed to its regulation of many of these proteins, and this is also likely to be true for the impact of selenium on cancer risk. There have been several excellent reviews that summarize the known functions of the 25 human selenoproteins[6,7,15] and many of these proteins have anti-oxidant properties, such the members of the glutathione peroxidase (GPX) and thioredoxin reductase (TXRND) families of proteins. The enzymatic activity of these proteins could potentially reduce the incidence of oxidative lesions in DNA that could be mutagenic, or attenuate ROS signaling pathways that contribute to carcinogenesis. Several of these selenoproteins have been implicated in contributing to cancer risk due to the association of specific polymorphisms in the corresponding genes for these proteins and cancer incidence.[16–18] All of the forms of selenium obtained from foods, be they organic or inorganic, can funnel into the selenoprotein synthesis pathway *via* the eventual conversion to selenide.[19] The metabolism of ingested selenium has been reviewed[20,21] as have the unique steps that result in the incorporation of selenium as selenocysteine into the elongating selenoprotein.[2,6] Regulation of selenoprotein levels by selenium is different for each of these proteins, with some being more responsive than others to varying selenium availability. The impact of selenium on selenoprotein levels is regulated at both the cellular and tissue levels, and this has been referred to as the "selenoprotein hierarchy", where, for example, the levels of GPX are very sensitive to selenium status while TXRND is not.[3,20] In spite of the common convergence of these pathways, the different forms of selenium provided to the media of cultured mammalian cells often result in different cellular responses with regard to molecular changes and phenotype.

Selenoprotein levels are also differentially responsive to various forms of selenium, and there are very distinct responses. For example, human Caco-2 colon carcinoma-derived cells differentially transport and take up seleno compounds with different efficiencies, with SeMet, MSC and digested selenium-enriched yeast being transported more efficiently than selenite.[22,23]

Using the ability of added seleno compounds to increase the levels of GPX1 to investigate the efficiency of different selenium forms to stimulate selenoprotein synthesis, lower levels of selenite can better increase GPX1 levels as compared to SeMet, and do so in a shorter period of time.[23–25]

A comprehensive evaluation of the efficacy of different forms of selenium to induce several selenoproteins was conducted in culture using HepG2 human liver carcinoma and HT-29 colon adenocarcinoma-derived cell lines, as well as immortalized colon cells derived from mice.[26] The forms of selenium evaluated were selenite, SeMet, methylseleninic acid (MSA) and MSC and the four selenoproteins evaluated were GPX1, SELENOW (SepW1), SELENOH (SelH) and SELENOM (SelM). In this study, selenite and MSA were the most efficient in increasing the levels of the selected proteins, with stronger effects being seen in the immortalized cells as compared to the tumor cell lines. As the authors concluded, these data support other published reports in concluding that ability of selenium to enhance selenoprotein levels depends on the form of selenium used, the particular cells being evaluated and the individual protein being examined.[26]

Smith *et al.* compared the effects of different forms of selenium on the phosphorylation of p53, a tumor suppressor whose inactivation represents one of the most common mutations that occur in most cancer types, using human lung carcinoma cells.[27] In addition to selenite and SeMet, the efficacy of MSA, a derivative seleno compound that can only minimally be metabolized for use in selenoprotein synthesis[28] was evaluated. In these studies MSA resulted in phosphorylation of p53 at threonine residues, selenite caused phosphorylation at serine residues while SeMet did not induce the phosphorylation of any amino acids in p53.[27] The difference in the post-translational modification of p53 observed after exposure of cells to the different forms of selenium may account for the differences in the ability of these forms to induce apoptosis.[29] Other processes that might contribute to the chemopreventive properties of selenium and may respond differently to selenium compounds include an impact on immunity, the cell cycle, cell motility and apoptosis.[30]

16.3.3 Selenium and the Repair of DNA Damage

One potential process affected by selenium that might be particularly relevant to cancer prevention is its impact on the levels of DNA damage. This might occur, for example, if selenium stimulates the production of antioxidant selenoproteins such as the GPXs that would reduce the levels of potentially mutagenic reactive oxygen species. Alternatively, selenium might impact signaling pathways that result in increased ability for the cellular DNA repair machinery to repair existing damage. The ability of selenium to reduce DNA damage might provide an explanation as to why higher selenium intake in the diet over the course of a lifetime may be beneficial, whereas providing selenium supplements to older individuals who have likely accumulated many of the necessary genetic changes for cancers to form is less effective.[31] Reports of the ability of selenium to enhance DNA damage repair using cultured cells have indicated the stimulation of several different repair pathways, including the repair of oxidative damage,[32] DNA mismatch repair[33] and base excision repair.[34,35] The mechanisms by which selenium might reduce genetic damage include the stimulation of DNA-repair associated proteins such as p53 and Brca1,[36] hMLH1 and ATM,[33] Gadd45, APE1/REF1,[34] H2AX and Chk1[37] and Fpg.[35] It can be difficult to distinguish activities associated with the prevention of DNA damage, perhaps due to the induction of anti-oxidant selenoproteins, from the repair of existing damage. However, pretreatment of primary mouse embryonic fibroblasts with SeMet resulted in protection from both UV and UV-mimetic drugs[36] and selenium provided either in the form of selenite or SeMet increased the repair of oxidative DNA lesions in LNCaP human prostate cells.[32]

16.3.4 Animal Studies

Enthusiasm for the development of selenium compounds for cancer prevention was greatly stimulated by a large number of studies using rodents to evaluate the effects of selenium supplementation on tumor induction. Experimental evidence frequently indicated the efficacy of various selenium compounds at non-toxic concentrations to reduce the development of several different cancer types following the inoculation of immunocompromised mice with human tumor cells in xenograft models.[38–40] In these studies, various forms of selenium selectively inhibit tumor growth and plausible mechanisms of action include the targeting of oncogenes, the enhancement of DNA repair or detoxification pathways, effects on the control of the cell cycle and epigenetics and the inhibition of angiogenesis, to name a few.[41] Studying the consequences of the form of selenium provided to mice, Zhang *et al.* examined the proteome of tumors that developed in mice due to the injection of PC3 human tumor cells after supplementing their diets with SeMet, selenite, MSA or MSC and described very different signatures for each compound.[42] Similarly, high doses of selenite or SeMet resulted in distinct proteomic signatures in mouse liver.[43] The differences in the efficacy of these different forms and their possible mechanisms of action in the anti-tumor effects of selenium has been reviewed.[21,44,45]

Far less research has been conducted in animal models directly addressing the ability of selenium compounds to prevent cancers from developing from non-cancerous cells, as would be the case in mouse models where the animals are genetically predisposed to cancer. Such studies have been conducted using genetic models of prostate cancer. Using a mouse model for prostate cancer where the animals were engineered to express the SV40 T antigen oncogene in the prostate, monomethylated forms of selenium that are precursors to methylselenol, MSA or MSC were effective in attenuating cancer development.[46,47] Similarly, MSA was shown to prevent cancer in a mouse prostate cancer model where the Pten tumor suppressor was specifically knocked out in the prostate.[48] SeMet supplementation failed to prevent carcinogenesis in a rat model of hormone-induced prostate cancer.[49] Selenium in the form of selenized yeast prevented the development of virus-induced mammary gland tumors[50] while selenite could prevent the spontaneous development of the same tumor type.[51]

Whether or not the chemopreventive properties of selenium in animal models is due to an impact on selenoproteins is controversial. It has been argued that selenoproteins are not involved in the observed protection, as GPX activity in the serum of selenium supplemented animals did not rise, although protection was observed.[30] However, the activity detected in serum may not reflect the impact of supplementation on selenoprotein levels in the various organs, which has not been thoroughly investigated. Data specifically implicating selenoproteins in the prevention of cancer have come from studies using transgenic animals engineered to be devoid of selenoproteins in either the mammary gland or the prostate due to organ-specific deletion of the gene for selenocysteine tRNA, which is essential for all selenoprotein synthesis. In these experiments, global inhibition of *de novo* translation of selenoproteins resulted in enhanced carcinogenesis in the targeted organ.[52,53] These experiments imply a role for selenoproteins in cancer prevention, and indirectly implicate selenium as well. In this regard, different forms of selenium, including SeMet, selenite, selenocysteine or selenized yeast were shown to be similar in their ability to enhance the levels of representative selenoproteins within a variety of tissues in both mice[54] and rats.[55]

16.4 Human Studies

By the 1990s, sufficient preclinical data derived from both cultured cells and animal work encouraged clinical studies to investigate the potential of selenium to prevent cancer. One of the more promising studies involved providing individuals who had previously been treated for skin cancer with a dose of 200 $\mu g\,day^{-1}$ of selenium in the form of selenized yeast and comparing the incidence of recurrence as compared to a control population given a placebo.[8] Selenized yeast was chosen as the form of selenium for its lower toxicity as compared to non-organic forms, and in this supplement, selenium was mostly in the form of SeMet, although detailed analysis of the selenium-containing species indicated several other minor forms, not all of which have been identified.[21] This patient population was selected for the high frequency of recurrence that occurs following skin cancer resection, making a study of cancer incidence feasible in a relatively short term. The trial, called the National Prevention of Cancer (NPC) trial was a double-blind randomized study with 1312 participants provided with either 200 $\mu g\,day^{-1}$ of selenized yeast or placebo, with the primary endpoint being the recurrence of squamous cell carcinoma and total non-melanoma skin cancer. In 1996, the first results were released, including 8 years of follow-up.[8] While no protection was evident regarding the reduction in skin cancers, secondary analyses indicated significant reductions in cancer mortality and incidence, as well as reduced incidence of three of the more prevalent cancer types: lung, colon and prostate. Efficacy of selenium to reduce cancer incidence was affected by the baseline selenium levels of the participants, with men who had the lowest baseline levels being those who benefited, while participants in the selenium group who had the highest baseline either showed no benefit, or in the case of basal and total non-melanoma skin cancer, an increased risk with selenium supplementation.[56–58] There were several weaknesses in this study, including the small numbers of cancers that were recorded and the observed beneficial effects not being in organs included in the primary study design. However, to a large degree due to the encouraging data from the NPC trial, several other placebo-controlled clinical studies were initiated. Two were conducted on men at high risk of developing prostate cancer, either using 200 $\mu g\,day^{-1}$ SeMet[59] or 200 $\mu g\,day^{-1}$ or 400 $\mu g\,day^{-1}$ of selenized yeast.[60] Another study investigated the efficacy of 200 $\mu g\,day^{-1}$ selenized yeast to prevent the recurrence of non-small cell lung cancer in patients who had been treated for their cancer by lung resection.[61] None of these studies indicated a benefit from the selenium supplementation.

The definitive study examining whether selenium supplementation could prevent cancer began with the accrual of participants in 2001 and was named the Selenium and Vitamin E Cancer Prevention Trial (SELECT).[62] The trial contained four arms: selenium at 200 $\mu g\,day^{-1}$, vitamin E at 400 IU day^{-1}, both supplements and placebo randomly provided to >32 000 men aged >55 years or >50 years in the case of African–American men, who are at greater risk of developing and dying from prostate cancer compared to Caucasians. Planning of the form and dosage of selenium to be used preceded the implementation of the study by several years, with SeMet being selected for several reasons.[63] SeMet was considered to be less toxic than selenite and previous data had indicated a better pharmacological profile when considering body retention and the impact on the body's selenium stores.[64] There was consideration of the use of the monomethylated forms of selenium shown to have greater efficacy in animal studies; however there were safety and regulatory concerns that eventually discouraged the panel from recommending these compounds.[63] In these discussions, selenized yeast, the form used in the NPC trial, was not selected, in part due the presence of uncharacterized organic forms of selenium and batch variations. The results of the trial, initially published in 2009[9] and updated in 2011,[65] indicated no reduction in prostate cancer incidence among those men receiving selenium as compared to those in the placebo arm. In this study, men in the vitamin E arm saw a significant increase in prostate cancer incidence and that was not apparent in the group of men receiving selenium in addition to vitamin E. Thus, there was a potential benefit of the selenium supplementation with regard to reducing the increased risk that was observed in the vitamin E-supplemented arm.

16.5 The Impact of SELECT on Evaluating Chemoprevention with Selenium

The results of SELECT, indicating that SeMet was ineffective in preventing prostate cancer, have significantly reduced enthusiasm for the further investigation of the potential of selenium in protecting against cancer. Moreover, ancillary studies of SELECT examining whether selenium was able to reduce the incidence of cancers of organs other than the prostate in the same population were performed, and similarly showed no benefit.[66,67] There have been several thoughtful commentaries offering possible explanations for the lack of efficacy of selenium, especially in the wake of the wealth of preclinical data supporting its development for that purpose. Among the explanations offered by these authors were the dosing, the possibility of reduced efficacy of selenium in a population with higher baseline levels than in the NPC trial and the form of selenium used.[68–73]

The genetics of individuals may impact whether selenium supplementation can reduce prostate cancer incidence and mortality. Statistically significant interactions were observed between selenium and the genotype of several important proteins with regard to prostate cancer risk. NKX3.1 is an androgen-regulated prostate specific homeobox protein considered a prostate cancer tumor suppressor.[74] Using a mouse model in which NKX3.1 was genetically removed, it was shown that these mice exhibited enhanced prostate epithelial cell proliferation when they were provided a with supplement of the anti-oxidant *N*-acetylcysteine.[75] When samples obtained from SELECT were stratified by a functional polymorphism in NKX3.1 associated with lower protein levels, a statistically significant interaction between that polymorphism and selenium supplementation was observed for prostate cancer risk.[76] Additional statistically significant interactions were found for the risk of high-grade prostate cancer between selenium supplementation and polymorphisms in five antioxidant genes: CAT, SOD2, PRDX6, SOD3 and TXNRD2.[77] These data indicate that at least for a subset of individuals identified by genotype, SeMet might be beneficial in cancer prevention.

16.6 Which forms of Selenium Should be Investigated for Cancer Prevention?

The forms of selenium that have most often been used to study cancer prevention are presented in Table 16.1.

There have been numerous efforts to examine the physiological consequences of administering selenium in different forms and in different species, including rats, birds, ruminants and humans, with several of these examining a wide range of parameters in the blood and other tissues. Readers are directed to a recent compilation and discussion of these data.[78] The strongest data to date on the ability of selenium to prevent cancer is

Table 16.1 The most commonly used forms of selenium used to investigate the element's chemopreventive actions.

Selenium compound	Cultured cells	Animal studies	Human trials
Sodium selenite Na_2SeO_3	✓	✓	
Sodium selenite Na_2SeO_4	✓	✓	
Methylseleninic acid (MSA) CH_3SeO_2H	✓	✓	
Se-methylselenocysteine (MSC) $CH_2Se\text{-}CH_2\text{-}CH(NH_2)\text{-}COOH$	✓	✓	
Selenomethionine (SeMet) $CH_2\text{-}Se\text{-}CH_2\text{-}CH_2\text{-}CH(NH_2)\text{-}COOH$	✓	✓	✓
Selenized yeast	✓	✓	✓

derived from observational studies indicating likely effects of dietary selenium on prostate cancer risk, and the study by Bostwick and co-workers indicating that supplementing the diet of dogs (the only animals to spontaneously develop prostate cancer) with selenium in the forms of SeMet or selenized yeast failed to yield differences in toenail or prostatic selenium levels, nor differences in multiple markers associated with the risk of prostate cancer.[79] In addition, a central tenet of cancer prevention is that any intervention should do no harm, as it will be administered to healthy individuals, or at least there should be minimal chances of harmful side effects if the targeted population is at an elevated risk of disease. Therefore, the use of non-organic forms of selenium such as selenite and selenate with potential toxicity, or less characterized methylselenol precursors shown to have the highest efficacy of reducing tumor growth in animal models may be less desirable.

16.7 Conclusions

The accumulated data indicating the potential of selenium as a chemopreventive agent were derived from a large number of animal and cell culture studies, as well as human observational studies indicating an inverse association between dietary selenium intake and cancer incidence. These results ultimately led to human supplementation trials that have not established a benefit of providing individuals with a selenium supplement as a means to protect against cancer. One possible explanation for the apparent discrepancy between the preclinical and supplementation results could lie in the particular form of the selenium used. Initially, promising results were obtained providing selenium in the form of selenized yeast, but the enthusiasm for using selenized yeast to prevent cancer diminished greatly when additional studies failed to detect any benefits. Moreover, SELECT represented a huge and comprehensive endeavor, and the results of this trial not only indicated that SeMet did not reduce the incidence of prostate cancer, but failed to reduce the incidence of other cancer types. While many have suggested that different forms of selenium should be investigated in human trials, it should be considered that there are fundamental differences between providing older adults with a selenium supplement and consuming relatively higher levels of selenium in the diet over the course of a lifetime where the current data generally support a benefit. A plausible explanation remains that the chemical form of selenium is not critical in humans as long as the selenium consumed can channel into the selenoprotein biosynthesis pathway, but these benefits need to be realized over a much longer period of time than could be achieved by giving older individuals the supplement. Mechanisms of protection by selenium compounds that would be consistent with both the observational and supplementation data would include the reduction of DNA damage that could be achieved by reduced oxidative stress or enhanced repair. Additional research into both the mechanism of protection seen in model systems, as well the impact of the dietary selenium levels at the molecular level, will be required to determine whether selenium, either as a supplement or in the diet can be used for cancer prevention.

Acknowledgements

AMD's research in this area is supported by grants from the National Institutes of Health (RO1CA127943, R21CA182103 and a Research Supplement to Promote Diversity in Health-Related Research) and a UIC Cancer Center Pilot Grant. The author would also like to thank Dr Dede Ekoue for the critical reading of the chapter.

References

1. G. N. Schrauzer, D. A. White and C. J. Schneider, *Bioinorg. Chem.*, 1977, **7**, 23.
2. A. L. Bulteau and L. Chavatte, *Antioxid. Redox Signaling*, 2015, **23**, 775.
3. D. M. Driscoll and P. R. Copeland, *Annu. Rev. Nutr.*, 2003, **23**, 17.
4. L. V. Papp, J. Lu, A. Holmgren and K. K. Khanna, *Antioxid. Redox Signaling*, 2007, **9**, 775.
5. G. V. Kryukov, S. Castellano, S. V. Novoselov, A. V. Lobanov, O. Zehtab, R. Guigo and V. N. Gladyshev, *Science*, 2003, **300**, 1439.
6. V. M. Labunskyy, D. L. Hatfield and V. N. Gladyshev, *Physiol. Rev.*, 2014, **94**, 739.
7. S. Kurokawa and M. J. Berry, *Met. Ions Life Sci.*, 2013, **13**, 499.
8. L. C. Clark, G. F. Combs Jr., B. W. Turnbull, E. H. Slate, D. K. Chalker, J. Chow, L. S. Davis, R. A. Glover, G. F. Graham, E. G. Gross, A. Krongrad, J. L. Lesher Jr., H. K. Park, B. B. Sanders Jr., C. L. Smith and J. R. Taylor, *J. Am. Med. Assoc.*, 1996, **276**, 1957.

9. S. M. Lippman, E. A. Klein, P. J. Goodman, M. S. Lucia, I. M. Thompson, L. G. Ford, H. L. Parnes, L. M. Minasian, J. M. Gaziano, J. A. Hartline, J. K. Parsons, J. D. Bearden 3rd, E. D. Crawford, G. E. Goodman, J. Claudio, E. Winquist, E. D. Cook, D. D. Karp, P. Walther, M. M. Lieber, A. R. Kristal, A. K. Darke, K. B. Arnold, P. A. Ganz, R. M. Santella, D. Albanes, P. R. Taylor, J. L. Probstfield, T. J. Jagpal, J. J. Crowley, F. L. Meyskens Jr., L. H. Baker and C. A. Coltman Jr., *J. Am. Med. Assoc.*, 2009, **301**, 39.
10. A. P. Fernandes and V. Gandin, *Biochim. Biophys. Acta*, 2015, **1850**, 1642.
11. M. Vinceti, G. Dennert, C. M. Crespi, M. Zwahlen, M. Brinkman, M. P. Zeegers, M. Horneber, R. D'Amico and C. Del Giovane, *Cochrane Database System. Rev.*, 2014, **3**, CD005195.
12. Y. Wei, X. Cao, Y. Ou, J. Lu, C. Xing and R. Zheng, *Mutat. Res.*, 2001, **490**, 113.
13. M. Kotrebai, M. Birringer, J. F. Tyson, E. Block and P. C. Uden, *Analyst*, 2000, **125**, 71.
14. C. Ip and D. J. Lisk, *Adv. Exp. Med. Biol.*, 1996, **401**, 179.
15. C. D. Davis, P. A. Tsuji and J. A. Milner, *Annu. Rev. Nutr.*, 2012, **32**, 73.
16. C. Meplan and J. Hesketh, *Cancer Treat. Res.*, 2014, **159**, 145.
17. P. Zhuo and A. M. Diamond, *Biochim. Biophys. Acta*, 2009, **115**, 227.
18. C. Meplan, *Nutrients*, 2015, **7**, 3621.
19. A. A. Turanov, X. M. Xu, B. A. Carlson, M. H. Yoo, V. N. Gladyshev and D. L. Hatfield, *Adv. Nutr.*, 2011, **2**, 122.
20. R. F. Burk and K. E. Hill, *Annu. Rev. Nutr.*, 2015, **35**, 109.
21. C. M. Weekley and H. H. Harris, *Chem. Soc. Rev.*, 2013, **42**, 8870.
22. H. Zeng, M. I. Jackson, W. H. Cheng and G. F. Combs Jr., *Biol. Trace Elem. Res.*, 2011, **143**, 1209.
23. Y. Wang and L. Fu, *Biol. Trace Elem. Res.*, 2012, **149**, 110.
24. R. Brigelius-Flohe, K. Lotzer, S. Maurer, M. Schultz and M. Leist, *Biofactors*, 1995, **5**, 125.
25. P. Zhuo, M. Goldberg, L. Herman, B. S. Lee, H. Wang, R. L. Brown, C. B. Foster, U. Peters and A. M. Diamond, *Cancer Res.*, 2009, **69**, 8183.
26. A. P. Kipp, J. Frombach, S. Deubel and R. Brigelius-Flohe, *Methods Enzymol.*, 2013, **527**, 87.
27. M. L. Smith, J. K. Lancia, T. I. Mercer and C. Ip, *Anticancer Res.*, 2004, **24**, 1401.
28. C. Ip, H. J. Thompson, Z. Zhu and H. E. Ganther, *Cancer Res.*, 2000, **60**, 2882.
29. M. Suzuki, M. Endo, F. Shinohara, S. Echigo and H. Rikiishi, *Cancer Chemother. Pharmacol.*, 2010, **66**, 475.
30. J. K. Wrobel, R. Power and M. Toborek, *IUBMB Life*, 2016, **68**, 97.
31. S. Bera, V. De Rosa, W. Rachidi and A. M. Diamond, *Mutagenesis*, 2013, **28**, 127.
32. V. de Rosa, P. Erkekoglu, A. Forestier, A. Favier, F. Hincal, A. M. Diamond, T. Douki and W. Rachidi, *Free Radical Res.*, 2012, **46**, 105.

33. Y. Qi, N. W. Schoene, F. M. Lartey and W. H. Cheng, *J. Biol. Chem.*, 2010, **285**, 33010.
34. H. J. Jung, H. L. Kim, Y. J. Kim, J. I. Weon and Y. R. Seo, *Oncol. Rep.*, 2013, **30**, 1581.
35. H. Blessing, S. Kraus, P. Heindl, W. Bal and A. Hartwig, *Eur. J. Biochem.*, 2004, **271**, 3190.
36. J. L. Fischer, J. K. Lancia, A. Mathur and M. L. Smith, *Anticancer Res.*, 2006, **26**, 899.
37. A. Jerome-Morais, S. Bera, W. Rachidi, P. H. Gann and A. M. Diamond, *Biochim. Biophys. Acta*, 2013, **1830**, 3399.
38. K. El-Bayoumy, *The Role of Selenium in Cancer Prevention*, ed. V. T. De Vita and S. A. Rosenberg, J. B. Lippincott Co., Philadelphia, 1991, pp. 1–15.
39. G. F. Combs Jr. and W. P. Gray, *Pharmacol. Ther.*, 1998, **79**, 179.
40. J. Lu, J. Zhang, C. Jiang, Y. Deng, N. Ozten and M. C. Bosland, *Nutr. Cancer*, 2016, **68**, 1.
41. K. El-Bayoumy and R. Sinha, *Mut. Res.*, 2004, **551**, 181.
42. J. Zhang, L. Wang, G. Li, L. B. Anderson, Y. Xu, B. Witthuhn and J. Lu, *Nutr. Cancer*, 2011, **63**, 778.
43. C. Lennicke, J. Rahn, A. P. Kipp, B. P. Dojcinovic, A. S. Muller, L. A. Wessjohann, R. Lichtenfels and B. Seliger, *Biochim. Biophys. Acta*, 2017, **1861**, 3323.
44. C. Ip, Y. Dong and H. Ganther, *Cancer Metastasis Rev*, 2002, **21**, 281.
45. J. Fleming, A. Ghose and P. R. Harrison, *Nutr. Cancer*, 2001, **40**, 42.
46. L. Wang, M. J. Bonorden, G. X. Li, H. J. Lee, H. Hu, Y. Zhang, J. D. Liao, M. P. Cleary and J. Lu, *Cancer Prev. Res.*, 2009, **2**, 484.
47. H. L. Nakken, E. D. Lephart, T. J. Hopkins, B. Shaw, P. M. Urie and M. J. Christensen, *Prostate*, 2016, **76**, 588.
48. L. Wang, X. Guo, J. Wang, C. Jiang, M. C. Bosland, J. Lu and Y. Deng, *Cancer Prev. Res.*, 2016, **9**, 35.
49. N. Ozten, L. Horton, S. Lasano and M. C. Bosland, *Cancer Prev. Res.*, 2010, **3**, 371.
50. G. N. Schrauzer, *Biol. Trace Elem. Res.*, 2006, **109**, 281.
51. G. N. Schrauzer, J. E. McGinnis and K. Kuehn, *Carcinogenesis*, 1981, **1**, 199.
52. T. S. Hudson, B. A. Carlson, M. J. Hoeneroff, H. A. Young, L. Sordillo, W. J. Muller, D. L. Hatfield and J. E. Green, *Carcinogenesis*, 2012, **33**, 1225.
53. H. A. Luchman, M. L. Villemaire, T. A. Bismar, B. A. Carlson and F. R. Jirik, *Am. J. Pathol.*, 2014, **184**, 871.
54. J. L. Barger, T. Kayo, T. D. Pugh, J. A. Vann, R. Power, K. Dawson, R. Weindruch and T. A. Prolla, *Genes Nutr.*, 2012, **7**, 155.
55. J. T. Deagen, J. A. Butler, M. A. Beilstein and P. D. Whanger, *J. Nutr.*, 1987, **117**, 91.
56. A. Duffield-Lillicoe, M. Reid, J. Turnbull, G. J. Combs, E. Slate, L. Fischbach, J. Marshall and L. Clarke, *Cancer Epidemiol., Biomarkers Prev.*, 2002, **11**, 630.

57. A. J. Duffield-Lillico, B. L. Dalkin, M. E. Reid, B. W. Turnbull, E. H. Slate, E. T. Jacobs, J. R. Marshall and L. C. Clark, *BJU Int.*, 2003, **91**, 608.
58. A. J. Duffield-Lillico, E. H. Slate, M. E. Reid, B. W. Turnbull, P. A. Wilkins, G. F. Combs Jr., H. K. Park, E. G. Gross, G. F. Graham, M. S. Stratton, J. R. Marshall and L. C. Clark, *J. Natl. Cancer Inst.*, 2003, **95**, 1477.
59. J. R. Marshall, C. M. Tangen, W. A. Sakr, D. P. Wood Jr., D. L. Berry, E. A. Klein, S. M. Lippman, H. L. Parnes, D. S. Alberts, D. F. Jarrard, W. R. Lee, J. M. Gaziano, E. D. Crawford, B. Ely, M. Ray, W. Davis, L. M. Minasian and I. M. Thompson Jr., *Cancer Prev. Res.*, 2011, **4**, 1761.
60. A. M. Algotar, M. S. Stratton, F. R. Ahmann, J. Ranger-Moore, R. B. Nagle, P. A. Thompson, E. Slate, C. H. Hsu, B. L. Dalkin, P. Sindhwani, M. A. Holmes, J. A. Tuckey, D. L. Graham, H. L. Parnes, L. C. Clark and S. P. Stratton, *Prostate*, 2013, **73**, 328.
61. D. D. Karp, S. J. Lee, S. M. Keller, G. S. Wright, S. Aisner, S. A. Belinsky, D. H. Johnson, M. R. Johnston, G. Goodman, G. Clamon, G. Okawara, R. Marks, E. Frechette, W. McCaskill-Stevens, S. M. Lippman, J. Ruckdeschel and F. R. Khuri, *J. Clin. Oncol.*, 2013, **31**, 4179.
62. E. A. Klein, I. M. Thompson, S. M. Lippman, P. J. Goodman, D. Albanes, P. R. Taylor and C. Coltman, *J. Urol.*, 2001, **166**, 1311.
63. S. M. Lippman, P. J. Goodman, E. A. Klein, H. L. Parnes, I. M. Thompson Jr., A. R. Kristal, R. M. Santella, J. L. Probstfield, C. M. Moinpour, D. Albanes, P. R. Taylor, L. M. Minasian, A. Hoque, S. M. Thomas, J. J. Crowley, J. M. Gaziano, J. L. Stanford, E. D. Cook, N. E. Fleshner, M. M. Lieber, P. J. Walther, F. R. Khuri, D. D. Karp, G. G. Schwartz, L. G. Ford and C. A. Coltaman Jr., *J. Natl. Cancer Inst.*, 2005, **97**, 94.
64. B. H. Patterson and O. A. Levander, *Cancer Epidemiol., Biomarkers Prev.*, 1997, **6**, 63.
65. E. A. Klein, I. M. Thompson Jr., C. M. Tangen, J. J. Crowley, M. S. Lucia, P. J. Goodman, L. M. Minasian, L. G. Ford, H. L. Parnes, J. M. Gaziano, D. D. Karp, M. M. Lieber, P. J. Walther, L. Klotz, J. K. Parsons, J. L. Chin, A. K. Darke, S. M. Lippman, G. E. Goodman, F. L. Meyskens Jr. and L. H. Baker, *J. Am. Med. Assoc.*, 2011, **306**, 1549.
66. Y. Lotan, P. J. Goodman, R. F. Youssef, R. S. Svatek, S. F. Shariat, C. M. Tangen, I. M. Thompson Jr. and E. A. Klein, *J. Urol.*, 2012, **187**, 2005.
67. P. Lance, D. S. Alberts, P. A. Thompson, L. Fales, F. Wang, J. San Jose, E. T. Jacobs, P. J. Goodman, A. K. Darke, M. Yee, L. Minasian, I. M. Thompson and D. J. Roe, *Cancer Prev. Res.*, 2017, **10**, 45.
68. H. L. Nicastro and B. K. Dunn, *Nutrients*, 2013, **5**, 1122.
69. M. C. Ledesma, B. Jung-Hynes, T. L. Schmit, R. Kumar, H. Mukhtar and N. Ahmad, *Mol. Med.*, 2011, **17**, 134.
70. D. L. Hatfield and V. N. Gladyshev, *Mol. Interventions*, 2009, **9**, 18.
71. K. El-Bayoumy, *Nutr. Cancer*, 2009, **61**, 285.
72. G. N. Schrauzer, *Crit. Rev. Biotechnol.*, 2009, **29**, 81.
73. M. J. Christensen, *Cancer Prev. Res.*, 2014, **7**, 781.

74. C. Abate-Shen, M. M. Shen and E. Gelmann, *Res. Biol. Div.*, 2008, **76**, 717.
75. E. E. Martinez, P. D. Anderson, M. Logan and S. A. Abdulkadir, *PLoS One*, 2012, **7**, e46792.
76. E. E. Martinez, A. K. Darke, C. M. Tangen, P. J. Goodman, J. H. Fowke, E. A. Klein and S. A. Abdulkadir, *Cancer Prev. Res*, 2014, **7**, 950.
77. J. M. Chan, A. K. Darke, K. L. Penney, C. M. Tangen, P. J. Goodman, G. S. Lee, T. Sun, S. Peisch, A. M. Tinianow, J. M. Rae, E. A. Klein, I. M. Thompson Jr., P. W. Kantoff and L. A. Mucci, *Cancer Epidemiol., Biomarkers Prev.*, 2016, **25**, 1050.
78. E. N. Bermingham, J. E. Hesketh, B. R. Sinclair, J. P. Koolaard and N. C. Roy, *Nutrients*, 2014, **6**, 4002.
79. D. J. Waters, S. Shen, S. S. Kengeri, E. C. Chiang, G. F. Combs Jr., J. S. Morris and D. G. Bostwick, *Nutrients*, 2012, **4**, 1650.

Subject Index